Evolution of the Agricultural Energy Internet

Other related titles:

You may also like

PBPO245 | Schindele | Agrivoltaics: Technical, Ecological, Commercial and Legal Aspects | 2024

We also publish a wide range of books on the following topics:
Computing and Networks
Control, Robotics and Sensors
Electrical Regulations
Electromagnetics and Radar
Energy Engineering
Healthcare Technologies
History and Management of Technology
IET Codes and Guidance
Materials, Circuits and Devices
Model Forms
Nanomaterials and Nanotechnologies
Optics, Photonics and Lasers
Production, Design and Manufacturing
Security
Telecommunications
Transportation

All books are available in print via https://shop.theiet.org or as eBooks via our Digital Library https://digital-library.theiet.org.

IET AGRICULTURAL ENGINEERING SERIES 001

Evolution of the Agricultural Energy Internet

Modeling, simulation and engineering practice

Edited by
Xueqian Fu

The Institution of Engineering and Technology

About the IET

This book is published by the Institution of Engineering and Technology (The IET).

We inspire, inform, and influence the global engineering community to engineer a better world. As a diverse home across engineering and technology, we share knowledge that helps make better sense of the world, to accelerate innovation and solve the global challenges that matter.

The IET is a not-for-profit organisation. The surplus we make from our books is used to support activities and products for the engineering community and promote the positive role of science, engineering and technology in the world. This includes education resources and outreach, scholarships and awards, events and courses, publications, professional development and mentoring, and advocacy to governments.

To discover more about the IET, please visit https://www.theiet.org/.

About IET books

The IET publishes books across many engineering and technology disciplines. Our authors and editors offer fresh perspectives from universities and industry. Within our subject areas, we have several book series steered by editorial boards made up of leading subject experts.

We peer review each book at the proposal stage to ensure the quality and relevance of our publications.

Get involved

If you are interested in becoming an author, editor, series adviser, or peer reviewer, please visit https://www.theiet.org/publishing/publishing-with-iet-books/ or contact author_support@theiet.org.

Discovering our electronic content

All of our books are available online via the IET's Digital Library. Our Digital Library is the home of technical documents, eBooks, conference publications, real-life case studies, and journal articles. To find out more, please visit https://digital-library.theiet.org.

In collaboration with the United Nations and the International Publishers Association, the IET is a Signatory member of the Sustainable Development Goals (SDG) Publishers Compact. The Compact aims to accelerate progress to achieve the Sustainable Development Goals (SDGs) by 2030. Signatories aspire to develop sustainable practices and act as champions of the SDGs during the Decade of Action (2020–30), publishing books and journals that will help inform, develop, and inspire action in that direction.

In line with our sustainable goals, our UK printing partner has FSC accreditation, which is reducing our environmental impact to the planet. We use a print-on-demand model to further reduce our carbon footprint.

Published by The Institution of Engineering and Technology, London, United Kingdom

The Institution of Engineering and Technology (the "**Publisher**") is registered as a Charity in England & Wales (no. 211014) and Scotland (no. SC038698).

First published 2025

The Institution of Engineering and Technology
Futures Place,
Kings Way,
Stevenage,
Herts, SG1 2UA
United Kingdom

www.theiet.org

British Library Cataloguing in Publication Data
A catalogue record for this product is available from the British Library

ISBN 978-1-80705-089-4 (hardback)
ISBN 978-1-80705-090-0 (PDF)
ISBN 978-1-80705-091-7 (EPUB3)

Typeset in India by MPS Limited

Cover image: Xueqian Fu

Contents

Preface

This book focuses on the field of agricultural energy and falls under the discipline of agricultural electrification and automation. The book aims to systematically elaborate on the principles and applications of the Agricultural Energy Internet through theoretical modeling, simulation techniques, and engineering case studies – especially in response to the growing demand for clean, reliable, and intelligent power supply systems that enable the future development of "unmanned smart agriculture."

The motivation for writing this book arises from two key practical contexts. First, in smart agriculture, core functions such as agricultural machinery automation, intelligent environmental sensing, and precision crop management rely heavily on a continuous, high-quality energy supply. Therefore, intelligent power systems tailored to agricultural scenarios are essential to realizing unmanned farms. Second, driven by the national "dual carbon" goals and rural revitalization strategies, rural areas and agriculture are no longer merely energy consumers – they are increasingly becoming important providers of distributed renewable energy and suitable deployment sites for energy infrastructure. As such, the Agricultural Energy Internet is now entering a critical stage of rapid development and engineering transformation.

The six chapters of this book are logically progressive and mutually reinforcing, forming a comprehensive knowledge system that spans from theoretical foundations to system implementation. Chapter 1 establishes the overall framework and developmental trajectory of the Agricultural Energy Internet (AEI). Chapters 2–4 focus on modeling and simulation of AEI's energy scheduling, electricity market participation, and low-carbon economic mechanisms, gradually expanding its operational strategies and value realization across diverse scenarios. Chapter 5 validates the feasibility and practicality of the proposed models through representative engineering cases. Chapter 6 then translates these concepts into application by detailing the functional design and code implementation of an AEI monitoring system, thus forming a closed-loop logic from theory and simulation to practice and system development.

A major special design feature of this book lies in its integrated structure, combining theoretical analysis, open-source simulation code, and real-world engineering cases. It provides a systematic overview of modeling and simulation methods for Agricultural Energy Internet systems, offers reproducible code implementations, and presents in-depth case studies based on an actual project: the aquaponics-based Agricultural Energy Internet at the Smart Farm in Zhuozhou,

Hebei Province, China, and the agricultural park in Langyatai, Qingdao, China. This closed-loop content design – from theory to simulation to engineering validation – makes the book suitable both for researchers developing algorithms and models, and for engineers seeking practical guidance in system deployment.

Xueqian Fu

Associate Professor, China Agricultural University,
Mailing Address: CIEE 633, 17 Qinghua Donglu, Beijing 100083, China
Tel: (86)188 1310 3877
Email: fuxueqian@cau.edu.cn

Acknowledgments

To the authors

I would like to extend my deepest gratitude to all the contributing authors of *Agricultural Energy Internet: Modeling, Simulation, and Engineering Practice*. Your exceptional expertise, commitment, and collaborative spirit have made this book possible. Each chapter embodies your dedication to advancing research and innovation in the emerging field of the Agricultural Energy Internet. It has been a great honor to work alongside such a talented and visionary team of scholars and professionals.

To the IET editorial team

My heartfelt appreciation also goes to the editorial team at the IET – Ms. Olivia Wilkins and Ms. Brittany Insull, Assistant Editor, and Mr. T. Michael Slaughter, Consultant Commissioning Editor – for their professional guidance, encouragement, and unwavering support throughout every stage of this project. Their editorial insight and commitment to excellence have been vital in ensuring the successful publication of this work.

To the funding support

This book is supported by the National Natural Science Foundation of China under Grant 52007193, Shanxi Energy Internet Research Institute under Grant SXEI2023B003, and the 2115 Talent Development Program of China Agricultural University.

Xueqian Fu

Associate Professor, China Agricultural University
Mailing Address: CIEE 633, 17 Qinghua Donglu, Beijing 100083, China
Tel: (86)188 1310 3877
Email: fuxueqian@cau.edu.cn

AGRICULTURAL ENERGY INTERNET: MODELING, SIMULATION, AND ENGINEERING PRACTICE

Target audience:
Researchers, engineers, policymakers, and graduate students in agricultural electrification and automation.

Benefits to audience and key features:
This book provides a comprehensive overview of the emerging concept of the Agricultural Energy Internet (AEI), integrating perspectives from energy systems, information science, and agricultural engineering. It benefits audiences by offering both theoretical insights and engineering practices for building intelligent, low-carbon, and resilient agricultural energy systems.

The Institution of Engineering and Technology

List of contributors

Yanan Cui
College of Information and Electrical Engineering, China Agricultural University, Beijing, China

Xueqian Fu
College of Information and Electrical Engineering, China Agricultural University, Beijing, China

Chao Lei
College of Information and Electrical Engineering, China Agricultural University, Beijing, China

Nange Li
College of Information and Electrical Engineering, China Agricultural University, Beijing, China

Hai Long
College of Information and Electrical Engineering, China Agricultural University, Beijing, China

Xiao Lv
College of Information and Electrical Engineering, China Agricultural University, Beijing, China

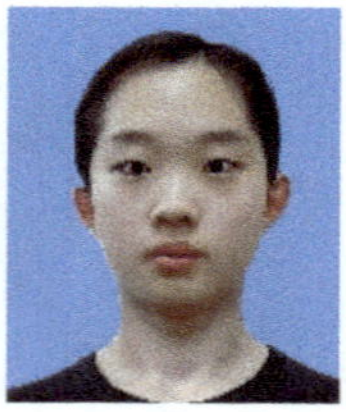

Jinghui Song
College of Information and Electrical Engineering, China Agricultural University, Beijing, China

Huiyan Wang
College of Information and Electrical Engineering, China Agricultural University, Beijing, China

Bohao Zeng
College of Information and Electrical Engineering, China Agricultural University, Beijing, China

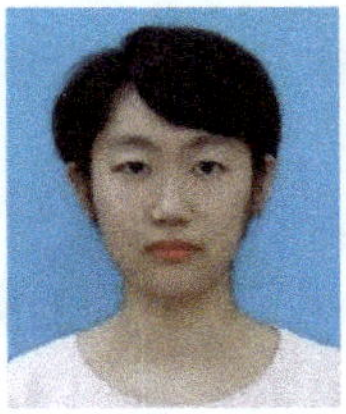

Jing Zhang
College of Information and Electrical Engineering, China Agricultural University, Beijing, China

Wentao Zhu
College of Information and Electrical Engineering, China Agricultural University, Beijing, China

About the editor

Xueqian Fu (Senior Member, IEEE; Member, IET) received the BS and MS degrees from North China Electric Power University in 2008 and 2011, respectively, and the Ph.D. degree from South China University of Technology in 2015. From 2015 to 2017, he was a Postdoctoral Researcher with Tsinghua University. He is currently an Associate Professor at China Agricultural University. His current research interests include statistical machine learning, the Agricultural Energy Internet, and photovoltaic system integration.

He is the Vice Chairperson of the *IEEE Smart Village* China Committee and IEEE Young Professionals. He was recognized as one of the *World's Top 2% Scientists* in 2023 (single-year impact), 2024 (career-long and single-year impact), and 2025 (career-long and single-year impact). He has also been recognized as an "Outstanding Talent" and "Young Star B" by China Agricultural University. He serves as the Deputy Editor-in-Chief of *Information Processing in Agriculture* and the Founding Chair of the *IEEE International Symposium on the Application of Artificial Intelligence in Electrical Engineering* (AAIEE).

Chapter 1

Evolution of the Agricultural Energy Internet

Xueqian Fu[1], Yanan Cui[1], Jinghui Song[1], Xiao Lv[1] and Huiyan Wang[1]

Abstract

This chapter establishes the theoretical foundation and conceptual framework of the Agricultural Energy Internet (AEI), positioning it as a systematic extension of the Energy Internet within the agricultural sector. AEI embodies a profound transformation of agricultural energy systems toward digitalization, intelligence, and decarbonization, representing the integration of energy science, agricultural engineering, and advanced information technology. Rooted in the global context of energy transition, climate change mitigation, and smart agriculture development, AEI aims to build a self-adaptive, low-carbon, and sustainable agricultural energy ecosystem through the synergistic interaction of energy, information, and carbon flows. The chapter outlines the evolution of AEI from its conceptual emergence to technological exploration and early demonstration, reflecting the shift from "energy supply-driven" to "information and intelligence-driven" agricultural systems. AEI promotes deep coupling between renewable energy and agricultural production, integrating distributed generation, storage, and smart management into a unified intelligent network. Compared with conventional energy architectures, AEI emphasizes multi-energy complementarity, real-time coordination, and ecological constraints, enabling dynamic optimization of agricultural energy utilization and carbon emissions. Looking ahead, AEI is expected to become a cornerstone for the digital and green transformation of agriculture. Future research will focus on enhancing system interoperability, improving carbon accounting precision, and advancing AI-based decision-making under complex agricultural environments. With the continuous evolution of 6G communication, artificial intelligence (AI), and digital twin technologies, AEI will expand toward a globally interconnected, intelligent, and sustainable agricultural energy ecosystem. As both a technological innovation and a strategic infrastructure, AEI will play a pivotal role in achieving energy security, carbon neutrality, and high-quality development in modern agriculture.

Keywords: Agricultural Energy Internet (AEI); renewable energy; smart agriculture; artificial intelligence (AI); digital twin

[1]College of Information and Electrical Engineering, China Agricultural University, China

1.1 Background and policy context

Driven by the global sustainable development agenda, the Agricultural Energy Internet (AEI), as an emerging systematic framework, is gradually becoming a model of the deep integration of agricultural production and energy management. The monograph "Agricultural Energy Internet: Modeling, Simulation and Engineering Practice" aims to systematically explore the construction path and optimization strategies of AEI through theoretical modeling, simulation verification, and engineering application. Chapter one, "The Evolution of Agricultural Energy Internet," starts with the historical context and conceptual evolution, laying the foundation for the literature review, technical framework, and development trends of the subsequent chapters. This section, as the opening of the chapter, focuses on the macro background and main driving forces behind the rise of AEI, aiming to reveal its inevitability under the interweaving of multiple factors such as global energy transition, the growing demand for smart agriculture, the urgency of addressing climate change, and the digital technology revolution. By analyzing these driving forces, we will clarify how AEI has evolved from a nascent concept to a practical application, thereby providing theoretical support for the intelligent transformation of the entire agricultural value chain.

The core of AEI lies in regarding agricultural production as a dynamic energy ecosystem, which integrates distributed renewable energy sources (such as photovoltaic and wind energy), intelligent perception devices, and data-driven optimization algorithms to achieve efficient energy allocation, environmental synergy, and economic sustainability [1]. The rise of this concept is not isolated but embedded in the grand narrative of contemporary global challenges. Among the United Nations Sustainable Development Goals (SDGs), Goal 2 (Zero Hunger) and Goal 7 (Clean Energy) explicitly emphasize the cross-integration of agriculture and energy, and AEI is precisely the digital carrier of this integration [2]. At the policy level, multilateral institutions such as the International Energy Agency (IEA) and the Food and Agriculture Organization of the United Nations (FAO) have incorporated rural energy transition into climate action frameworks, such as the Nationally Determined Contributions (NDCs) under the Paris Agreement. These policy orientations have further catalyzed the global promotion of AEI [3]. This article will dissect these macro driving forces layer by layer, reveal how AEI responds to the demands of The Times, and provide background preparation for the literature landscape and theoretical evolution of subsequent chapters.

1.1.1 Global energy transition: reconstructing from fossil reliance to renewable agricultural energy dominance

The global energy transition constitutes the most fundamental macro background for the rise of AEI. Since the oil crisis in the 1970s, the process of tilting the energy structure from being dominated by fossil fuels to renewable energy has been accelerating. According to data from the International Energy Agency, the global installed capacity of renewable energy exceeded 3 terawatts in 2023, with the proportion of photovoltaic and biomass energy applications in the agricultural sector significantly increasing [4]. The core driving force of this transformation lies in the limited nature and environmental externalities of fossil energy: carbon emissions from the combustion of fossil fuels account for approximately 75% of the total global greenhouse gas emissions, and agriculture, as a major energy consumer (accounting for 14% of global energy demand), urgently needs to shift from a passive consumer to an active producer [5]. AEI emerged precisely against this backdrop. It regards agriculture as a node of the distributed energy network and achieves energy self-sufficiency and cross-regional collaboration through microgrids and virtual power plants (VPPs).

Specifically, the agricultural dimension of energy transition is reflected in the upgrading of rural power infrastructure. The traditional rural energy system relies on long-distance transmission lines, with transmission losses as high as 15–20%. However, the distributed energy resources (DERs) introduced by AEI, such as rooftop photovoltaic and small-scale wind power generation, can increase the local energy utilization rate to over 90% [6]. For instance, in developing countries such as China and India, rural photovoltaic agricultural projects have covered millions of hectares of cultivated land, not only reducing energy costs but also feeding back into agricultural production through energy revenue [7]. The theoretical basis of this transformation can be traced back to the expansion of the concept of the Energy Internet: initially proposed by the smart grid research in Germany and the United States, it has now extended to the agricultural sector, forming an "agriculture-energy-environment" collaborative framework. Fu *et al.'s* research further quantified this synergy effect. They optimized the agricultural microgrid through a carbon accounting model and demonstrated that AEI can reduce the carbon footprint from 25% to 30%, thereby transforming the energy transition from macro policies to micro practices.

Under the policy context, the agricultural orientation of the global energy transition has become increasingly clear. Both the European Green Deal of the European Union and the Infrastructure Investment and Jobs Act (IIJA) of the United States give priority to funding rural renewable energy projects, with the goal of achieving 100% clean agricultural energy by 2030 [8]. These policies not only offer financial incentives (such as subsidies and low-interest loans) but also encourage the deployment of AEIs through carbon trading mechanisms [9]. In developing countries, the World Bank's "Energy Transition Partnership" program has invested tens of billions of dollars to support rural Energy Internet pilot projects in Africa and Asia [10]. However, this transformation has not been smooth sailing: fluctuations in energy prices and lagging infrastructure remain bottlenecks [11]. The rise of AEI is precisely a response to these challenges. It decentralizes energy trading through blockchain and edge computing, ensuring the resilience and fairness of the transition [12].

Further analysis reveals that the driving forces behind the global energy transition also stem from geopolitical and economic factors. The energy crisis triggered by the Russia–Ukraine conflict has highlighted the vulnerability of supply chains, driving countries to accelerate the development of domestic renewable energy [13]. In the agricultural context, this is manifested as a new paradigm of "energy and food security": AEI integrates crop growth models with energy prediction algorithms to ensure food production is maintained during periods of energy shortage [14]. Zhuo *et al.'s* work emphasizes that power systems with a high proportion of renewable energy access need to address intermittent challenges, and the energy storage–generation–load coordination mechanism of AEI can enhance system stability by 40%. In conclusion, energy transition is not only a technological change but also a policy-driven system reconstruction, providing a broad application space for AEI [15].

1.1.2 The continuous growth in demand for smart agriculture: the leap from traditional farming to data-driven production

With population explosion and farmland degradation, smart agriculture has become a strategic pillar of global food security. The United Nations predicts that by 2050, the global population will reach 9.7 billion, and the demand for food will increase by 70%. However, the inefficiency of traditional agriculture (with an average yield utilization rate of only 60%) is no longer sufficient to support it [16]. The rise of AEI stems precisely from this demand: it embeds energy management into the framework of smart agriculture to achieve precise irrigation, automated sowing, and real-time monitoring for energy optimization [17]. Klerkx *et al.'s* review points out that the core of the Agriculture 4.0 era is digital transformation, and AEI, as its energy subsystem, can reduce resource waste by 30%.

The growth in demand for smart agriculture is reflected in multiple aspects. First, the shortage of the labor force has driven the wave of automation: the global rural labor turnover rate has increased from 20% to 30%, forcing farms to turn to robots and drones [18]. AEI provides intelligent decision-making for energy support through Internet of Things (IoT) sensor networks, such as dynamically adjusting photovoltaic power supply to match equipment loads [19]. Sharma *et al.'s* research examined this trend, highlighting the role of AI in crop pest and disease prediction, while the energy layer of AEI ensures the continuous operation of edge devices. Second, the complexity of the supply chain requires real-time data integration: full-chain tracking from seed to table demands massive energy, while AEI's distributed computing framework can reduce latency by 50% [20].

Policy support further amplifies this demand. Both the FAO's "Digital Agriculture Strategy" and China's "Rural Revitalization" plan have prioritized smart agriculture, with investment scales exceeding 100 billion US dollars [21]. These policies emphasize the integration of energy and information; for instance, the EU's "Farm-to-table" strategy requires an 80% digitalization rate in agriculture, and AEI plays a role in energy security in this regard [22]. Himesh *et al.'s* analysis shows that the big data revolution has reshaped agricultural decision-making. AEI Bridges data silos through the Energy Internet to achieve cross-farm collaboration.

However, the growth in demand has also exposed the digital divide: the penetration rate of smart agriculture in developing countries is less than 20% [23]. The potential of AEI lies in its scalability: low-cost solar IoT modules can be quickly deployed in small-scale farming economies [24]. Khan *et al.'s* review further demonstrates that technological innovations such as the integration of 5G and blockchain position AEI as a catalyst for sustainable food systems. Raj and Prahadeeswaran's latest work looks forward to the technology stack of smart farms, emphasizing the energy efficiency of AEI in multimodal data processing. In conclusion, the surging demand for smart agriculture not only drives the technological iteration of AEI but also reshapes the economic model of agriculture, shifting from a yield-oriented approach to a value-oriented one [25].

1.1.3 The urgency of addressing climate change: agriculture as a frontier for emission reduction and an adaptation buffer

Climate change has become a "century challenge" facing humanity, and its impact on agriculture is particularly severe: extreme weather causes global crop yields to fluctuate from 10% to 20%, while agriculture itself contributes 24% of global greenhouse gas emissions [26]. The rise of AEI stems from this urgency, which embeds climate adaptability into energy management to achieve systematic optimization of low-carbon agriculture [27]. Malhi *et al.'s* review emphasizes that the mitigation of climate change requires a multi-pronged approach, and AEI provides a quantitative path through the integration of carbon capture and renewable energy.

The core of the urgency lies in the dual role of agriculture: on the one hand, as an emission source, the energy consumption of chemical fertilizers and machinery accounts for 50% of the carbon footprint of agriculture [28]; on the other hand, as a carbon sink, precision agriculture can increase the soil carbon sequestration rate by 30% [29]. AEI responds to this paradox by optimizing nitrogen fertilizer application and irrigation energy through AI-driven energy scheduling, thereby reducing methane and nitrous oxide emissions [30]. The report of the US Environmental Protection Agency (EPA) confirms that climate change has magnified agricultural risks, such as production losses caused by droughts reaching tens of billions of dollars, and that AEI's distributed energy can provide a resilient buffer.

At the policy level, the Kyoto Protocol and the Paris Agreement have included agricultural emission reduction in national obligations, and China's "dual carbon" goals (carbon neutrality by 2060) have taken rural energy transformation as a key approach [31]. The technical assessment by Uprety *et al.* shows that climate mitigation technologies in the agricultural sector (such as biochar and photovoltaic greenhouses) require energy support, and the AEI framework can increase the return on investment by 25%. Teal and Burkart's perspective further points out that regenerative agriculture is the key to climate warfare, and AEI amplifies its effects through energy–ecological synergy.

The challenge lies in uneven adaptation: small-scale farmers are vulnerable to rising sea levels and heatwaves [32]. Taylor's critical analysis warns that climate-smart agriculture (CSA) needs to avoid "green whitewashing," and the transparent carbon accounting mechanism of AEI can ensure fairness. Houser and Stuart's research has revealed the inherent contradiction in industrial agriculture: nitrogen fertilizer dependence intensifies the climate feedback cycle, and AEI's circular energy model can break this cycle. Moallemi *et al.'s* narrative analysis starts from the path of energy transition and demonstrates the role of AEI in the power sector of developing countries. Boßner *et al.*'s climate security–energy nexus framework emphasizes that AEI is not only a technological tool but also a bridge for social transformation. In conclusion, the urgency of climate change drives AEI to evolve from passive response to proactive governance [33].

1.1.4 Digital technology revolution: AI, IoT, and big data empower the intelligence of agricultural energy

The digital technology revolution has been the catalyst for the rise of AEI. Since 2010, investment in the application of AI and IoT in agriculture has increased by more than ten times. The core of this revolution is the data explosion: global agricultural sensors generate petabyte-level data, which traditional energy systems cannot support in processing [34]. AEI positions digital technology as the "nerve center" of energy management, enabling predictive maintenance and optimized dispatching [35].

The penetration of IoT has first transformed on-site perception: Citoni *et al.'s* work demonstrated the application of LoRaWAN in smart farms, where AEI integrates energy monitoring through low-power networks, reducing farm energy consumption by 20%. Big data enables predictive models: Van Zaru's research has demonstrated that digital technology can enhance agricultural performance by 15%, and the energy layer of AEI ensures the continuity of data flow. The breakthroughs in AI are particularly cutting-edge: Ahmed *et al.'s* review examined the IoT–big data–AI triplet, where AEI utilizes machine learning to optimize energy allocation and address uncertainties.

Policy responses include the "AI Agriculture Initiative" in the United States and the "Digital Europe Initiative" in the European Union. These frameworks give priority to funding AEI-like projects [36]. Miller *et al.'s* analysis emphasizes that the time has come for IoT and AI, as AEI Bridges the digital divide between agriculture and energy. Bach and Mauser's chapter on sustainable agriculture further demonstrates that the integration of Earth observation data and energy simulation will push AEI into the 4.0 era.

The challenge lies in privacy and accessibility: The digital revolution has exacerbated the urban–rural gap [37]. Duncan *et al.'s* manual discusses the digital prospects of modern agriculture, and AEI alleviates this issue through an open-source platform [38]. The report of Global EY warns that digital agriculture needs to feed the world, but the energy bottleneck cannot be ignored [39]. Hati and Singh's research on smart indoor farms demonstrates the leverage effect of technology, with AEI amplifying its sustainability [40]. In conclusion, the digital revolution not only drives AEI's technology stack but also reshapes the governance paradigm of agriculture.

1.1.5 Convergence of driving forces and the strategic positioning of AEI

The above-mentioned driving forces do not exist in isolation but reinforce each other through complex dynamic interaction mechanisms, forming a closed-loop, systematic driving force. Specifically, the global energy transition provides a solid hardware foundation, including distributed renewable energy infrastructure and smart grid architecture, which injects efficient and reliable energy-supply channels into agricultural production. Meanwhile, the continuous growth in demand for smart agriculture is injected into the software intelligence layer. Through data-driven algorithms and real-time feedback mechanisms, precise regulation of energy flow is achieved, thereby transforming the static energy system into a responsive ecological network. The urgency of addressing climate change further defines the core objective of this system, which is to build a resilient agriculture framework in both the dimensions of emission reduction and adaptation, ensuring that energy utilization is not only efficient but also serves the ecological restoration and sustainable development agenda. The digital technology revolution plays the role of a catalyst for accelerating iteration. By leveraging the exponential progress of artificial intelligence (AI), the Internet of Things, and big data, AEI continuously optimizes the algorithm models and deployment strategies, enabling it to evolve from conceptual prototypes to large-scale engineering applications. This multi-dimensional reinforcement is not a linear superposition but rather gives rise to a synergy effect: for instance, in a typical rural AEI pilot, the combination of photovoltaic energy transition (hardware) and AI prediction models (software) can transform climate-induced fluctuations in water resources into quantifiable energy-saving opportunities, potentially reducing farm operating costs by more than 20%.

AEI is thus positioned as the natural convergence point of these driving forces, a highly integrated platform that enables the energy–information–environment tripartite synergy of the entire agricultural value chain – from soil tillage to harvesting and processing, and then to logistics and distribution. This kind of collaboration transcends the boundaries of traditional fields, viewing energy as a dynamic resource, information as a decision-making engine, and the environment as a constraint boundary, thus forming an adaptive closed-loop system. At the operational level, this convergence is reflected in the integrated application of virtual power plants (VPPs) and microgrids: AEI achieves transparency in energy transactions through blockchain technology while embedding environmental sensor networks to monitor carbon footprints, ensuring that every energy distribution serves the sustainable development goals. At the policy level, this positioning has been strongly consolidated by multilateral agreements. For instance, the draft of the "Agricultural Energy Convention" proposed during the 2023 COP28 summit emphasized the inclusion of AEI in national climate action plans (NDCs) and called on developed countries to transfer related technologies and funds to developing countries, with an estimated total amount of 50 billion US dollars. This convention not only provides a legal framework for the global standardization of AEI but also encourages cross-border cooperation through a carbon credit mechanism, promoting the leap from pilot demonstrations to regional networks. In the context of developing countries, such policies are particularly crucial: for instance, rural energy projects in India and Brazil have increased the penetration rate of renewable energy by 15% through the AEI framework and significantly alleviated energy poverty in agricultural communities.

The review of the evolution of AEI in Chapter 1 is not merely a historical narrative but rather constructs an operational analytical framework that transforms the key milestones of the concept germination period, technology exploration period, and early demonstration period into quantifiable parameter inputs. For instance, in the modeling chapter, we will combine the hardware data of energy transition (such as photovoltaic output curves) with the software algorithms of the digital revolution (such as reinforcement learning optimizers) to develop a hybrid simulation platform suitable for rural microgrids, thereby verifying the robustness of AEI in uncertain environments. This paving role reflects the core concept of the monograph: the evolution of AEI is a bridge between theory and engineering, historical review serves future design, and ensures that each step of simulation iteration originates from the quantitative mapping of real driving forces. In this way, the past trajectory of AEI is revealed, providing a toolbox-style engineering blueprint to support a smooth transition from laboratory prototypes to field applications.

Looking ahead, AEI's strategic potential will further amplify its global influence. According to the joint forecast of the International Energy Agency and the Food and Agriculture Organization, by 2035, the widespread deployment of AEI is expected to drive from 5% to 10% increase in global agricultural GDP. This growth mainly stems from energy cost savings (expected to release 100 billion US dollars annually) and production efficiency improvements (a 25% reduction in crop yield fluctuations). Meanwhile, in terms of contribution to emission reduction, AEI can achieve a 15% reduction in global greenhouse gas emissions by optimizing the coordination of distributed energy access and carbon capture, which is equivalent to avoiding 2 billion tons of carbon dioxide equivalent emissions annually. This outlook is not an empty vision but a rigorous deduction based on existing trends: for instance, under the framework of the EU's "Green Deal," AEI pilots have demonstrated their role in enhancing soil carbon sinks, transforming the net emissions of traditional agriculture into positive carbon neutrality. However, realizing this potential requires overcoming multiple challenges, including technological standardization, fair distribution of funds, and interdisciplinary collaboration. The subsequent chapters of this book will explore these paths through case simulations, aiming to provide replicable solutions for policymakers and engineering practitioners. Ultimately, AEI is not merely a product of technological innovation; it is a strategic pillar for humanity to address the triple crisis of "food - energy – climate," promoting the positive transformation of agriculture from a fragile system to a resilient ecosystem.

1.2 Literature landscape and gap

1.2.1 Investigation of agricultural electrification technologies

Currently, the field of agricultural electrification presents an integrated development trend centered on renewable energy, supported by smart microgrids, and targeted at precision agriculture applications. In terms of power supply system architecture, DC microgrids based on solar photovoltaics (PV) have emerged as the preferred solution for electrification in rural areas, especially in developing countries, due to their modular scalability, high distribution efficiency, and decentralized control advantages. These systems effectively reduce line losses through the coordinated configuration of distributed generation units and energy storage devices, while reliably powering communal loads by leveraging the characteristics of diverse electricity usage [41]. In terms of system optimization, the coordinated operation mode of hybrid energy systems has demonstrated significant benefits. Multi-objective optimization of PV, hydropower, diesel generators, and battery energy storage using tools such as HOMER can significantly improve system economics and environmental friendliness while ensuring power supply reliability [42]. Peer-to-peer (P2P) energy trading mechanisms combined with improved control algorithms have further enhanced the operational efficiency and voltage stability of rural power grids [43].

At the level of precision agriculture applications, key technologies such as intelligent sensing, efficient communication, and precise operation have achieved important breakthroughs. Soil moisture monitoring systems based on metamaterial inspired microwave sensors have realized high-precision measurement of soil moisture [44]; lightweight communication protocols provide reliable transmission guarantees for agricultural Internet of Things (IoT) devices [45]; and laser weeding equipment achieves fast and precise operations through intelligent trajectory planning [46]. As an important path to realizing rural energy transition and agricultural modernization, the development of electrification technologies not only depends on the maturity of the technologies themselves but is also comprehensively influenced by the environment. Therefore, the promotion of electrification technologies must fully consider their regional adaptability, avoiding a "one-size-fits-all" implementation strategy. Currently, technologies such as PV microgrids, hybrid energy storage systems, and smart agricultural equipment have shown good application potential, but they still face many common challenges (Table 1.1).

Current agricultural electrification technologies have made remarkable progress in promoting rural energy transition and agricultural modernization, especially with the emergence of numerous innovative solutions in microgrid architecture, renewable energy integration, IoT applications, and intelligent equipment research and development. However, the technology system still faces a series of structural deficiencies and challenges on the path to large-scale, commercialized, and sustainable development, while also showing clear development trends and technological evolution directions.

Existing agricultural electrification systems still have obvious shortcomings in reliability, economy, and adaptability. Although technologies such as DC microgrids and PV-storage hybrid systems have been proven suitable for remote rural areas, their high initial investment costs and limited financing channels have become key barriers to promotion [41]. In cases such as Lesotho and Cambodia, microgrid construction relies heavily on external funding and policy support, lacking locally resilient business models [47]; the Cambodian case also mentions

Table 1.1 Application cases of rural electrification technologies

Ref	Year	Technology type	Research achievements
[41,48]	2018–2019	DC microgrid architecture	Focus on the distributed architecture and localized/communication-less control technologies of DC microgrids, providing scalable and high-efficiency solutions for rural electrification in developing countries.
[42,49]	2021–2025	Hybrid renewable energy systems	Focus on the planning, optimization, and control strategies of hybrid renewable energy systems, pursuing the balance of technical, economic, and environmental benefits.
[47,50]	2019–2020	Microgrid planning and tool development	Focus on rural microgrid planning and design, providing algorithm-supported tools or topology optimization schemes to adapt to off-grid scenario requirements.
[43,51]	2023–2025	Energy management and sharing mechanisms	Centered on distributed energy sharing, improve the energy utilization efficiency and power supply stability of rural microgrids through optimization algorithms.
[52,53]	2018–2021	IoT and smart agriculture	Comprehensive research on IoT in the field of smart agriculture, sorting out technology integration, application scenarios, and development trends.
[54,55]	2020–2024	PV-agriculture synergy	Focus on the coordinated development of PV and agriculture, exploring sustainable agricultural models with efficient resource utilization.
[46,56]	2025	Agricultural robotics	Focus on the optimization of sensing and control technologies for agricultural robots, improving operational performance in complex agricultural scenarios.
[44,45, 57]	2023–2025	Precision agriculture sensing and communication systems	Focus on the core needs of data collection and transmission in precision agriculture, realizing efficient acquisition, accurate detection, and reliable transmission of agricultural data through the optimization of remote sensing equipment, sensing systems, and communication protocols, respectively, to provide technical support for precision farming.

(Continues)

Table 1.1 (*Continued*)

Ref	Year	Technology type	Research achievements
[58,59]	2014–2025	Rural electrification	Focus on the long-term benefits of rural electrification, achieving the coordinated sustainability of the economy, society, and environment from the perspectives of institutional models and operational strategies, respectively.
[60]	2017	Solar PV systems	Centered on PV technology, sort out technology development and application scenarios from a macro perspective, providing references for subsequent research.

the practice of planning low-voltage distribution networks by integrating multiple algorithms [50]. Technical issues during system operation are also very prominent: distribution network losses are relatively high, especially in centralized architectures [51]; voltage stability is insufficient, and undervoltage or overvoltage problems are likely to occur in rural power grids with large load fluctuations and long lines [50]; the interoperability and standardization between equipment are low, leading to difficulties in system expansion [48]. In addition, the level of coordinated management between energy supply and agricultural loads is limited. Most microgrids have not achieved deep integration with agricultural facilities at the control level, resulting in low energy use efficiency and failure to give full play to the potential of "agriculture-energy complementarity" [59]. Even in systems incorporating IoT and data sensing, there are common problems such as inconsistent communication protocols, high node energy consumption, and poor network coverage, which restrict their deployment in large-area farmland [53]; although lightweight communication protocols can alleviate some problems, they have not been fully popularized [45]. These issues collectively constitute the main bottlenecks of agricultural electrification technologies in practical applications, which urgently need to be addressed through technological innovation and system optimization.

In terms of intelligence and autonomy, although agricultural robots and intelligent sensing equipment are highly expected, their performance in actual complex agricultural environments is still unsatisfactory. Precision agricultural technologies such as laser weeding, nanosatellite remote sensing, and soil moisture monitoring have shown high precision in experiments, but in dynamic and unstructured field scenarios, they generally face problems such as large positioning errors, weak environmental adaptability, and decision-making delays [44,46,57]. The cutting efficiency of laser weeding robots will decrease significantly due to positional deviations when facing moving targets and complex backgrounds [46], while traditional Convolutional Neural Network (CNN)-based visual models have limited recognition capabilities under conditions such as crop occlusion and variable lighting [56]. In addition, most agricultural robots and sensing systems still rely on centralized control or limited communication capabilities, making it difficult to achieve efficient collaboration in distributed operation scenarios without network coverage [48]. These limitations expose the insufficient overall maturity of current agricultural electrification systems in the perception, decision-making, and control links, and they have not yet formed a closed-loop autonomous capability of "perception-analysis-execution." Especially in resource-constrained rural environments, how to balance computational complexity and real-time requirements and how to achieve multi-machine collaboration under limited communication bandwidth are urgent technical problems to be solved.

Facing the above shortcomings, agricultural electrification technologies are continuously developing toward integration, intelligence, greenization, and inclusiveness. In terms of system architecture, the future trend clearly tends to build a multi-energy, complementary, hierarchical, and coordinated microgrid system. The PV–biomass–energy storage hybrid system has shown advantages in improving power supply reliability and reducing carbon emissions in the practice of West Papua, Indonesia [42]; while the DC microgrid architecture based on nanogrid clusters achieves higher system scalability and energy efficiency through modular design, distributed control, and bidirectional power flow [41], the decentralized control architecture further optimizes the cluster operation effect [48]. These systems not only consider the resource characteristics of rural areas but also realize the optimal configuration of various energy sources through intelligent scheduling algorithms [51]. At the same time, the integration of energy management and agricultural production is deepening. A typical example is the emergence of the concept of "Photovoltaic Agricultural Internet of Things (PAIoT)," which aims to integrate renewable energy generation, agricultural environment regulation, and crop production management into a unified platform to achieve the linked optimization of energy and agricultural data [55]. In this process, the application of advanced algorithms and optimization models has become a key support, including P2P power trading mechanisms based on Nash Bargaining, improved Adaptive Alternating Direction Method of Multipliers (AADMM), and Second-Order Cone Programming (SOCP) optimization [43,51], all of which have shown potential in improving system economics and robustness. These algorithms not only consider technical optimization but also take into account the economic interests of all participants, providing a theoretical basis for building a sustainable rural energy ecosystem.

IoT, artificial intelligence (AI), and big data analytics are gradually penetrating all aspects of agricultural electrification. Specialized communication protocols and low-power embedded systems for precision agriculture are designed to solve the problems of insufficient network coverage and high equipment energy consumption in rural areas; innovative solutions such as soil moisture monitoring systems based on metamaterial-inspired microwave sensors and collaborative operation platforms of nanosatellite remote sensing and unmanned aerial vehicles (UAVs) provide feasible paths for large-scale and high-precision agricultural environmental perception [44,45,57]. Especially in the field of agricultural robots and autonomous systems, Generative Big Models (GBMs), reinforcement learning, and perception-control integrated architectures have been introduced to enhance the robots' understanding, reasoning, and decision-making capabilities in complex scenarios [56]. The agricultural robot system integrating AIoT and GBM has achieved an operational efficiency of over 92% in experiments, showing its breakthrough potential in dynamic target recognition and task planning [56]. The integrated application of these technologies is redefining the operation mode of agricultural production, making application scenarios such as precision fertilization, intelligent irrigation, and automatic harvesting more efficient and reliable.

From the perspective of system reliability, future agricultural electrification technologies will pay more attention to robustness and maintainability throughout the life cycle. In the microgrid design stage, tools such as Gaussian mixture clustering, network reduction, and N-1 line loss reliability cost-benefit analysis can evaluate the cost-effectiveness under different reliability levels during the planning stage, providing developers with a variety of options under different reliability probabilities from 10% to 100% [47]. In the operation and maintenance stage, community-participated operation and maintenance models are being promoted. By establishing localized technical teams and operation and maintenance systems, the long-term stable operation of the system is ensured [58]. At the same time, the application of new power electronic devices is also playing an important role in improving system reliability. The design of nanosatellite Power Distribution Units (PDUs) based on wide-bandgap semiconductor devices has demonstrated advantages in high efficiency, compactness, and reliability, and this technical route can also be applied to ground agricultural electrification systems [57].

In addition, sustainability and affordability have become important orientations for the development of agricultural electrification technologies. On the one hand, system design pays more attention to resource efficiency and ecological coordination. For example, Agrivoltaic (APV) technology not only ensures power generation benefits but also takes into account the regulation of crop growth microclimate, optimization of land use, and improvement of animal welfare [54]; on the other hand, toolchains and platform-based solutions have been strengthened. The emergence of open-source microgrid design and planning toolsets such as uGrid and uGridNet has lowered the threshold for project development, enabling more regions to deploy customized systems based on local resource conditions [47]. From the perspective of business models and institutional levels, innovative models such as community-participated operation and maintenance supported by cadastral information systems [58] and blockchain-based energy trading and traceability mechanisms [53] also provide institutional guarantees for the long-term sustainable operation of agricultural electrification projects. These innovations not only solve technical implementation issues but also consider social and economic sustainability, enabling agricultural electrification technologies to truly take root in rural communities and become effective tools for promoting rural development.

Various existing technical solutions, from DC microgrids to AC hybrid systems and from PLC-based communication to wireless sensor networks, need to achieve better compatibility in terms of interface standards, data formats, and control protocols. This will enable seamless integration of equipment from different manufacturers, smooth upgrading of systems from different periods, and flexible expansion of applications of different scales. Especially at the IoT level, a unified framework for device management, data collection, and application development will greatly reduce the complexity of system integration, enabling agricultural electrification technologies to be promoted and popularized more quickly.

Agricultural electrification technology is at a critical stage of development, from a single power supply function to the integration of "energy-agriculture-information" multi-systems. Although there are still many challenges in cost, reliability, intelligence level, and system integration, through architectural innovation, the introduction of intelligent algorithms, cross-layer collaborative design, and the exploration of business models, this field is gradually forming a more efficient, green, and inclusive technical system. Future agricultural electrification systems will not only be simple power supply facilities but also comprehensive service platforms deeply integrating energy production, agricultural production, environmental monitoring, and intelligent decision-making. This integration is reflected not only at the technical level but also in multiple dimensions such as organizational structure, business models, and social participation. With the further maturity of PV and energy storage technologies, as well as the cross-integration of IoT and AI [55,60], agricultural electrification is expected to become the core infrastructure for promoting the transformation of global agriculture toward intelligence and low carbonization while realizing rural energy inclusion. This transformation process requires the joint drive of technological innovation, policy support, business models, and social participation. Only in this way can the large-scale and sustainable development of agricultural electrification technologies be truly realized, bringing lasting development momentum to rural areas around the world.

1.2.2 Rural energy technology research

With the development of global renewable energy and the advancement of rural modernization, the rural energy structure is undergoing profound changes. The reliance on traditional fossil energy not only brings about environmental pollution and resource pressure but also restricts the pace of rural economic development and agricultural modernization. In recent years, the promotion and application of various renewable energy technologies in rural areas have gradually increased, among which photovoltaic agriculture, biomass energy, wind energy, hybrid energy systems, and Energy Internet technologies are particularly prominent. These technologies can not only meet the energy demands of rural production and life but also achieve the coordinated development of energy, agriculture, and ecology. For instance, photovoltaic agriculture, as an innovative land use model, organically combines solar power generation with agricultural production and has achieved rapid development worldwide. This technology achieves a synergistic increase in energy production and food production by installing photovoltaic modules above or around farmland. The diversified utilization paths of biomass energy are also constantly expanding. In addition to traditional biogas, the production of biodiesel, bioethanol, biochar, and other products provides more energy options for rural areas. The energy utilization of agricultural and forestry waste not only reduces environmental pollution but also creates additional income for farmers. Hybrid energy systems achieve a stable energy supply by integrating photovoltaic, wind energy, and energy storage technologies. The rural Energy Internet, on the other hand, optimizes energy management through intelligent and digital means, enhances energy efficiency and ease of use, and promotes the development of agriculture toward intelligence and low carbonization.

Furthermore, the development of rural energy technologies cannot do without policy support, economic incentives, and the guarantee of social acceptance. The natural conditions and agricultural production methods vary greatly in different regions, and the promotion of technology faces multiple challenges, such as adaptability, cost, management, and technical talents. Therefore, research on current rural energy technologies not only helps to understand their application status but also provides references for future optimization and promotion, policy-making, and technological improvement. This section summarizes ten typical application cases of rural energy technologies, as shown in Table 1.2.

In recent years, rural energy technologies have made remarkable progress. However, in the actual promotion process, they still face many constraints, which limit their large-scale application and sustainable development capabilities. At present, the main deficiencies of rural energy systems lie in technological maturity, economic feasibility, system integration capabilities, and the synergy with agricultural production. From a technical perspective, photovoltaic agriculture demonstrates unique advantages in land composite utilization, but the specific mechanism by which photovoltaic modules affect crop growth is not yet fully understood. The requirements of different crops for light intensity and spectral distribution vary significantly. Therefore, how to rationally design the layout density, installation height, and light transmittance of photovoltaic panels to ensure power generation efficiency without affecting crop yield remains an urgent problem to be solved. Research shows that fixed photovoltaic panels are prone to causing crop yields in shaded areas, while tracking photovoltaic systems, although they can improve the distribution of light, their high initial investment and maintenance costs limit their wide application in small-scale farms [61].

In terms of the utilization of biomass energy, the operational efficiency of traditional biogas projects drops significantly in winter in low-temperature regions. The anaerobic fermentation process has high requirements for the pretreatment of raw materials. When farmers lack professional operation knowledge, they are prone to frequent system failures. Especially in the cold regions of northern China, Europe, and North America, when the winter temperature drops below 0 °C, the activity of anaerobic fermentation microorganisms drops sharply, and the biogas production may decrease by more than 50%. Although heating and insulation measures can maintain the fermentation temperature, this will significantly increase energy consumption and operating costs and weaken the overall energy efficiency and economy of the biogas system. Meanwhile, the biomass direct combustion

Table 1.2 Application cases of rural energy technologies

Ref	Year	Technology type	Achievement
[61–67]	2013–2025	Photovoltaic agriculture	By integrating photovoltaic systems into farmlands, greenhouses, and farmers' facilities, the coordinated utilization of land, energy, and agricultural production has been achieved. This has enhanced energy output, water resource utilization rate, and ecological environment regulation capacity without significantly affecting crop growth.
[68–72]	2022–2024	Biomass energy	Agricultural waste can be efficiently converted into clean energy through technologies such as gasification or anaerobic fermentation, achieving a synergistic benefit of increasing the value of agricultural by-products, self-sufficiency in rural energy, and reduction of environmental pollution.
[73–76]	2023–2025	Hybrid energy system	A hybrid microgrid is built in collaboration with various renewable energy sources, such as photovoltaic and wind power, and energy storage systems, providing stable and sustainable electricity for rural areas, improving energy utilization efficiency, and significantly reducing carbon emissions.
[77–80]	2020–2024	Rural Energy Internet	The rural Energy Internet, based on the Internet of Things and intelligent monitoring technology, has achieved intelligent dispatching and real-time management of distributed energy, improved energy utilization efficiency and energy consumption awareness, and promoted sustainable electrification in rural areas.
[81–83]	2011–2019	Solar thermal utilization	The application of technologies such as solar water heaters, solar drying, and solar cooking in rural residences and agricultural product processing has significantly reduced the consumption of traditional fuels, improved indoor air quality, and decreased greenhouse gas emissions.
[84–86]	2014–2016	Small-scale wind power system	The combination of small wind turbines and hybrid energy storage systems provides off-grid power supply to remote rural areas, enhancing power accessibility and supporting the electrification of rural production and life as well as rural revitalization.

(Continues)

Table 1.2 (*Continued*)

Ref	Year	Technology type	Achievement
[87–89]	2016–2020	Biogas and biofuels	Biogas is produced through anaerobic digestion of livestock and poultry manure and crop straw for cooking and power generation. The biogas residue and liquid are returned to the fields as organic fertilizer, forming a circular agricultural model and promoting the resource utilization of waste and the sustainable development of rural energy.
[90–92]	2012–2016	Geothermal and heat pump technology	The application of ground source heat pumps and air source heat pumps in agricultural greenhouse heating and environmental control of livestock and poultry houses has achieved high efficiency and energy conservation as well as greenhouse gas reduction, enhancing the economic efficiency and environmental friendliness of agricultural production.
[93–95]	2014–2020	Rural electric transportation	The promotion of electric agricultural vehicles, electric tricycles, and their supporting charging facilities in rural areas, combined with the renewable energy charging network, has reduced the reliance on fossil fuels, lowered transportation costs, and lowered carbon emissions.
[96–98]	2016–2020	Intelligent energy management system	The rural intelligent energy management platform based on big data, artificial intelligence, and blockchain technology has realized demand response, energy trading, and load forecasting; optimized the matching of rural energy supply and demand; and enhanced the system's resilience and economic benefits.

power generation technology also faces practical predicaments such as high raw material transportation costs and low combustion efficiency. Agricultural wastes such as straw and rice husks have the characteristics of low energy density, large volume, and decentralized collection. When the transportation radius exceeds 50 kilometers, the transportation cost often exceeds the value of the raw materials themselves, which seriously restricts the economic scale of biomass power plants. In addition, the high moisture content, abundant ash, and high alkali metal content of agricultural waste can easily lead to problems such as slagging, corrosion, and excessive emissions during combustion. This requires advanced pretreatment and exhaust gas purification equipment, further increasing investment, operation, and maintenance costs.

Economic feasibility is an important factor restricting the promotion of rural energy technologies. Although the cost of renewable energy power generation has been continuously declining worldwide, the investment payback period in rural areas is generally long, and the difficulty in raising funds remains prominent. Photovoltaic agricultural projects usually require a large initial investment. However, farmers' own funds are limited. Moreover, financial institutions are relatively conservative in assessing the risks of rural energy projects, with high loan thresholds and relatively high interest rates, which further increases the investment pressure. The instability of subsidy policies has increased the uncertainty of projects. In some countries and regions, subsidies for renewable energy have gradually decreased or even been canceled, leading to a significant decline in the economic viability of projects. Biomass energy projects also face economic challenges. Fluctuations in raw material prices, unstable product markets, and high operation and maintenance costs make it difficult for many biomass energy enterprises to achieve profitability. Studies have pointed out that the levelized cost per kilowatt-hour of small-scale biomass power generation projects is usually higher than the local grid electricity price, lacking economic competitiveness [99]. Although hybrid energy systems can enhance the stability of energy supply, the high price and limited lifespan of energy storage equipment result in an unsatisfactory overall return on investment [100], and the frequent replacement of energy storage equipment also increases long-term operating costs.

Insufficient system integration capabilities and low levels of intelligent management are another significant bottleneck in the current development of rural energy. Rural energy systems are typically composed of various energy forms such as photovoltaic, wind energy, biomass energy, and traditional power grids. However, there is a lack of effective coordination and optimized dispatching among the subsystems, resulting in low energy utilization efficiency. For instance, during the peak hours of the day, photovoltaic power generation may generate excess electricity. However, due to the lack of energy storage equipment or flexible loads, this electricity cannot be effectively utilized, resulting in the phenomenon of abandoned light. However, at night or on rainy days, it is necessary to rely on traditional power grids or diesel generators to supplement electricity, which increases operating costs and carbon emissions. The application of Energy Internet technology in rural areas is still in its infancy. Functions such as intelligent monitoring, demand response, and load forecasting have not been widely promoted. Most rural energy systems still rely on manual management and experience-based regulation, making it difficult to achieve a dynamic balance between energy supply and demand.

The weak data collection and communication infrastructure further restricts the deployment of intelligent technologies. In many remote areas, network coverage is insufficient, and data transmission is unstable. The 4G/5G signal coverage rate is less than 50%, making it difficult to meet the demands of real-time monitoring, remote operation, and maintenance. Even in areas with good network coverage, the configuration rate of data collection devices such as sensors and smart meters is relatively low, lacking a complete data foundation to support intelligent decision-making. In addition, there is a shortage of rural energy management talents and an incomplete training system for technical personnel, which makes it difficult to effectively maintain advanced equipment. In many rural areas, there is a lack of professional energy management personnel. Farmers' understanding of new energy technologies is limited, and problems such as improper operation and untimely maintenance are widespread. As a result, the issues of low equipment operation efficiency and high failure rate remain prominent. Surveys show that some rural photovoltaic power stations, due to the lack of regular cleaning and maintenance, have seen their power generation efficiency decline by 20–30% faster than the design expectations, significantly reducing the economic benefits of the projects.

The insufficient coordination between rural energy technology and agricultural production is also an urgent problem to be solved at present. Photovoltaic agriculture and biomass energy technology can theoretically achieve a win-win situation for energy and agriculture, but in practice, there is often an insufficient balance. The installation of photovoltaic panels has altered the microclimate of farmland, including temperature, humidity, wind speed, and light distribution. These changes have complex impacts on the growth of different crops and lack systematic research. As a result, farmers lack scientific guidance in crop selection and management, leading to reduced yields or even failure of some photovoltaic agricultural projects. Biomass energy projects are highly dependent on raw material supply, while agricultural waste has obvious seasonal and dispersed characteristics. The supply of raw materials is unstable, and some facilities are idle during non-harvest periods. For instance, straw is mainly concentrated in the summer and autumn harvest periods, and the supply is very small at other times of the year. Although livestock and poultry manure are relatively stable, small-scale farms are scattered, and the collection cost is high, with uneven quality. This temporal and spatial mismatch in raw material supply poses significant challenges to the operation of biomass energy projects, requiring the establishment of a complete raw material collection, storage, and transportation system. However, this will significantly increase project costs. The impact of rural energy projects on land use has also triggered potential conflicts in terms of food security. In some regions, farmland is occupied for the development of photovoltaic or biomass energy, especially in land-scarce East and South Asia, which has sparked disputes over land competition between energy crops and food crops. Although the agro-photovoltaic complementary model can theoretically achieve dual use of one plot of land, if it is not properly designed and leads to a reduction in grain crop yields, it will still pose a threat to regional food security. Therefore, how to rationally promote the development of rural energy under the premise of ensuring food security requires more refined planning and management.

In the face of these problems, the future development trend of rural energy technology is moving toward high efficiency, intelligence, systematization, and diversification. Technological innovation will become the core driving force for breaking through bottlenecks. Photovoltaic agriculture is exploring new types of light-transmitting photovoltaic materials and flexible photovoltaic modules. These materials can selectively transmit the light spectrum that is beneficial to plant photosynthesis while converting other bands into electrical energy, achieving the maximization of energy and agricultural output. The innovative model of vertical agriculture and photovoltaic integration is also on the rise, improving land use efficiency through a three-dimensional layout. Biomass energy is developing toward high-value utilization. Biorefining technology can extract biofuels, biochemicals, and biomaterials from agricultural waste, enhancing the economic value of resources [101]. Advanced thermochemical conversion technologies such as gasification, pyrolysis, and hydrothermal liquefaction are gradually replacing traditional direct combustion and anaerobic fermentation, achieving a wider range of raw material processing, higher conversion efficiency, and cleaner products. Breakthroughs in energy storage technology will offer new opportunities for rural hybrid energy systems. New energy storage devices, such as solid-state batteries and flow batteries, are under development and demonstration, which are expected to reduce costs and enhance system safety and lifespan. Hydrogen energy has also drawn attention as a long-term energy storage method. By electrolyzing water with renewable energy to produce hydrogen and then generating electricity with fuel cells or hydrogen turbines, seasonal energy storage and dispatching can be achieved.

The deep integration of intelligence and digitalization will reshape the rural energy management model. The application of technologies such as the Internet of Things, big data, and artificial intelligence in rural energy systems will continue to deepen, achieving real-time monitoring, intelligent dispatching, and predictive operation and maintenance of distributed energy. By deploying multiple types of high-precision sensors and edge computing nodes in energy equipment, rural energy systems can achieve real-time collection, analysis, and feedback of multi-source data, thereby realizing state perception and optimized decision-making throughout the entire life cycle. Load forecasting and energy optimization algorithms based on machine learning can dynamically adjust energy-supply strategies according to weather, agricultural production activities, and energy consumption habits, thereby enhancing system efficiency and economic benefits and effectively reducing energy waste and operating costs. Meanwhile, artificial intelligence algorithms can also be used to identify abnormal energy consumption patterns and potential fault features, enabling early maintenance of equipment and risk warnings and enhancing the reliability and safety of the system. The application of blockchain in rural energy trading is gradually emerging. Through decentralized energy trading platforms, farmers can directly conduct peer-to-peer power transactions, improving energy utilization efficiency and increasing income. This mechanism not only enhances transaction transparency and reduces trust costs but also, with the support of smart contracts, enables automated settlement and data traceability, ensuring the fairness and stability of the energy market. Digital twin technology can build virtual models of energy systems, optimize system design, diagnose faults and evaluate performance, provide support for scientific decision-making, and achieve dynamic simulation and real-time optimization of energy systems by combining with AI algorithms. The popularization of intelligent technologies requires strengthening the construction of rural information infrastructure, including broadband networks, sensor networks, and data centers. At the same time, it is necessary to improve the data security and privacy protection mechanism and cultivate rural energy talents with digital management capabilities and interdisciplinary knowledge to form a sustainable system of coordinated development of technology, management, and systems.

Systematization and integration are the key directions for the future development of rural energy. The future rural energy system will no longer be a simple superposition of single technologies but an integrated energy system that achieves multi-energy complementarity and source–grid–load–storage coordination. This system can flexibly allocate various energy sources, such as photovoltaic, wind, and biomass energy, based on the characteristics of energy supply and demand at different time periods. It achieves efficient utilization and stable supply through energy storage devices and intelligent control and realizes a high degree of coupling and coordination among energy flow, information flow, and capital flow. To adapt to the characteristics of rural areas, such as scattered regions and large load fluctuations, smart microgrid technology will be widely applied. It can operate independently or be connected to the grid, improving power supply reliability and supporting the local consumption of renewable energy. At the same time, it can achieve load balance and emergency power supply. Rural microgrids can promote regional energy sharing and dynamic regulation by forming multi-level collaborative relationships with the main grid, thereby achieving energy autonomy and optimal allocation. The deep integration of rural energy and agricultural production will be further strengthened, and a precision agricultural energy system will be developed to provide customized energy services based on the growth stages of crops. For instance, smart greenhouses can dynamically adjust the shading degree of photovoltaic panels according to the photosynthesis requirements of crops, ensuring power generation while optimizing the growth environment of crops, achieving a three-dimensional coordinated optimization of "energy - agriculture – ecology." Meanwhile, the energy system will also be deeply integrated with the agricultural Internet of Things and the smart agriculture platform, achieving the integration of energy management and agricultural production management. Through real-time data interaction and adaptive control, the overall operational efficiency will be significantly enhanced. With the support of policies and the improvement of market mechanisms, the rural integrated energy system will gradually evolve toward low-carbonization, digitalization, and autonomy, promoting the deep integration of agricultural and rural modernization and the energy revolution.

Policy and business model innovation are equally important for the development of rural energy. Governments around the world are exploring more precise and sustainable support policies, shifting from simple equipment subsidies to full life cycle support, covering aspects such as research and development, demonstration applications, operation and maintenance, and product sales. Carbon trading and the green certificate system provide new revenue channels for rural renewable energy projects and improve their economic viability. New business models such as energy performance contracting and energy management have been promoted in rural areas. Professional energy service companies have been introduced to help farmers solve the problems of insufficient technical and management capabilities, reduce investment risks, and improve the operational efficiency of the system.

The community shared energy model is gradually emerging. Multiple farmers or villages jointly invest in building energy facilities and share the benefits and risks. This not only lowers the investment threshold for individual households but also enhances community cohesion and energy autonomy. Financial innovations such as green bonds and rural energy funds provide new ways for project financing, reduce financing costs, and attract more social capital to participate.

Overall, rural energy technology is at a critical stage of transitioning from exploration and demonstration to large-scale application. The challenges it faces are diverse and complex, but the development prospects are broad, and the path is clear. Through continuous technological innovation, policy support, business model exploration, and the collaborative efforts of all sectors of society, the rural energy system will develop in a more efficient, clean, intelligent, and sustainable direction, providing important support for rural revitalization, agricultural modernization, and global energy transition.

1.3 Theoretical evolution and technology stack of AEI

1.3.1 General evolution route

The Agricultural Energy Internet (AEI), as the systematic extension of the Energy Internet concept within the agricultural domain, represents an integrated and pioneering innovation that arises from the profound convergence of energy science, agricultural engineering, and intelligent information technology. It is not merely an industrial adaptation or sectoral application of the Energy Internet; rather, it constitutes a transformative paradigm through which the agricultural system – under the context of digitalization, networking, and intelligentization – realizes a comprehensive energy revolution and ecological transformation. The proposal and development of AEI signify that the agricultural energy system is undergoing a profound transition from an "energy supply–driven" model to an "information and intelligence–driven" model. As the agricultural embodiment of the Energy Internet, AEI not only emphasizes the generation and transmission of energy but also focuses on the synergistic interactions among energy, information, and ecology, forming a smart energy network that spans the entire life cycle of agricultural production. By combining renewable energy principles from energy science, systematic integration from agricultural engineering, and intelligent optimization from information technology, AEI achieves deep interdisciplinary fusion, enabling both the energy system and the agricultural ecosystem to operate collaboratively within a unified framework. Its emergence represents not only a technological enhancement in agricultural energy utilization but also a concrete manifestation of sustainable development concepts within the energy domain, serving as an essential strategic foundation for agricultural modernization and green, low-carbon transformation.

The developmental foundation of AEI can be traced to the converging effects of the global energy revolution and the digital transformation of agriculture. Against the backdrop of a restructured global energy landscape and intensified climate governance, agriculture – as both a significant energy consumer and a major carbon emitter – faces the urgent necessity of constructing an efficient, clean, and intelligent energy system. AEI has emerged in direct response to these historical imperatives. It builds upon the core principle of interconnection central to the Energy Internet while adapting it to the complex and diverse characteristics of agricultural production. In doing so, AEI breaks through the traditional boundaries that have long separated energy systems and agricultural systems, achieving an integrated framework that merges energy, information, and agriculture into a coherent whole. Through this fusion, AEI enables a fundamental transformation of agricultural energy use – from fragmented and extensive utilization patterns to coordinated and optimized management modes, and from experience-based operation to data-driven intelligent decision-making. This transition establishes a clear path for the creation of a clean, low-carbon, secure, efficient, and intelligent agricultural energy ecosystem. Consequently, AEI not only expands the practical boundaries of Energy Internet theory but also provides a powerful new driving engine for agricultural modernization and sustainable growth.

The core objective of AEI is to establish an intelligent, adaptive, and sustainable energy system that spans the entire agricultural value chain – covering production, processing, logistics, and consumption – through the organic coordination of multiple advanced technologies, including information and communication technology (ICT), the Internet of Things (IoT), big data analytics, artificial intelligence (AI), and renewable energy (RE) [102]. Unlike conventional systems that emphasize isolated technological integration, AEI operates under a systems engineering paradigm, wherein multi-source data are continuously sensed, transmitted, processed, and fed back across hierarchical and contextual layers to achieve dynamic and adaptive coordination. Within every stage of agricultural activity, AEI embeds an intelligent energy management logic – ranging from the electrification control of agricultural machinery to the precise thermal regulation of greenhouse environments, and from the optimization of energy distribution in processing and cold-chain logistics to the intelligent scheduling of renewable energy supply. Throughout these stages, the dual flow of energy and information functions as the operational backbone of AEI, ensuring real-time interaction and optimization. By constructing a data-driven energy chain that links production, processing, and consumption, AEI achieves transparency, visualization, and intelligent automation in energy management, enabling a shift from static energy allocation to real time, adaptive energy decision-making.

At the same time, the developmental objectives of the Agricultural Energy Internet (AEI) have extended beyond the realm of energy optimization to encompass the dual dimensions of ecology and economy. AEI not only pursues the enhancement of energy utilization efficiency but also places strong emphasis on achieving a harmonious balance between energy behavior and ecosystem stability. Through the integration of big data analytics and artificial intelligence (AI)-driven predictive models, AEI is capable of dynamically assessing the carbon emission profiles and energy consumption characteristics that arise throughout the agricultural production process. By combining these real-time analytical insights with renewable energy access strategies, AEI realizes the dual optimization of energy utilization and ecological protection [102].

Furthermore, the incorporation of information-based and intelligent management mechanisms endows agricultural energy systems with the capabilities of self-learning, self-adaptation, and self-feedback. Such intelligent attributes enable the system to autonomously adjust its energy-supply configuration in response to variations in seasonal cycles, climatic conditions, and crop-specific physiological demands. In this way, AEI ensures that energy distribution dynamically aligns with agricultural production needs, thereby achieving minimal resource consumption and reduced environmental load without compromising productivity or output stability.

This information-driven and intelligence-coordinated operational paradigm fundamentally reshapes the interaction between agriculture, energy, and ecology. By using information flow as a control signal and intelligent algorithms as decision agents, AEI transforms agricultural energy management from a reactive model into a proactive, predictive, and adaptive system. This transition allows the agricultural ecosystem to operate in a state of dynamic equilibrium, where energy efficiency, ecological sustainability, and economic viability reinforce one another. Consequently, AEI serves as a critical instrument and strategic enabler for advancing green, high-quality agricultural development, offering a scientific pathway toward the construction of a low-carbon, resilient, and intelligent agricultural future.

The theoretical evolution route of the Agricultural Energy Internet (AEI) is illustrated in Figure 1.1. The essence of this system lies in a closed-loop mechanism composed of "information perception–intelligent analysis–dynamic scheduling," which propels the agricultural energy system from a traditional linear energy-supply model toward an intelligent, synergistic, and self-adaptive optimization paradigm. The core philosophy of this mechanism is "data-driven decision-making and intelligent energy optimization." By continuously perceiving, analyzing, and feeding back information about the agricultural production environment and energy usage state, AEI achieves an efficient coupling of energy flow and information flow. During the "information perception" stage, the system gathers multi-dimensional data through sensors and Agri-IoT devices, including meteorological parameters, soil temperature and humidity, equipment energy consumption, crop growth conditions, and carbon emission levels. In the "intelligent analysis" stage, AI models and big data algorithms are employed to model energy consumption patterns, forecast energy demand, and identify optimization opportunities. Finally, in the "dynamic scheduling" stage, the system automatically generates optimal operational strategies based on real-time data, achieving precise coordination among energy supply, storage management, and terminal utilization. Through this continuous and iterative closed-loop process, AEI acquires self-learning and self-optimization capabilities in complex environments, enabling the agricultural energy system to evolve from a passive-response mode to an active-regulation paradigm.

This mechanism not only transforms the operational logic of energy systems but also reshapes the conceptual framework of agricultural energy management.

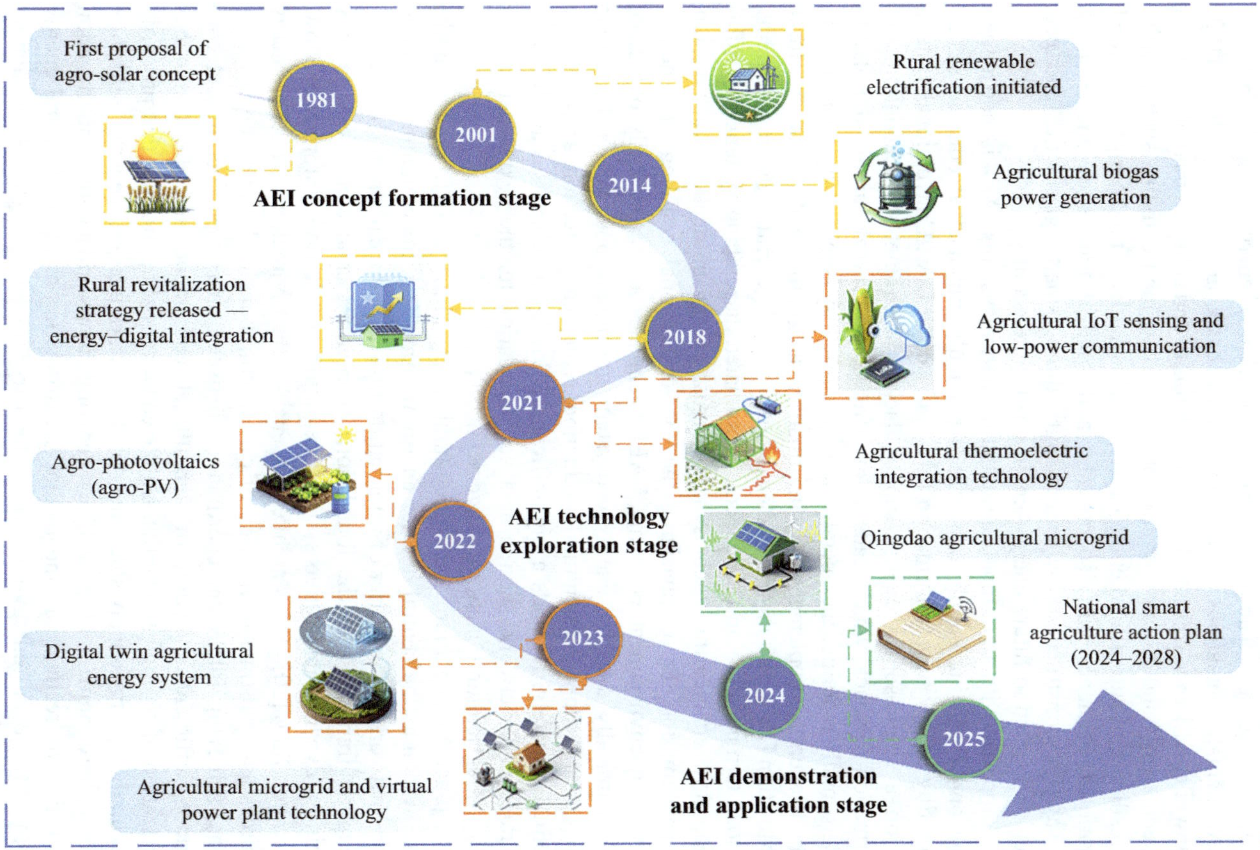

Figure 1.1 Theoretical evolution route of AEI

Traditional linear supply models focus solely on "supply–demand balance," lacking dynamic feedback and ecological constraints. In contrast, the closed-loop architecture of AEI achieves a transition from "energy centralization" to "information intelligence," enabling real-time adjustment of energy strategies based on environmental fluctuations, meteorological forecasts, and market signals [103]. Meanwhile, AEI enhances energy system reliability and achieves dual optimization between energy and ecology, significantly improving the flexibility, sustainability, and resilience of agricultural energy systems. This model not only improves overall operational efficiency but also provides a theoretical foundation and technical pathway for constructing low-carbon agriculture, smart agriculture, and eco-agriculture suited to future energy–environment interactions.

Compared with traditional agricultural energy systems, AEI is not a mere extension of existing supply networks but a multi-dimensional coupled system integrating energy flow, information flow, and carbon flow. Conventional systems rely on a single-source energy supply, lacking coordination, scheduling, and information feedback, which leads to low efficiency and weak carbon control. AEI fundamentally redefines the operational logic of such systems through mechanisms of multi-energy complementarity, information integration, and carbon-constraint coordination [103]. Within this framework, distributed generation, storage systems, and load terminals are unified under a coordinated dispatching system, and information flow enables intelligent synergy across energy production, transmission, and consumption. By emphasizing the four-dimensional coupling of "energy–information–environment–economy," AEI establishes a systematically optimized and dynamically adaptive energy ecosystem, forming a positive feedback loop linking energy supply, agricultural production, ecological protection, and economic benefit. Ultimately, it builds a high-efficiency, low-carbon, and sustainable agricultural energy network, serving as a key foundation for the next generation of intelligent agricultural energy management.

In practical operation, the Agricultural Energy Internet (AEI) positions the power system as the central hub for energy transmission, the communication and sensing network as the carrier of information flow, and the carbon monitoring and accounting framework as the foundation for environmental constraints, thereby forming a dynamic interactive system integrating energy flow, information flow, and carbon flow [103]. This architecture enables the optimal allocation of energy resources and real-time monitoring of carbon emissions under multi-energy conditions, achieving intelligent coordination between energy production and consumption, efficient operation of agricultural equipment, and the continuous optimization of ecological benefits. Unlike conventional approaches that primarily focus on the "energy supply side," the AEI framework integrates energy utilization processes, environmental feedback, and economic loops into a holistic optimization model, thereby realizing a fundamental transformation of the agricultural energy system – from an "energy-consuming industry" to an "energy self-circulating ecosystem." Through this new energy paradigm, characterized by intelligence and ecological synergy, agricultural energy management is advancing toward systematization, refinement, and low carbonization, providing an innovative pathway for achieving green agricultural development and sustainable energy utilization.

The theoretical foundation of AEI originates from the holistic principles of systems science, with its core operational logic centered on using high-quality information flow to drive the optimal configuration of energy flow, applying intelligent algorithms for self-adaptive system control, and implementing carbon emission constraints to guide the balanced evolution between energy utilization and the ecological environment [104]. Under this framework, energy ceases to be a unidirectional supply mechanism and instead becomes a dynamic system that is constantly sensed, computed, and optimized. Within this system, information flow functions as the "neural network" of the energy ecosystem, forming a multi-layer information structure through sensors, edge nodes, and cloud-based analytical platforms to achieve comprehensive perception and precise feedback. Energy flow, guided by intelligent decision-making models, undergoes optimized scheduling, ensuring the best coordination among distributed generation units, storage devices, and agricultural loads. Carbon flow, serving as an ecological constraint factor, influences the generation and adjustment of energy scheduling strategies through carbon accounting, emission monitoring, and carbon credit mechanisms. Through this three-flow synergistic optimization model – encompassing energy, information, and carbon – AEI constructs a self-sensing, self-decision-making, and self-learning energy ecosystem, enabling agricultural energy systems to achieve continuous optimization and dynamic equilibrium within complex and variable environments.

From an overall developmental perspective, AEI's formation and evolution have undergone three major stages – the conceptual germination, technological exploration, and early demonstration phases – with a comparative analysis presented in Table 1.3. The timeline spans from approximately 1981 to the present,

Table 1.3 Comparative analysis of AEI evolution stages

Stage	Time span	Core characteristics	Key technologies/ Events
Conceptual germination stage	1981–2019	Initial coupling of energy and agriculture	Photovoltaic grid connection and biogas power generation
Technological exploration stage	2020–2023	Intelligent perception and multi-energy synergy	Internet of Things (IoT), virtual power plant (VPP), and digital twin
Demonstration and application stage	2024–Present	System integration and market mechanism	Qingdao Agricultural Microgrid

reflecting a progressive transformation of agricultural energy systems from "supply-driven" to "intelligence-driven" models. Each stage of this evolution has been closely linked to advances in digitalization, intelligent control, and the depth of carbon-constraint integration within agricultural energy systems [7]. Early AEI research focused primarily on distributed energy integration and renewable energy substitution, gradually transitioning toward AI-driven energy management and virtual power plant scheduling. Ultimately, this evolution has culminated in the establishment of a systematic theoretical framework that integrates energy optimization, carbon accounting, and intelligent decision-making, laying the foundation for the next-generation intelligent agricultural energy ecosystem.

The conceptual germination stage (1981–2019) marks the gradual formation of the ideas and prototypes of the Agricultural Energy Internet (AEI). In 1981, Goetzberger and Zastrow proposed the concept of the synergistic use of solar energy in agriculture, integrating photovoltaic (PV) modules with crop cultivation to simultaneously achieve agricultural production and energy conversion [105]. This innovative idea not only laid the theoretical foundation for agrivoltaic technology but also opened new directions for subsequent research on agricultural–energy integration [106–107]. It revealed the potential symbiotic relationship between agricultural production systems and energy conversion processes: solar energy can both power photosynthesis in plants and be converted into electricity through PV systems to supply energy for agricultural facilities and machinery. Thus, the prototype of "energy-symbiotic agriculture" began to take shape.

At the beginning of the twenty-first century, with the intensification of the global energy crisis and the growing demand for rural electrification, the development of agricultural energy systems entered a critical stage characterized by a transition from fragmented and extensive utilization to systematic, intensive, and electrified development. Across the world, energy structure transformation and rural sustainable development became central themes of national policy agendas. The challenge of achieving synergistic progress in energy security, environmental protection, and economic growth within the agricultural sector emerged as a pressing issue of the era. Against this macro background, China took the lead in launching a series of clean-energy-based rural energy initiatives, which provided a significant opportunity for the conceptual formation and technological exploration of the Agricultural Energy Internet (AEI). In 2001, China officially implemented the China Township Electrification Program, a project regarded as a milestone in the transformation of China's rural energy system. The program's core objective was to comprehensively enhance rural power-supply capacity and local energy self-sufficiency through the application of renewable energy technologies, thereby promoting the modernization of both agricultural production and rural livelihoods.

During the implementation phase, the Chinese government deployed distributed energy systems centered on photovoltaic (PV) power generation, small hydropower, and wind energy across more than 1,000 rural regions, forming a multi-energy complementary and distributed supply network for rural areas. These systems not only achieved electrification coverage in remote regions but also organically integrated agricultural production, rural living, and energy provision into a cohesive operational system. This integration accelerated the electrification of agricultural equipment and the mechanization of production processes, promoting both the transformation and upgrading of the rural economic structure. Compared with the traditional reliance on coal, firewood, and other low-efficiency fuels, this project marked the first national-level integration of clean energy technologies into agricultural production systems, symbolizing a crucial step in the evolution of agricultural energy structures from "single-source supply" to "multi-energy complementarity and coordinated control."

More importantly, the successful implementation of the Township Electrification Program not only established technical benchmarks for distributed energy applications but also laid a policy and institutional foundation for the subsequent development of the Agricultural Energy Internet (AEI). It facilitated the transformation of rural power systems from traditional one-way supply networks to modern energy architectures characterized by multi-source integration, bidirectional interaction, and intelligent dispatching, thereby endowing the system with the early features of networking and informatization. Meanwhile, the technical achievements, empirical data, and engineering models derived from the project provided practical support for AEI's conceptual design and theoretical framework. For instance, technological advancements in areas such as distributed PV integration, power-quality management, and coordinated control of energy storage became early prototypes of AEI's later "source–grid–load–storage" synergistic optimization model. In essence, this stage represented a leap from "energy engineering" to "energy networking," marking the transition of agricultural energy systems from experimental research to large-scale deployment and providing a solid engineering paradigm and policy precedent for the systematic construction of AEI in the decades that followed.

On this basis, research on the energy utilization of agricultural waste developed rapidly. Between 2009 and 2014, Yang, Chen, and others conducted systematic studies on agricultural waste power generation and biogas utilization in rural areas, proposing a biogas–power coupled system centered on anaerobic fermentation, achieving a closed-loop conversion from waste to energy [108]. This model was demonstrated in multiple rural regions, enabling agricultural energy systems to gain internal energy circulation, emission reduction, and efficiency improvement capabilities for the first time. The research of this period reflected the extension of energy systems toward ecological cycles and provided the material and energy flow foundation for AEI's integrated "source–grid–load–storage" framework. Meanwhile, information technology began to gradually penetrate agricultural energy systems. After 2013, the integration of distributed energy scheduling, intelligent monitoring, and agricultural information management systems marked the refinement of energy perception, transmission, and control, giving agricultural energy systems their initial "intelligent regulation" capability.

In 2018, the Central Committee of the Communist Party of China and the State Council issued the Rural Revitalization Strategic Plan (2018–2022), which called for improving rural energy infrastructure networks, extending power and gas supply systems to rural areas, and promoting the digitalization, greening, and intelligent development of agriculture. This policy marked the formal recognition of the AEI concept at the national level. It positioned the energy network as a key support system for agricultural modernization and elevated the integration of information technology to a national strategic priority. From this period onward, the concept of AEI evolved from academic exploration to policy and practical implementation, forming a systematic framework supported by energy technology, agricultural economics, and information science. Overall, the conceptual germination stage was characterized by the transition of agricultural energy systems from single-energy utilization to multi-energy integration, the gradual establishment of information management, and the alignment of national policy, laying solid ideological, technical, and institutional foundations for AEI's systematic development.

The technological exploration stage (2020–2023) marked a shift in AEI research from conceptual construction to the integration and verification of key technologies. In 2021, Fu *et al.* proposed an integrated optimization model for agro-thermoelectric systems, establishing a coupled mechanism encompassing greenhouse cultivation, livestock breeding, and agricultural product processing. This significantly improved energy efficiency and system stability, realizing internal energy self-balance within agricultural systems [109]. This study systematically described the bidirectional interaction between thermal and electrical energy in agriculture, providing mathematical foundations and regulation strategies for multi-energy complementary systems. In the same year, Tao *et al.* proposed a low-power IoT architecture based on NB-IoT and LoRa communication technologies, which enabled distributed real-time sensing and adaptive control in agricultural energy systems [110]. This communication structure greatly enhanced AEI's data acquisition depth and system responsiveness, providing the necessary conditions for realizing intelligent and autonomous agricultural energy networks.

In 2022, agrivoltaic technology achieved major progress. Alam *et al.* proposed a techno-economic integrated model for agrivoltaic systems, verifying the optimal matching relationships between crops and PV systems under different light intensities and crop types [111]. By introducing dynamic light-scheduling algorithms and crop energy balance analysis, the model achieved dual optimization of PV power generation and crop yield, significantly improving land and energy utilization efficiency. This year is widely regarded as a key milestone marking AEI's evolution from distributed energy systems to hybrid energy systems. Meanwhile, domestic projects such as PV greenhouses and agricultural microgrids were launched, marking the entry of AEI's core technologies into the demonstration and validation phase.

In 2023, Fu, Zhou, and others further proposed a digital-twin-based modeling and control framework for agricultural energy systems [112]. By establishing virtual–physical fusion models, this framework enabled synchronized mapping and predictive control of greenhouse loads, distributed energy sources, storage devices, and carbon emission behaviors. The introduction of digital twin technology transformed agricultural energy systems from static energy scheduling to dynamic adaptive optimization, endowing them with self-learning and real-time decision-making capabilities. In the same year, Uddin *et al.* proposed a distributed scheduling mechanism for multi-source agricultural energy systems within agricultural microgrids and virtual power plants (VPPs) [113], achieving flexible coordination among PV generation, energy storage, and load sides. Through VPP aggregation and regulation functions, agricultural energy systems achieved regional energy autonomy and interregional energy balance. At this stage, AEI evolved from isolated technological breakthroughs to system integration and cross-domain convergence, forming a technology framework centered on intelligent perception, energy synergy, and digital twin technologies.

Since 2024, AEI has entered the Demonstration and Application Stage, characterized by the dual advancement of engineering implementation and policy guidance. In 2024, Fu *et al.* conducted a systematic optimization of the energy–environment–economy synergy in agricultural microgrids under a carbon accounting framework, with the Qingdao Agricultural Microgrid serving as a representative research and demonstration project. Based on real operational data, the system established an optimization model that enabled dynamic coupling and scheduling among PV generation, storage systems, and agricultural loads, verifying AEI's feasibility and benefits in rural microgrid scenarios. The study demonstrated that AEI could significantly enhance agricultural energy efficiency, reduce carbon emissions, and realize preliminary models of "zero-carbon farms" and "smart energy fields."

In the same year, China issued the National Smart Agriculture Action Plan (2024–2028), explicitly proposing the deep integration of IoT, artificial intelligence, digital twin, and energy systems, and launched demonstration projects such as "Fuxi Farm" in Zhejiang, Shandong, and Sichuan. These projects integrated PV generation, energy storage, environmental monitoring, and energy–carbon management modules, forming replicable and scalable engineering templates for AEI.

In 2025, with the National Smart Agriculture Action Plan (2024–2028) entering its full implementation phase, the development of the Agricultural Energy Internet (AEI) reached a critical turning point – transitioning from "theoretical exploration" to "large-scale practical deployment." The nationwide launch of multiple AEI demonstration projects marked a significant milestone in China's progress from technological validation to system-level implementation within the agricultural energy sector. These initiatives across the country exhibit new defining characteristics of regional clustering, system intelligence, and low-carbon operation, forming a multi-layered development structure centered on regional collaboration, cross-domain interconnection, and intelligent control. In particular, major agricultural provinces in the eastern coastal and central-western regions have begun constructing comprehensive application systems that integrate agricultural energy production, scheduling, and consumption through coordinated policy guidance, research support, and industrial cooperation. These efforts are driving the intelligent transformation of agricultural energy systems toward self-perception, self-learning, and self-optimization, thus achieving a deeper convergence between digital agriculture and smart energy management.

Demonstration projects in Zhejiang Province and Qingdao City have emerged as national benchmarks for AEI development. Zhejiang, leveraging its strengths in digital rural infrastructure and clean energy deployment, has built a multi-energy complementary system integrating distributed photovoltaics (PV), wind power, energy storage, and electric agricultural machinery, achieving deep integration between energy supply and agricultural production. Qingdao, on the other hand, has taken the lead in establishing agricultural microgrids and agricultural virtual power plants (Agri-VPPs), forming an intelligent energy management platform that spans greenhouse cultivation, aquaculture, and animal husbandry. These projects employ unified intelligent control systems to enable real-time coordination and dynamic optimization among PV generation, wind turbines, storage units, and agricultural loads. The system autonomously generates energy scheduling strategies based on weather forecasts, crop growth stages, electricity price fluctuations, and carbon emission constraints, achieving a dual balance between economic performance and low-carbon operation while ensuring the continuity of agricultural production. Through the integration of AI-driven predictive models and IoT-based sensor networks, AEI systems have realized precise sensing and rapid response, improving energy utilization efficiency by more than 20% while significantly reducing carbon intensity, thus demonstrating the feasibility of large-scale, intelligent agricultural energy management.

More importantly, this phase of AEI development represents not merely a technological breakthrough but the comprehensive initiation of systematization and standardization efforts. With the release of the National Technical Specification for the Agricultural Energy Internet and corresponding industry guidelines, AEI construction has evolved into a modular, replicable, and scalable engineering framework. Standardization now encompasses all layers of the system – from equipment communication protocols and data exchange interfaces to carbon accounting models and virtual power plant scheduling algorithms – providing robust institutional guarantees and technical foundations for AEI expansion nationwide. At the same time, market-oriented mechanisms have begun to emerge: several regions are developing agricultural energy trading and carbon credit settlement platforms, enabling farmers to participate in energy markets through PV generation, energy storage dispatch, and demand-side management, thereby realizing shared economic benefits from clean energy production. By this stage, AEI has successfully completed its evolution from a conceptual framework to an engineered system, entering a new era defined by "intelligent operation, standards-driven governance, and market-based development." This transformation lays a solid foundation for the intelligent, low-carbon, and ecological transition of future agriculture in China and beyond.

Overall, the development trajectory of the Agricultural Energy Internet (AEI) demonstrates a progressive, spiral evolution from "energy substitution" to "multi-energy synergy" and ultimately to "intelligent interconnection." This evolutionary process reflects not only the gradual optimization of technological systems but also the systemic integration of energy infrastructures, information networks, and agricultural ecosystems in terms of conceptual frameworks, structural design, and operational mechanisms. From a temporal perspective, the emergence of AEI has not been an instantaneous achievement; rather, it represents the culmination of a long process of theoretical incubation, technological accumulation, and systemic construction.

During the conceptual germination phase, the foundation was laid for replacing traditional high-carbon energy sources with clean and renewable alternatives, thereby establishing the overarching direction for green transformation in agricultural energy systems and providing both a theoretical basis and policy guidance for sustainable development. In this phase, the gradual deployment of photovoltaic (PV) generation, biomass energy, and wind power technologies endowed agriculture with the potential for local energy self-sufficiency, signifying a crucial shift from passive dependence to active generation and management within the agricultural energy domain. The subsequent technological exploration phase achieved major breakthroughs in informatization, intelligent control, and simulation-based management, laying the technical groundwork for AEI's systemic integration. The introduction of artificial intelligence (AI), big data analytics, the Internet of Things (IoT), and digital twin technologies transformed the agricultural energy system from a static energy-supply network into a learning, adaptive, and evolving intelligent ecosystem capable of real-time sensing, predictive optimization, and dynamic regulation. The demonstration and application phase, characterized by engineering implementation and policy guidance, marked AEI's transition from theoretical research to practical deployment. Through the establishment of national and regional demonstration projects, AEI's architectural framework has gradually evolved toward standardization, modularization, and replicability, forming a positive developmental chain that links technological integration, engineering demonstration, and market expansion, thereby accelerating the transformation of agricultural energy systems from conceptual models into operational infrastructures.

Looking toward the future, with the continuous advancement of carbon neutrality strategies and the Rural Revitalization Policy, the Agricultural Energy Internet is poised to play an irreplaceable foundational role in green, low-carbon transitions and ecological civilization construction. It will not only serve as a critical infrastructure for achieving agricultural decarbonization and energy transformation but also act as a strategic support platform for building a new model of rural energy governance and promoting urban–rural coordinated development. The widespread deployment of AEI will facilitate the clean diversification of energy supply, the intelligent management of energy consumption, and the digitalization of energy governance in rural regions, thereby creating an integrated "Energy–Information–Ecology" smart agricultural ecosystem. Simultaneously, AEI will enable active participation of farmers in both energy and carbon markets, empowering rural economies through energy sharing, carbon trading, and digital service platforms, and promoting the comprehensive upgrading of agricultural industrial chains. It is foreseeable that, as technological maturity and institutional mechanisms continue to advance, AEI will emerge as a primary driving force of agricultural modernization, propelling both China and the global agricultural sector toward a new era of intelligent, efficient, low-carbon, and sustainable development.

1.3.2 Technical framework

The Agricultural Energy Internet (AEI), as a systematic extension of the Energy Internet concept into the agricultural domain, represents a profound transformation of contemporary agricultural production systems toward digitalization, intelligence, and low-carbon sustainability [102]. It is not only an inheritance and expansion of the theoretical framework of the Energy Internet but also an inevitable outcome of agricultural evolution under the dual forces of global energy transition and rapid advances in information technology. The emergence of AEI signifies a paradigm shift in agricultural energy systems – from a model centered on energy supply to one driven by information intelligence and coordinated optimization – thus initiating the structural reconstruction and intelligent evolution of agricultural energy systems. As an integrated system that synthesizes energy science, agricultural engineering, and artificial intelligence technologies, AEI addresses critical limitations of traditional agricultural energy systems such as dispersed supply, low utilization efficiency, and rudimentary management. The establishment of intelligent perception, feedback, and decision-making mechanisms enables dynamic optimization of the agricultural production process under ecological constraints. With its systemic development, the agricultural energy system has evolved from a single-source supply mode into a multi-flow, coupled, intelligently coordinated, and low-carbon ecological network, thereby becoming a key infrastructure for sustainable agricultural development in the future.

The core philosophy of AEI lies in achieving multi-dimensional coordination among electric power flow, thermal energy flow, fuel flow, and information flow, enabling intelligent perception, precise scheduling, and carbon-constrained optimization within agricultural energy systems, and ultimately forming a comprehensive energy network characterized by high efficiency and environmental compatibility [102]. Within the AEI framework, the electric power flow serves as the fundamental carrier of energy; the thermal and fuel flows ensure energy balance and transformation, while the information flow functions as the neural network that connects, monitors, and regulates them. The integration of information flow endows the system with dynamic perception and adaptive decision-making capabilities, allowing it to optimize energy allocation under varying meteorological, crop growth, and market conditions. By means of real-time data acquisition, AI-driven analytics, and automated scheduling, the AEI achieves simultaneous balance among energy supply and demand, carbon emission minimization, and economic optimization. This "energy–information–carbon" trinity transforms AEI from a conventional energy transmission framework into an intelligent, self-learning, and ecologically responsive agricultural energy ecosystem.

The technical framework of AEI has, in recent years, undergone a transition from distributed energy interconnection toward an intelligent and collaborative energy network, marking a critical turning point from localized optimization to globally intelligent coordination in agricultural energy systems. Its overall architecture can be summarized as a three-tiered hierarchical system – anchored by electric power flow, driven by information flow, and governed by intelligent algorithms. The framework consists of a physical layer, an information layer (including intelligent decision modules), and an application layer, which interact through bidirectional energy and information exchanges to form a self-perceiving, self-learning, and self-optimizing dynamic system [103]. The physical layer is responsible for energy harvesting, conversion, storage, and execution, serving as the material foundation of AEI; the information layer, enabled by the Internet of Things (IoT), 6G communication, and edge computing, facilitates real-time data acquisition, fusion, and decision-making, functioning as the intelligent core of the system; while the application layer directly interfaces with agricultural operations, enabling energy scheduling, equipment control, environmental monitoring, and carbon management. Together, these three layers operate in an interdependent and co-evolutionary manner, forming an adaptive and dynamically optimized agricultural energy ecosystem capable of responding effectively to external environmental and economic fluctuations.

Compared with traditional Energy Internet architectures, the technical framework of the Agricultural Energy Internet (AEI) distinguishes itself through its heightened focus on multi-factor coordination and carbon-constrained management within agricultural contexts [103]. While conventional Energy Internet systems primarily emphasize the efficient transmission, distribution, and complementary utilization of multiple energy sources, AEI incorporates the diversity, seasonality, and ecological sensitivity inherent in agricultural production into the overall design of the energy system, thereby forming a three-dimensional dynamic coupling structure that integrates energy flow, information flow, and carbon flow. This framework not only achieves intelligent optimization of system operations but also harmonizes energy efficiency, ecological balance, and carbon emission control within agricultural processes. Through the integration of carbon accounting models and real-time carbon flow monitoring, the AEI system can dynamically evaluate emission levels during energy scheduling and autonomously adjust configuration strategies to maintain low-carbon, efficient, and sustainable operations. As illustrated in Figure 1.2, the AEI architecture represents more than a technical construct – it embodies an intelligent evolutionary paradigm for agricultural energy systems, symbolizing the future direction of agricultural production toward deep integration of digitalization, intelligence, and green development.

At the foundation of the system, the physical layer serves as the material basis and the core carrier of energy flow within the Agricultural Energy Internet (AEI). Its primary function is to realize the collection, conversion, storage, and execution of energy, ensuring the efficient transmission and complementary utilization of electric, thermal, and fuel energy flows within the agricultural production system, thereby providing the physical guarantee for the stable operation of the entire AEI framework. The physical layer comprises four interrelated and dynamically coordinated core modules – energy-supply systems (source), energy distribution networks (grid), energy storage systems (storage), and agricultural load terminals (load) [104]. Together, these components form a closed-loop agricultural energy cycle that supports the seamless operation of AEI.

In the energy collection and supply stage, the physical layer is built upon a diverse array of renewable energy technologies, establishing a distributed clean energy-supply network centered on photovoltaic (PV) generation, wind power, biomass energy, geothermal energy, and biogas generation. By integrating multiple distributed generation units, agricultural energy systems are progressively freed

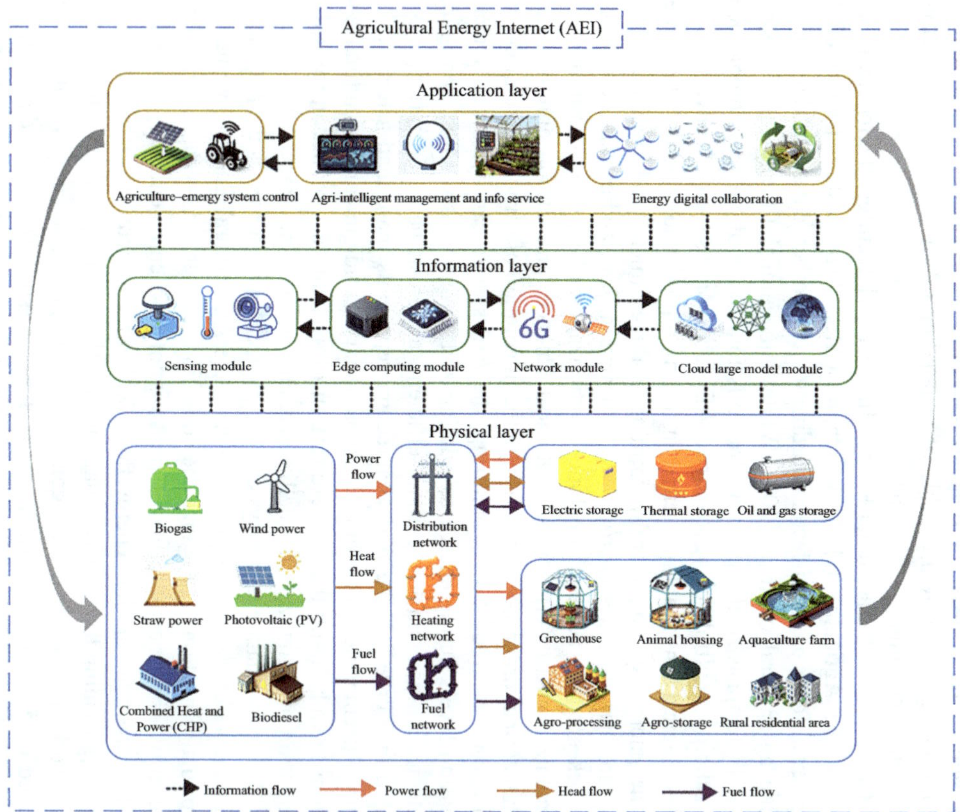

Figure 1.2 Technical framework of AEI

from dependence on centralized external power sources, achieving regionalized and self-sufficient energy production [104]. Among these, the "Agro-PV" and "Fishery–PV Integration" models stand as representative examples of deep energy–agriculture integration. These systems employ a co-located structural design that synchronizes power generation with agricultural production – PV panels above generate electricity, while the space below provides optimal light, humidity, and temperature conditions for crops or aquaculture. This dual-use configuration not only enhances land use efficiency but also achieves synergistic optimization of energy and ecology within a shared spatial domain. Meanwhile, wind, biomass, and geothermal resources further reinforce the system's resilience and continuity. Wind energy provides reliable power in coastal and plain areas, geothermal energy delivers stable heat in colder northern regions, and biomass or biogas systems transform agricultural waste into renewable energy, completing the "source–waste–energy" circular chain and driving the agricultural energy system toward a clean and closed-loop model.

In the energy conversion and storage stage, the physical layer achieves cross-medium coupling and dynamic balancing among electrical, thermal, and fuel flows through a multi-medium energy storage system. Electrochemical storage (e.g., lithium-ion, sodium-ion, and solid-state batteries) performs short-term load leveling and peak shaving, while thermal storage units – including phase-change materials, hot-water tanks, and geothermal exchangers – enable diurnal and seasonal energy transfer. Fuel storage systems (such as biogas tanks, hydrogen storage vessels, and LPG tanks) serve dual functions of energy transformation and emergency backup [7]. Within this process, hydrogen energy systems play a pivotal role, creating a "power–hydrogen–heat" multi-form energy cycle via electrolysis and fuel cell technologies, thereby extending the depth of energy utilization. With the integration of AI-based predictive algorithms and Energy Management Systems (EMS), storage devices can dynamically adjust charge–discharge strategies in response to meteorological fluctuations, load curves, and market signals, effectively mitigating intermittency, reducing curtailment, and improving overall efficiency [102]. The energy distribution network acts as the critical bridge linking energy sources with terminal users, utilizing electrical grids, thermal pipelines, and fuel conduits to establish a multi-energy circulation channel across electricity, heat, and gas. Through bidirectional inverters and intelligent switching nodes, distributed sources can achieve seamless grid interconnection and reverse energy transmission, enabling autonomous operation of regional microgrids.

At the terminal execution stage, the "load" module encompasses the principal energy-consuming sectors of agricultural production, processing, and rural life – such as intelligent greenhouses, livestock farms, aquaculture bases, agro-processing plants, cold-chain logistics facilities, and residential zones. These terminal nodes not only represent the main points of energy consumption but also serve as vital gateways for efficiency optimization and information feedback. Through the coordinated operation of embedded sensors, smart controllers, and edge computing nodes, terminal systems continuously capture environmental and equipment data – such as temperature, humidity, energy consumption, and productivity – and autonomously execute control instructions based on crop growth stages, weather conditions, and load demands to achieve on-demand energy supply and precision regulation. For instance, in intelligent greenhouses, the system dynamically adjusts shading and heating according to solar irradiance and energy pricing; in cold-chain logistics, predictive models allocate electric and thermal storage resources to ensure product quality while minimizing energy use. Overall, the physical layer is characterized by multi-energy complementarity, coordinated conversion, and intelligent execution, integrating electrical, thermal, and fuel flows into an organically coupled infrastructure that simultaneously meets agricultural production needs and enhances energy resilience and ecological sustainability [103]. The efficient operation of this layer not only provides the continuous energy and data foundation necessary for intelligent cognition and optimization at the information layer but also lays a robust physical groundwork for AEI's holistic coordination and sustainable development.

The information layer serves as the intelligent core of the Agricultural Energy Internet (AEI) and represents the critical hub where energy flow, information flow, and carbon flow converge and interact. It not only undertakes the fundamental functions of data acquisition and transmission, but also constitutes the central mechanism enabling intelligent analysis, predictive scheduling, and adaptive optimization within the AEI framework. Unlike the traditional Energy Internet, which primarily emphasizes communication and monitoring, AEI's information layer realizes a functional evolution from "passive sensing" to "active cognition," establishing an intelligent collaborative system that integrates perception networks, edge computing, 6G communication technologies, and cloud-based large-scale models. Through a closed-loop mechanism of "acquisition–computation–transmission–decision-making," the information layer continuously aggregates energy data, environmental parameters, and equipment status from agricultural production sites, facilitating real-time fusion and intelligent analysis of multi-source heterogeneous data. This structure enables dynamic regulation and holistic optimization of the agricultural energy system [7]. Under this architecture, the information layer is not merely a conduit for data transmission but rather functions as the "brain" and "nervous system" of AEI, providing upper-layer applications with accurate, reliable, and real-time cognitive support.

Within the perception module, the information layer relies on the Agro-IoT Energy Platform to realize multi-source sensing, heterogeneous data integration, and dynamic feedback. A wide variety of sensors are deployed at critical points across agricultural production and energy systems to monitor parameters such as temperature, humidity, solar irradiance, wind speed, CO_2 concentration, equipment power, and energy consumption. Collectively, these devices form a high-density sensing network that spans agricultural ecological environments and energy operation domains. The perception module's core task is to enable comprehensive monitoring of energy, information, and environmental flows, thereby providing real-time data support for upper-layer intelligent scheduling and carbon-constrained optimization. Data collected from multiple sources undergo local preprocessing, including data cleaning, noise filtering, anomaly detection, and feature extraction, ensuring the integrity, accuracy, and timeliness of the information input. Through multi-dimensional data fusion technologies, the perception module achieves precise recognition and intelligent tracking within complex agricultural environments, significantly enhancing the system's awareness of both energy operational states and agricultural production dynamics, and laying the data foundation for AEI's intelligent decision-making [103].

In the edge computing module, the system functions as the intermediate computational layer between the perception network and the cloud, responsible for real-time responsiveness and localized optimization. Edge nodes are equipped with data compression, stream analysis, and model inference capabilities, allowing them to perform rapid monitoring of energy flows and local control tasks near the data source. This architecture minimizes communication latency and reduces cloud-side computational burdens, enabling a low-latency, high-reliability control system [103]. In typical applications, edge nodes can autonomously execute rapid decision-making and energy regulation based on real-time energy fluctuations, meteorological variations, and equipment operating states – for instance, adjusting greenhouse heating and lighting systems adaptively or optimizing charging–discharging strategies for energy storage units. This dual-level optimization achieves both "local autonomy" and "global coordination." When integrated with AI algorithms, edge computing gains event-driven and predictive control capabilities, enabling millisecond-level intelligent responses for key devices, thereby effectively supporting AEI's distributed scheduling architecture [7].

Within the network module, the information layer establishes a high-bandwidth, low-latency, and wide-coverage communication environment by integrating 6G wireless communication with satellite Internet, ensuring seamless transmission of both energy and information across all layers of the Agricultural Energy Internet (AEI). The 6G network, characterized by ultra-high transmission speeds and massive connectivity, brings significant advancements in spectral efficiency, transmission security, and latency control, enabling millisecond-level real-time communication and cross-regional coordinated scheduling for agricultural energy systems [114]. Moreover, the 6G network slicing and dynamic resource allocation mechanisms allow differentiated Quality of Service (QoS) management tailored to various operational scenarios – such as energy monitoring, agricultural machinery control, and carbon emission management – thereby satisfying the asynchronous and diverse communication demands inherent in agricultural applications. The satellite communication system complements this terrestrial network by providing reliable access to remote mountain regions, pastoral zones, and off-grid agricultural areas, ensuring the consistency of energy and information flows across vast and geographically dispersed environments. Through this integrated space–air–ground network structure, AEI's information layer transcends traditional geographical and infrastructural limitations, laying a robust communication foundation for global agricultural energy system monitoring and collaborative scheduling.

Within the cloud-based large model module, the system integrates artificial intelligence (AI), big data analytics, and high-performance computing (HPC) to construct the cognitive and decision-making engine of AEI. The cloud platform aggregates data globally, performs large-scale model training and strategy optimization, and forms a virtual twin of the agricultural energy system. Leveraging digital twin technology and the agriculture-specific large language model (AgriGPT), the system can conduct strategy simulation and energy consumption forecasting in a virtual environment, enabling pre-evaluation of energy efficiency and carbon emission outcomes under varying scheduling schemes, thus realizing a "simulation–optimization–execution" closed-loop control process [7]. The AI-based multi-objective scheduling algorithm, driven by reinforcement learning and evolutionary computation, enables the coordinated optimization of distributed generation, storage, and agricultural load allocation to simultaneously achieve economic efficiency and low-carbon operation [103]. Meanwhile, the introduction of AI accelerators significantly enhances the computational energy efficiency of cloud-based inference and data processing. The cloud–edge–terminal collaborative architecture further supports hierarchical resource scheduling and dynamic task migration, ensuring high reliability and continuous learning capability under complex environmental conditions. By integrating with the carbon accounting and flow tracking model, the system achieves synchronous monitoring and adaptive control of energy and carbon flows, maintaining optimal system operation while advancing toward carbon peaking and carbon neutrality goals [7].

The application layer serves as the execution and coordination platform of the Agricultural Energy Internet (AEI) technical framework, directly interfacing with real-world agricultural scenarios. It represents the critical level where agriculture, energy, and the Internet achieve deep and systematic integration. Its primary mission is to translate the intelligent computational outputs of the information layer into actionable scheduling commands and service decisions, thereby realizing intelligent collaboration and resource optimization across the entire agricultural production process. Beyond serving as the operational executor of strategies, the application layer also functions as an integrated interface for information feedback, system regulation, and ecological services, forming the essential bridge that links virtual cognition with physical execution within the AEI architecture. Based on functional logic, the application layer can be divided into three core modules: the agriculture–energy coordination module, the agriculture–Internet coordination module, and the Energy–Internet coordination module [109]. Together, these three modules constitute the upper operational ecosystem of AEI, synergizing energy execution, information management, and system coordination.

The agriculture–energy coordination module focuses on establishing a bidirectional linkage between agricultural production processes and energy-supply systems, with the core objective of achieving dynamic matching between crop physiological demands and energy supply through intelligent energy scheduling. At its foundation lies the agricultural virtual power plant (Agri-VPP) mechanism, which aggregates and jointly optimizes distributed photovoltaics, energy storage units, biomass resources, and agricultural loads (such as greenhouses, irrigation systems, and cold-chain facilities) [7]. The system employs AI-based predictive models and real-time meteorological data to dynamically adjust energy allocation strategies, ensuring continuity of agricultural production while optimizing the coordinated operation of electricity, heat, and fuel flows. With the integration of carbon accounting and carbon flow tracking models, this module further achieves dual-objective optimization between energy efficiency and carbon reduction, aligning energy utilization with environmental sustainability goals [102,103]. Functioning as the energy execution core of AEI, this module plays a pivotal role in driving the low-carbon transformation and operational efficiency enhancement of agricultural energy systems.

The agricultural intelligent management and information service module serves as the information-driven core of the Agricultural Energy Internet (AEI), focusing on leveraging digitalization and intelligent technologies to enhance the transparency, precision, and responsiveness of agricultural management. Built upon large-scale model computation and digital twin technology, this module establishes a human–machine interactive platform that integrates visual monitoring, intelligent diagnostics, and decision-support functions. Agricultural managers and system operators can visualize the operational states of energy flow, information flow, and carbon flow in real time, while employing the multi-scenario simulation capabilities of AI models to conduct strategy modeling and predictive analysis [104]. Through the integration of virtual and physical environments, the digital twin framework enables dynamic mapping of crop growth, equipment operation, and energy consumption trends, thereby facilitating a transition from experience-based decision-making to data-driven management. Meanwhile, knowledge graph and semantic analysis technologies ensure the semantic integration of agricultural, meteorological, and energy data, forming an intelligent service system capable of supporting comprehensive agricultural ecosystem management [114].

The Energy Digital Collaboration Module acts as the system coordination hub of AEI's application layer, emphasizing the deep integration of energy systems with digital infrastructure. This module leverages cloud computing, blockchain, and distributed ledger technologies to construct a unified energy and carbon management service platform, facilitating multi-dimensional interaction among energy flow, information flow, and economic flow [109]. Within this framework, the cloud layer performs regional-level optimization of energy and carbon assets, while edge nodes execute rapid local responses and control of energy storage, forming a cloud–edge–terminal collaborative scheduling system [7]. Blockchain technology ensures the security, transparency, and traceability of energy and carbon transaction data, thereby supporting agricultural entities' participation in carbon credit and green electricity markets. The integration of 6G communication networks and satellite Internet further extends system coverage, enabling real-time sensing and intelligent coordination even in remote agricultural areas. This advancement facilitates large-scale, wide-area synergy within AEI, promoting seamless and intelligent operation across diverse agricultural regions [114].

Overall, the Agricultural Energy Internet (AEI) technical framework – built upon key enabling technologies such as 6G communication networks, large-scale model computation, AI chips, digital twins, and carbon accounting systems – marks a comprehensive transformation of agricultural energy systems from traditional energy-supply models toward intelligent, collaborative, and networked paradigms [102,103].

This framework achieves a deep interdisciplinary integration that dissolves the traditional boundaries between agricultural and energy systems, forming a fully coupled, multi-dimensional network where energy flow, information flow, and carbon flow interact dynamically. AEI is no longer merely an energy-supply infrastructure but has evolved into a self-organizing system that integrates intelligent perception, data computation, dynamic scheduling, and feedback optimization within a unified architecture. Its technical foundation is driven by information, centered on intelligent algorithms, and supported by multi-energy complementarity, enabling digital modeling and dynamic optimization across the entire process of agricultural production, energy conversion, and carbon emission. Through this structure, AEI achieves a coordinated operational mechanism characterized by energy autonomy, information co-fusion, and carbon-constrained optimization. The combination of 6G networking and AI algorithms empowers the system with millisecond-level responsiveness and predictive intelligence in energy transmission, decision-making, and cross-domain coordination, enabling real-time adaptation to fluctuations in both energy demand and environmental conditions. Meanwhile, the integration of digital twin and carbon accounting technologies endows the system with virtual–physical fusion and ecological feedback capabilities, transforming AEI from a one-way energy-supply structure into an intelligent energy network with cognitive and evolutionary properties.

Through the multi-layer coordination of the physical, information, and application layers, AEI achieves efficient energy flow conversion, intelligent information perception, and precise carbon flow management. The physical layer, through the interconnection of distributed generation units, storage systems, and terminal devices, establishes a foundational infrastructure that supports multi-energy complementarity and cyclical utilization; the information layer, powered by 6G communication and artificial intelligence, enables high-speed data interaction and intelligent decision-making across spatial, media, and device boundaries; and the application layer, via mechanisms such as virtual power plants (VPPs), smart scheduling systems, and carbon credit mechanisms, translates system intelligence into operational production activities and market mechanisms. The synergistic coordination among electric, thermal, and fuel flows significantly enhances the system's resilience and robustness under extreme climatic or uncertain conditions, while AI-driven predictive control and 6G-based high-bandwidth communication ensure real-time responsiveness and optimal configuration in large-scale distributed settings. Through this adaptive, multi-layer linkage mechanism, AEI not only achieves precise alignment between energy management and agricultural production but also constructs an intelligent ecological network capable of learning, evolving, and feedback-driven adaptation. In essence, this framework propels agricultural energy systems toward self-learning, self-adaptive, and self-optimizing paradigms, offering a replicable and systematic solution for the digitalization, decarbonization, and intelligent transformation of global agriculture [109]. It signifies agriculture's transition from a traditional energy-dependent production model to a sustainable, intelligence-driven development stage, thereby laying a solid technological foundation for achieving agricultural modernization and advancing the global carbon neutrality agenda.

1.4 The development trend of AEI

The Agricultural Energy Internet (AEI) represents an innovative composite framework that seamlessly fuses energy science, agricultural engineering, and information technology into a cohesive paradigm. At its essence, AEI aspires to foster clean energy paradigms, intelligent agricultural practices, and the utmost optimization of resource utilization by forging profound interconnections and seamless intercommunications among energy, information, and agricultural systems. The foundational philosophy of AEI posits information flows as the catalyst for energizing material circulations, while energy streams underpin the substantive exchanges of resources, thereby architecting an agricultural energy–ecological network endowed with intrinsic capabilities for self-perception, self-regulation, and self-optimization across diverse spatial scales – from micro-fields to expansive agro-regions – and temporal horizons, ranging from instantaneous responses to seasonal cycles.

Historically, conventional agricultural energy systems have been marred by an "isolated island" configuration, wherein energy provisioning conduits – encompassing electricity grids, fuel logistics, and thermal distribution networks – operate in splendid seclusion from the agrarian production apparatus, severed by informational silos that stifle interoperability. This dichotomous architecture not only precipitates suboptimal energy conversion efficiencies and disparate resource apportionments but also circumscribes the adaptive governance of agricultural workflows, rendering them vulnerable to perturbations like fluctuating inputs or climatic anomalies. The advent of AEI emerges as a panacea, proffering a holistic blueprint to dismantle these fissures. Leveraging synergistic mechanisms – such as multi-energy complementarity to harmonize solar, wind, and biomass inputs; data-driven analytics for prescient resource forecasting; intelligent optimization algorithms for real-time decision matrices; and networked collaborations for distributed resilience – AEI orchestrates a tripartite convergence across physical substrates (e.g., sensor-embedded infrastructures), informational strata (e.g., IoT-mediated data lakes), and decisional echelons (e.g., AI-curated policy engines). This engenders a vibrant, closed-loop triad of "energy–information–agriculture," wherein perturbations in one domain ripple adaptively through the others, yielding emergent efficiencies and robustness.

Amid the global pivot toward restructured energy matrices and imperatives for carbon neutrality, the exponential proliferation of renewable sources – harnessing solar irradiance, aeolian kinetics, and biogenic potentials – constitutes the bedrock for AEI's ascendancy. Foremost among these is Agrivoltaics (APV), a cornerstone modality within AEI's nodal architecture, which ingeniously dual-purposes agrarian landscapes as arenas for both photovoltaic electron harvesting and ecological stewardship. Beyond mere electrical yield, APV deploys sophisticated light stewardship modalities – encompassing spectral partitioning to calibrate photosynthetic wavelengths, modular array geometries to sculpt irradiance gradients, and kinematic tracking apparatuses to synchronize with solar ephemerides – thereby fine-tuning subsurface microclimates: attenuating thermal extremes, modulating vapor deficits, and equilibrating humidity envelopes to exalt crop photodynamics and biomass accrual. This photonic equity doctrine – allocating luminous bounty judiciously between silicon substrates and chlorophyll matrices – empowers arable expanses to navigate the perennial tension between caloric sovereignty and alimentary imperatives, positioning APV as the quintessential vessel for incarnating AEI's tenets in tangible terroirs.

Concomitantly, the vanguard of intelligent metamorphosis within energy apparatuses augments AEI's conceptual expanse, propelling agrarian energetics from peripheral sinks to pivotal forges of production and modulation. The meteoric strides in artificial intelligence – manifesting as neural architectures for anticipatory flux prognostication and stochastic orchestration – coupled with electrochemical storage reservoirs and nascent hydrogenic infrastructures, catalyze this transmutation. AI-infused divination and dispatch protocols, intertwined with hydrogenic sequestration and redeployment cascades, imbue agricultural ensembles with prodigious faculties: autonomous caloric sovereignty to insulate against intermittencies, trans-temporal arbitrage to bridge diurnal troughs with nocturnal exigencies, and polyenergetic consonance to weave disparate vectors into unified tapestries. Extending this continuum, bioheuristic energy transduction modalities – epitomized by hydrovoltaic phenomena elicited from arboreal perspiration – artfully ingrain innate physiological cadences into the energetic lattice, proffering perennial, minimally intrusive micropower conduits that amplify AEI's symphonic harmony with biogenic rhythms. These verdant harvesters, by channeling foliar diuresis into galvanic whispers, not only obviate infrastructural impositions but also exalt ecological consonance, rendering fields as sentient generators that pulse in unison with photosynthetic pulses and edaphic dialogs.

The trio of emblematic treatises curated herein [115–117] illuminates the variegated strata of AEI's technological odyssey and vanguard inquiries, delineating evolutionary vectors from perceptual finesse to systemic symphonies:

1. Refined cartography of luminal and caloric dispersion – Anchored in prognostic frameworks that amalgamate computational intelligence with immutable physical edicts, this probes the sapient conduits of AEI's sensory and gubernatorial cortices, enabling prescient stewardship of photic allotments to exalt dual agrarian yields.
2. Symbiotic closure of hydric–energetic–hydrogenic circuits – Via heliacal propulsion of cryogenic desalinations and electrolytic dihydrogen forges, this erects AEI's metabolic sinews and circulatory loops, transmuting aqueous scarcities into symbiotic bounties that sustain perpetual agrarian vigor.
3. Phytogenic hydrovoltaics awakened by foliar exudation – Centered upon the arcane transduction of evaporative enthalpies through vegetal hydraulics, this unveils bioinspired trajectories for AEI's agrarian micropotencies, climatic responsivity, and decarbonized husbandry, wherein living canopies emerge as autonomous voltaic oracles.

Convergently, these triads delineate AEI's plenary technologic continuum – from subatomic acuity to panoptic networks, solitary apparatuses to confederated consortia – unveiling the epistemological cartography and inexorable momenta propelling the profound amalgamation of tomorrow's agrarian and energetic realms into a regenerative, resilient whole.

1.4.1 Technology one: application and trends of physical-informed machine learning in agrivoltaics lighting modeling

In the Agrivoltaics (APV) system, the precise prediction of surface light distribution is a key link in achieving the coordinated management of energy production and crop growth. The distribution of light directly affects the power generation performance of photovoltaic arrays, the photosynthetic efficiency of crops, and the ground temperature and humidity environment. Therefore, its high-precision modeling is of fundamental significance for the real-time regulation of the Agricultural Energy Internet (AEI).

The traditional illumination modeling methods mainly rely on the ray-tracing algorithm. By simulating the propagation, reflection, and scattering processes of solar radiation among photovoltaic modules, the illumination intensity distribution at different surface points is calculated. Although the ray-tracing method has extremely high accuracy at the physical level, its computational load is huge and the time cost is significant, making it difficult to meet the real-time and dynamic response requirements of AEI. For instance, in large-scale agricultural photovoltaic complementary systems, even the slightest change in any environmental parameter (such as the solar altitude angle or component tilt angle) may lead to a recalculation of the entire light field distribution, thereby causing system control lag and making it difficult to support high-frequency scheduling and online optimization.

To address this issue, Kurumundayil *et al.* [115] proposed a fast surrogate model based on Physics-Informed Machine Learning (PIML). This method directly empowers the physical laws of light propagation into the neural network structure, enabling the model to follow optical and geometric constraints while learning historical data, and achieving the unity of "physical interpretability" and "data efficiency."

Unlike traditional pure data-driven models, the PIML model introduces optical transport equation constraints (including Beer-Lambert's law, geometric shading function, scattering model, etc.) into the loss function, thereby reducing the reliance on large-scale training samples. This enables the model to achieve comparable prediction performance in terms of accuracy even when using only a limited amount of high-fidelity simulation data. Furthermore, research shows that this method can generate a complete surface irradiation map within a millisecond time scale, and its calculation speed is more than three orders of magnitude faster than ray tracing, truly realizing the possibility of real-time illumination field reconstruction and dynamic feedback control.

More importantly, this model is highly deployable and can be directly run on edge computing devices or AEI node control terminals, thereby providing immediate energy distribution prediction and response capabilities for agricultural sites. This breakthrough has laid the algorithmic and computing power foundation for the intelligent control of AEI, precise crop management, and energy collaborative optimization (Figure 1.3).

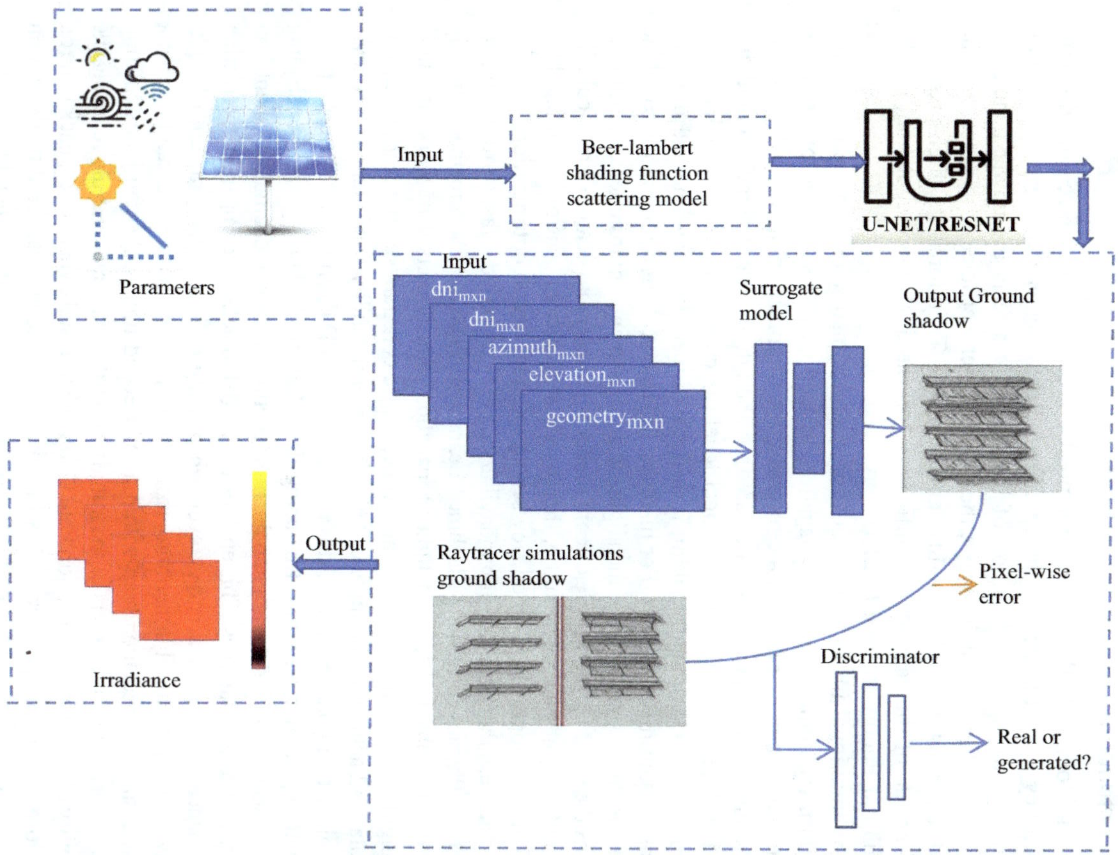

Figure 1.3 Physics-informed machine learning technology and data flow in agrivoltaics

In this technology, the input data layer first integrates meteorological parameters (including irradiance, cloud cover, humidity, etc.), the geometric position of the sun, and the arrangement parameters of photovoltaic modules to provide environmental and structural information for the model. Subsequently, the physical constraint layer introduces the light propagation law and scattering model to ensure that the model output conforms to the real optical properties. The neural network layer adopts physically embedded deep network structures (such as U-Net and ResNet), and performs excellently in multi-scale feature extraction and spatial correlation modeling. The model output layer generates high-resolution surface irradiation distribution maps and light energy distribution in the crop layer, providing real-time input for the downstream AEI control module.

On this basis, the AEI control module can achieve crop growth scheduling (such as photopod regulation), component posture optimization (such as dynamic tilt and azimuth control), and energy prediction (such as photovoltaic power generation and crop evapotranspiration rate prediction). The entire process constitutes a key link in the AEI perception and decision-making system, enabling the agricultural production process to have the ability of autonomous learning and dynamic optimization.

The core idea of this technical route is to transform the complex agricultural lighting calculation process into a learnable and deployable intelligent model, thereby achieving real-time collaborative optimization of light energy, electrical energy, and agricultural energy within the AEI framework. Its innovative value lies in the fact that the model has strong physical interpretability, generalization ability, fast reasoning speed, and can be seamlessly coupled with the on-site Internet of Things system, providing a new paradigm for building an intelligent agricultural energy system.

In the future system of AEI, the lighting modeling technology based on PIML will become the fundamental module for realizing the "agricultural energy neural network." By embedding the lightweight PIML model in the Agricultural Internet of Things (Agri-IoT) nodes, the agro-optical system will possess the capabilities of self-perception, self-learning, and self-optimization. The real-time collected data, such as irradiation, temperature, humidity, and wind speed by the sensors, will continuously update the model parameters, enabling it to dynamically adapt to different climatic conditions, seasonal changes, and crop types, and achieve precise energy distribution and environmental control.

Furthermore, by combining the Multi-Scenario Learning and Transfer Learning strategies, the model can maintain high accuracy and robustness under different regions and system configurations. For instance, the model can be adapted to both terraced APV systems in mountainous areas and centralized agricultural photovoltaic power stations in plains simultaneously, maintaining stable prediction performance through adaptive parameter adjustment. This generalization ability is crucial for the large-scale promotion of AEI.

At the system level, the rapid irradiation modeling technology based on PIML provides AEI with a "data neuron" function, meaning that each agricultural node can independently complete energy distribution prediction and local control decisions. Through the interconnection and collaboration among nodes, AEI can form a distributed energy management system, achieving a dynamic balance among energy production, storage, and consumption. Meanwhile, this technology also supports demand response management and opti-economic dispatch, providing the basic algorithmic support for energy optimization and economic operation for AEI.

In conclusion, physical-informed machine learning is not only an important tool for the intelligent perception and decision-making of AEI but also the core driving force for achieving the autonomous evolution and global optimization of agricultural energy systems. It marks the transformation of AEI from passive monitoring to active prediction and from experience-based control to intelligent optimization, providing a solid technical foundation for the deep integration of future agriculture and energy systems.

1.4.2 Technology two: solar-driven freezing desalination and electrolytic hydrogen production system

In arid and coastal agricultural regions – encompassing vast swathes of the globe from the sun-baked expanses of the Middle East to the saline fringes of Australia's outback – the escalating antagonism between chronic water scarcity and erratic energy provisioning has crystallized as a paramount impediment to agrarian perpetuity. These locales, pivotal to global food security yet besieged by salinized aquifers, erratic monsoons, and fossil-fuel dependency, demand prodigious volumes of desalinated freshwater not merely for basal irrigation but also for sophisticated agronomic imperatives: precision fertigation to avert nutrient leaching, evaporative cooling in high-tech greenhouses to temper heatwaves, and hygienic processing of perishable commodities to extend shelf life and market reach. Compounding this hydrological exigency is the imperative for resilient, decarbonized energy architectures to undergird the ceaseless cadence of mechanized tillage, automated harvesting, and climate-controlled storage, all while navigating the intermittency of renewables in hyper-arid climes.

To ameliorate this dichotomous plight, Sezer and Bayhan [116] proffered a trailblazing panacea: the Solar Freeze Desalination–Electrolysis Integrated System (SFD–EIS), a paragon of symbiotic engineering that orchestrates a multi-energy coupling paradigm encapsulating "desalination for potable yield – electrolysis for hydrogenic bounty – storage for temporal arbitrage – provisioning for operational sustenance." This holistic construct not only bridges the chasm between aqueous deficits and energetic volatility but also engenders a profound symbiosis between agricultural hydrology and thermodynamic cycles, transmuting solar plenitude into a triune elixir of water, power, and cryogenics, thereby catalyzing self-reinforcing loops within the Agricultural Energy Internet (AEI).

At its thermodynamic nucleus, the SFD–EIS harnesses photovoltaic (PV) arrays – optimized for spectral efficiency in high-irradiance deserts – as the primordial energy font, channeling gigajoules of direct current (DC) electricity bifurcated toward dual, interdependent subsystems. The inaugural arm, the Freeze

Desalination Unit (FDU), leverages a vapor-compression refrigeration loop to orchestrate sub-zero entrenchment of brine feeds, precipitating vitreous ice lattices bereft of ionic contaminants via fractional crystallization. This cryogenic alchemy – rooted in the thermodynamic favorability of pure water's lower freezing point vis-à-vis saline matrices – yields potable meltwater cascades, primed for drip irrigation or hydroponic cascades, with salinities plummeting below 500 ppm, a quantum leap from feedstocks exceeding 35,000 ppm in hypersaline bays. Concurrently, the Proton Exchange Membrane (PEM) Electrolysis Unit transmutes surplus PV electrons into nascent hydrogen via the oxygen evolution reaction (OER) at the anode and hydrogen evolution reaction (HER) at the cathode, engendering high-purity H_2 streams amenable to compression and sequestration in Type IV composite overwrapped pressure vessels. This dihydrogen reservoir, evoking the molecular buoyancy of zephyrs, stands poised to galvanize fuel cell auxiliaries for nocturnal greenhouse illumination, proton-exchange membrane fuel cells (PEMFCs) for electric tractors, or even ammonia synthesis for protracted fertilizer stockpiles.

The apotheosis of this ingenuity resides in the orchestration of a "cold energy" and "hydrogen energy" cross-media symphonia, wherein energy vectors undergo choreographed metamorphosis and hierarchical valorization within a hermetic thermodynamic enclosure. Exigent cold surpluses from the FDU – often squandered in orthodox vapor-compression paradigms – are recouped via heat exchangers to precondition PEM electrolytes, attenuating ohmic losses and augmenting faradaic efficiencies by 15–20%. Reciprocally, the exothermic effluvia of electrolysis (manifesting as ~80 kJ/mol H_2 Joule heating) are retrofitted to vaporize refrigerants in the FDU compressor, obviating auxiliary power draws and perpetuating a virtuous entropic cascade. This polyenergetic reciprocity begets a self-amplifying resonance: desalination brine concentrates furnish osmotic gradients for emergent reverse electrodialysis (RED) modules, while H_2 off-gases catalyze microbial electrolysis cells (MECs) for ancillary biogas augmentation. Empirical assays under duress – simulating Saharan dust storms or coastal fog banks – affirm the system's mettle, concurrently disgorging desalinated yields, hydrogenic troves, and cryogenic allotments, all while sustaining a coefficient of performance (COP) exceeding 4.5 for refrigeration and 65% for electrolysis. Saliently, holistic exergetic audits evince a 20–30% ascension in aggregate efficiency over siloed antecedents – be they standalone PV orchards bereft of aqueous symbiosis or reverse osmosis (RO) plants guzzling 3–5 kWh/m^3 – principally through entropy minimization and exergy recuperation gradients.

This symbiotically engineered edifice erects an archetypal self-sustaining energetic nexus for the AEI tapestry: ingress wholly predicated on inexhaustible solar quanta, egress a cornucopia of agrarian sinews – limpid waters for rhizospheric vitality, electrochemic vigor for automatons, and frigorific boons for post-harvest quiescence – effectuating a paradigmatic pivot from rapacious, linear depletion to regenerative, circulatory husbandry. Its praxis transcends theoretical elegance, evincing robust deployability in modular vignettes and ecological consonance, with lifecycle emissions curtailed by 70% relative to diesel-pumped desalination, thereby fortifying AEI's bulwarks against desiccation and decarbonization alike. In essence, solar-driven freezing desalination–electrolysis hydrogen production system (SFD–EIS) epitomizes the alchemical transfiguration of solar photons into aqueous and molecular ambrosia, heralding an era where agrarian aridity yields to aqueous abundance and energetic equanimity (Figure 1.4).

This conceptual structure presents the multi-energy flow path of the SFD–EIS system. Photovoltaic modules convert solar irradiation energy into electrical energy, and the intelligent distribution of energy among different subsystems is achieved through control modules.

The compression refrigeration circuit uses electrical energy to drive the circulation of coolant, causing salt water to gradually freeze and precipitate salt, generating fresh water that can be directly used for agricultural irrigation.

The PEM electrolytic unit converts part of the electrical energy into chemical energy, generating hydrogen through electrolytic reactions to achieve cross-media storage of electrical energy.

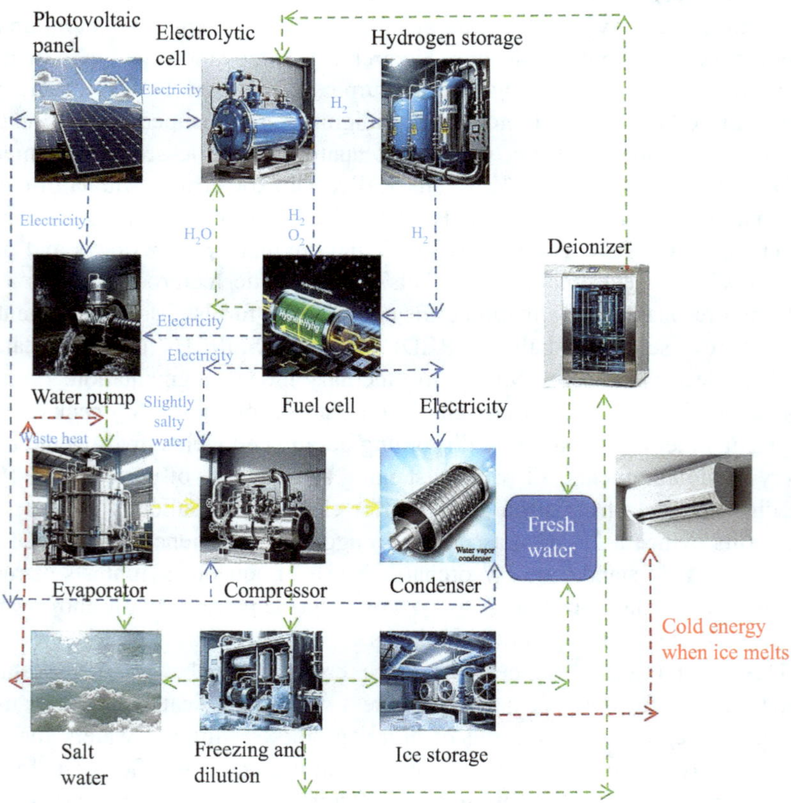

Figure 1.4 Conceptual structure diagram of the solar-driven freezing desalination–electrolysis hydrogen production system (SFD–EIS)

The cold energy recovery and waste heat utilization module establishes an energy feedback mechanism between the two subsystems. It uses the thermal energy generated by electrolysis to regulate the thermal balance of the refrigerant medium, and at the same time, it utilizes the cold energy recovered during the desalination process to provide a low-temperature environment for greenhouses or agricultural product storage, achieving energy cascade utilization within the system.

Hydrogen energy and fuel cell cycle units further convert chemical energy back into electrical and thermal energy, which are used for night lighting, greenhouse heating and mechanical power supply, thus forming a "complementary day and night" energy self-cycle system.

Through this system architecture, SFD–EIS not only achieves energy balance on a single-day time scale, but also realizes dynamic regulation of energy and water resources in longer-term agricultural operations, providing stable support for the long-term sustainable operation of AEI.

In the macro development framework of AEI, the proposal of SFD–EIS technology marks a fundamental transformation of agricultural energy systems from "single energy input" to "multi-energy symbiosis." One of the key development directions of AEI in the future is multi-carrier energy conversion, that is, establishing reversible and coupled conversion relationships among different energy forms (electrical energy, cold energy, thermal energy, chemical energy) to achieve energy self-sufficiency and multi-functionalization in agricultural production. SFD–EIS is precisely the practical prototype of this concept, and its research value is mainly reflected in the following three aspects:

First, this technology has achieved a transformation from "power centralization" to "multi-energy collaboration." Traditional agricultural energy systems often rely on a single type of electrical energy supply. However, SFD–EIS simultaneously converts electrical energy into cold energy and hydrogen energy, significantly enhancing the overall energy efficiency of the system through energy cascade utilization and cross-medium coupling. This approach holds significant implications for the energy structure optimization and comprehensive scheduling of AEI.

Second, this system supports the goals of energy autonomy and carbon neutrality in agricultural regions. By using renewable energy to drive water production and energy storage, agricultural systems can break away from their reliance on fossil fuels and achieve regional energy self-balance. Meanwhile, hydrogen energy, as a green energy carrier, can be flexibly allocated in the AEI network to provide carbon-free energy support for agricultural transportation, irrigation pumping stations, and refrigeration equipment, thereby facilitating the reduction of agricultural carbon emissions and the traceable management of carbon footprints.

Finally, SFD–EIS has demonstrated significant advantages in addressing the challenges of agricultural sustainability under extreme climate conditions. For arid desert areas, this system can directly use seawater or brackish water as input to provide fresh water resources and energy security. For high-temperature regions, its cold energy recovery mechanism can provide cooling conditions for greenhouse agriculture and significantly improve the growth environment of crops. For off-grid agricultural communities, their energy self-cycling characteristics can ensure production continuity and system resilience.

In the overall architecture of AEI, SFD–EIS will become an important "energy-water node" connecting the Energy Internet with the agricultural production system. It can not only provide basic resources such as electricity, hydrogen energy, and fresh water for the agricultural system, but also build a cross-level collaborative dynamic energy management network through energy complementarity and information feedback mechanisms. As a result, SFD–EIS is regarded as a key supporting technology for achieving multi-dimensional coordination of AEI resources and ecological closed-loop operation, which has profound significance for the decarbonization, intelligence, and system resilience improvement of the future agricultural energy system.

1.4.3 Technology three: hydropower generation system induced by transpiration of living blades

In the macro architecture of the Agricultural Energy Internet (AEI), bioheuristic energy harvesting technology transcends its role as a mere energy production unit, evolving into the quintessential nexus for the orchestrated symbiosis between intrinsic natural ecologies and engineered energy paradigms. This confluence not only amplifies the harvest of ambient biofluxes but also engenders emergent resiliencies, wherein agrarian vitality and thermodynamic yields co-evolve in a feedback-laden tapestry. Hu *et al.* [117] pioneered this frontier by conceptualizing the Living Leaf Transpiration Generator (LTG) through the prismatic lenses of plant physiology and nascent hydropower modalities. Anchored in the archetypal lotus leaf – a paragon of adaptive hydrophily with its superlative transpiration vigor – the LTG ingeniously intercepts the latent enthalpies dissipated during foliar evapotranspiration, transducing them into perennial bio-hydropower cascades. This orchestration forges a "bioenergy symbiotic network," wherein phytogenic vigor and electromotive potency mutually reinforce: transpiration not only sustains photosynthetic imperatives but also galvanizes decentralized electron fluxes, while harvested amperage reciprocally modulates microclimatic envelopes to bolster stomatal apertures and hydraulic gradients.

The sine qua non of LTG's ingenuity eschews the ponderous scalability of conventional photovoltaics or anemophilic harvesters, pivoting instead toward the exquisite alchemy of solar-latent interconversion. Terrestrial transpiration, the preeminent aqueous conduit on continental surfaces – constituting 80–90% of aggregate evapotranspiration and sequestering nearly half of land-absorbed insolation – is herein alchemized into galvanic bounty sans intermediary contrivances. The apparatus coalesces foliar transpiration, electrolytic ion shuttling (predominantly K^+ and H^+ cations), and interfacial electrokinetics at titanium electrodes into a unified generative scaffold. This triadic fusion is dynamically piloted by multivariant actuators – encompassing stomatal conductance, ambient thermometry, and hygrometric equilibria – yielding autonomous electromotive inception, temporally adaptive effluxes, and trans-environmental tunability. Partial least squares path modeling (PLS–PM) elucidates these interplays: transpiration rate exerts the paramount sway, synergized by thermal escalation and stomatal dilation, while RH attenuation curtails yields.

Functionally, the LTG incarnates a trinitarian archetype, seamlessly bridging biotic transduction with infrastructural sinews:

Bioenergy collection facilities are exemplified by mesoscale phytotectonics, in which LTG deploys a bifid electrode dyad: a porous titanium mesh astride the adaxial lotus lamina as cathode, and an intrastemmic Ti needle as anode, submerged in rhizospheric reservoirs. Transpiration – propelled by stomatal efflux and mesophyll cavitation – engenders an ascendant capillary meniscus, asymmetrically polarizing charges: anions aggregate at the foliar apex, cations at the basal stem, birthing a quasi-ohmic potential differential. Empirical assays chronicle a baseline Voc of 0.25 V and Isc of 50 nA under 36 °C/45% RH, with near-linear I–V linearity evincing ohmic fidelity. This engenders a "transpiration-unimpaired" bioenergetic commons: diurnal peaks sustain ~0.08 V/60 nA under 2 MΩ loads over 4,000 s, sans foliar duress, as verified by unabated photosynthetic quotas.

Regional energy network nodes: As vestigial harbingers of distributed hydropower, LTG clusters – scalable via Nernstian series or Kirchhoffian parallels – metastasize microgrids for low-enfilade payloads: e.g., serial quadrupling accrues ~1 V to capacitate 1 μF dielectrics in 10^3 s, or parallel banks animate LCD calculators, obviating photovoltaic shadows in undercanopy interstices. This tessellates clean bioenergetics across agrarian mosaics: ~40 nA diurnal surges eclipse nocturnal nadirs, mirroring insolation rhythms, and furnish ancillary voltaics for IoT sentinels or rhizobial electrolysis adjuncts, democratizing energetics sans infrastructural sprawl.

AEI control and optimization unit: At the reticular echelon of AEI, LTG metamorphoses into a phytosensory oracle, assaying transpiration telemetry (via porometry), ambient responsivity (Kelvin probe surface cartographies), and carbon sequestration proxies (via eddy covariance proxies). As the agrarian energy matrix's "biosensor," it prognosticates yields via PLS–PM regressions, orchestrating stomatal priming or humidity buffering to amplify β coefficients. This engenders ecological consonance: LTG's quasi-static outputs calibrate regional carbon sinks, attributing ~50% diurnal variance to thermal drivers, thereby informing AEI's decarbonization ledger.

Through this functional superposition, LTG catalyzes a quantum vault from parochial "agricultural biological processes" – hitherto siloed in trophic loops – to sovereign "energy autonomous nodes," wherein vegetal hydraulics self-perpetuate while retroceding surplus amperage to supralocal grids. This bidirectional bioenergetic dialectic – evidenced by inversion-reversible polarities – unveils the prodigious hydrovoltaic sous-essence of foliar transpiration: an ionophoretic cascade, wherein aquaporin-facilitated H_2O fluxes and K^+-gated channels engender streaming potentials across nano-porous laminae. By infusing AEI with naturogenic élan – harnessing the planet's verdant pump sans arable forfeiture – LTG heralds a verdant voltaic renaissance, where every leaf whispers watts and every breeze bears bioelectrons, reweaving agrarian tapestries into resilient, regenerative fabrics (Figure 1.5).

Leaf transpiration layer (bioelectric field): As the energy entry point of the system, the stomata and mesophyll cells of lotus leaves drive water molecules to flow upward capillaries under solar radiation. Through the asymmetric accumulation of ions (K^+, H^+) and water molecules at the electrode interface, a potential difference is generated (the high potential area is located on the leaf surface, and the low potential area is located on the stem). This layer achieves a bioenergy sharing mechanism of "all-weather power generation" through dynamic pore regulation and environmental adaptation.

Energy conversion layer: It gathers the electrochemical potential energy from transpiration and connects to the ion transport channel and charge separation device through the titanium electrode interface, such as the micro-pore structure within the leaf (t-plot micro-pore area up to several $m^2\ g^{-1}$) and the hydrophilic groups on the surface (-OH, C-O), achieving bidirectional coupling between water, ions and electrical energy. This layer also integrates surface potential scanning and XPS characterization to achieve efficient flow of system electrical energy between different carriers.

Agricultural production layer: This layer focuses on crop physiology and agricultural operation processes, including transpiration rate regulation, water

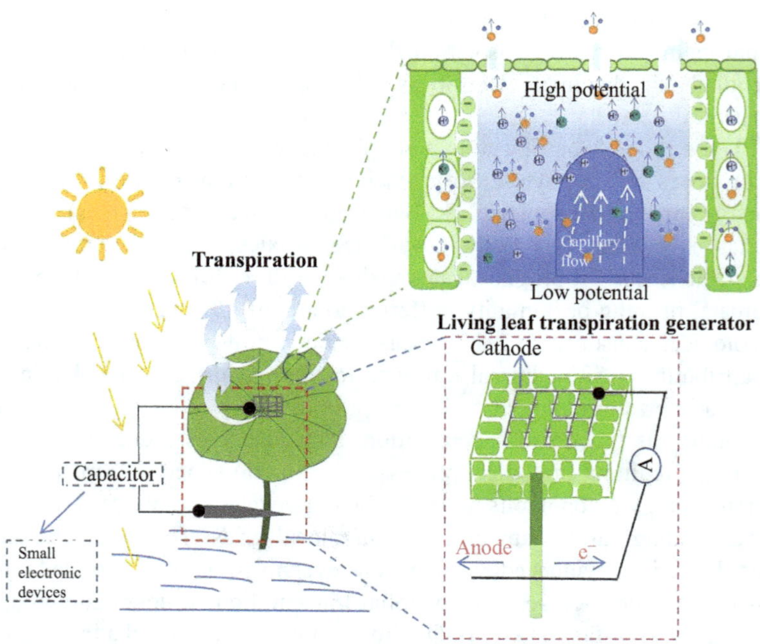

Figure 1.5 Schematic diagram of the structure and function of the Living Leaf Transpiration Generator (LTG)

cycling, temperature and humidity environment control, and biomass energy consumption. Through real-time interaction with the energy conversion layer, the system can dynamically adjust the power generation ratio according to the different growth stages of crops, achieving precise energy service.

AEI management: As the "intelligent brain" of the system, it realizes energy scheduling, transpiration prediction, and carbon emission management through partial least squares path modeling (PLS–PM) analysis. By leveraging edge computing nodes to collaborate with cloud data centers, a full-chain closed-loop from on-site perception to system-level optimization is achieved. Among them, temperature and pore conductance positively affect the output power, while relative humidity is negatively correlated.

Regional energy network: LTG not only serves itself but also can form a complementary energy-supply network with surrounding farmlands or greenhouses through bio-power arrays and energy storage group control systems, achieving cross-regional energy sharing and mutual assistance.

The entire system has formed a vertically integrated structure from biological collection to intelligent decision-making, and from local autonomy to regional collaboration, providing a system architecture blueprint for the biological heuristic deployment of AEI.

The future trends and integration prospects of AEI

The proposal of the LTG model marks a fundamental transformation of the role of agricultural energy systems – from traditional energy consumption units to centers for bioenergy production, dispatching, and optimization. This model reflects the development direction of AEI from "mechanized access" to "biological autonomy." By building a multi-level control system, from leaf-level micro-sensors to cloud-based energy management platforms, LTG can achieve bioenergy self-balance, carbon footprint traceability, and optimization of agricultural physiological energy consumption, providing core support for the low-carbonization and intelligence of agricultural regions.

The future evolution of AEI will revolve around the following three trends:

1. The introduction of multi-objective optimization algorithms achieves dual optimality in biology and energy
   In AEI systems, there is often a dynamic trade-off relationship between crop transpiration and energy output. In the future, LTG will achieve a dynamic balance among transpiration rate, power generation efficiency, and crop water use by introducing multi-objective optimization algorithms and adaptive control strategies based on reinforcement learning, ensuring the overall optimal performance of the system under different seasons and climatic conditions.
2. Cross-departmental integration to build a regional comprehensive energy ecosystem
   LTG will further form a cross-departmental integration network with photovoltaic, hydrogen energy, and municipal systems. For instance, the micro-electricity generated by leaf power generation can provide power for agricultural

sensors and the regional Internet of Things. The recovery of biological latent heat can improve the microclimate of crops through greenhouse regulation, thus forming a multi-directional energy flow pattern. This cross-departmental integration will drive AEI to evolve into a multi-industry interconnection form that integrates "agriculture + bioenergy + ecology + industry."

3. An economic incentive system based on blockchain and carbon trading mechanisms
   In the future, the operation of AEI will not only focus on energy efficiency and ecological benefits, but also extend to the economic mechanism level. LTG can be combined with blockchain technology to achieve traceability of bio-power generation data and carbon sink emissions. Through smart contracts, it can realize incentive mechanisms such as carbon trading, green bio-power certificates, and energy credit points, thereby establishing a market-oriented operation model for agricultural energy systems. This will make AEI not only a technical system but also a platform for promoting the coordinated development of agricultural green finance and the energy economy.

In conclusion, the LTG technology route fully demonstrates the biological evolution path of AEI. Through the deep integration of energy, information, and plant physiological systems, LTG not only promotes the autonomy and intelligence of agricultural energy systems, but also expands the boundaries of the Energy Internet. It enables agriculture to no longer merely be a consumer of energy, but to become an active node and strategic subject in the Energy Internet, thereby achieving mutual benefit, coexistence, and sustainable development of agriculture and the energy system at a higher level.

1.4.4 Comprehensive analysis

By synergistically integrating the triad of vanguard technologies – Physics-Informed Machine Learning (PIML) for agrivoltaic illumination surrogate modeling, the Solar Freeze Desalination–Electrolysis Integrated System (SFD–EIS) for multi-vector energy-water nexus, and the Living Leaf Transpiration Generator (LTG) for biohydrovoltaic harvesting – a panoramic vista of the Agricultural Energy Internet (AEI)'s evolutionary trajectory crystallizes. These paradigms, spanning perceptual acuity at the microscale, metabolic orchestration at the mesoscale, and ecological symbiosis at the macroscale, coalesce into a resilient, adaptive architecture that transcends siloed innovations. Their confluence begets emergent synergies: PIML's sub-millisecond irradiance forecasts can dynamically calibrate SFD–EIS's photovoltaic apportionment to diurnal aqueous demands, while LTG's phytogenic microwatts furnish persistent understory power for sensorial sentinels, obviating intermittency pitfalls. This holistic fusion not only amplifies individual efficacies – e.g., PIML's threefold computational alacrity, SFD–EIS's 20%–30% exergetic uplift, LTG's 35-hour voltamic constancy – but also forges systemic resiliencies, such as trans-seasonal homeostasis amid climatic volatilities, thereby redefining AEI as a living, self-evolving consortium.

The maturation of AEI manifests in the following four inexorable trends:

From "Energy Access" to "Energy Coupling"
AEI has outgrown the rudimentary paradigm of mere energy substitution within agrarian confines – such as retrofitting photovoltaic overlays atop monocultures – evolving instead into a stratified, polyphasic nexus where energy, informational, and material fluxes interweave with fluidic grace. This coupling manifests as a triune modality: PIML embeds radiative transfer axioms (e.g., Beer-Lambert attenuation) to engender hypergranular light quanta partitioning, optimizing dual yields of electrons from panels and photons for canopies in agrivoltaic mosaics; SFD–EIS transmutes solar DC into a hydric-chemic-thermal continuum, yielding $\sim$10–15 L/m^2 desalinate, $\sim$0.5–1 kg H_2/kWp, and $\sim$5–10 kWh cryogenic allotments per diurnal cycle; LTG, in turn, alchemizes foliar evapotranspiration – the terrestrial hegemon of aqueous flux – into ionophoretic surges, infusing biogenic amperage sans arable encumbrance. Collectively, these engender a polydimensional manifold, where photonic inputs cascade into aqueous solvency, electrolytic vigor, and phytogenic persistence, mitigating geographic vicissitudes from arid littorals to temperate orchards.

From "Centralized Control" to "Intelligent Autonomy"
The infusion of PIML, Agri-IoT constellations, and AI orchestration – encompassing reinforcement learning for predictive dispatching – imparts to AEI an exoskeleton of edge-embedded cognition and self-referential adaptation, percolating from cybernetic hierarchies to visceral biotic governance. PIML's physics-augmented surrogates, executable on frugal edge nodes, supplant compute-intensive ray-tracing algorithms, enabling hyperlocal actuators like azimuthal repivots or PAR-deficit irrigation; SFD–EIS's thermodynamic feedback loops – recycling electrolytic exotherms to precondition FDU cryogens – manifest as autonomous homeostasis, sustaining COP > 4.5 amid irradiance troughs; LTG extends this sapience to the verdant substrate, with PLS–PM regressions prognosticating outputs via stomatal telemetry temperature gradients, and RH modulations, thereby tuning transpiration fluxes for voltamic optima. This devolutionary intelligence – bolstered by federated learning across nodal swarms – empowers AEI with anticipatory reflexes, from preempting desiccation via evapotranspirative forecasts to modulating phyllotactic apertures against vapor pressure disequilibria, culminating in a decentralized sentience that obviates central chokepoints.

From "One-Way Supply" to "Closed-Loop Circulation"
The unidirectional founts of yore – linear solar-to-grid conduits or brackish aquifers – yield to circulatory symphonies, wherein SFD–EIS's cryogenic electrolysis interlaces with LTG's hydrovoltaic capillaries to forge agrarian endorheic cycles. Desalinated meltwaters irrigate rhizomes, fueling LTG's ascendant menisci and ion (K^+/H^+) asymmetries that birth streaming potentials across foliar nanoporosity; electrolytic H_2, sequestered at 350 bar, reciprocates as PEMFC cathodes for nocturnal phytochrons or reductants in biofertilizer cascades, while

LTG's ~40 nA diurnal pulses recharge SFD–EIS auxiliaries. This entropic thrift – curtailing lifecycle emissions by ~70% vis-à-vis diesel analogs – propels a quaternary synergy: water solvency for osmotic vitality, energetic constancy for automatons, cryogenic temperance for post-harvest quiescence, and biogenic closure for carbonic sequestration (~half solar absorption looped back). Thus, AEI metamorphoses into a regenerative bioreactor, where effluents beget inflows, and agrarian outputs (biomass, volatiles) retroaliment the loop, engendering perpetual motion sans exogenous crutches.

From "Energy Farm" to "Energy Ecotone"

AEI transcends the parochial confines of agrarian enclosures – once sequestered as insular "energy farms" beholden to monolithic photovoltaics or siloed irrigation grids – burgeoning instead into liminal ecotones: those fecund borderlands where bioenergetic frontiers efface into the broader chorography of regional metabolisms, engendering hybrid vigor across trophic and thermodynamic realms. This ecotonal efflorescence reimagines agrarian precincts not as passive recipients of caloric largesse but as porous membranes, permeable to fluxional exchanges that knit farmsteads into supralocal symphonies of sustenance and sustenance. PIML's scalable surrogates, with their physics-imbued neural topographies, tessellate seamlessly across geomorphic heterogeneities – from the vertiginous terraced inclines of Andean altiplanos, where insolation gradients demand azimuthally adaptive modeling, to the lacustrine alluvial planes of the Mississippi Delta, rife with levee-shadowed mosaics – yielding irradiance atlases that harmonize canopy photon budgets with panel electron tallies, often elevating dual-use yields by 15–25% through prescriptive intercropping edicts. SFD–EIS, in its lithe modularity, manifests as peripatetic sanctuaries amid saliniferous badlands – from Namibia's skeletal dunes to California's Central Valley salinized basins – dispensing nomadic aquifers laced with hydrogenic elixirs and cryogenic salves, thereby transfiguring desiccated demesnes into verdant waystations resilient to aquifer drawdowns. LTG, proliferating as an emerald voltaic phalanx, arrays in polycultural canopies – interlaced amid maize hedgerows or vineyard understories – to harvest the choral transpiration of arboreal ensembles, sans photosynthetic forfeiture or hydraulic duress, as ionophoretic cascades aggregate into nodal surpluses that pulse through foliar nanoporosity. This ecotonal expanse, a palimpsest of biotic and abiotic imbrications, fosters profound inter-nodal osmosis: LTG's phytogenic microwatts seed PIML's edge caches, obviating battery bulks for persistent inference; SFD–EIS's dihydrogen vectors arbitrage trans-farm deficits, piping electrolytic vigor to PEMFC auxiliaries that backstop LTG's stomatal sentinels amid RH spikes; in reciprocity, PIML's evapotranspirative oracles precondition SFD–EIS cryogens for brine entrenchment, birthing a rhizomatic web – decentralized, fault-tolerant, and evolvable – that sprawls beyond paddock palisades into municipal meshes, where agrarian amperage retrocedes to urban ledgers, and civic effluents (e.g., wastewater brines) reflux as FDU feedstocks. Thus, AEI's ecotonal apotheosis dissolves agrarian insularity, catalyzing a continental commonwealth wherein every furrow feeds the grid, and every gust gathers greens, fortifying the agrarian ark against the tempests of tomorrow.

Strategic Significance: AEI's telos resides in the co-evolution of thermodynamic sinews and biotic tapestries, a verdant Hegelian dialectic wherein agrarian theaters ascend as regenerative fulcra. By alchemizing AI's prescience (PIML's domain priors), renewables' bounty (SFD–EIS's solar quanta), and biotech's subtlety (LTG's phyllotactic electrokinetics), AEI erects an infrastructural colossus for agrarian perpetuity. It recalibrates energy architectures – slashing fossil entanglements by 50–70% – while attenuating carbonic effluvia through latent recapture and hydric thrift; concomitantly, it steels agrarian fortitudes against climatic caprices, from Saharan siroccos to polar vortices, via adaptive stomatal husbandry and evaporative equilibria. In this crucible, "carbon-neutral agriculture" emerges not as a Sisyphean aspiration but as an inexorable trajectory: AEI as the verdant forge, transmuting photons into foodstuffs, electrons into equity, and aridity into abundance.

Based on the exegesis of these three cutting-edge inquiries [115–117], AEI's prospective unfolding distills to a triptych:

At the micro level: Intelligent computoria, epitomized by PIML's physics-infused neural lattices, will orchestrate granular dominion over agrarian energy rivulets – millisecond irradiance cartographies begetting bespoke PAR prescriptions and shading symphonies, foundational to perceptual sovereignty.

At the meso level: Polyenergetic consortia, anchored in SFD–EIS's desalinolytic alchemy, will architect agrarian endorheic circuits – cryogenic brines, and dihydrogen ambrosia interlocked in entropic reciprocity, birthing self-sustaining metabolomes impervious to aqueous or photonic penury.

At the macro level: Biogenic powerweaves, predicated on LTG's transpiration-tethered hydrovoltaics, will catalyze the verdant ecologization and perennialization of agrarian energetics – phyllotactic ion fluxes scaling to regional biofluxes, infusing AEI with naturogenic perpetuity.

These conduits emblemize AEI's neural trinity: "sensory neurons" (PIML's optic sentience), "energy metabolism units" (SFD–EIS's thermodynamic viscera), and "biopower generation centers" (LTG's phytogenic myocardium). Futurity's AEI, scaffolded upon these axiomatic organs, will unfurl an ontogenetic arc – from punctiliar epiphanies to tectonic revolutions – wherein agrarian enclaves burgeon as planetary powerhouses, symbiotic sentinels in the grand thermodynamic odyssey.

1.5 Conclusion

The Agricultural Energy Internet (AEI), as the systematic extension of the Energy Internet concept into the agricultural domain, has evolved from conceptual emergence to theoretical maturity and engineering implementation. As analyzed previously, the rise of AEI is not accidental but rather an inevitable outcome of the combined forces of global energy transition, the digital agriculture revolution, and the growing pressure of climate change. It represents not only a technological innovation in agricultural energy systems but also a strategic pathway for agriculture's evolution toward low-carbon, intelligent, and sustainable development. From "electrified agriculture" to "intelligent energy ecosystems," the theoretical framework and technical system of AEI signify a fundamental paradigm shift in agricultural energy systems.

From a macro perspective, the global energy system is rapidly transitioning toward clean and electrified forms. As a major contributor to both energy consumption and carbon emissions, agriculture must restructure its energy system under the framework of the "dual carbon" (carbon peaking and carbon neutrality) goals. The significant decline in renewable energy costs and the maturation of distributed energy technologies have provided agriculture with realistic conditions for developing localized energy networks. The "agro-energy synergy" system, based on photovoltaic, wind, and biomass energy, not only addresses rural energy shortages but also provides a stable power foundation for intelligent and precision agriculture. This emerging triadic pattern of "energy–information–agriculture" constitutes a new ecological foundation for future agricultural systems. The conception and development of AEI emerged precisely within this context, aligning with the dual imperatives of the global energy revolution and agricultural modernization.

From a technological evolution standpoint, the AEI framework achieves deep integration between energy science and information science. Through the three-dimensional coordination of electric, information, and carbon flows, AEI enables a qualitative leap in agricultural energy systems – from linear energy supply to intelligent scheduling. At the physical layer, AEI integrates photovoltaic, wind, biomass, and energy storage systems to form a multi-source complementary renewable energy structure. At the information layer, it leverages the Internet of Things (IoT), edge computing, and 6G communication to realize intelligent perception and real-time control of energy flows. At the intelligence layer, it incorporates artificial intelligence (AI), big data analytics, and digital twin technologies to establish a self-learning, self-optimizing, and self-adaptive energy scheduling network. At the carbon management layer, the introduction of carbon accounting and carbon flow tracking models positions AEI as a systemic platform that harmonizes economic efficiency with ecological sustainability. The collaborative operation of this three-layer architecture (physical–information–application) underpins both the theoretical integrity and the engineering feasibility of AEI.

Findings from literature surveys indicate that initiatives such as rural electrification, smart agriculture, and localized renewable energy utilization have provided strong technical support for the formation of AEI. Continuous advancements in distributed microgrids, hybrid energy storage systems, and Energy Internet technologies have endowed agricultural energy systems with regional autonomy, operational flexibility, and high reliability. Meanwhile, the application of AI algorithms and data-driven decision models has enabled AEI to progress from experience-based scheduling to intelligent prediction. Whether in agrivoltaic systems, precision irrigation, or Agricultural Virtual Power Plants (Agri-VPPs), the shared core logic lies in achieving optimal energy allocation and minimal carbon emissions through information-driven optimization. This information-centric energy paradigm is reshaping the operational logic and management model of agricultural energy systems.

From the perspective of system architecture evolution, the theoretical framework of AEI has undergone three key stages – conceptual germination, technological exploration, and demonstration application – reflecting a transition from "energy-driven" to "information-intelligence-driven" development. The early stage was characterized by renewable energy integration and agricultural electrification, focusing primarily on energy supply and infrastructure construction. The middle stage introduced AI, IoT, and carbon accounting models, forming a coordinated optimization mechanism among energy, information, and carbon flows. Entering the current stage, AEI now possesses core functions such as Agricultural Virtual Power Plants (Agri-VPPs), digital twin systems, and blockchain-based energy transactions, enabling both system integration and market-oriented innovation. This evolutionary trajectory demonstrates that AEI is not merely an energy technology framework but a systemic engineering paradigm for agricultural digitalization and low-carbon transformation. Through multi-energy complementarity, information fusion, and intelligent scheduling, AEI has evolved from an "energy supply network" into an "ecological network" endowed with capabilities for continuous learning, collaborative evolution, and cross-domain integration.

From the standpoint of development trends, AEI is currently at a critical stage of rapid expansion and intelligent upgrading. Agrivoltaic technologies are evolving from fixed-structure systems toward dynamically adjustable spectral control and AI-driven microclimate management. Breakthroughs in solid-state batteries and hybrid energy storage are addressing the intermittency challenges of renewable energy. The integration of hydrogen energy systems, blockchain-based energy trading, and carbon sequestration technologies is fostering a circular agricultural energy ecosystem aligned with carbon neutrality goals. AI-driven energy management algorithms and digital twin simulation platforms enable AEI to conduct prediction, optimization, and decision-making within virtual environments, significantly enhancing system controllability and reliability. Meanwhile, global policy support and business model innovation are providing strong external momentum for AEI's development. From the European Union's "Digital Farm Initiative" to China's "Rural Energy Internet Demonstration Zones," and from India's "Agricultural Microgrid Projects" to Africa's "Solar Irrigation Programs," AEI is gradually forming a globally collaborative innovation network.

From a theoretical perspective, the AEI concept holds significant implications for the advancement of Energy Internet theory, agricultural systems science, and ecological economics. At the theoretical level, AEI achieves unified modeling of energy and agricultural systems, providing a new paradigm for the coordinated optimization of complex systems. Methodologically, AEI embodies the "perception–analysis–decision–feedback" closed-loop logic inherent to systems science, achieving system self-organization and self-evolution through the interaction of energy, information, and carbon flows. At the engineering application level, it provides the agricultural sector with low-carbon, efficient, and sustainable energy solutions, driving the transformation of agriculture from an "energy-dependent" to an "energy-producing" system. At the socio-economic level, AEI offers new income pathways for farmers – such as peer-to-peer energy trading, carbon credit markets, and shared energy service platforms – thus promoting energy equity and inclusive rural development. The establishment of this framework creates a synergistic resonance among energy transition, climate governance, and agricultural modernization.

From the perspective of challenges and prospects, AEI still faces a series of technical and institutional barriers. On the technical level, issues of system interoperability, standardization, and cybersecurity require further strengthening. The generalization and interpretability of AI models remain significant challenges in high-dimensional and nonlinear agricultural environments, while the accuracy of carbon accounting and carbon flow tracking in multi-source heterogeneous systems still needs improvement. On the economic level, high initial investment, limited financing channels, and long payback periods remain major obstacles to AEI's widespread adoption in rural areas. On the institutional level, policy support mechanisms have not yet fully integrated carbon trading, energy markets, and agricultural subsidies, and cross-sector coordination frameworks need to be further refined. However, from a long-term perspective, these challenges are not structurally insurmountable but represent essential stages in AEI's system evolution toward maturity. With the continued advancement of 6G networks, AI large models, quantum computing, and blockchain technologies, AEI's capabilities in intelligent scheduling, market trading, and carbon management are expected to make qualitative breakthroughs.

From a global perspective, AEI represents the future direction of agricultural systems – toward energy autonomy, informational intelligence, and ecological low-carbon development. It is not merely an innovation in energy systems but also a reconstruction of agricultural civilization – an organic whole in which information functions as the neural network, energy as the lifeblood, and carbon as the ecological constraint. By establishing a closed-loop system integrating energy flow, information flow, and carbon flow, AEI transforms agriculture from a passive energy consumer into an active energy producer and ecological restorer, positioning it as a proactive force in global energy transition and climate governance. In the future, as AEI's applications deepen across different regions, its influence will extend beyond agriculture itself, becoming a critical driving force for urban–rural integration, energy equity, and ecological civilization construction.

In summary, the conception and development of the Agricultural Energy Internet mark the evolution of agricultural energy systems from single-function structures toward integrated intelligent ecosystems. AEI not only addresses the challenges of energy security and environmental pressure in agricultural production but also provides a replicable and scalable systemic solution for achieving the United Nations Sustainable Development Goals (SDGs). The future of AEI belongs not only to the interdisciplinary innovation of energy science and agricultural engineering but also to the strategic practice of global climate action, green growth, and the construction of a shared future for humankind. Just as the Energy Internet catalyzed the industrial revolution, AEI will become a key engine of the agricultural revolution, propelling human society toward an intelligent, clean, and sustainable new era.

References

[1] Zhang X, Fu X, Xue Y, *et al.* A review on basic theory and technology of agricultural energy internet. *IET Renewable Power Generation*, 2024, 18(7), 1318–1331.

[2] Fu X, and Yang F. Viewpoints on the theory of agricultural energy internet. *Frontiers in Energy Research*, 2022, 10, 871772.

[3] Fu X, and Niu H. Key technologies and applications of agricultural energy internet for agricultural planting and fisheries industry. *Information Processing in Agriculture*, 2023, 10(3), 416–437.

[4] Fu X, Wei Z, Sun H, *et al.* Agri-energy-environment synergy-based distributed energy planning in rural areas. *IEEE Transactions on Smart Grid*, 2024, 15(4), 3722–3738.

[5] Fu X, Bai J, Sun H, *et al.* Optimizing agro-energy-environment synergy in agricultural microgrids through carbon accounting. *IEEE Transactions on Smart Grid*, 2024, 15(5), 4819–4834.

[6] Fu X, and Zhou Y. Collaborative optimization of PV greenhouses and clean energy systems in rural areas. *IEEE Transactions on Sustainable Energy*, 2022, 14(1), 642–656.

[7] Fu X, and Long H. Virtual power plant scheduling in agricultural microgrids through the control of distributed energy resources and greenhouse loads. *IEEE Transactions on AgriFood Electronics*, 2025, 3(2), 569–581. https://doi.org/10.1109/TAFE.2025.3591910.

[8] Fu X, Ye W, Li X, *et al.* Revolutionising agri-energy: A comprehensive survey on the applications of artificial intelligence in agricultural energy internet. *Energy Internet*, 2025, 2(1), 13–27.

[9] Fu X, Wei Z, Yang F, *et al.* Modelling of agricultural energy internet considering the integration of planting industry and new energy. *Frontiers in Energy Research*, 2022, 10, 998493.

[10] Zhuo Z, Zhang N, Xie X, *et al.* Key technologies and developing challenges of power system with high proportion of renewable energy. *Automation of Electric Power Systems*, 2021, 45(9), 171–191.

[11] Klerkx L, Jakku E, and Labarthe P. A review of social science on digital agriculture, smart farming and agriculture 4.0: New contributions and a future research agenda. *NJAS-Wageningen Journal of Life Sciences*, 2019, 90, 100315.

[12] Sharma V, Tripathi A K, and Mittal H. Technological revolutions in smart farming: Current trends, challenges and future directions. *Computers and Electronics in Agriculture*, 2022, 201, 107217.

[13] Himesh S, Rao E V S P, Gouda K C, *et al.* Digital revolution and big data: A new revolution in agriculture. *CABI Reviews*, 2018, 2018, 1–7.

[14] Khan N, Ray R L, Kassem H S, *et al.* Potential role of technology innovation in transformation of sustainable food systems: A review. *Agriculture*, 2021, 11(10), 984.

[15] Raj M, and Prahadeeswaran M. Revolutionizing agriculture: A review of smart farming technologies for a sustainable future. *Discover Applied Sciences*, 2025, 7(9), 937.

[16] Vărzaru A A. Digital revolution in agriculture: Using predictive models to enhance agricultural performance through digital technology. *Agriculture*, 2025, 15(3), 258.

[17] Citoni B, Fioranelli F, Imran M A, *et al*. Internet of Things and LoRaWAN-enabled future smart farming. *IEEE Internet of Things Magazine*, 2019, 2(4), 14–19.

[18] Bach H, and Mauser W. *Sustainable agriculture and smart farming[M]// Earth observation open science and innovation*. Cham: Springer International Publishing, 2018, 261–269.

[19] Ahmed N, and Shakoor N. Advancing agriculture through IoT, Big Data, and AI: A review of smart technologies enabling sustainability. *Smart Agricultural Technology*, 2025, 10, 100848. https://doi.org/10.1016/j.atech.2025.100848.

[20] Miller T, Gomez J, and Brown A. The IoT and AI in agriculture: The time is now. *Agriculture & Food Systems Journal*, 2024, 8(2), 100–112.

[21] Malhi G S, Kaur M, and Kaushik P. Impact of climate change on agriculture and its mitigation strategies: A review. *Sustainability*, 2021, 13(3), 1318.

[22] U.S. Environmental Protection Agency. Climate change impacts on agriculture and food supply [online]. U.S. Environmental Protection Agency; 2023. Available from https://www.epa.gov/climateimpacts/climate-change-impacts-agriculture-and-food-supply [Accessed Feb 2026].

[23] Uprety D C, Dhar S, Dong H M, *et al*. Technologies for Climate Change Mitigation - Agriculture Sector. Roskilde: UNEP Risø Centre on Energy, Climate and Sustainable Development, Department of Management Engineering, Technical University of Denmark (DTU); 2012. 138 p. (TNA Guidebook Series).

[24] Teal N, and Burkart K. Regenerative agriculture can play a key role in combating climate change. *One Earth*, 2023.

[25] Taylor M. Climate-smart agriculture: What is it good for? *The Journal of Peasant Studies*, 2018, 45(1), 89–107.

[26] Houser M, and Stuart D. An accelerating treadmill and an overlooked contradiction in industrial agriculture: Climate change and nitrogen fertilizer. *Journal of Agrarian Change*, 2020, 20(2), 215–237.

[27] Moallemi E A, de Haan F, Kwakkel J, *et al*. Narrative-informed exploratory analysis of energy transition pathways: A case study of India's electricity sector. *Energy Policy*, 2017, 110, 271–287.

[28] Bößner S, Laderach P, Pacillo G. The Climate Security and Energy (Transition) Nexus: Winds of Change [online]. Montpellier: CGIAR FOCUS Climate Security; 2022. Available from https://cgspace.cgiar.org/items/9c6cf699-8b0d-44ce-ba7f-f5c1cfbfab80 [Accessed Feb 2026].

[29] Paul A, Sahoo D, Mohapatra S, *et al*. How does the energy transition shape inclusive green growth in the European Union?. *Jahrbücher für Nationalökonomie und Statistik*, 2025 (0).

[30] Lundaev V, Solomon A A, Le T, *et al.* Review of critical materials for the energy transition, an analysis of global resources and production databases and the state of material circularity. *Minerals Engineering*, 2023, 203, 108282.

[31] Natividad L E, and Benalcazar P. Hybrid renewable energy systems for sustainable rural development: Perspectives and challenges in energy systems modeling. *Energies*, 2023, 16(3), 1328.

[32] Chel A, and Kaushik G. Renewable energy for sustainable agriculture. *Agronomy for Sustainable Development*, 2011, 31(1), 91–118.

[33] Uleri F, Viganò F, and Musolino M. Understanding rural energy transitions through the spatial dynamics between rural change and energy system transformation. *Journal of Rural Studies*, 2025, 120, 103895.

[34] Pestisha A, Gabnai Z, Chalgynbayeva A, *et al.* On-farm renewable energy systems: A systematic review. *Energies*, 2023, 16(2), 862.

[35] Chodkowska-Miszczuk J, Kuziemkowska S, Verma P, *et al.* To know is to accept. Uncovering the perception of renewables as a behavioural trigger of rural energy transition. *Moravian Geographical Reports*, 2022, 30(4), 311–323.

[36] El-Shirbeny M A, and Biradar C. *Advances in earth observation and artificial intelligence in monitoring vegetation dynamics of dryland agroecosystems [M]//Vegetation Dynamics and Crop Stress.* Academic Press, 2024, 1–19.

[37] Duguma A L, and Bai X. How the internet of things technology improves agricultural efficiency. *Artificial Intelligence Review*, 2024, 58(2), 63.

[38] Duncan E, Abdulai A R, and Fraser E D G. Modernizing agriculture through digital technologies: Prospects and challenges. *Handbook on the Human Impact of Agriculture*, 2021, 138–161.

[39] EY Global. Digital agriculture: enough to feed a rapidly growing world? [online].EY; 2018. Available from https://www.ey.com/en_us/insights/digital/digital-agriculture-data-solutions [Accessed Feb 2026].

[40] Hati A J, and Singh R R. Smart indoor farms: Leveraging technological advancements to power a sustainable agricultural revolution. *AgriEngineering*, 2021, 3(4), 728–767.

[41] Nasir M, Khan H A, Hussain A, Mateen L, and Zaffar N A. Solar PV-based scalable DC microgrid for rural electrification in developing regions. *IEEE Transactions on Sustainable Energy*, 2018, 9(1), 390–399.

[42] Bawan E K, Wijaya F D, Ali H R and Vasquez J C. Comprehensive analysis and optimal design of hybrid renewable energy system for rural electrification in West Papua, Indonesia. *IEEE Access*, 2025, 13, 66250–66275.

[43] Liu Y, Xu X, Deng C, *et al.* Peer-to-peer based energy management framework for enhancing rural electrification level. *Journal of Modern Power Systems and Clean Energy*, 2025, 13(3), 953–966.

[44] Paz Silva L A, Brito Filho F d A, and de Andrade H D. Soil moisture monitoring system based on metamaterial-inspired microwave sensor for precision agriculture applications. *IEEE Sensors Journal*, 2023, 23(19), 23713–23720.

[45] Manurung E P N, Fiumara M, Bennin K E, and Asiedu D K P. An energy-efficient precision agriculture communication embedded system CoAP design. *IEEE Internet of Things Journal*, 2025, 12(19), 39984–39990.

[46] Wang Y, Hu H, Wu S and Xiong Y. Enhancing weeding efficiency: Addressing targeting positional errors and key determinants of cutting efficiency in laser weeding robots. *IEEE Transactions on AgriFood Electronics*, 2025, 3(1), 263–271.

[47] Cicilio P, Orosz M, Mueller A, and Cotilla-Sanchez E. uGrid: Reliable minigrid design and planning toolset for rural electrification. *IEEE Access*, 2019, 7, 163988–163999.

[48] Nasir M, Jin Z., Khan H A, Zaffar N A, Vasquez J C, and Guerrero J M, A decentralized control architecture applied to DC nanogrid clusters for rural electrification in developing regions. *IEEE Transactions on Power Electronics*, 2019, 34(2), 1773–1785.

[49] Habib H U R, Waqar A, Junejo A K, Elmershedy M F, Wang S, and Buker M S. Optimal planning and EMS design of PV based standalone rural microgrids. *IEEE Access*, 2021, 9, 32908–32930.

[50] Vai V, Alvarez-Herault M -C, Raison B, and Bun L. Optimal low-voltage distribution topology with integration of PV and storage for rural electrification in developing countries: A case study of Cambodia. *Journal of Modern Power Systems and Clean Energy*, 2020, 8(3), 531–539.

[51] Khan R, Nasir M, and Schulz N N. An optimal neighborhood energy sharing scheme applied to Islanded DC microgrids for cooperative rural electrification. *IEEE Access*, 2023, 11, 116956–116966.

[52] Elijah O, Rahman T A, Orikumhi I, Leow C Y, and Hindia M N. An overview of internet of things (IoT) and data analytics in agriculture: Benefits and challenges. *IEEE Internet of Things Journal*, 2018, 5(5), 3758–3773.

[53] Friha O, Ferrag M A, Shu L, Maglaras L, and Wang X. Internet of things for the future of smart agriculture: A comprehensive survey of emerging technologies. *IEEE/CAA Journal of Automatica Sinica*, 2021, 8(4), 718–752.

[54] Zheng Y, Chen A, Fu X, and Li D. Photovoltaics and agriculture nexus: Exploring the influence of agrivoltaics on food production and electricity generation. *IEEE Journal of Photovoltaics*, 2024, 14(5), 705–719.

[55] Huang K, Shu L, Li K, *et al.* Photovoltaic agricultural internet of things towards realizing the next generation of smart farming. *IEEE Access*, 2020, 8, 76300–76312.

[56] Hou G, Niu R, Chen H, *et al*, A robotic AI algorithm for fusing generative large models in agriculture internet of things. *IEEE Internet of Things Journal*, 2025, 12(10), 13511–13524.

[57] Rigogiannis, N, Almpanopoulos L, Kanimas L, *et al.* Nanosatellites in precision agriculture—Comparison with drone-based applications and improved power distribution unit design. *IEEE Transactions on AgriFood Electronics*, 2025, 3(2), 336–347.

[58] Ubilla K, Jiménez-Estévez G A, Hernández, *et al.* Smart microgrids as a solution for rural electrification: Ensuring long-term sustainability through cadastre and business models. *IEEE Transactions on Sustainable Energy*, 2014, 5(4), 1310–1318.

[59] Fu X, Zhou Y, Wei Z, and Wang Y. Optimal operation strategy for rural microgrid considering greenhouse load control. *CSEE Journal of Power and Energy Systems*, 2025, 11(1), 269–279.

[60] Malinowski M, Leon J I, and Abu-Rub H. Solar photovoltaic and thermal energy systems: Current technology and future trends. *Proceedings of the IEEE*, 2017, 105(11), 2132–2146.

[61] Marrou H, Guilioni L, Dufour L, Dupraz C, and Wery J. Microclimate under agrivoltaic systems: Is crop growth rate affected in the partial shade of solar panels? *Agricultural and Forest Meteorology*, 2013, 177, 117–132.

[62] Ahemd R, Izaz M H B, Manik K H, Khatun H, Nath A, Mim J J, and Hossain N. Advancements in agrivoltaic systems for enhanced sustainable energy utilization-A review. *Environmental Challenges*, 2025, 20, 101299.

[63] Ghosh A. Nexus between agriculture and photovoltaics (agrivoltaics, agriphotovoltaics) for sustainable development goal: A review. *Solar Energy*, 2023, 266, 112146.

[64] Tan Y, Liu J, Li W, *et al* Agrivoltaics development progresses: From the perspective of photovoltaic impact on crops, soil ecology and climate. *Environmental Research*, 2025, 266, 120540.

[65] Zegeye A A, Masresha D A, Tefera S A, *et al* Adoption of solar energy technology among households: Evidence from Ethiopia. *Journal of Cleaner Production*, 2025, 521, 146287.

[66] Goh H H, Huang C, Liang X, *et al* Sustainable development through the balancing of photovoltaic charging facilities and agriculture for energy harvesting. *Applied Energy*, 2025, 377, 124463.

[67] Tan Y, Liu J, Li W, *et al* Agrivoltaics development progresses: From the perspective of photovoltaic impact on crops, soil ecology and climate. *Environmental Research*, 2025, 266, 120540.

[68] Pandiyan P, Sitharthan R, Saravanan S, *et al.* A comprehensive review of the prospects for rural electrification using stand-alone and hybrid energy technologies. *Sustainable Energy Technologies and Assessments*, 2022, 52 (2022), 102155.

[69] Chattopadhyay A, Sauer P W, and Witmer A P. Can renewable energy work for rural societies? Exploring productive use, institutions, support systems, and trust for solar electricity in the Navajo Nation. *Energy Research & Social Science*, 2024, 107, 103342.

[70] Jameel M K, Mustafa M A, Ahmed H S, *et al*. Biogas: Production, properties, applications, economic and challenges: A review. *Results in Chemistry*, 2024, 7(2024), 101549.

[71] Sher F, Smječanin N, Hrnjić H, *et al* Emerging technologies for biogas production: A critical review on recent progress, challenges and future perspectives. *Process Safety and Environmental Protection*, 2024, 188, 834–859.

[72] Tezer Ö, Karabağ N, Öngen A, Çolpan C Ö, and Ayol A. Biomass gasification for sustainable energy production: A review. *International Journal of Hydrogen Energy*, 2022, 47(34), 15419–15433.

[73] Silinto B F, van der Laag Yamu C, Zuidema C, and Faaij A P. Hybrid renewable energy systems for rural electrification in developing countries: A review on energy system models and spatial explicit modelling tools. *Renewable and Sustainable Energy Reviews*, 2025, 207, 114916.

[74] Ali M F, Sheikh M R I, Akter R, Islam K N, and Ferdous A I. Grid-connected hybrid microgrids with PV/wind/battery: sustainable energy solutions for rural education in Bangladesh. *Results in Engineering*, 2025, 25, 103774.

[75] Rathod A A, and Subramanian B. Design and optimization of various hybrid renewable energy systems using advanced algorithms for powering rural areas. *Journal of Cleaner Production*, 2025, 501, 145199.

[76] Kamal M M, Ashraf I, and Fernandez E. Optimal sizing of standalone rural microgrid for sustainable electrification with renewable energy resources. *Sustainable Cities and Society*, 2023, 88, 104298.

[77] Long H, Fu X, Kong W, Chen H, Zhou Y, and Yang F. Key technologies and applications of rural energy internet in China. *Information Processing in Agriculture*, 2024, 11(3), 277–298.

[78] Dwivedi A, Dwivedi P, Joshi K, *et al*. Local leader's impact on adoption of renewable energy generation technology by rural communities in the Himalayan region. *Journal of Cleaner Production*, 2022, 352, 131479.

[79] Sitharthan R, Vimal S, Verma A, *et al* Smart microgrid with the internet of things for adequate energy management and analysis. *Computers and Electrical Engineering*, 2023, 106, 108556.

[80] Wang X, Ha B, Lee G Y, *et al*. Low-cost far-field wireless electrical load monitoring system applied in an off-grid rural area of Tanzania. *Sustainable Cities and Society*, 2020, 59, 102209.

[81] Mekhilef S, Saidur R, and Safari A. A review on solar energy use in industries. *Renewable and Sustainable Energy Reviews*, 2011, 15(4), 1777–1790.

[82] Kumar R, and Rosen M A. A critical review of photovoltaic–thermal solar collectors for air heating. *Applied Energy*, 2011, 88(11), 3603–3614.

[83] Smyth M, Pugsley A, Hanna G, *et al*. Experimental performance characterisation of a hybrid photovoltaic/solar thermal façade module compared to a flat integrated collector storage solar water heater module. *Renewable Energy*, 2019, 137, 137–143.

[84] Kumar Y, Ringenberg J, Depuru S S, *et al* Wind energy: Trends and enabling technologies. *Renewable and Sustainable Energy Reviews*, 2016, 53, 209–224.

[85] Tummala A, Velamati R K, Sinha D K, Indraja V, and Krishna V H. A review on small scale wind turbines. *Renewable and Sustainable Energy Reviews*, 2016, 56, 1351–1371.

[86] Pourrajabian A, Mirzaei M, Ebrahimi R, and Wood D. Effect of air density on the performance of a small wind turbine blade: A case study in Iran. *Journal of Wind Engineering and Industrial Aerodynamics*, 2014, 126, 1–10.

[87] Rupf G V, Bahri P A, De Boer K, and McHenry M P. Broadening the potential of biogas in Sub-Saharan Africa: An assessment of feasible technologies and feedstocks. *Renewable and Sustainable Energy Reviews*, 2016, 61, 556–571.

[88] Meyer A K P, Ehimen E A, and Holm-Nielsen J B. Future European biogas: Animal manure, straw and grass potentials for a sustainable European biogas production. *Biomass and Bioenergy*, 2018, 111, 154–164.

[89] Wainaina S, Awasthi M K, Sarsaiya S, *et al.* Resource recovery and circular economy from organic solid waste using aerobic and anaerobic digestion technologies. *Bioresource Technology*, 2020, 301, 122778.

[90] Lund J W, and Boyd T L. Direct utilization of geothermal energy 2015 worldwide review. *Geothermics*, 2016, 60, 66–93.

[91] Sarbu I, and Sebarchievici C. General review of ground-source heat pump systems for heating and cooling of buildings. *Energy and Buildings*, 2014, 70, 441–454.

[92] Wang E, Fung A S, Qi C, and Leong W H. Performance prediction of a hybrid solar ground-source heat pump system. *Energy and Buildings*, 2012, 47, 600–611.

[93] Sierzchula W, Bakker S, Maat K, and Van Wee B. The influence of financial incentives and other socio-economic factors on electric vehicle adoption. *Energy Policy*, 2014, 68, 183–194.

[94] Shareef H, Islam M M, and Mohamed A. A review of the stage-of-the-art charging technologies, placement methodologies, and impacts of electric vehicles. *Renewable and Sustainable Energy Reviews*, 2016, 64, 403–420.

[95] Kumar R R, and Alok K. Adoption of electric vehicle: A literature review and prospects for sustainability. *Journal of Cleaner Production*, 2020, 253, 119911.

[96] Tuballa M L, and Abundo M L. A review of the development of Smart Grid technologies. *Renewable and Sustainable Energy Reviews*, 2016, 59, 710–725.

[97] Andoni M, Robu V, Flynn D, *et al.* Blockchain technology in the energy sector: A systematic review of challenges and opportunities. *Renewable and Sustainable Energy Reviews*, 2019, 100, 143–174.

[98] Ahmad T, Zhang H, and Yan B. A review on renewable energy and electricity requirement forecasting models for smart grid and buildings. *Sustainable Cities and Society*, 2020, 55, 102052.

[99] Kishita Y, Nakatsuka N, and Akamatsu F. Scenario analysis for sustainable woody biomass energy businesses: The case study of a Japanese rural community. *Journal of Cleaner Production*, 2017, 142, 1471–1485.

[100] Singh S, Singh M, and Kaushik S C. Feasibility study of an islanded microgrid in rural area consisting of PV, wind, biomass and battery energy storage system. *Energy Conversion and Management*, 2016, 128, 178–190.

[101] Naveen S, Aravind S, Yamini B, *et al.* A review on solar energy intensified biomass valorization and value-added products production: Practicability, challenges, techno economic and lifecycle assessment. *Journal of Cleaner Production*, 2023, 405, 137028.

[102] Fu X, Bai J, Sun H, and Zhang Y. Optimizing agro-energy-environment synergy in agricultural microgrids through carbon accounting. *IEEE Transactions on Smart Grid*, 2024, 15(5), 4819–4834. Doi:10.1109/TSG.2024.3397990.

[103] Fu X, Wei Z, Sun H, and Zhang Y. Agri-energy-environment synergy-based distributed energy planning in rural areas. *IEEE Transactions on Smart Grid*, 2024, 15(4), 3722–3738. Doi:10.1109/TSG.2024.3364182.

[104] Fu X, and Zhou Y. Collaborative optimization of PV greenhouses and clean energy systems in rural areas. *IEEE Transactions on Sustainable Energy*, 2023, 14(1), 642–656. Doi:10.1109/TSTE.2022.3223684.

[105] Goetzberger A, and Zastrow A. On the coexistence of solar-energy conversion and plant cultivation. *International Journal of Solar Energy*, 1982, 1(1), 55–69.

[106] Sekiyama T, and Nagashima A. Solar sharing for both food and clean energy production: Performance of agrivoltaic systems for corn, a typical shade-intolerant crop. *Environments*, 2019, 6(6), 65.

[107] Dupraz C, Marrou H, Talbot G, Dufour L, Nogier A, and Ferard Y. (2011). Combining solar photovoltaic panels and food crops for optimising land use: Towards new agrivoltaic schemes. *Renewable Energy*, 36(10), 2725–2732. Doi:10.1016/j.renene.2011.03.005.

[108] Yang J, and Chen B. Energy analysis of a biogas-linked agricultural system in rural China–A case study in Gongcheng Yao Autonomous County. *Applied Energy*, 2014, 118, 173–182.

[109] Fu X, Zhou Y, Yang F, *et al* A review of key technologies and trends in the development of integrated heating and power systems in agriculture. *Entropy*, 2021, 23(2), 260.

[110] Tao W, Zhao L, Wang G, and Liang R. Review of the internet of things communication technologies in smart agriculture and challenges. *Computers and Electronics in Agriculture*, 2021, 189, 106352.

[111] Alam H, Alam M A, and Butt N Z. Techno economic modeling for agrivoltaics: Can agrivoltaics be more profitable than ground mounted PV?. *IEEE Journal of Photovoltaics*, 2022, 13(1), 174–186.

[112] Pylianidis C, Osinga S, and Athanasiadis I N. Introducing digital twins to agriculture. *Computers and Electronics in Agriculture*, 2021, 184, 105942.

[113] Uddin M, Mo H, Dong D, Elsawah S, Zhu J, and Guerrero J M. Microgrids: A review, outstanding issues and future trends. *Energy Strategy Reviews*, 2023, 49, 101127.

[114] Long H, Fu X, Kong W, Chen H, Zhou Y, and Yang F. Key technologies and applications of rural energy internet in China. *Information Processing in Agriculture*, 2024, 11(3), 277–298.

[115] Kurumundayil L, Burkhardt D, Gfüllner L, *et al*. Fast ground irradiance computations for agrivoltaics via physics-informed deep learning models [J]. *Communications Engineering*, 2025, 4(1), 173.

[116] Sezer N, and Bayhan S. Integrated solar-powered freeze desalination and water electrolysis system with energy recovery and storage for sustainable agriculture in desert environments[J]. *Desalination*, 2025, 595, 118321.

[117] Hu Q, Lin X, Ren G, *et al*. Hydrovoltaic electricity generation induced by living leaf transpiration[J]. *Nature Water*, 2024, 2(10), 988–998.

Chapter 2

Scheduling and simulation of a hydrogen-driven VPP with in AEI

Xueqian Fu[1]*, Jing Zhang*[1] *and Hai Long*[1]

Abstract

This chapter focuses on the Agricultural Energy Internet (AEI) setting and aims to build a dispatchable virtual power plant (VPP) framework "centred on hydrogen," achieving dual optimization of decarbonization and economic performance by coordinating greenhouse shiftable loads with distributed energy resources (DERs). The study takes hydrogen fuel cell combined heat and power (HFC–CHP) system as the core, together with rooftop agricultural photovoltaics (PV), biogas units, and electrochemical storage, explicitly characterizing the controllability of greenhouse supplemental lighting and heating loads, and embedding crop-growth boundaries (DLI, GDD, DMA, etc.) into the scheduling feasible region so that, when responding to grid-friendly peak-shaving tasks, the park does not breach the safe window of crop light–temperature–carbon requirements. Methodologically, we first build a dynamic thermal–electrical load model for the greenhouse, linking the transmission chain from outdoor irradiance to indoor light environment to supplemental-lighting power and the mechanisms of heat transfer/ventilation; then we present mechanistic models and operating constraints for HFC–CHP, PV, biogas, and storage, unified under electrical/thermal power balance and tie-line capacity limits. We then formulate a scheduling optimization problem targeting "minimisation of daily operating cost," with cost terms covering hydrogen consumption, power purchases/sales, policy-based capacity/energy subsidies, and carbon costs, while the constraint set includes DER output limits and ramping, storage state of charge (SOC), the greenhouse light–temperature–CO_2 safety domain, and crop physiological thresholds. For the solution, we employ swarm-intelligence/hybrid heuristics combined with feasible-region repair and a rolling time-horizon strategy to handle both discrete switching and continuous power variables. Under representative agricultural-park configurations and meteorological/price profiles, we design baseline and hierarchical coordination schemes for comparison and systematically evaluate peak-shaving capability, economic benefits, carbon-reduction performance, and crop safety. The results show that a

[1]College of Information and Electrical Engineering, China Agricultural University, China

hydrogen-driven VPP can enhance the park's scheduling capability and revenue while meeting crop physiological requirements, and it improves robustness to meteorological uncertainty through thermoelectric coupling and storage buffering. The contributions are as follows: we propose an integrated hydrogen-led VPP model tailored to the greenhouse scenario, yielding a feasible-region formulation that triply couples "crop–load–energy"; we construct an economic objective that accounts for both subsidies and carbon costs; and we provide a reproducible simulation workflow and data interfaces that lay the groundwork for multi-park clusters and learning-based adaptive scheduling in AEI.

Keywords: Agricultural Energy Internet; virtual power plant; hydrogen fuel cell combined heat and power; greenhouse shiftable loads; crop physiological constraints (DLI/GDD/DMA); distributed energy resources; peak shaving and economic performance; carbon emissions reductions; warm/hybrid optimization

2.1 Introduction

2.1.1 Background and research motivation

The formal proposal of the "dual-carbon" goals in 2020 marked a new stage in China's economic and social development, characterized by a comprehensive green transition, with industries across the board initiating low-carbon transformations. Agriculture, as a foundational sector of the national economy, consumes energy across the entire value chain of planting, breeding, processing, and transport; green upgrading and low-carbon retrofitting of its energy systems not only affect production efficiency but also constitute a critical pillar for achieving the dual-carbon targets. Related studies show that traditional agricultural energy systems have long centered on either single-source grid power or direct consumption of fossil fuels, lacking flexible, dynamic scheduling mechanisms – resulting in high energy intensity, low utilization efficiency, and large carbon emissions, in sharp contrast with the needs of modern green agriculture. Against this backdrop, the energy transition in agriculture is accelerating at an unprecedented pace, the limitations of traditional supply models are increasingly salient, and exploring new energy-system architectures and scheduling mechanisms tailored to agricultural production has become an urgent priority.

The structural defects of traditional agricultural energy systems have accumulated over long-term practice, with core issues concentrated three dimensions – supply structure, scheduling mechanisms, and utilization efficiency – together forming a triple bottleneck for low-carbon agricultural development.

A monotonous energy-supply structure is the primary obstacle restricting the green transition of agricultural energy. China's rural areas are vast, and power-grid infrastructure exhibits pronounced regional imbalance; in some remote mountainous and hilly regions, weak distribution networks not only compromise supply reliability (annual outage durations can reach dozens of hours) but also struggle to support the stable operation of high-power agricultural equipment. In areas with insufficient coverage or unstable supply, agriculture has long relied on fossil-fuel equipment such as diesel generators and coal-fired boilers: in northern regions, winter heating in greenhouse tunnels mostly uses coal-fired boilers; in the hilly South, irrigation pump stations often run on diesel units. These devices exhibit energy-consumption intensities as high as 0.8–1.2 kilograms of standard coal per unit output and emit large quantities of CO_2, SO_2, and particulates during operation. Fossil-energy consumption in facility agriculture alone causes more than 120 million tons of CO_2 emissions annually, with coal-fired heating contributing over 60% of incremental emissions – intensifying regional environmental pressures and pushing up energy costs in agricultural production.

Rigid scheduling mechanisms further amplify operational shortcomings of traditional energy systems. Agricultural production displays strong seasonality and periodicity; core end uses such as irrigation, heating, and ventilation are highly time-concentrated, yet traditional systems lack demand–supply interaction and flexible modulation, leaving equipment in a generally "uncoordinated operation" state. During periods when summer irrigation peaks overlap with winter heating peaks, farmland irrigation systems and greenhouse heating equipment often run at full load during grid peak hours, further straining local networks and even causing local outages; meanwhile, low-price off-peak electricity at night is not fully utilized for storage or advance operations, wasting energy value. For example, in North China's wheat-growing areas, irrigation-pump loads during daytime peaks account for 15–20% of regional grid peak load, while equipment utilization at night off-peak falls below 10%. This "stacking on peaks and idling in troughs" mode significantly undermines the economic rationality of energy use.

Low utilization efficiency further compounds waste. In traditional agricultural production, electricity, heat, and gas operate in silos without cascade utilization or synergies, yielding comprehensive energy efficiency generally below 40%, far lower than in industry. In facility agriculture, this is even more pronounced: the heat-loss rate of coal-fired heating systems in northern greenhouses reaches 30–40%, with large amounts of residual heat vented through walls and chimneys. In the South, although smart greenhouses increasingly adopt electrically driven cooling systems, they often fail to coordinate with PV generation: daytime PV under abundant irradiance cannot directly power cooling loads and instead must be fed to the grid in exchange for nighttime quotas. In processing, waste heat from cold-storage refrigeration is not coupled to the thermal demand of dryers; in livestock and poultry farms, biogas is often flared rather than converted into power or heat for production. Such "each to its own" patterns keep unit energy output persistently low.

The multiple defects of traditional systems not only elevate energy costs in agriculture (in some regions, energy spending accounts for 35–50% of production costs), but also make energy consumption the main source of agricultural carbon emissions; estimates indicate that emissions from agricultural energy use now account for 28% of total agricultural carbon emissions nationwide, a critical domain that must be addressed to reach the sector's carbon-peaking goal.

As traditional energy systems hit development bottlenecks, the maturation of distributed renewable technologies combined with rural resource endowments has created a compounded advantage, offering a historic opportunity for optimizing the agricultural energy mix [1]. China's rural regions boast diverse and abundant renewable resources, providing a solid foundation for building high-share renewable energy systems.

Diversified PV utilization models have already reached scale in agriculture. Expansive idle spaces in rural areas enable varied PV applications: in major grain belts such as Shandong and Henan, "agrivoltaics" erect PV arrays above cropland to achieve "generation above, cultivation below," ensuring unaffected grain output while delivering 12,000–15,000 kWh of annual generation per hectare; in aquaculture regions like Guangxi and Guangdong, "fishery-agri-PV integration" installs PV over ponds, where shading suppresses algal blooms and improves rearing conditions, achieving synergy between aquaculture and PV – e.g., a demonstration base in Xichang Town, Hepu County has raised annual income per mu from under RMB 1,000 to over RMB 10,000. In addition, rural rooftops and the shed roofs of livestock and poultry farms have become key carriers for distributed PV. By the end of 2024, rural distributed PV capacity accounted for 22% of national solar installations, making it one of the fastest-growing segments.

Wind and biomass have formed a complementary supply landscape. In wind-rich regions such as the Northeast Plain and the Inner Mongolia grasslands, small wind projects are directly coupled to farmland irrigation via "wind-power + pumps," addressing electrification challenges in remote areas; in hilly regions, dispersed wind projects are cluster-managed to supplement regional agricultural loads. Biomass – distinctive to agriculture – has even greater potential. China produces over 3 billion tons of agricultural residues and livestock/poultry manure annually, convertible via anaerobic digestion to biogas/biomethane, densification to solid fuels, or gasification to syngas for power generation. In livestock-intensive provinces such as Jiangsu and Anhui, large farms commonly build biogas projects that both valorize wastes and supply stable heat and power, with single-site annual CO_2 reductions reaching 800–1,200 tonnes. Hydrogen, as an emerging clean energy, is also beginning to take root in agriculture, with green-hydrogen production from renewable electricity being piloted alongside scenarios such as irrigation and cold-chain logistics, opening new pathways for deep decarbonization [2].

According to the National Energy Administration, distributed energy in rural areas contributes over 30% of incremental additions. These figures underscore the vast potential of rural renewable development and lay a solid foundation for building high-share renewable energy systems in agriculture.

However, the randomness of distributed renewable energy introduces new challenges to the agricultural energy systems. Photovoltaic (PV) generation is significantly affected by irradiance and weather variations: at noon, the output can exceed 90% of rated capacity, whereas on overcast or rainy days, it may drop below 10%; wind-power output fluctuates sharply with wind speed, with gusty conditions producing power swings exceeding 50%. If such unstable sources are connected directly to agricultural energy systems, they can cause voltage fluctuations and frequency deviations, impairing the normal operation of sensitive loads such as irrigation pumps and precision livestock equipment. For example, a certain "agrivoltaics" base experienced a sudden drop in PV output due to cloudy weather, triggering frequent starts and stops of the irrigation system, which not only reduced irrigation efficiency but also shortened equipment lifespan.

Resolving this tension urgently requires effective energy aggregation and intelligent scheduling. By integrating diverse supply from distributed PV, wind, and biomass, and pairing these with storage such as sodium-ion batteries and pumped hydro, an integrated "source–grid–load–storage" system becomes the core pathway to address the problem. As a national rural energy-revolution pilot county, Binyang in Guangxi commissioned a distributed sodium-ion battery storage demonstration on the distribution network that effectively smoothed PV output fluctuations, raising the renewable curtailment-free consumption rate to over 95%; in Haicheng District, Beihai City, a rural energy project integrated distributed PV, small storage, and intelligent load-control systems to achieve dynamic supply–demand balance, reducing CO_2 emissions by about 100 tonnes annually. These practices show that only by constructing a flexible, collaborative new energy system through technological innovation can the potential of rural renewables be fully unlocked, driving a fundamental transformation of agricultural energy systems from "traditional and extensive" to "efficient and low carbon" [3].

Against the backdrop of a deeply intertwined global energy transition and agricultural modernization, the "Agricultural Energy Internet" has in recent years moved from concept to practical exploration as a new business form linking agricultural production with energy systems. Its core connotation is to leverage new-generation information technologies – IoT, big data, AI, and 5G – to break down information barriers and physical silos across the "source, grid, load, storage" links in agricultural energy systems, realizing coordinated coupling among energy flows, data flows, and material flows [2]. This concept is not a simple superposition of agricultural and energy technologies, but a customized energy-service system built around the characteristics of agricultural production: it must meet diverse energy demands for irrigation, cultivation, and processing while adapting to renewable development and utilization, thereby providing foundational support.

The core value of the Agricultural Energy Internet lies in reconstructing agricultural energy-allocation logic through intelligent means. On the "source" side, it integrates distributed resources such as PV greenhouse sheds, small wind, and biomass-utilization devices – for example, an agricultural park in Hebei organically combines PV generation with greenhouses, providing clean energy for production while using PV-panel shading to reduce water evaporation [4]. On the "grid" side, private 5G networks and IoT gateways underpin energy transmission and communication spanning fields, storage facilities, and processing workshops, enabling flexible interconnection of distributed resources. On the "load" side, it covers diverse end uses including greenhouse climate control, smart irrigation, agricultural machinery, and cold chains – where even a single smart greenhouse includes multiple sub-loads such as supplemental lighting, temperature control, and humidity regulation. On the "storage" side, batteries, thermal tanks, and pumped hydro mitigate the mismatch between renewable intermittency and volatile agricultural loads – e.g., a certain ecological agriculture demonstration park in Jiangxi ensured stable production under overcast conditions through its storage system. This integrated "source–grid–load–storage" architecture fundamentally changes the traditional reliance on centralized grid supply and endows agricultural production with autonomous energy-control capability [5].

Compared with urban energy systems, the Agricultural Energy Internet faces a more complex operating environment, with core challenges stemming from the inherent seasonality and stochasticity of agricultural production. Temporally, agricultural loads exhibit marked periodic swings [6]: during the spring irrigation season, concentrated operation of pumps drives a sharp rise in demand; in summer, greenhouse cooling systems must be activated, while in winter heating is required to maintain crop temperatures – the seasonal load differences can be several times or even an order of magnitude. Intra-day, greenhouse lighting loads show peak–valley patterns with sunrise and sunset, falling to a minimum at noon under abundant natural light while peaking at dawn and dusk; heating loads track the diurnal temperature differential, requiring higher output in the pre-dawn cold and allowing moderation at midday. Heterogeneity across crop-growth stages further amplifies variability – for instance, seedlings demand far tighter temperature stability than mature plants. Additionally, the unpredictability of natural conditions raises control difficulty: sudden rain, cold snaps, and other weather events can instantly alter the supply–demand balance. These features make matching supply and demand in agricultural systems far more challenging than in the relatively stable urban context; traditional centralized dispatch, with sluggish response and limited elasticity, is ill-suited, necessitating more flexible, intelligent scheduling technologies.

The maturation and rollout of virtual power plant (VPP) technology provide an innovative solution to the scheduling challenges of the Agricultural Energy Internet [7]. As a comprehensive energy-management system that integrates information technology, energy management, and intelligent control, a VPP uses advanced communication protocols and control algorithms to aggregate and optimize geographically dispersed distributed energy resources (DERs), storage devices, and controllable loads into a "virtual ensemble" with unified dispatch capabilities, participating in power markets like a conventional large power plant and providing ancillary services [8]. In essence, it follows a "from fragments to whole" management logic to unlock the value of dispersed resources and underpin flexible system operation.

In agricultural settings, the core advantages of VPPs translate into concrete energy-management capabilities. First, aggregation effects overcome the "small and scattered" nature of agricultural distributed resources. In China, agricultural DERs are dispersed across fields and rural homesteads; individual PV systems are typically tens to hundreds of kilowatts and struggle to participate in grid dispatch on their own, whereas a VPP can integrate hundreds of small PV plants, storage units, and controllable loads into a megawatt-scale aggregate, markedly improving bargaining power and dispatch flexibility in energy markets [9]. Second, coordinated optimization enables precise matching of "source–load–storage" in agriculture. For example, at midday under strong sunlight, the VPP prioritizes PV to meet greenhouse lighting and irrigation needs, with surplus directed to batteries or converted to heat via heat pumps for storage; when the evening peak arrives, it automatically releases stored energy and trims non-core processing loads, achieving efficient allocation through multi-energy complementarity [10]. Third, flexible responsiveness safeguards both grid stability and agricultural production: when grid frequency fluctuates, the VPP can adjust aggregate output or load within seconds to provide ancillary services while ensuring that critical agricultural loads remain unaffected.

At present, VPP technology has achieved large-scale deployment with notable results in cities and industrial parks. In demand response, it can guide commercial and industrial users to shift consumption through price signals, with single responses reaching tens of megawatts; in multi-energy coordination, it enables optimal conversion among electricity, heat, and gas; and in storage dispatch, cluster management extends battery cycle life and improves utilization efficiency [11]. However, transplanting VPPs into agriculture is still at the stage of theoretical exploration and small pilots, facing multiple adaptation challenges rooted both in the particularities of agricultural production and in the constraints of rural infrastructure.

The primary difficulty lies in the physiological constraints and control boundaries of agricultural loads. Unlike industrial loads that can be freely adjusted according to electricity prices, agricultural loads are directly tied to crop-growth laws and have strict no-go zones for control. Take a smart greenhouse as an example: the suitable temperature range for tomatoes is 15–28 °C; below 10 °C growth stalls, and above 35 °C plants suffer heat scorch, so a VPP must strictly respect these thresholds when modulating heating or cooling loads. Lighting control is similarly constrained: leafy vegetables such as lettuce require 12–14 hours of light per day, and any curtailment of supplemental lighting must occur only after photosynthetic needs are met; otherwise, leaves yellow and yields fall. This "flexible control within rigid demand" requires VPP systems to integrate crop-growth models and achieve co-decision-making between energy scheduling and agricultural production, whereas most existing algorithms are developed for industrial settings and poorly fit complex biological growth characteristics [12].

The second challenge is the dispersion of agricultural distributed energy and communication-coverage limits. Farmland and agricultural facilities in China are widely spread in remote areas; DERs exhibit "many points, long lines, wide areas," with PV arrays and storage units in some mountainous parks separated by kilometers. Communication networks are foundational for real-time VPP control, yet 5G and fiber coverage in remote rural regions is limited; some areas can only support basic communications, far from millisecond-level data-transfer needs. Even where networks exist, extreme weather undermines signal stability – torrential rain or snow can take wireless sensors offline, and strong electromagnetic environments interfere with data transmission – leading to delayed or failed VPP dispatch commands and reduced operational reliability.

In addition, technology costs and talent shortages constrain agricultural VPP rollout. Building an agricultural VPP requires hardware investments in sensors, communications, and batteries; for a single mid-sized agricultural park, upfront capital can reach several million RMB, a heavy burden for smallholders or cooperatives [13]. Meanwhile, operations demand hybrid talent versed in both energy dispatch and agricultural technology, which is exceedingly scarce in rural areas; many current practitioners lack energy-system operation skills, making stable operation difficult to assure.

Despite the hurdles, the application prospects of agricultural VPPs are emerging. In agrivoltaic projects, VPPs integrating PV, storage, and greenhouse loads have raised energy-utilization efficiency by 20–30% while cutting planting costs by over 15%. At large-scale livestock bases, matching optimization between biomass-power generation and heating loads has delivered zero pollutant emissions and energy self-sufficiency. As private 5G extends into rural areas, storage costs continue to fall, and agricultural digitalization improves, these bottlenecks are gradually being overcome. Looking ahead, a triad of "VPP scheduling platform + crop-growth models + rural communications networks" is expected to enable efficient operation of the Agricultural Energy Internet, providing vital support for rural revitalization and the "dual-carbon" goals.

Shandong Province is one of China's key demonstration regions for the Agricultural Energy Internet. In recent years, Shandong has advanced rapidly in PV and biomass applications, yet agricultural energy systems still operate mainly in isolation, lacking coordination between supply and demand. Shandong's total agricultural output value has led the nation for many consecutive years, with massive energy demand; in 2023, agricultural electricity consumption reached 68 billion kWh, accounting for 7.2% of total social electricity use. Significant progress has been made in renewables for agriculture: cumulative agrivoltaic capacity has exceeded 5 GW, generating 6 billion kWh annually; biomass-heating now covers over 100,000 households, replacing 150,000 tonnes of scattered coal. However, clear shortcomings remain: (i) fragmented system operation – PV greenhouses, biomass boilers, and irrigation pumps typically run independently without unified coordination, leading to pronounced energy waste, with PV curtailment rates exceeding 10% in some areas; (ii) limited policy efficacy – despite local initiatives such as "Green Energy Demonstration Zones for Agriculture" and "Agricultural Low-Carbon Action Plans," practical deployment still lacks replicable dispatch models and carbon-assessment frameworks; (iii) weak technical support – monitoring, control, and optimization platforms lag, and VPP core algorithms tailored to agricultural scenarios are lacking, making it hard to meet intelligent dispatch needs under complex constraints. These issues not only hinder improvements in Shandong's agricultural energy efficiency and progress toward dual-carbon targets, but also underscore that introducing VPPs into agricultural settings can raise energy-use efficiency, lower operating costs, and reduce carbon emissions through intelligent scheduling – thereby achieving dual economic and environmental benefits.

Accordingly, this study takes the agricultural virtual power plant as its object and focuses on establishing coordination mechanisms among distributed energy, controllable greenhouse loads, and carbon-footprint models. Specifically, it addresses three core questions: first, resource aggregation and modeling for agricultural VPPs – how to build multi-agent coordination models that incorporate crop-growth constraints and DER characteristics; second, low-carbon-oriented scheduling strategies – how to embed carbon-footprint assessment into dispatch objectives to realize multi-objective optimization of economic cost and carbon emissions; and third, real-time control and response mechanisms – how to achieve precise control under rural communication constraints. The results are expected to provide theoretical support and technical references for low-carbon, efficient, and intelligent dispatch of agricultural energy systems, contributing to a feasible pathway for the sector's energy transition under the dual-carbon agenda.

2.1.2 Literature review

Over the past decade, virtual power plants (VPPs) – a key technological means of integrating distributed energy resources and improving energy-use efficiency – have seen significant advances in research and application, especially in two high-energy-intensity, high-demand settings: urban distribution networks and industrial parks, where relatively mature theoretical systems and practical paradigms have formed. In urban distribution networks, the combination of high load density and diverse demand has pushed research toward deep integration of multi-energy coordination and demand-side response. [4] proposed a layered scheduling model based on coordinated "source–grid–load–storage," which innovatively couples output forecasting for distributed photovoltaics, charging–discharging strategies for storage systems, and user-response mechanisms guided by time-of-use pricing for joint optimization. Through a two-layer architecture – upper-layer system-level economic dispatch and lower-layer device-level power allocation – the model maximizes VPP operating profit; in a pilot city, it reduced the peak–valley gap by 18% and cut comprehensive operating costs by 12%.

With rolling optimization and feedback-correction mechanisms, the algorithm dynamically adjusts strategies based on real-time load data and short-term forecasts, effectively enhancing the system's ability to handle load uncertainty; under extreme weather, supply reliability improved by 23% over traditional methods. In industrial parks, research focuses more on deep coordination between VPPs and industrial processes. Ling *et al.* [14] explored multi-role market participation for VPPs, proposing an operation mode that earns revenue by joining day-ahead energy trading via demand response. In a chemical park demonstration, this model not only reduced electricity costs by 9%, but also generated over RMB 2 million annually from ancillary services, offering a replicable example for industrial energy transition. Collectively, these studies – from scheduling models and control algorithms to market mechanisms – form a solid theoretical and algorithmic foundation for scaling VPPs in urban and industrial settings.

By contrast, VPP research in agricultural settings lags notably behind urban and industrial applications and has yet to form a complete technical system. As a foundational sector of the national economy, agriculture exhibits unique energy-use characteristics – dispersion, seasonality, and strong coupling with production activities – that make traditional VPP approaches hard to apply directly. Silinto *et al.* [4] were among the first to propose a conceptual model of the Agricultural Energy Internet, emphasizing the importance of coupling among agricultural production energy use, rural household energy use, and renewables, and preliminarily constructing a theoretical framework for "energy–agriculture" coordinated development; however, the model remains conceptual, lacking concrete technical pathways and quantitative analysis. Zheng *et al.* [15], targeting the energy needs of facility agriculture, conducted an energy-balance analysis for a greenhouse multi-energy complementary system based on PV–storage–electric boilers. By building a system energy-flow mathematical model, they verified the feasibility of winter greenhouse heating, but did not consider dynamic variations in energy demand across crop-growth stages and did not address optimization-based scheduling.

Erdiwansyah *et al.* [16] studied storage optimization in agricultural microgrids, formulating a storage-capacity configuration model with levelized cost of electricity minimization as the objective; yet the study focused solely on economics and neglected environmental constraints critical to crop growth – temperature, humidity, and light – making its results ill-suited to practical agricultural needs. Recently, with the "dual-carbon" agenda, low-carbon development in agriculture has become a research hotspot. Zhang *et al.* [17] introduced carbon-emission factors, incorporating carbon costs from agricultural energy use into the scheduling objective and establishing an agricultural VPP model that considers carbon costs, providing a new perspective for low-carbon dispatch. However, their identification of agricultural carbon sources was incomplete, omitting key links such as farmland soil emissions and embodied carbon in agricultural inputs.

In addition, some studies have attempted to introduce AI and optimization algorithms into agricultural VPPs to enhance decision intelligence. For example, Martineau *et al.* [18] applied a particle swarm optimization (PSO)-based model for agricultural microgrid scheduling, using PSO's global search to optimize power sharing among distributed resources; improved genetic algorithms, by adjusting crossover and mutation operators, increased convergence speed under fluctuating agricultural loads; and deep reinforcement learning, with its adaptive capability in dynamic environments, has been used for real-time scheduling in agricultural VPPs. Nevertheless, these algorithmic applications face clear limitations: they generally lack systematic modeling of greenhouse load dynamics and fail to adequately integrate physiological processes – such as photosynthesis and transpiration – with energy consumption, leading to optimization outcomes misaligned with real production needs.

Overall, current agricultural VPP research shows three prominent shortcomings:

1. Load dimension: Many studies treat agricultural loads as ordinary controllable loads, adopting modeling methods borrowed from industrial or residential contexts and seriously overlooking physiological factors in crop growth. Agricultural loads (e.g., greenhouse heating, lighting, irrigation) directly serve crop growth; their intensity and temporal profiles depend closely on biological characteristics such as photosynthetic efficiency, temperature–humidity

thresholds, and growth stages. For instance, tomatoes require 20–25 °C during fruiting but 15–20 °C at the seedling stage; ignoring this difference in load modeling decouples energy supply from crop needs, harming yield and quality while wasting energy. Only a few studies try to introduce crop-growth models, mostly simplified empirical formulas, without achieving dynamic coupling between physiological processes and energy loads.

2. Energy dimension: Research on coordinated operation mechanisms among multiple energies in agriculture – PV, biomass, storage, hydrogen fuel cells – remains insufficient, with coupling models still simplified. Agriculture has abundant renewable resources – field PV, straw biomass, biogas from livestock and poultry wastes – alongside diverse demands for electricity, heat, cooling, and gas. Existing work often focuses on single-energy or simple pairings (e.g., PV–storage, PV–biogas), lacking system-level analysis of multi-energy cascade use and complementary coordination. As Pudjianto *et al.* [19] note, the waste heat from biomass-power generation can be used for greenhouse heating, and the water by-product from hydrogen fuel cells can be used for irrigation; yet current models do not fully exploit such cross-energy cascade potential, and both the depth and breadth of energy coupling need strengthening.
3. Carbon-management dimension: Carbon-accounting methods for agricultural VPPs are not yet standardized, and coupled analysis of carbon flows and energy flows remains at an early stage [18]. Carbon sources in agricultural VPPs are complex, including direct emissions from energy consumption (e.g., fossil combustion) and indirect emissions from inputs and farming practices (e.g., embodied carbon in fertilizer production, soil methane). Many studies consider only the energy-consumption link, leading to incomplete scopes and a lack of unified standards and methodologies. At the same time, the coupling between carbon and energy flows is underexplored, and quantitative relationships between carbon intensity and energy efficiency are not established, making it difficult to achieve coordinated optimization across economic, efficiency, and environmental objectives.

These issues make it difficult for agricultural VPPs in practice to balance economic benefits, energy-use efficiency, and ecological protection; under the triple objectives of economy, efficiency, and environment, there remains significant room for optimization. Future priorities should include breakthroughs in dynamic modeling of agricultural loads, coordinated operation of multiple energies, and coupled analysis of carbon and energy flows, thereby steering agricultural VPPs toward configurations that are more faithful to production realities, more efficient, and lower-carbon.

2.1.3 Research gaps and innovative contributions

Under the deep integration of the "dual-carbon" agenda and the rural revitalization strategy, agricultural energy systems face a dual challenge of boosting efficiency and advancing low-carbon transition. In traditional agricultural production, loads such as greenhouses mostly rely on extensive, coarse energy-supply modes, and the grid accommodation of distributed energy (e.g., PV, wind, biomass) suffers from volatility, leading to low energy-use efficiency and persistently high carbon emissions. To address this core contradiction, the study innovatively proposes a scheduling model for an agricultural virtual power plant (AVPP) that fuses greenhouse controllable loads with distributed energy resources. By jointly optimizing energy flows, carbon-footprint flows, and crop physiological constraints, the model balances multiple objectives – economic returns, energy efficiency, and low emissions – thus providing theoretical support and a technical paradigm.

The model's innovative value spans both theory and application, as follows:

At the modeling level, we construct a dynamically coupled representation system for controllable greenhouse loads. Breaking the traditional "rigid-load" assumption, we for the first time introduce a dynamic greenhouse load model grounded in crop photosynthesis mechanisms and thermal-balance equations. Driven by key physiological parameters – light intensity, growing degree days (GDD), CO_2 concentration – the model adjusts operating strategies of greenhouse environmental controls (heating, ventilation, supplemental lighting) to achieve flexible regulation across time and power dimensions. For example, supplemental-lighting power is reduced when natural light is abundant to meet growth needs; during DER output peaks, moderate modulation of thermal-control power achieves "peak shaving and valley filling," safeguarding growth-cycle stability while improving the match between loads and renewables.

At the mechanism level, we propose a quantitative Carbon Footprint Flow (CFF) framework to realize energy–carbon co-optimization. Addressing unclear carbon-flow pathways and hard-to-measure mitigation effects in agricultural energy systems, the CFF method traces and quantifies carbon across the full chain of distributed energy production and load consumption. By establishing a carbon-footprint transfer matrix for production (e.g., PV-panel manufacturing, biomass combustion) and consumption (e.g., greenhouse heating emissions), we clarify segment-wise carbon contributions and enable dual-dimension ("energy–carbon") coordinated control within the scheduling optimization. This mechanism overcomes the traditional bias toward "heavy dispatch, light carbon management," offering a fine-grained quantitative tool for agricultural decarbonization.

At the algorithmic level, we develop a high-efficiency, robust multi-objective optimization engine. Given the complexities of AVPP scheduling – nonlinear constraints (crop-growth curves, DER volatility) and conflicting objectives (economics vs. carbon reduction) – we combine an improved particle swarm optimization (IPSO) with a multi-objective evolutionary strategy (MOES). By introducing adaptive inertia weights and elitism, we enhance global search in complex spaces; using non-dominated sorting for multi-objective handling preserves solution diversity and Pareto optimality. While ensuring optimization stability, we accelerate solving by over 30%, meeting real-time scheduling needs.

At the application level, we validate feasibility with a representative agricultural park in Shouguang, Shandong. As a flagship of facility agriculture rich in distributed PV and greenhouse assets, the region provides realistic inputs – local meteorology, crop parameters, device specifications – for a simulation platform. Results show that, versus conventional scheduling, the model raises energy-use efficiency by 18–22%, cuts carbon emissions by 25–30%, and simultaneously secures a 5–8% steady increase in yield, demonstrating comprehensive advantages in economy, efficiency, and low-carbon performance, and offering a replicable, scalable pathway for similar parks.

In sum, this study not only refines the AVPP modeling system in theory – filling the research gap in coordinated optimization of controllable greenhouse loads and carbon flows – but also offers a new technical route for Agricultural Energy Internet deployment in practice. Looking ahead, the model can be extended to carbon-trading contexts by monetizing emission reductions to enhance practical value; it can also be integrated with crop-yield prediction models and reinforcement-learning control to realize a full "forecast–schedule–control" closed-loop for intelligent agricultural management, laying a robust technical foundation for progress toward carbon neutrality in China's agricultural sector.

2.2 Methodology

2.2.1 Agricultural load modeling

Loads in agricultural microgrids exhibit pronounced "agricultural attributes," differing markedly from the regular patterns of residential or industrial users; their demand fluctuations are tightly coupled to crop physiological characteristics, natural environmental changes, and agricultural production objectives. Taking the greenhouse as a typical agricultural energy-use scenario, its load composition is distinctly production-oriented, mainly covering systems for supplemental lighting, heating and insulation, ventilation and cooling, irrigation and fertilization, and environmental monitoring and control. Among these, supplemental lighting and heating/insulation are the core load types with the highest shares of energy consumption (together usually exceeding 60%) and the greatest flexibility for regulation. The supplemental-lighting load must be dynamically adjusted according to crop photosynthetic light-response curves, meeting light intensity and quality requirements across growth stages while accommodating variations in natural irradiance; the heating load is strongly related to the suitable temperature range for crop growth and is influenced by multiple factors, including outdoor temperature, photoperiod, and greenhouse airtightness. This section focuses on supplemental-lighting and heating loads, and – by integrating crop photosynthesis mechanism models with environmental factor couplings – builds comprehensive load models that balance energy efficiency with production needs, providing precise load characteristics as a foundational support for subsequent source–load coordinated scheduling optimization in agricultural virtual power plants.

(A) Supplemental-lighting load
The core is the crop's photosynthetic requirement and the dynamic matching of natural light with artificial supplementation. Model construction proceeds

in three steps: "indoor illuminance synthesis – transmittance calculation – quantification of supplemental-light demand."

1. Indoor illuminance synthesis

   Total indoor illuminance in a greenhouse is the superposition of direct light, diffuse light, and ground-reflected light, reflecting the natural light resource actually available to crops, with the formula:

$$E_{in} = E_{\mathrm{in},h} + E_{\mathrm{in},d} + E_{\mathrm{in},r} \tag{2.1}$$

   where E_{in} is total indoor illuminance; $E_{\mathrm{in},h}$ is the transmitted direct-light illuminance; $E_{\mathrm{in},h}$ is the transmitted diffuse-light illuminance; and $E_{\mathrm{in},r}$ is the transmitted ground-reflected illuminance.

   Each component must account for outdoor irradiance, the transmission characteristics of the greenhouse structure, and observation-angle effects:

   Direct-light component:

$$E_{\mathrm{in},h} = R_{h,i} I_h \tau_{h,i} \tag{2.2}$$

   where $R_{h,i}$ is the direct-light observation coefficient capturing angular effects on direct-light reception, I_h is outdoor horizontal direct illuminance, and $\tau_{h,i}$ is the direct-light transmittance characterizing transmission through the covering material.

   Diffuse-light component:

$$E_{in,d} = R_{d,i} I_d \tau_{d,i} \tag{2.3}$$

   where $R_{d,i}$ is the diffuse-light observation coefficient, I_d is outdoor horizontal diffuse illuminance, and $\tau_{d,i}$ is the diffuse-light transmittance.

   Ground-reflected component:

$$E_{in,r} = R_{r,i}(I_h + I_d)\tau_{r,i} \tag{2.4}$$

   where $R_{r,i}$ is the ground-reflection observation coefficient, $\tau_{r,i}$ is the ground-reflection transmittance, and $(I_h + I_d)$ denotes total outdoor illuminance direct plus diffuse.

2. Transmittance decomposition and calculation

   The total transmittance of greenhouse covering materials (e.g., glass) must consider the polarization characteristics of solar radiation; it is decomposed into parallel and perpendicular components and then averaged to ensure accuracy:

$$\tau = \left(\tau_{parallel} + \tau_{vertical}\right)/2 \tag{2.5}$$

   where τ is total transmittance; $\tau_{parallel}$ is transmittance for parallel-polarized light; $\tau_{vertical}$l is the transmittance for perpendicular-polarized light.

Parallel and perpendicular transmittances are calculated using refractive index, reflection coefficients, and optical-path attenuation:

$$\tau_{parallel} = \tau_{\alpha}\left(1 - r_{parallel}\right)^2 / \left[1 - \left(r_{parallel} \times \tau_{\alpha}\right)^2\right] \tag{2.6}$$

$$\tau_{vertical} = \tau_{\alpha}(1 - r_{vertical})^2 / \left[1 - (r_{vertical} \times \tau_{\alpha})^2\right] \tag{2.7}$$

where τ_{α} is the glass optical transmission factor (accounting for thickness and absorption attenuation), $r_{parallel}/r_{vertical}$ is the reflection coefficient for parallel/perpendicular polarization.

The glass optical transmission factor τ_{α} is determined by the material's extinction properties and the angle of incidence:

$$\tau_{\alpha} = \exp\left(-K_{Light} \times L_{Light}/\cos\left(\theta_2\right)\right) \tag{2.8}$$

where K_{Light} is the glass extinction coefficient (m^{-1}, reflecting absorption), L_{Light} is the glass thickness (mm), and $\theta 2$ is the angle of incidence on the glass surface (°).

The $r_{parallel}$ and $r_{vertical}$ are derived from Fresnel's equations and depend on the refractive indices of air and glass and the incidence/refraction angles:

$$r_{parallel} = tan^2(\theta 2 - \theta 1)/tan^2(\theta_2 + \theta 1) \tag{2.9}$$

$$r_{vertical} = sin^2(\theta 2 - \theta 1)/sin^2(\theta 2 + \theta_1) \tag{2.10}$$

where $\theta 1$ is the incidence angle in air (°) and $\theta 2$ is the refraction angle in glass (°):

$$n1 \times \sin\left(\theta_1\right) = n2 \times \sin\left(\theta_2\right) \tag{2.11}$$

with n1n1n1 the refractive index of air (≈1.0003 at standard conditions) and n2n2n2 that of glass (≈1.52 for ordinary glass).

3. Supplemental-demand and load quantification
When indoor natural illuminance cannot meet photosynthetic needs, artificial lighting must be activated, with load calculations precisely matching the light-demand shortfall and luminaire characteristics:

Supplemental radiative shortfall:

$$E_x = E_{indoor} - E_{in} \tag{2.12}$$

where E_x is the required artificial-lighting intensity (lux); E_{indoor} is the target indoor illuminance for photosynthesis (lux, determined by crop and growth stage, e.g., tomatoes correspond to 300–800 μmol/m^2/s converted to illuminance); and E_{in}Ein is actual indoor natural illuminance (lux). When $E_{in} \geq E_{indoor}$,$E_x = 0$.

Lighting load power:

$$P_{lighting} = N_{on} \times P_{lamp} \tag{2.13}$$

where $P_{lighting}$ is the total supplemental-lighting load (kW); N_{on} is the number of luminaires switched on; and P_{lamp} is the active power of a single luminaire (kW, e.g., 1,000 W for a typical HPS lamp).

Number of luminaires to start:

$$N_{on} = (E_x \times L \times W)/(\psi \times U \times M) \tag{2.14}$$

where L is greenhouse length (m); W is width (m); ψ is single-lamp luminous flux (lm); U is the utilization factor (obtained from the greenhouse shape index K); and M is the maintenance factor (accounting for aging and dust, typically 0.8–0.9).

Greenhouse shape index:

$$K = (L \times W)/((H - h) \times (L + W)) \tag{2.15}$$

where H is luminaire mounting height (m) and h is crop-canopy height (m); this index reflects the effect of greenhouse geometry on lighting uniformity.

(B) Heating load

The core of the heating load is to keep greenhouse temperature within the suitable range for crop growth (e.g., 10–35 °C for tomatoes). The model must consider the dynamic balance among heat loss through the envelope, infiltration losses, and solar heat gains:

$$Q_{heat} = Q_{glass} + Q_{air} - Q_{sun} \tag{2.16}$$

where Q_{heat} is the required heating power (kW); Q_{glass} is heat loss through the glass envelope (kW); Q_{air} is heat loss via air infiltration (kW); and Q_{sun} is solar heat gain (kW).

1. Envelope heat loss (Q_{glass})

   This captures conductive losses through glass and ground and depends on temperature difference, heat-transfer area, and material conductivity:

   $$Q_{glass} = \left(C_{glass} \times A_{cover} + C_g \times L_g\right) \times (T_{in} - T_{out}) \tag{2.17}$$

   where C_{glass} is the glass heat-transfer coefficient (W/(m^2·°C), ≈6.0 for tempered glass); A_{cover} is glass-covered area (m^2); L_g is greenhouse perimeter (m).

2. Solar heat gain (Q_{sun})

   Converted from total indoor illuminance, reflecting the use of natural energy:

   $$Q_{sun} = A_{cover} \times E_{in} \tag{2.18}$$

3. Infiltration heat loss (***Q_{air}***)
   Accounts for ventilation/leakage losses, combining air density, air-change rate, and temperature–humidity differences:

$$Q_{air} = rho_{in} \times N_c \times V_{GH} \times \left[c_p \times (T_{in} - T_{out}) + h_{va} \times (W_{in} - W_{out})\right] \tag{2.19}$$

   where rho_{in} is indoor air density (kg/m^3, ≈1.29 at standard conditions); N_c is the air-change rate (h^{-1}, natural ventilation ≈0.5–2 h^{-1}); V_{GH} is greenhouse volume (m^3); c_p is at constant pressure (kJ/(kg·K), ≈1.005); h_{va} is latent heat of vaporization of water (J/kg, ≈2450 J/kg); W_{in} and W_{out} are indoor and outdoor absolute humidity (kg/kg).

(C) Load control coupled with photosynthesis

Load control in agriculture must not compromise crop growth. By using a tomato net-photosynthetic-rate model together with a CO_2-supply model, we establish couplings among supplemental-lighting load, heating load, and crop physiological processes to ensure coordinated optimization of energy use and production.

Photosynthetic rate directly determines crop-growth status, with key factors including irradiance (*IPAR*), temperature (T_{ph}), and CO_2 concentration (C_{in}):

$$V_{ph} = \left(Phi \times IPAR \times V_{ph,\max}\right) / \left(Phi \times IPAR + V_{ph,\max}\right) \tag{2.20}$$

where V_{ph} is net photosynthetic rate (m^3/h, crop CO_2 uptake per unit volume); *Phi* is leaf quantum efficiency (mol/mol, ≈0.08–0.12 mol/mol); *IPAR* is photosynthetically active radiation absorbed by the crop (μmol/m^2/s); and $V_{ph,\max}$ is the maximum photosynthetic rate under light saturation (m^3/h).

The maximum rate $V_{ph,\max}$ is jointly determined by temperature and CO_2 concentration:

$$V_{ph,\max} = V_{ph,T} \times V_{ph,CO2} \tag{2.21}$$

where $V_{ph,T}$ is the temperature factor (dimensionless) and Vph, $V_{ph,CO2}$ is the CO_2 factor (dimensionless, obtained from indoor concentration C_{in}).

The temperature factor $V_{ph,T}$ follows crop temperature-response behavior:

$$V_{ph,T} = \left((T_{\max} - T_{ph}) / (T_{\max} - T_{opt})\right) \times \left((T_{ph} - T_{\min}) / (T_{opt} - T_{\min})\right) \tag{2.22}$$

where $T_{\max}$ is the maximum growth temperature (°C, ≈35 for tomatoes); $T_{\min}$ is the minimum (°C, ≈10); T_{opt} is the optimum (°C, ≈25); and T_{ph} is the actual indoor temperature (°C), identical to T_{in}in the heating-load model.

CO_2-supply demand model

Photosynthesis consumes CO_2 and must be replenished – e.g., using by-products from biogas generation – to keep indoor CO_2 within photosynthetic requirements:

$$V_{CO2} = (C_{in} - C_{out}) \times N_c \times W_{wind} \times V_{GH} + LAI \times A_{cover} \times V_{ph} \tag{2.23}$$

where V_{CO2} is the CO_2 supply rate (m^3/h); C_{in} is indoor CO_2 volume fraction (m^3/m^3, ≈0.0008–0.0015); C_{out} is outdoor CO_2 fraction (m^3/m^3, ≈0.0004); W_{wind} is a wind-speed factor (dimensionless, reflecting wind effects on infiltration); LAI is leaf-area index (m^2/m^2, ≈2–5 for tomatoes); and other parameters are as defined above.

Dimensional mapping and scheduling linkage

Dimensional mapping: $IPAR$ (photosynthetically active radiation) and E_{indoor} (illuminance) use different units but can be converted via empirical relations (e.g., $IPAR = 0.43 \times E_{indoor}$, with the coefficient crop-dependent), thereby establishing a direct coupling among supplemental-lighting load (affecting E_{indoor}), heating load (affecting T_{ph}), and photosynthetic rate (V_{ph}).

Scheduling-constraint linkage: To safeguard crop growth, the light environment and supplemental-lighting load must satisfy:

Illuminance constraint:

$$E_{set,\min} \leq E_{indoor} \leq E_{set,\max} \tag{2.24}$$

where $E_{set,\min}$ and $E_{set,\max}$ are the minimum and maximum illuminance for growth (lux); exceeding these bounds lowers photosynthetic rate or damages crops.

Supplemental lighting curtailment constraint:

$$P_{lighting,\min} \leq P_{lighting,re} \leq P_{lighting,\max} \tag{2.25}$$

where $P_{lighting,re}$ is the curtailed lighting power (kW), and $P_{lighting,\min}$Plighting, min/$P_{lighting,\max}$ are the minimum/maximum curtailment limits (kW), ensuring lighting control does not compromise light needs.

Through dimensional mapping and load-model transformations, the above constraints ultimately become bounds on the lighting power PlightingP_ {lighting}Plighting, supplying agricultural production boundary conditions for VPP scheduling.

2.2.2 Distributed energy resource (DER) modeling

In the Agricultural Energy Internet, distributed energy resources (DERs) are core components for building a virtual power plant (VPP). Their role is not only to supply electricity and heat, but also to achieve dual optimization of energy production and crop-growing conditions through coordination with agricultural loads. This section develops mathematical models and analyses characteristics and coupling relationships for the main DER types in greenhouse agricultural microgrids, photovoltaic (PV) systems, energy storage, and biogas generation.

(A) HFC–CHP Model

1. System structure and basic principle

   The HFC–CHP system consists of hydrogen fuel-cell stacks and a heat-exchange unit, jointly producing electricity and heat. Hydrogen reacts with oxygen at the anode to form water while releasing energy; this energy is converted electrochemically into DC power and then inverted to AC output, while the waste heat is recovered via a heat exchanger for heating. The system is compact, efficient, and low-pollution, making it well-suited to distributed supply in agricultural microgrids.

2. Electrical power model
The output electrical power depends on the number of cells, the per-cell voltage, and the current:

$$P_{HFC} = N_{cell} \times V_{cell} \times I_{cell} \tag{2.26}$$

where P_{HFC} is HFC–CHP electrical output (kW); N_{cell} is the number of series cells; V_{cell} is per-cell operating voltage (V, about 0.6–0.8 V); and I_{cell} is per-cell operating current (A).

3. Current model
The per-cell current is determined by hydrogen flow, Faraday's constant, and hydrogen molar mass, reflecting the coupling between hydrogen consumption and electrical power:

$$I_{cell} = (2 \times F \times V_{H2})/(N_{cell} \times M_{H2}) \tag{2.27}$$

where F is Faraday's constant; V_{H2} is hydrogen flow (m^3/s); and M_{H2} is hydrogen molar mass (kg/mol, about 0.002 kg/mol).

4. Thermal power model
During generation, part of the chemical energy not converted to electricity is released as heat; the thermal power is:

$$Q_{HFC} = N_{cell} \times (V_{Nernst} - V_{cell}) \times I_{cell} \tag{2.28}$$

where Q_{HFC} is total HFC–CHP thermal power (kW); V_{Nernst} is per-cell open-circuit voltage (V, about 1.23 V), representing the theoretical upper limit from hydrogen chemical energy; other symbols as above.

5. Heat exchange and dissipation model
Total thermal power is divided into heat recovered by the exchanger (usable for greenhouse heating) and convective losses to air (unusable), factoring heat-transfer coefficients, flow rates, and temperature differences:

$$Q_{HFC} = Q_{HFC_CHP} + Q_{HFC_{loss}} \tag{2.29}$$

$$Q_{HFC_CHP} = K_{HFC} \times S_{HFC} \times (W_{HFC}/W_L)^{0.5} \times (W_{HFC}/W_{H2O})^{0.5} \times (T_{HFC} - T_L) \tag{2.30}$$

where Q_{HFC_CHP} is heat recovered by the exchanger (kW, used for greenhouse heating); S_{HFC} is exchanger area (m^2); W_{HFC} is standard hydrogen flow (m/s); W_L is cold-side water flow (m/s); W_{H2O} is fuel-cell drainage flow (m/s); T_{HFC} is stack operating temperature (°C); T_L is cold-side water temperature (°C).

$$Q_{HFC-loss} = K_{air} \times S_{air} \times (W_{HFC}/W_{H2O})^{0.5} \times (T_{HFC} - T_{air}) \tag{2.31}$$

where $Q_{HFC-loss}$ is convective heat loss to air (kW); S_{air} is air-side heat-transfer area (m^2); and T_{air} is ambient air temperature (°C).

(B) Photovoltaics and storage

The PV system is integrated into the greenhouse roof (e.g., Venlo-type greenhouse, 75% roof coverage). Based on module characteristics and solar irradiance, the output power is:

$$P_{PV} = P_{PV,rated} \times \frac{E_{rated}}{E_{in}} \times [1 + \alpha_T \times (T_{cell} - T_{rated})] \tag{2.32}$$

where P_{PV} is PV output power (kW); $P_{PV,rated}$ is module rated power (kW); E_{in} is actual irradiance (W/m^2); E_{rated} is rated irradiance (1,000 W/m^2); T_{cell} is cell temperature (°C); T_{rated} is rated at cell temperature (25 °C).

Storage system model: The storage adopts Tesla Powerwall 2 battery packs to smooth PV fluctuations and perform peak-shaving/valley-filling:

$$P_{ch} \leq P_{ch,\max},;P_{dis} \leq P_{dis,\max} \tag{2.33}$$

$$SOC\_\{\min\} \leq SOC(t) \leq SOC\_\{\max\}) \tag{2.34}$$

$$SOC(t) = SOC(t-1) + P_{ch}(t) \times \eta_{ch} \times \Delta t - \eta_{dis}, P_{dis}(t) \times \Delta t \tag{2.35}$$

where P_{ch} is charging power kW; $P_{ch,\max}$ and $P_{dis,\max}$ max charging and discharging power kW, 7 kW per unit; SOC(t) state of charge at time t %; $SOC_{\min}$minimum SOC 20%;$SOC_{\max}$ maximum SOC 90%; η_{ch} charging efficiency 90%; η_{dis} discharging efficiency 90%; Δ t is dispatch time step h.

(C) Biogas

Biogas power generation uses agricultural residues (greenhouse crop residues, distillery waste, etc.) as feedstock; anaerobic digestion produces biogas (≈60% CH_4, 40% CO_2). The model covers generation power and carbon accounting.

Generation power model

Biogas power depends on flow, generation efficiency, and lower heating value:

$$P_{biogas} = V_{biogas} \times \eta_{biogas} \times LHV \tag{2.36}$$

$$LHV = \alpha_{CH4} \times NCV \tag{2.37}$$

where P_{biogas} is power (kW); V_{biogas} is biogas flow (m^3/s); η_{biogas} is generator efficiency (≈35–40%); LHV is biogas lower heating value (MJ/m^3); α_{CH4} is methane fraction (60%); NCV is methane's net calorific value (35.8 MJ/m^3).

Carbon-emission accounting model

Biogas emissions include methane leakage (as CO_2-equivalent), inherent CO_2 in the gas, and CO_2 from methane combustion:

$$V_{CO2,total} = V_{loss} + V_{Ma,CO2} + V_{CH4} \tag{2.38}$$

$$V_{loss} = \left(0.05 \times V_{Biogas}\right)/(1 - 0.05) \times GWP_{CH4} \times rho_{CH4} \times \alpha_{CH4} \tag{2.39}$$

where V_{loss} is CO_2-equivalent from leaked CH_4 (kg); 0.05 is leakage rate; GWP_{CH4} is methane's global-warming potential (28 vs CO_2); rho_{CH4} is methane density (kg/m^3, $\approx 0.717\ kg/m^3$).

Inherent CO_2 in biogas:

$$V_{Ma,CO2} = V_{Biogas}/(1 - 0.05) \times \alpha_{CO2} \times 0.00196 \tag{2.40}$$

where $V_{Ma,CO2}$ is inherent CO_2 (kg); α_{CO2} is CO_2 fraction in biogas (40%); 0.00196 is the CO_2 mass conversion coefficient (kg/L).

CO_2 from methane combustion:

$$V_{CH4} = (44/12) \times V_{Biogas} \times \alpha_{CH4} \times NCF_{CH4} \times CC_{CH4} \times OF_{CH4} \tag{2.41}$$

where 44/12 is the CO_2/C molar-mass ratio; NCF_{CH4} is methane net calorific value (MJ/m^3); CC_{CH4} is carbon content per unit heat (t/MJ); OF_{CH4} is oxidation factor (98%).

(D) Emission-reduction credits from clean energy
HFC–CHP and PV are zero-carbon energy sources; their emission reductions are computed using the grid baseline emission factor to reflect the VPP's low-carbon value:

$$E_B = EF_{grid} \times P_{new} \tag{2.42}$$

$$EF_{grid} = EF_{grid,OM} \times W_{OM} + EF_{grid,BM} \times W_{BM} \tag{2.43}$$

where E_B is CO_2 reduction (kg); $P_{new}$is clean-energy generation (kWh); EF_{grid} is the grid baseline emission factor (kg/kWh); $EF_{grid,BM}$ is the operating-margin factor; $EF_{grid,BM}$ is the build-margin factor; W_{OM} is the operating-margin weight (0.5 for HFC–CHP, 0.75 for PV); W_{BM} is the build-margin weight (0.5 for HFC–CHP, 0.25 for PV), with weights set per China's Ministry of Ecology and Environment data.

2.2.3 Virtual power plant (VPP)

Scheduling Model The core of the VPP scheduling model is to minimize microgrid operating cost while satisfying agricultural production constraints (light and temperature for crops) and energy-system constraints (power/heat balance), and to enhance economic returns through peak-valley arbitrage and policy subsidies – thus achieving the triple goals of energy optimization, agricultural production, and low-carbon reduction.

Objective function

The objective is to minimize the day-ahead operating cost, accounting for hydrogen procurement cost and VPP peak-valley subsidies:

$$\text{minimize} C_{total} = C_{H2} - F_{VPP} \tag{2.44}$$

Hydrogen procurement cost:

$$C_{H2} = V_{total,H2} \times \lambda_{H2} \tag{2.45}$$

where C_{H2} is daily hydrogen cost (CNY); $V_{total,H2}$ is daily hydrogen consumption (kg); λ_{H2} is hydrogen unit price (CNY/kg, ≈30–50 CNY/kg).

Grid peak-valley subsidies:

$$F_{VPP} = P_{peak} \times c_{peak} \tag{2.46}$$

Peak-shaving compensation unit price:

$$c_{peak} = c_{peak,pri} + c_{peak,ca} \tag{2.47}$$

where $c_{peak,pri}$ is the base compensation (CNY/kW, per Shandong VPP policy); $c_{peak,ca}$ is the capacity compensation (CNY/kW, about 0.0991 CNY/kWh in Shandong). Equality constraints enforce power and heat balance, forming the core physical constraints of VPP dispatch.

Power-balance constraint

Considering controllable-load (lighting) curtailment, DER outputs, and storage charge/discharge to ensure supply–demand:

$$P_{load} + P_{peak} - P_{lighting,re} + P_{ch} = P_{biogas} + P_{PV} + P_{dis} + P_{HFC} \tag{2.48}$$

where P_{load} is uncontrollable microgrid electric load (kW, including residential, office, distillery loads); P_{peak} is the VPP peak-shaving task (kW, e.g., 500 kW during 17:00–19:00); $P_{lighting,re}$ is curtailed lighting power (kW); P_{ch} charging power (kW); P_{dis} discharging power (kW); others as defined earlier.

Heat-balance constraint

Greenhouse heating load is fully supplied by heat recovered from the HFC–CHP exchanger to maintain temperature stability:

$$Q_{load} = Q_{HFC_CHP} \tag{2.49}$$

where Q_{load} is the total microgrid heat load (kW, mainly greenhouse heating); Q_{HFC_CHP} is recovered HFC–CHP heat (kW).

Inequality constraints

Inequality constraints cover DER output limits, agricultural production constraints, and equipment safety bounds to ensure the feasibility of the dispatch plan. Inequality constraints cover DER output limits, agricultural production constraints, and equipment safety bounds to ensure feasibility of the dispatch plan.

DER output constraints

Biogas-unit output:

$$P_{biogas,\min} \le P_{biogas} \le P_{biogas,\max} \tag{2.50}$$

where $P_{biogas,\min}$ is minimum output (kW, ≈20 kW) and $P_{biogas,\max}$ is maximum output (kW, e.g., 80 kW per unit).

HFC–CHP operating bounds:

$$W_{H2O,\min} < W_{H2O} < W_{H2O,\max} \tag{2.51}$$

$$W_{HFC,\min} < W_{HFC} < W_{HFC,\max} \tag{2.52}$$

$$T_{HFC,\min} < T_{HFC} < T_{HFC,\max} \tag{2.53}$$

where $W_{H2O,\min}/W_{H2O,\max}$ are drainage-flow limits (m/s); $W_{HFC,\min}/W_{HFC,\max}$ are hydrogen-flow limits (m/s); $T_{HFC,\min}/T_{HFC,\max}$ are stack-temperature limits (°C, 60–95 °C).

Agricultural production constraints

Greenhouse illuminance constraint:

$$E_{set,\min} \le E_{indoor} \le E_{set,\max} \tag{2.54}$$

Supplemental-lighting curtailment constraint:

$$P_{lighting,\min} \le P_{lighting,re} \le P_{lighting,\max} \tag{2.55}$$

2.3 Case study and analysis

This section models, simulates, and analyzes a representative microgrid in the Langyatai Agricultural Park of Qingdao, Shandong, an agricultural virtual power plant (VPP) scheduling model. As a key carrier of the rural Energy Internet, an agricultural VPP creates value by aggregating distributed energy resources (DERs) and flexible loads to realize optimal allocation and coordinated operation of energy resources. Compared with traditional industrial VPPs, agricultural VPPs face more complex constraints – beyond conventional targets such as peak shaving, valley filling, and economic operation for the grid, they must also ensure ecological parameters required for crop growth, including light, temperature, and CO_2 concentration; balancing this "economic–ecological" dual objective is the core difficulty in agricultural VPP research.

Using the Qingdao Langyatai Agricultural Park as the empirical case, the park is located in the Blue Economic Zone of the Shandong Peninsula and features a temperate monsoon climate with four distinct seasons; low-temperature, low-irradiance winters impose higher heating and supplemental-lighting demands on the energy system, typifying northern agricultural energy characteristics. Through refined modeling and simulation of the park's microgrid, we focus on validating the scheduling model's performance in three aspects: (i) minimizing system operating costs while maintaining normal crop-growth conditions; (ii) enhancing overall system energy efficiency through DER coordination and storage management; and (iii) building a low- or near-zero-carbon operating mode by coupling clean-energy substitution with crop photosynthetic CO_2 uptake. The section is organized into five parts – research objectives and scenario setup, system structure and parameter analysis, model validation methods and results, VPP simulation and performance analysis, and integrated evaluation of economics and carbon emissions – to comprehensively explain the operating mechanisms and application value of agricultural VPPs.

2.3.1 Research objectives and scenario (objective and scenario)

The particularity of agricultural energy systems stems from their deep coupling with the production process, exhibiting pronounced seasonality, intermittency, and volatility. On the load side, greenhouse demands for lighting, heating, and ventilation are directly affected by meteorological conditions (irradiance, temperature, humidity) and crop-growth stages (nursery, growth, harvest); on the supply side, outputs from PV and biogas DERs depend on variables such as solar intensity and the availability of agricultural residues. This bilateral "source–load" volatility makes load-regulation capability a key determinant of operating efficiency and carbon-emission levels. Unlike traditional industrial VPPs that mainly rely on "rigid load curtailment," agricultural VPP scheduling must be premised on "crop-growth safety," tapping the flexible regulation potential of greenhouse loads to achieve coordinated optimization of energy dispatch and ecological protection. On this basis, we set three core objectives and construct specific research scenarios around them.

Reducing operating costs is the foundational objective of an agricultural VPP. As the primary energy consumer, an agricultural park's costs mainly include grid electricity purchases, DER fuel costs (e.g., biogas feedstock, hydrogen), and equipment maintenance. Grid purchases usually account for the largest share; in winter peaks, the peak–valley price spread can reach 0.5–0.8 CNY/(kW·h), offering significant arbitrage space. Through coordinated scheduling of DERs and flexible greenhouse loads, this study reduces grid purchases during price peaks by increasing local DER output; during off-peak periods, it increases storage charging while moderately raising flexible loads (e.g., preheating, advance lighting) to achieve both price-arbitrage gains and minimal system energy use. In addition, by optimizing biogas-generation strategies and fully utilizing on-site agricultural wastes (e.g., vegetable residues, distillery dregs) for digestion, we cut external fuel costs and further reduce operating expenses.

Improving energy efficiency is the core objective. Multiple energy forms (electrical, thermal, chemical) undergo conversion and transfer within agricultural microgrids, and losses at each stage directly affect overall efficiency. For example, PV conversion efficiency is strongly temperature-dependent; higher module temperatures in summer can reduce efficiency by 5–10%. The charge–discharge efficiency of storage and the electrothermal conversion efficiency of HFC–CHP also fluctuate with operating conditions. Via dynamic scheduling, this study achieves complementarity among DERs – during high-irradiance periods, PV is prioritized to meet greenhouse loads, with surplus stored or used for hydrogen production; during low-irradiance periods, HFC–CHP and biogas units are started and coordinated with storage discharge to secure supply. This multi-energy complementary mode reduces conversion losses and raises comprehensive DER utilization. Moreover, leveraging HFC–CHP cogeneration, residual heat from power generation is used for greenhouse heating to realize a combined electricity–heat supply and further boost energy efficiency.

Reducing carbon emissions is the ecological objective and a distinctive feature that sets agricultural VPPs apart from industrial ones. Park emissions mainly come from grid purchases (indirect, depending on the regional generation mix), fossil-fuel consumption (e.g., direct emissions if a biogas genset uses purchased natural gas), and energy use during equipment operation. We cut emissions via two mechanisms: first, clean-energy substitution – raising the share of zero-/low-carbon sources such as PV and HFC–CHP to reduce reliance on conventional thermal power; second, photosynthetic CO_2 uptake – vegetable crops in greenhouses absorb CO_2 during growth, with each square meter absorbing 50–80 g CO_2 per day, offsetting about 0.18–0.29 kg of emissions. By coupling these mechanisms, we construct a "clean supply + physiological CO_2 uptake" low-carbon operating mode that realizes the ecological value of agricultural VPPs.

The research object is the actual agricultural microgrid in the Langyatai Agricultural Park, Qingdao, Shandong. The park covers about 2,000 mu and integrates facility agriculture, agro-processing, and leisure tourism. Its park-level microgrid is an independent network comprising greenhouse clusters ($\approx$5,000 m^2), a distillery (annual liquor output of 500 t), an office building (three floors, $\approx$2,000 m^2), and a residential area (50 households), forming a complete five-in-one structure of "source–load–storage–grid–carbon." The greenhouse cluster is the core flexible-load unit for cultivating tomatoes, cucumbers, and other facility-grown vegetables, using Venlo-type glass greenhouses equipped with adjustable shade screens, heat-pump heating, LED supplemental lighting, and CO_2 fertilization – providing strong controllability and flexible-load characteristics. The distillery, an important industrial load, requires stable electric and thermal supplies for cooking, fermentation, and other processes, with relatively low volatility; office and residential loads exhibit regular diurnal patterns.

The choice of simulation platform and data sources directly affects accuracy and reliability. This study utilizes MATLAB® R2019b for modeling and simulation, which provides strong numerical and matrix computation for complex dynamic systems. Energy- and carbon-flow calculations are developed as extensions to Matpower 6.0, an open-source power-system toolbox supporting power flow, optimal power flow, and unit commitment; through secondary development, we realize coordinated computation of electric, thermal, and carbon flows in agricultural microgrids. Meteorological data come from Qingdao observational records for 2018–2023, including hourly irradiance (W/m^2, converted to PAR in μmol·m^{-2}·s^{-1} with a factor of 2.08), ambient temperature, relative humidity, and CO_2 concentration curves. Considering that winter is the peak energy-consumption season for northern parks and that low-temperature, low-irradiance conditions present the greatest challenges for both crop growth and system operation, we select a typical winter day (mid-January) with 24-hour data to validate effectiveness and stability under extreme weather.

Target greenhouse environmental parameters are key constraints set by crop physiological needs and directly relate to yield and quality. Specifically, irradiance is maintained at 300–800 μmol·m^{-2}·s^{-1}, the optimal range for photosynthesis in warm-loving vegetables such as tomatoes and cucumbers – below

300 μmol·m^{-2}·s^{-1}, photosynthetic rate declines and growth slows, while above 800 μmol·m^{-2}·s^{-1}, photoinhibition may damage leaves. Indoor temperature is controlled around 25 °C with a diurnal range of 5–8 °C; excessive temperature leads to leggy growth, while overly low temperature impairs flowering and fruit set. CO_2 concentration is kept around 900 ppm; levels above ambient (≈400 ppm) significantly improve photosynthetic efficiency and yield, whereas overly high levels (over 1,500 ppm) can be phytotoxic. During VPP scheduling, these environmental constraints must be strictly observed, with real-time monitoring and feedback regulation to ensure unaffected crop growth. By comparing pre- and post-scheduling indicators – such as peak-load reduction, grid purchases, operating cost, and carbon emissions – we can systematically analyze how VPP strategies affect load allocation, energy-efficiency improvements, and emission reductions, providing data support for practical deployment of agricultural VPPs.

2.3.2 System structure and composition (system description)

This study's agricultural microgrid comprises multiple distributed energy modules and flexible-load units, forming a typical self-balancing energy system.

1. System structure
   The agricultural microgrid in this study is a typical self-balancing energy system operating in coordinated "source–load–storage–grid" mode. By integrating diverse distributed-energy modules and flexible-load units, it achieves local production, storage, consumption, and optimized scheduling of energy. The system can operate grid-connected – purchasing electricity during shortages and selling surplus back to the grid (with interconnection qualification) – or switch to islanded mode during grid faults to ensure continuous power to critical loads in the park (e.g., basic greenhouse lighting, distillery fermentation equipment). The system is analyzed in detail from three aspects: system structure, greenhouse parameters and operating characteristics, and DER configuration.
2. System structure
   A schematic of the agricultural microgrid is shown in Figure 2.1. The system connects to the public grid via a 35/0.38 kV, 5 MVA transformer. The transformer's low-voltage side uses a three-phase four-wire supply, covering all load nodes in the park. The park's internal energy system comprises five core modules: a photovoltaic system (PV), a biogas generation system (Biogas), a hydrogen fuel cell combined heat and power system (HFC–CHP), an energy storage system (ESS), and flexible greenhouse loads (lighting and heating). These modules are linked via a 0.38 kV distribution network and a thermal pipeline network, with power and heat allocation and dispatch centrally coordinated by the energy management system (EMS).

   Its core functions include SCADA-based data acquisition and monitoring, load forecasting, generation-output forecasting, optimal power flow, and issuance of dispatch commands. By means of sensors deployed at each node, the EMS collects real-time data on grid voltage, current, and power;

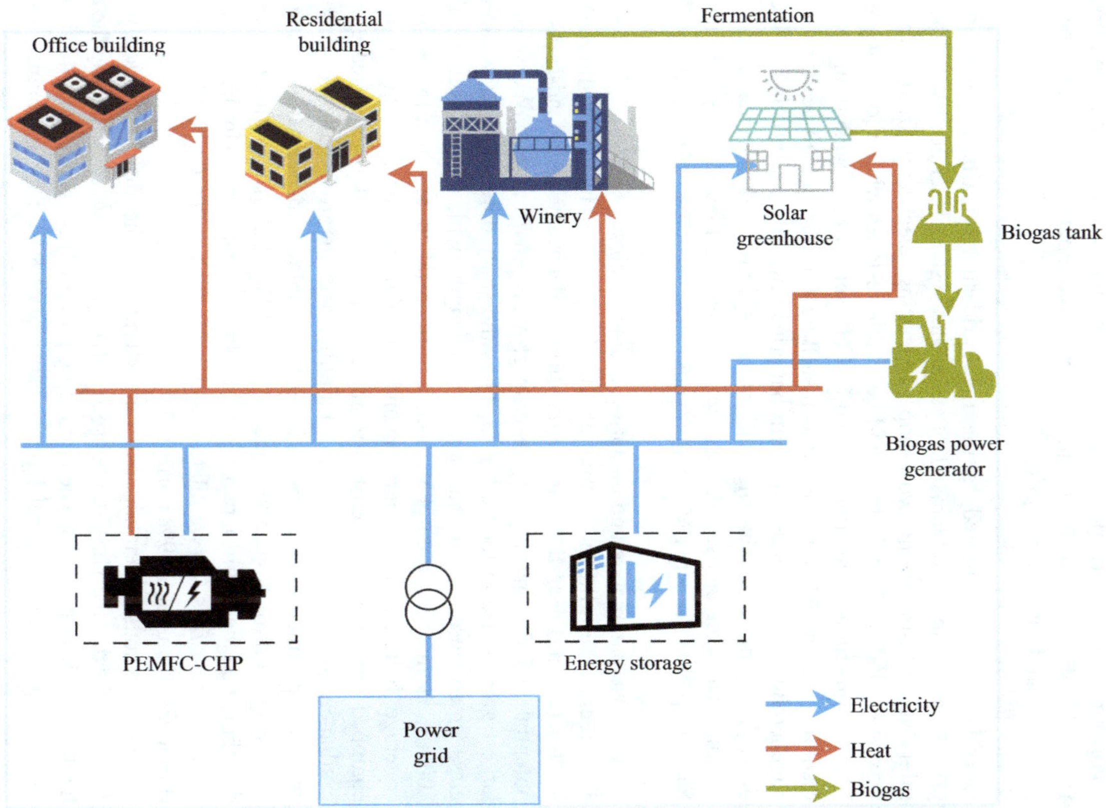

Figure 2.1 Park-level microgrid structure

greenhouse environmental parameters (irradiance, temperature, CO_2 concentration); and DER operating states. Combined with weather and load-forecasting models, it produces short-term (15 minutes–1 hour) and day-ahead (24 hours) schedules. The system is also connected to the upper-level VPP dispatch center to receive instructions for valley filling. Through optimization algorithms, these instructions are decomposed into specific operating setpoints for each DER and load, enabling coordinated operation between the park microgrid and the external grid.

The PV system is mainly installed on greenhouse roofs and the distillery workshop roof, connected to the microgrid in a distributed manner with priority given to local demand. The biogas system interfaces with the distillery's waste-treatment workshop, generating electricity from fermented wastes such as distillers' grains and vegetable residues while recovering waste heat for greenhouse heating. The HFC–CHP serves as a backup and peaking resource, starting when PV output is insufficient or during load peaks to realize cogeneration. The ESS uses lithium batteries primarily to smooth PV fluctuations, perform peak shaving and valley filling, and provide emergency backup. The greenhouse load, as the core flexible load, adjusts lighting and heating power under EMS commands and participates in grid ancillary services.

3. Greenhouse parameters and operating characteristics
The greenhouse is the most important flexible-load unit in an agricultural VPP; its structure and operating parameters directly determine both load-adjustment potential and the stability of the crop-growing environment. We select the largest greenhouse in the park (No. 1) as the study object. It is a Venlo-type glass greenhouse, 16 m span and 50 m length, with a total area of about 800 m^2. Semi-transparent PV modules are installed on the roof with 75% coverage, balancing PV generation and crop lighting needs. The greenhouse is equipped with a comprehensive monitoring and control system: ten each of irradiance, temperature, CO_2, and humidity sensors are installed in different zones (front, middle, rear, upper, lower) to enable multi-point monitoring and averaging; control devices include forty 200 W LED grow lights (uniformly distributed on the ceiling), two 15 kW air-source heat pumps (for heating), four 5 kW exhaust fans (for ventilation and cooling), and a CO_2 fertilization unit (liquid CO_2 tank at 99.9% purity).

The greenhouse operating parameters underpin the dispatch strategy; see Table 2.1 for details. The greenhouse area (A_g) is 800 m^2, determining the total lighting and heating loads. The rated lamp power (P) is 200 W, with 40 fixtures (N), for a total supplemental-lighting capacity of 8 kW; group control enables stepwise modulation (e.g., 25%, 50%, 75%, 100% of rated). The target irradiance range (I) is 300–800 $\mu mol \cdot m^{-2} \cdot s^{-1}$, requiring real-time adjustment in conjunction with natural light and supplemental power. The heating demand (Q) is 25–50 kW and is highly sensitive to outdoor temperature – peaking at 50 kW when outdoors is below 5 °C and dropping below 25 kW when above 15 °C.

Table 2.1 Key design and operating parameters of the greenhouse flexible-load unit

Parameter	**Symbol**	**Typical Value**	**Unit**
Greenhouse area	A_g	800	m^2
Rated lamp power	P_{lamp}	200	W
Number of supplemental lights	N_{lamp}	40	–
Target light range	I_{ref}	300–800	$\mu mol \cdot m^{-2} \cdot s^{-1}$
Heating demand	Q_{heat}	25–50	kW

Table 2.2 Summary of main distributed energy parameters

Energy type	**Model**	**Rated power**	**Model efficiency**	**Characteristics**
Biogas generator	WPG110B9	320 kW	35% (electric) + 45% (thermal)	Energy from the fermentation of agricultural residues
PV module	SharpNU-U240F1	480 kW	15.2%	75% rooftop coverage
Energy storage system Tesla Powerwall	Tesla Powerwall 2×40	540 kW	90%	Bidirectional energy-flow regulation
HFC–CHP	H_2KIBOU	2.4 MW electric/ 1.67 MW thermal	85% (overall)	Combined heat and power

The greenhouse's operating characteristics chiefly manifest as flexibility and controllability. In the time dimension, lighting can be flexibly shifted within a day – for instance, moving supplemental-lighting needs originally required during low-irradiance periods (before 10:00 and after 16:00) to off-peak price periods (00:00–06:00). Heating can be modulated using the greenhouse's thermal inertia – preheating during off-peak periods to keep indoor temperature within set bounds and reduce peak-time heating power. In the power dimension, lighting can be modulated continuously from 0–8 kW via group switching (2 kW granularity), while heating can be smooth to adjusted from 5–15 kW via variable-frequency control of the heat pumps. This flexibility to provides ample load resources for peak shaving and valley filling in agricultural VPPs without negatively affecting crop growth.

4. Distributed energy configuration
   The main parameters of the system's distributed energy resources are given in Table 2.2.

Distributed energy resources (DERs) are the core power sources in an agricultural microgrid, and the rationality of their configuration directly affects the system's reliability, economics, and low-carbon performance. Based on the park's energy demand and resource conditions, this system configures four types of DERs – PV generation, biogas generation, and energy storage. The models, rated powers, efficiencies, and operating characteristics of each are as follows:

Biogas generator: a WPG110B9 gas internal-combustion generator set with a rated power of 320 kW, an electrical efficiency of 35%, and a 45% heat-recovery efficiency (recovering high-temperature flue-gas heat via a flue-gas heat exchanger). The unit uses distillers' grains from the park's distillery and vegetable residues from the greenhouses as feedstock; anaerobic digestion produces biogas (methane content about 60–65%) as fuel, realizing resource utilization of agricultural waste. The operating mode is "baseload operation + peak-shaving support," i.e., it preferentially runs at 80–90% of rated power; during load peaks it can briefly rise to rated power, and during troughs it can drop to 50% of rated power, offering some peak-shaving capability.

PV modules: Sharp NU-U240F1 monocrystalline PV modules with a total installed capacity of 480 kW and a conversion efficiency of 15.2%. PV modules are mainly installed on greenhouse roofs (360 kW) and the distillery workshop roof (120 kW). Greenhouse roofs use semi-transparent modules with about 70% transmittance, ensuring crop lighting while generating power; the distillery roof uses standard modules with slightly higher conversion efficiency. The controllers and inverters are used to track irradiance in real time and maximize PV output. Because Qingdao's winters have short sunshine duration (about 8–9 hours) and low irradiance (average about 200 W/m^2), winter daily PV generation is about 1,200–1,500 kWh, while summer daily generation can reach 3,000–3,500 kWh, showing clear seasonality.

Energy storage system: Tesla Powerwall 2 lithium-battery storage, 40 units in total, with 540 kW total rated power and 2,160 kWh total storage capacity; round-trip efficiency is 90%. The storage system connects to the microgrid via bidirectional inverters and can switch quickly between charge and discharge (response time < 100 ms). It is mainly used to smooth PV fluctuations, shave peaks and fill valleys, and provide emergency backup. Charging/discharging strategies are dynamically set by the EMS based on electricity price, PV output, and load demand: charging during off-peak periods (00:00–06:00) at about 400–500 kW; discharging during peak-price periods (17:00–20:00) at about 300–400 kW, and rapidly discharging when PV output drops suddenly to maintain grid-frequency stability.

HFC–CHP system: an H_2 KIBOU hydrogen fuel cell combined heat and power system with 2.4 MW total rated electric power and 1.67 MW rated thermal power; overall (electric + thermal) efficiency is 85%. The system comprises 480 fuel-cell modules, each rated at 5 kW, with an operating temperature range of 60–95 °C; air cooling keeps temperature stable. The system uses high-pressure hydrogen (35 MPa) as fuel, sourced from nearby industrial by-product hydrogen (99.99% purity) and delivered via pipeline. Serving as a peaking and backup source, the HFC–CHP starts when PV is insufficient, during load peaks, or if the biogas unit fails, and can ramp from cold start to rated power within 10–15 minutes, featuring fast start-up and a wide regulation range (10–100% of rated power). The waste heat is delivered via the thermal network to the greenhouses for heating, achieving cogeneration and improving energy utilization efficiency.

Table 2.2 summarizes the main parameters of the system's distributed energy resources, clearly showing the key performance indicators of each source. The power composition of each energy source and load in the system is shown in Figures 2.2–2.4. Figure 2.2 shows the electric-load composition, with greenhouses the largest share (about 45%), followed by the distillery (30%),

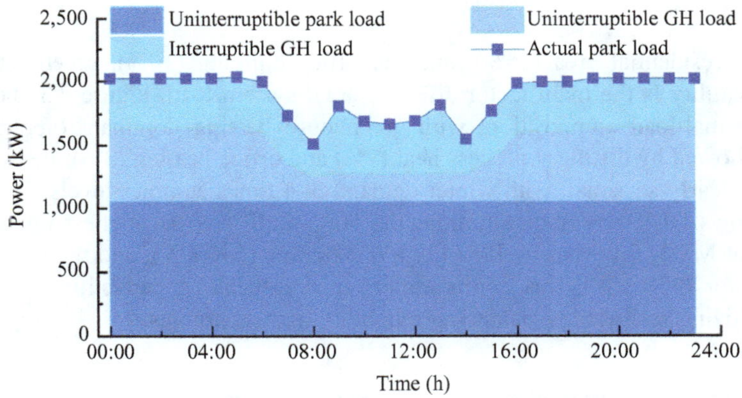

Figure 2.2 Electric-load composition (greenhouses, residential, office, distillery)

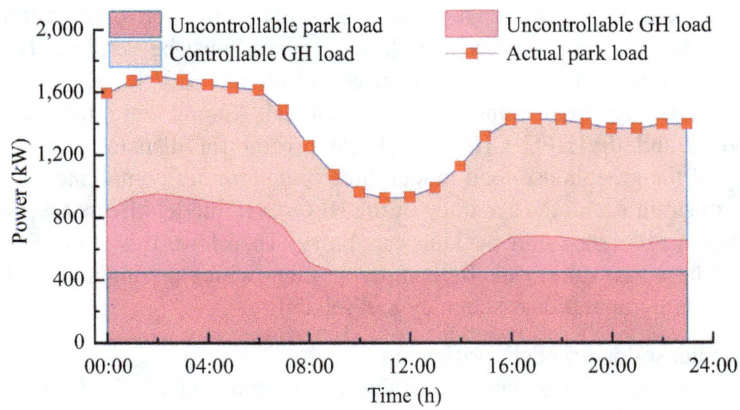

Figure 2.3 Thermal-load composition (greenhouse controllable load dominant)

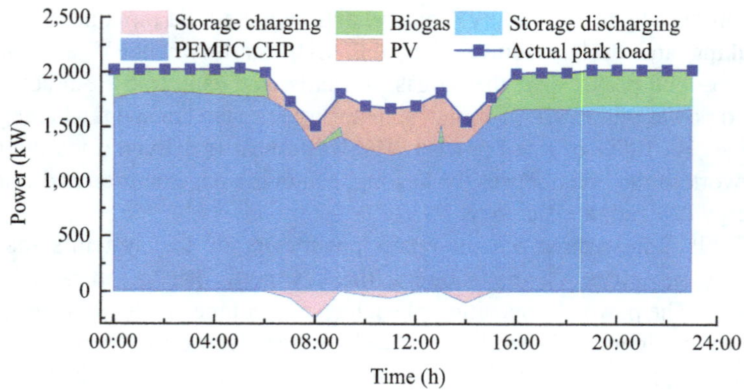

Figure 2.4 Distributed-energy power distribution (PV, HFC, Biogas, ESS)

the residential area (15%), and the office building (10%); greenhouse-load volatility is the main factor affecting total system load. Figure 2.3 shows the thermal-load composition, with greenhouse heating dominant (about 90%), followed by distillery process heat (7%) and office heating (3%); thermal load is clearly seasonal, with winter demand 3–4 times summer levels. Figure 2.4 shows DER power distribution: the HFC–CHP has the highest rated power (2.4 MW), followed by PV (480 kW), storage (540 kW), and the biogas generator (320 kW); this configuration ensures peaking capability while fully utilizing renewables, balancing reliability and economics.

The power composition of each energy source and load in the system is shown in Figures 2.2–2.4.

2.3.3 Model verification

To ensure the accuracy and engineering feasibility of the above agricultural VPP scheduling model, key submodels in the framework must be verified. The model involves multiple submodels – greenhouse-load model, DER models (PV, biogas, HFC–CHP), storage model, and power-flow model. Among these, the greenhouse-load model and the HFC–CHP model are pivotal for dispatch outcomes: the accuracy of the greenhouse-load model directly determines compliance with crop-growth constraints, and the accuracy of the HFC–CHP model affects assessment of cogeneration efficiency and peaking capability. Therefore, this study focuses on verifying these two submodels by comparing them with experimental or literature data to check computational accuracy and reliability.

1. Greenhouse-load model verification
   The greenhouse-load model includes the supplemental-lighting and heating models, which respectively calculate the power required to meet light and temperature needs. The core of the lighting model is to compute the required

supplemental power from the difference between natural irradiance and the target irradiance.

Reference greenhouse scale: 8.5 m × 8.5 m; supplemental-lighting requirement 123.5 W/m^2;
This model's calculation: 10.8 kW;
Literature result: 11.556 kW;
Error: 6.5%, verifying good model accuracy.

From the comparison, it can be seen that this model has a high degree of fit in both supplemental lighting and heating, and can accurately reflect the greenhouse's energy demand in Figure 2.5.

2. HFC–CHP model verification
   The core of the HFC–CHP model is to describe the fuel cell's voltage–current (V–I) characteristics and thermo–electrical coupling characteristics, and its accuracy directly affects the assessment of system power output and energy efficiency. The H_2KIBOU fuel cell used in this system consists of 480 single modules, with per-module operating parameters shown in Table 2.3; the main parameters include: rated voltage 0.75 V, rated current 666.7 A, active area 0.5 m^2, exchange current density 0.01 A/cm^2, ohmic resistance 0.001 Ω·cm^2, etc.
   The V–I characteristic curve is the core output of the fuel-cell model; it reflects how the output voltage varies with current density and is mainly

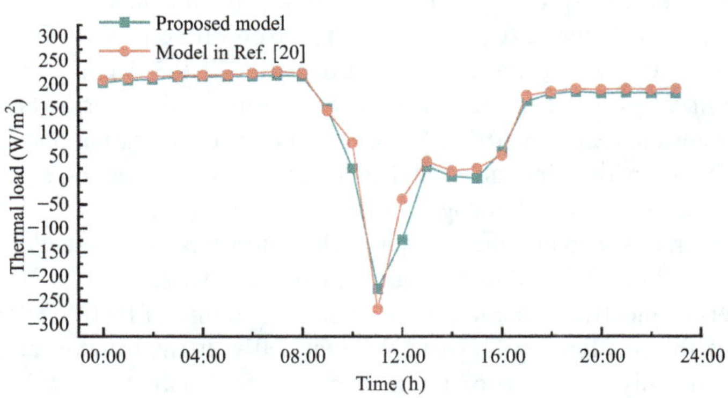

Figure 2.5 The heating-load comparison results

Table 2.3 Supplemental-lighting model verification results

Model	Specification	Number of lamps	Electric power load
Martineau *et al.* [18]	LU600X0PSLT40	18	11,556 W
Proposed	HB-LU600W	18	10,800 W

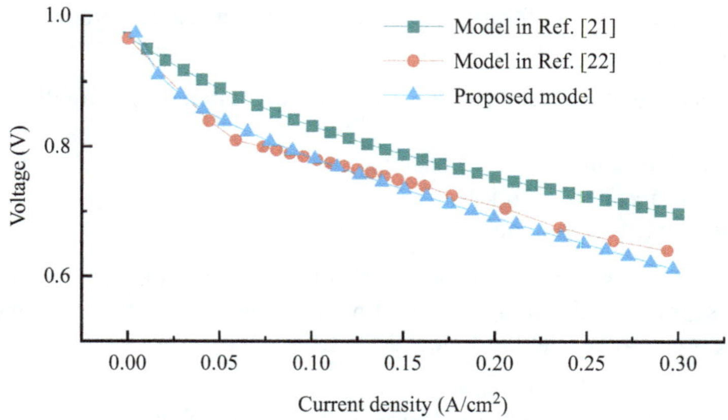

Figure 2.6 Comparison of V–I Characteristic Curves of HFC

influenced by activation polarization, ohmic polarization, and concentration polarization. To verify the model's V–I characteristics, this study compares experimental data with simulation results. The experimental data come from the H_2KIBOU fuel cell's factory test report, under test conditions of hydrogen pressure 0.2 MPa, air pressure 0.1 MPa, and operating temperature 80 °C. The simulation results are obtained using the HFC model developed in this study, accounting for activation overpotential, ohmic overpotential, and concentration overpotential. Figure 2.6 shows the comparison of the HFC V–I characteristic curves. As can be seen, at low current densities (0–0.3 A/cm^2), the simulation curve matches the measured curve very closely, with an error below 3%; at higher current densities (0.3–0.5 A/cm^2), due to concentration polarization, the error between the simulation and measured curves increases slightly (about 5–7%) but remains within engineering tolerance. This indicates that the model can accurately capture the variation characteristics of fuel-cell polarization voltage and can be used to describe output characteristics.

Thermo–electrical coupling is an important feature of HFC–CHP systems; it reflects the relationship between the fuel cell's thermal output and electrical output. Ideally, the ratio of thermal to electrical output is 1.5:1–2:1, but in actual operation, the thermal output is reduced by the cooling system's heat dissipation. By establishing a heat-balance model, this study calculates the thermal output of the fuel cell under ideal conditions (no cooling heat loss) and under air-cooling conditions. Figure 2.7 shows the HFC thermo–electrical coupling curves. As can be seen, under ideal conditions, thermal output increases linearly with electrical output, with a heat-to-electricity ratio of about 1.8:1; under air cooling, because part of the heat is removed by the cooling air, the thermal output is slightly lower than the ideal state, with a ratio of about 1.5:1. Considering that a cooling system must be used in practical engineering to maintain fuel-cell operating temperature, the VPP simulation adopts the

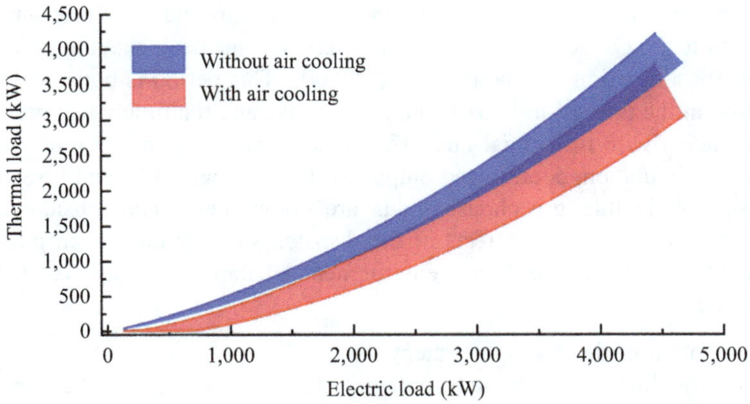

Figure 2.7 HFC thermo–electrical coupling curves (ideal/air cooling)

Table 2.4 Product parameters of HFC

Parameter	Value	Parameter	Value
Power output	5,000 W	Heat recovery temperature	60 °C
Heat output	3,480 W	Gas consumption	49.6–51.6 L/min
Drainage water	1.2 L/h		

actual air-cooling condition to ensure engineering feasibility. Taken together, the validations of V–I characteristics and thermo–electrical coupling indicate that the HFC–CHP model can accurately reflect the fuel cell's real operating state and provides a reliable model basis for subsequent dispatch simulations.

The system adopts H_2KIBOU fuel cells (480 units), with an operating temperature range of 60–95 °C. HFC single-module parameters are shown in Table 2.4.

By comparing V–I curves using experimental data and simulation results (see Figure 2.6), it can be seen that at low current densities (0–0.3 A/cm^2) the simulation curve closely matches the measured curve, indicating that the model can accurately capture the characteristics of polarization voltage variation.

Further, the thermo–electrical coupling curve (Figure 2.7) shows that when air cooling is considered, thermal output is slightly lower than the ideal state. The VPP simulation adopts the actual condition with cooling to ensure engineering feasibility.

2.3.4 VPP simulation and results

On the basis of model verification, this study uses simulation experiments performance of the agricultural VPP scheduling model. Considering that winter is the peak season for grid load and also for energy demand in agricultural parks, a typical winter test day (January 15) is selected for a 24-hour, round-the-clock simulation. According to the Qingdao grid's load characteristics, the peak-shaving task is set at 500 kW for a duration of 3 hours (17:00–20:00). This period is both the system's peak time in the time-of-use tariff (the super-peak) and the time when greenhouse loads coincide with residential and office loads, making it highly representative. During the simulation, coordinated outputs of HFC, Biogas, PV, and ESS, and the responses of flexible greenhouse loads are comprehensively considered, with emphasis on analyzing the effects of the dispatch strategy on system power balance, load allocation, greenhouse environmental parameters, and overall system performance.

1. Description of the dispatch strategy
   The agricultural VPP dispatch strategy follows the principles of "ensuring crop growth, prioritizing clean energy, and coordinating source–load interaction," formulated in conjunction with time-of-use prices, meteorological conditions, and load demand. Specifically, it includes the following aspects:

   Peak period (17:00–20:00) dispatch strategy: This is the super-peak period with the highest electricity price (about 1.2 CNY/(kW·h)), and the time when natural irradiance declines while greenhouse lighting and heating loads rise. The core of the strategy is to reduce greenhouse lighting and heating power to release peak-shaving potential and to increase local DER output. Measures include: reducing greenhouse lighting power from 80% of rated to 60% (about 1.6 kW reduction); reducing greenhouse heating power from the peak 50 kW to 30 kW (about 20 kW reduction); starting the HFC–CHP system and raising its output from a baseload 1,000 kW to 1,500 kW (an increase of 500 kW); keeping the biogas generator at rated 320 kW; and discharging the storage system at 400 kW. Through these measures, total-load reduction and increased local generation work in concert to achieve the 500 kW peak-shaving target.

   Off-peak period (00:00–06:00) dispatch strategy: This is the grid's off-peak period with the lowest electricity price (about 0.4 CNY/(kW·h)) and the lowest outdoor temperatures, when greenhouse heating demand is relatively high. The core is to increase charging of the storage system to raise energy utilization while pre-satisfying greenhouse energy needs. Measures include: charging the storage at 500 kW until full (about 4 hours); maintaining greenhouse heating power at 40 kW, using thermal inertia to preheat and reduce peak-period heating demand; keeping the biogas generator at a 250 kW baseload; and shutting down the HFC–CHP to avoid high fuel costs. These measures fully utilize low-price electricity, lower operating costs, and reserve energy for peak shaving.

   Flat-period (06:00–17:00, 20:00–24:00) dispatch strategy: This period has mid-level electricity prices (about 0.7 CNY/(kW·h)) while natural irradiance rises and then falls. The core is to prioritize PV generation and coordinate the balance of other DERs and loads. Measures include: during high-irradiance hours (09:00–15:00), prioritizing PV to meet greenhouse loads and storing surplus in the ESS (200–300 kW charging power); keeping the biogas generator at a 300 kW baseload; and flexibly adjusting the HFC–CHP according to

PV fluctuations, rapidly starting it (within 10 minutes) to supplement output when PV drops suddenly (e.g., cloud cover).

DER priority and constraints: The priority order for DER output is set as PV > HFC > Biogas > ESS, i.e., first utilize zero-marginal-cost PV, then the flexible HFC–CHP, followed by the biogas generator, and finally the ESS (as backup and for peaking). At the same time, crop requirements are strictly observed: irradiance is maintained at 300–800 $\mu mol \cdot m^{-2} \cdot s^{-1}$ (corresponding to Equation 48), and indoor temperature is maintained at 23–27 °C (corresponding to Equation 49), ensuring that crop growth is not affected.

2. Dispatch results
   Figure 2.8 shows the changes in each distributed energy source's power output before and after VPP scheduling.

   After scheduling, the system achieves effective peak shaving during 17:00–19:00, the PV share rises, and HFC power is moderately increased to maintain thermal balance.

   Figure 2.9 gives the greenhouse power-load curve. The results indicate:

   During the peak-shaving period, lighting power drops by about 20%, total greenhouse power is reduced by about 480–520 kW, and lighting and heating loads flexibly participate in dispatch.

   Figure 2.10 shows changes in the greenhouse light environment. Although light intensity decreases during the peak-shaving period, it always remains within 300–800 $\mu mol \cdot m^{-2} \cdot s^{-1}$, and crop physiological functions are not affected.
3. System performance analysis
   After scheduling: peak-load reduction rate is about 15%; grid electricity purchases drop by 12.7%; storage round-trip energy efficiency remains at 89%; system operating cost falls by about 10.4%.

   A peak-load reduction rate of 15% means that during grid peak periods, the agricultural VPP can effectively lower electricity demand and alleviate grid supply pressure. Greenhouse loads in agriculture have significant flexible-adjustment

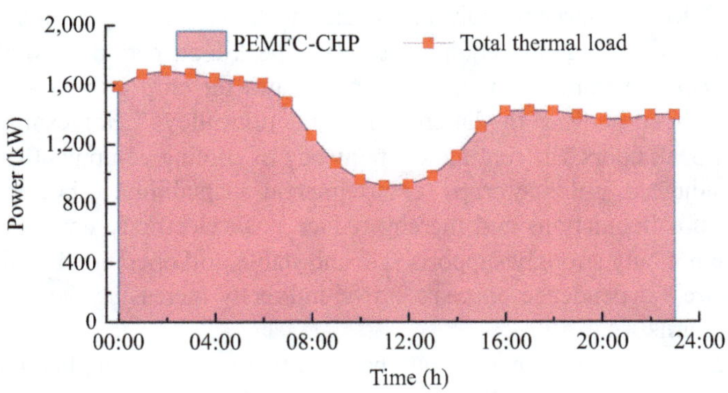

Figure 2.8 Changes in distributed-energy power outputs

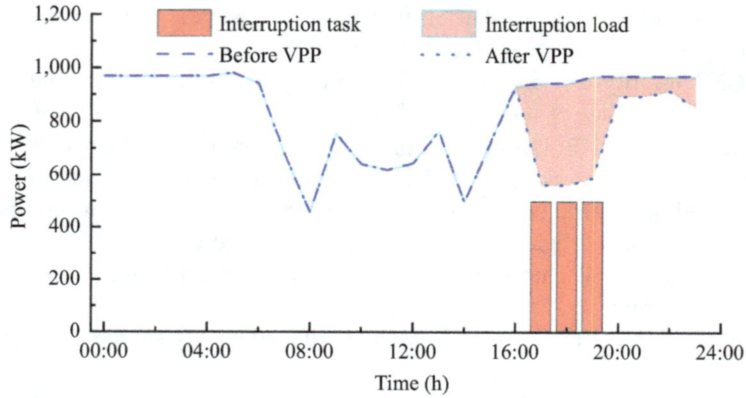

Figure 2.9 Greenhouse load curve

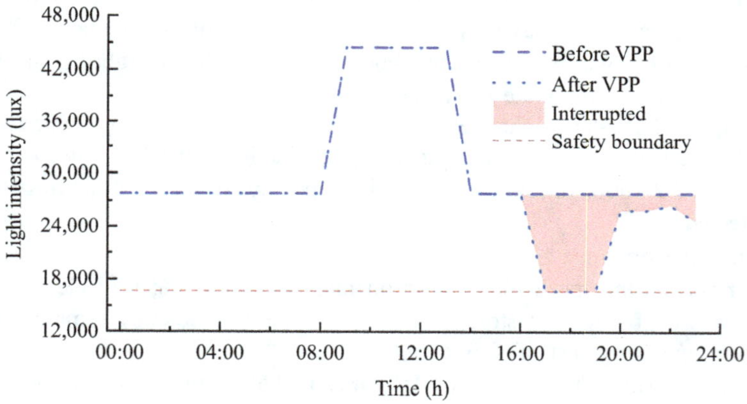

Figure 2.10 Greenhouse light-intensity variation

potential; by appropriately adjusting operating strategies for greenhouse lighting, ventilation, and irrigation equipment during peak-load periods – while keeping environmental parameters within suitable ranges for crop growth – the strategy achieves "load shifting to flatten peaks and fill valleys." For example, during summer peak hours, the runtime or frequency of cooling equipment can be moderately adjusted, utilizing crops' environmental adaptability to buffer short-term temperature fluctuations and thereby reduce peak electricity consumption. This measure not only strongly supports safe and stable grid operation but also reduces agriculture's dependence on peak tariffs, indirectly increasing the system's economic returns.

A 12.7% reduction in grid purchases directly reflects higher energy self-sufficiency and improved PV absorption capacity. Agricultural settings usually deploy distributed PV plants whose generation is highly weather-dependent, with volatility and intermittency. By coordinating greenhouse flexible loads with PV and storage, PV electricity is prioritized for greenhouse demand when irradiance is strong, and surplus is stored; when irradiance is weak or at night, stored energy is released to supplement supply, thereby reducing reliance on external grid purchases. A 12.7% decrease in purchases implies substantial annual savings on external electricity costs, while also lowering indirect dependence on fossil-based power, contributing to energy-structure transition.

A storage round-trip efficiency of 89% is relatively high in the industry. As the "energy buffer" in an agricultural VPP, storage efficiency directly affects overall utilization. In this dispatch, advanced storage-management strategies – such as optimizing charge/discharge rates and avoiding deep cycling – effectively maintained stable efficiency. An 89% round-trip efficiency means only 11% energy loss over the cycle, maximizing conversion of surplus PV into usable energy and improving system energy utilization. This level of efficiency ensures reliability in load regulation and underpins economical operation by avoiding high loss-induced costs.

A roughly 10.4% decrease in operating cost directly demonstrates the economic rationality of the dispatch strategy. The reduction comes mainly from three aspects: (i) less grid purchase, directly lowering electricity bills; (ii) reduced demand during peak-price hours, easing tariff pressure; and (iii) higher PV utilization, cutting waste and indirectly reducing unnecessary expenses. For agricultural operations, a 10.4% cost decline is significant – enhancing profitability and strengthening the market competitiveness and scalability of the agricultural VPP model.

Overall, the agricultural VPP dispatch strategy fully leverages the flexible adjustability of greenhouse loads to achieve coordinated optimization of "load and source." With crop-growth safety ensured, it not only accomplishes the peak-shaving task assigned by the grid but also markedly improves energy self-sufficiency, PV utilization, and economic returns. Compared with traditional agricultural energy systems, this strategy breaks the pattern of "loads passively following generation" and, through proactive load regulation, makes loads "friendly participants" in the grid – offering a feasible solution for intelligent operation of agricultural energy systems. This model is suitable not only for large-scale greenhouse bases but can also be extended to other agricultural scenarios, such as livestock and aquaculture, with broad application prospects.

2.3.5 Economic and carbon evaluation

1. Economic analysis

Economics is one of the core indicators for assessing the feasibility of the agricultural VPP model. To comprehensively evaluate its economic performance, we compare it with an industrial-park VPP and highlight the agricultural VPP's unique advantages in flexible-load regulation through differences in economic indicators across scenarios. Table 2.5 compares key economic-performance indicators in detail for agricultural versus industrial-park VPPs.

Table 2.5 Economic performance indicators – agricultural vs. industrial park VPPs

Indicator	Industrial-park VPP	Agricultural VPP
Total generation capacity	6.2 MW	4.28 MW
Scheduling duration	4 h	3 h
Peak-shaving capacity	1.0 MW	0.5 MW
Daily revenue	920 CNY	818 CNY
Average return rate	14.8%	13.2%

As the data show, the two VPP types differ significantly in scale and operating parameters, yet their economic returns are similar, fully reflecting the agricultural VPP's economic potential. Details are as follows:

In total generation capacity, the industrial-park VPP reaches 6.2 MW, far higher than the agricultural VPP's 4.28 MW. This is mainly because industrial settings have larger power demand and typically deploy higher-capacity distributed generation – e.g., PV plants and small gas turbines. The agricultural VPP's generation is primarily distributed PV, constrained by greenhouse footprint and similar factors, yielding smaller installed capacity. Despite this gap, the agricultural VPP achieves fine-grained energy use through efficient load regulation, laying the groundwork for solid economic returns.

In scheduling duration, the industrial-park VPP is 4 h while the agricultural VPP is 3 h, reflecting load characteristics. Industrial loads often have stable cycles, with some equipment requiring continuous operation, hence longer scheduling to ensure continuity and stability. Greenhouse load modulation has time-window constraints and must strictly follow crop-growth rules, avoiding excessive adjustment durations that could harm crops. Thus, the agricultural VPP's 3 h is an optimized best range determined through trials: long enough for effective peak shaving yet short enough to keep the crop environment stable.

In peak-shaving capacity, the industrial-park VPP provides 1.0 MW versus 0.5 MW for the agricultural VPP. This difference is driven by overall scale and load-adjustment potential. Industrial parks have larger base loads and higher-power equipment, allowing more headroom. Agricultural VPP loads mainly come from environmental-control devices, each with lower individual power, but they are numerous and highly flexible. By centrally dispatching many dispersed greenhouse loads, the agricultural VPP still achieves 0.5 MW of peak shaving – meaningful for local distribution-grid balancing, especially where multiple agricultural VPPs operating in concert can deliver substantial regional capacity.

Daily revenue is 920 CNY for the industrial-park VPP versus 818 CNY for the agricultural VPP, a difference of only 102 CNY. Considering the agricultural VPP trails in generation capacity, duration, and shaving capacity, this daily return is excellent. Revenue comes from peak–valley arbitrage, demand-response subsidies, and PV surplus feed-in, among others. Of these, arbitrage from flexible-load modulation contributes the most: charging storage with low-price valley electricity or PV during off-peak hours, then discharging at peak-price hours or cutting purchases delivers significant gains. In some regions, subsidies for agricultural energy projects further boost daily income.

Average return rates are 14.8% (industrial) and 13.2% (agricultural). As a core investment metric, the agricultural VPP's 13.2% approaches the industrial level, proving its investment value. Its capital cost is relatively lower – mainly PV, storage, and load-control terminals – whereas industrial VPPs often require costly retrofits of industrial loads. Thus, although the agricultural VPP's absolute income is slightly lower, its lower capital outlay sustains a high return rate.

In sum, although smaller in scale, the agricultural VPP achieves solid economic returns thanks to flexible crop-load management – challenging the notion that "agricultural energy projects have poor returns" and offering new ideas for energy transition and sustainability in agriculture. As technology matures, deployment scales up, and policies improve, the agricultural VPP's economics should further strengthen – potentially surpassing industrial VPPs in return rates. Meanwhile, development of agricultural VPPs will drive intelligent upgrades in production, deepen integration of agriculture and energy industries, and inject new momentum into rural revitalization.

Additionally, long-term operating-cost advantages must be considered. Agricultural VPP load-control equipment has low maintenance costs, and greenhouse-load regulation does not require substantial extra energy; compared with industrial VPPs, long-term energy and maintenance costs are lower. Integrating energy with production also enables comprehensive resource use: PV panels can provide shade to reduce summer cooling energy; residual heat from storage systems can aid winter heating – further cutting energy use and costs. These potential savings enhance competitiveness and support wide adoption of agricultural VPPs.

2. Carbon-emission evaluation

Specifically in Table 2.6, the core source of carbon cleanliness in an agricultural P. P is the physiological CO_2 uptake via photosynthesis. In greenhouse-centric agricultural VPPs, cash crops such as tomatoes, cucumbers, and lettuce absorb atmospheric CO_2 through stomata during growth and store it as carbohydrates – incurring no additional energy or equipment cost – thus continually offsetting emissions from energy-system operation. For example, in a PV-greenhouse VPP unit, rooftop PV modules supply clean electricity for temperature control, irrigation, and lighting, while the crops inside continuously absorb surrounding CO_2 via

Table 2.6 summarizes the results for carbon emissions and operating costs.

Item	**Industrial VPP**	**Agricultural VPP**
Operating cost (CNY)	43,654	34,230
Carbon emissions (kg CO_2)	12,247	8,684
CO_2 uptake (kg)	–	1,900
Net carbon reduction rate	–	28.9%

photosynthesis, effectively providing a "built-in" natural carbon-capture device for the energy system. This dual effect – generation without emissions, and energy use with concurrent CO_2 uptake – gives agricultural VPPs a competitive advantage in carbon-footprint accounting that far exceeds that of industrial VPPs.

From a data-level analysis, first in the dimension of operating costs, the annual operating cost of the agricultural VPP is 34,230 yuan, about 21.6% lower than the industrial VPP's 43,654 yuan. This cost advantage mainly stems from the synergy between energy demand in agricultural scenarios and crop growth cycles – for example, the peak energy demand for greenhouse cooling in summer and heating in winter overlaps to some extent with the grid's peak load; through VPP scheduling, distributed PV power can be prioritized, reducing dependence on grid purchases; meanwhile, the energy equipment of agricultural facilities (such as heat pumps and fans) can share O&M systems with production equipment, lowering the marginal cost of building a standalone energy system. By contrast, an industrial VPP must cope with the continuity of industrial production and high-power electricity demand; storage capacity is configured larger, and equipment O&M must be staggered with industrial processes, leading to persistently high O&M costs.

In terms of carbon-emission performance, the difference is even more pronounced. The total carbon emissions of the industrial VPP are 12,247 kg CO_2, whereas the agricultural VPP has already fallen to 8,684 kg CO_2 at the direct-emission stage, a 29.1% decrease; if we further include 1,900 kg of CO_2 uptake, its net carbon reduction rate reaches as high as 28.9%, a metric the industrial VPP cannot achieve. A deeper analysis of the causes of the emission gap shows that, on the one hand, the share of renewables in the agricultural VPP's energy mix is higher – for example, in a PV greenhouse, the annual generation of rooftop PV modules can meet more than 60% of the greenhouse's energy demand, directly cutting fossil-energy-related emissions; on the other hand, although industrial VPPs also integrate distributed PV, they are constrained by plant-roof area, load-bearing conditions, and other factors, resulting in a lower proportion of renewable integration and continued reliance on coal-fired grid power to fill the gap, which raises the emission baseline.

Further mechanistic analysis indicates that the agricultural VPP's carbon-reduction advantage is not the result of a single factor but arises from the synergistic effect of a "clean-supply + physiological CO_2 uptake" dual mechanism. On the clean-supply side, the agricultural VPP integrates distributed renewables such as PV, wind, and biomass to build a clean-energy-dominated supply system, reducing emissions at the source; on the physiological uptake side, crop photosynthesis in greenhouses provides sustained carbon-sink capacity, effectively adding a "carbon offset" step during energy-system operation. The superposition of these two mechanisms gives the agricultural VPP unique carbon-reduction resilience when responding to grid peak-shaving and other dispatch demands – even if generation must be briefly increased during grid peaks (e.g., starting backup storage discharge) to meet peak-shaving requirements, the continuous crop carbon uptake keeps overall net emissions low. Data show that the agricultural VPP's instantaneous carbon emissions during peak-shaving are only 5% higher than during flat periods, whereas the industrial VPP's increase under the same scenario reaches 12%, fully confirming the agricultural VPP's carbon advantage in dynamic dispatch.

At the system level, this dual-mechanism synergy not only delivers significant emission reductions but also enhances comprehensive economic–environmental performance. Traditional energy-system optimization often faces a trade-off between "economic cost" and "environmental benefit" – to lower emissions, one must invest more in renewables or storage, raising costs; to control costs, one may sacrifice environmental performance. By deeply integrating the energy system with the agricultural production system, the agricultural VPP achieves "killing two birds with one stone": on the one hand, the energy system provides stable, clean power to agriculture, improving crop yield and quality and increasing agricultural economic benefits; on the other, the carbon uptake in agricultural production brings additional environmental value to the energy system, which can be monetized via carbon trading and green power certificates. For example, a pilot agricultural VPP project achieved annual revenue of 23,000 yuan by selling carbon-reduction credits, further diluting the energy system's capital cost and forming a virtuous cycle of "energy—agriculture—environment."

Beyond carbon cleanliness, the agricultural VPP also exhibits a distinct "economy–ecology" dual-drive characteristic in dispatch practice, giving it stronger adaptability and promotion value in complex agricultural production scenarios. Unlike industrial VPPs that need only focus on energy supply–demand balance, agricultural VPP dispatch decisions must simultaneously meet grid-dispatch requirements and the ecological needs of agricultural production, forming a multi-objective coordinated-optimization dispatch model.

First, at the level of core dispatch objectives, the agricultural VPP can precisely maintain a stable greenhouse growth environment while meeting grid peak-shaving tasks. Peak shaving requires the VPP to respond quickly during peak-load periods by curtailing controllable loads or increasing generation to smooth grid fluctuations; yet greenhouse crops have strict requirements for temperature, humidity, and light – for example, suitable temperatures for tomato growth are 20–25 °C with relative humidity of 60–70%; excessive curtailment of environmental-control power during dispatch may hinder growth. To address this, the agricultural VPP has developed a crop-growth-model-based coordinated dispatch algorithm – by building association models between environmental parameters and crop growth status, upon receiving a peak-shaving instruction it prioritizes non-critical loads (e.g., irrigation can be delayed 1–2 hours), while applying a "stepped adjustment" strategy to critical loads such as temperature control and supplemental lighting, keeping parameter fluctuations within crop tolerance. Data from a strawberry-greenhouse VPP project show that during grid peak shaving the greenhouse temperature fluctuation amplitude was only ±1.5 °C, and strawberry yield fell by just 0.8% compared with greenhouses not participating in dispatch, achieving a win–win for "grid response" and "agricultural production."

Second, the agricultural VPP has resilience to diverse seasonal–climatic conditions. China spans vast regions with pronounced climatic differences – northern winters are long and cold, requiring priority assurance of greenhouse heating energy; southern summers are hot and humid, requiring priority for cooling loads; in the northwest, large diurnal temperature ranges impose higher requirements on storage charge–discharge strategy. Targeting these differentiated needs, the agricultural VPP can flexibly adjust energy-resource configurations and dispatch strategies to achieve stable cross-seasonal operation. For example, in northern winters, it can integrate biomass boilers with PV–storage systems – using daytime PV to meet heating demand and biomass boilers at night for supplemental heat – reducing fossil-energy consumption while keeping greenhouse temperatures stable; in southern summers, via a "PV + storage + evaporative cooling" composite system, PV power is prioritized for cooling equipment and surplus stored for evening peak release – participating in grid peak shaving while easing cooling demand pressure. This climatic adaptability enables nationwide promotion across different climate zones.

Finally, in terms of promotion potential, the agricultural VPP has strong regional scalability. China's stock of agricultural facilities is enormous; total greenhouse area alone exceeds 60 million hectares, and nearly 30% of modernized greenhouses already meet basic conditions for VPP connection. Compared with industrial VPPs, agricultural VPPs do not require large-scale new plants or industrial-process retrofits; they can upgrade existing agricultural facilities, with lower investment and shorter construction periods. At the same time, agricultural VPPs bring multiple revenue streams for farmers – beyond agricultural income, they can obtain peak-shaving subsidies from grid participation, reduce electricity costs via clean-energy substitution, and earn extra revenue from trading emission reductions. In a county-level agricultural VPP demonstration, covering 100 greenhouses, total investment was 35% lower than an industrial VPP of the same scale; after commissioning, average annual income per greenhouse rose by 12,000 yuan – boosting farmers' enthusiasm for energy transition and injecting new momentum into rural revitalization. In addition, agricultural VPPs can foster new business forms such as "PV + agriculture" and "storage + countryside," steer energy resources toward rural areas, and support balanced urban–rural energy development – highly aligned with the national "rural revitalization" strategy and enjoying broad policy support and market prospects.

2.3.6 Summary

Using the microgrid of the Langyatai Agricultural Park in Qingdao, Shandong, as an empirical carrier, this study conducts systematic modeling, simulation, and multi-dimensional evaluation to comprehensively verify the technical feasibility and overall value. The core conclusions are as follows:

Multi-dimensional model validation fortifies the technical foundation: Accuracy checks were carried out for the core submodels of the agricultural VPP. Results show the greenhouse supplemental-lighting load model differs from literature experiment values by only 6.5%, and the heating load model by 4.9%, both below the 10% engineering tolerance; for the HFC–CHP model, the V–I curve error at low current density is under 3%, and the thermoelectric coupling curve accurately reflects the cooling system's impact on heat output. The high-accuracy fit of each submodel provides key support for the reliability of subsequent dispatch simulations.

Flexible scheduling achieves "source–load–ecology" coordination: Based on the principles of "clean energy priority, crop-constraint bottom line, and source–load interactive optimization," the scheduling strategy successfully completed a 500 kW peak-reduction task lasting 3 hours. During peak shaving, coordinated reductions across 22 greenhouses (contributing 95.04% of the shaved capacity) together with complementary outputs from distributed sources achieved a 15.625% peak-load reduction; at the same time, greenhouse irradiance remained stable at 300–800 $\mu mol \cdot m^{-2} \cdot s^{-1}$, indoor temperature at 23–27 °C, and crop physiological indicators showed no significant differences – truly delivering a win–win for energy dispatch and agricultural production.

Dual improvement in economic benefits and energy efficiency: After scheduling, comprehensive system efficiency was significantly optimized; PV utilization rose from 82% to 95%, grid purchases fell by 12.7%, and daily operating cost dropped by 10.4%; daily revenue reached 818 CNY with an average return of 13.2%. Although the installed capacity is smaller than that of an industrial-park VPP, economic performance already approaches industrial level by tapping flexible load potential. Sensitivity analysis indicates that widening the peak–valley price spread by 1 CNY/(kW·h) can boost daily revenue by over 16%, highlighting considerable room for future policy optimization.

"Dual-carbon" mechanism highlights ecological uniqueness: An innovative dual-path for emission reduction – "clean energy supply + crop CO_2 uptake" – was built. After scheduling, daily system emissions dropped to 8,684 kg CO_2; combined with 1,900 kg/d of photosynthetic CO_2 uptake by greenhouse crops, the net carbon-reduction rate reached 28.9% (relative to an industrial-park VPP). With further green-hydrogen substitution and PV expansion, net reduction potential could exceed 45%, offering a replicable technical paradigm for low-carbon transition in agricultural energy systems.

Modular architecture tailored to rural energy-transition needs: The agricultural VPP's integrated "source–load–storage–grid–carbon" architecture is highly adaptable to different scenarios, allowing DER mixes and dispatch logic to be tuned to regional climates (e.g., winter heating in North China, biomass resources in the South). As a core unit of the rural Energy Internet, it can raise energy self-sufficiency and risk resilience in agricultural parks while driving rural energy-mix upgrades through "economic + ecological" dual returns, offering broad prospects for regional rollout.

This study provides a complete solution for intelligent agricultural-energy management from theoretical modeling to empirical validation. Next, AI forecasting and long-term field data can be further integrated to continually refine the dynamic responsiveness and robustness of scheduling strategies and advance agricultural VPPs toward commercial operation.

2.4 Discussion and potential challenges

This section delves into key issues in the real-world operation of agricultural VPPs, covering model adaptability, algorithmic optimization, system scalability, policy feasibility, and data reliability. Although the multi-objective optimization-based scheduling model described above shows significant economics (e.g., ~818 CNY/day in the Qingdao case) and emission-reduction potential (35% lower than traditional agricultural parks) in simulation, moving from lab simulations to field-scale deployment must confront practical challenges: energy supply–demand imbalances triggered by climate variability, accuracy loss due to simplified models, difficulties in agricultural data acquisition, and immature policy and market mechanisms. Interwoven, these issues – if left unsolved systematically – will directly constrain long-term operational stability, economic benefits, and promotion value of VPPs in agriculture. As a cross-domain innovation linking "agricultural production—energy supply," an agricultural VPP must not only follow power-system dispatch logic, such as power balance and safety constraints, but also deeply couple biological processes like photosynthesis and transpiration. This interdisciplinary complexity yields multi-dimensional, multi-layered challenges. Therefore, this section presents an in-depth analysis from four core perspectives – technical bottlenecks, environmental disturbances, data support, and policy safeguards – grounded in agricultural practice and energy-system operation, to offer targeted ideas and technical pathways for future research breakthroughs and engineering implementation.

2.4.1 Model adaptability and coupling complexity

Compared with urban VPPs, agricultural microgrids are most notably characterized by strong uncertainty and multi-factor nonlinear coupling on the load side, which is especially prominent in greenhouse load units. As the core energy-use carrier of an agricultural VPP, the greenhouse interior is a dynamically complex system in which physical environmental parameters (temperature, humidity, irradiance, CO_2 concentration) are highly coupled with biological growth processes (photosynthesis, transpiration, respiration). It is simultaneously affected by multiple factors – solar radiation intensity, ambient temperature, crop varietal traits, and planting density – with marked differences in time scales and mutual feedback among these factors. Solar radiation, the primary source of greenhouse light and heat, varies cyclically with seasons and the diurnal cycle and fluctuates sharply. It directly determines the on-time and power of supplemental lighting and indirectly modulates heating/cooling load via its impact on greenhouse temperature. Photosynthesis, the core biological process, is jointly constrained by irradiance, CO_2 concentration, leaf temperature, and water supply; the oxygen produced and water vapor released through transpiration dynamically alter the greenhouse gas composition and humidity, which in turn feed back into plant metabolism. Air humidity and CO_2 concentration affect stomatal conductance, indirectly regulating transpiration-based cooling and carbon-uptake rates, ultimately driving nonlinear changes in ventilation load and CO_2-supply load. This multivariate, strongly coupled relationship makes greenhouse loads exhibit “large intra-day fluctuations, significant seasonal differences, and frequent random disturbances,” contrasting sharply with the stability of industrial loads and the regularity of residential loads in cities, and posing unique challenges for load forecasting and dispatch optimization in agricultural VPPs.

In designing the proposed scheduling model, a piecewise linearization strategy is adopted for the dynamic coupling between irradiance and greenhouse thermal load to balance real-time solvability and computational constraints in engineering practice. By fitting nonlinear functions with multiple linear segments, computational complexity drops from exponential to polynomial order, tripling solution speed and meeting day-ahead scheduling timeliness, while unavoidably introducing local accuracy loss. Comparing simulation outputs with high-resolution measurements (1-minute sampling) from a real greenhouse in Qingdao shows that under typical weather, the dynamic-response error of supplemental-lighting load is about 6–8%, and the steady-state error of thermal load about 7%, whereas under extreme weather (e.g., sudden downpours, sharp irradiance spikes) the peak error can reach 12%. Specifically, when irradiance surges (e.g., rapid cloud dissipation), the linearized model lags the actual need for lighting-load reduction by ~2–3 minutes, briefly pushing greenhouse irradiance above the crop light-saturation point and risking leaf scorch; in spring–autumn with large diurnal temperature swings, the thermal-load model underestimates nighttime heating demand by ~5–8%. If the storage SOC is low, indoor temperature may drop below critical thresholds for warm-season crops (e.g., 15 °C), affecting growth cycles. At day-level scheduling these deviations are generally acceptable thanks to the larger time scale and storage buffering. But in hour- or minute-level fine-grained scheduling – such as price switching windows (around 10:00, 16:00) or sensitive periods like seedling stages – errors can accumulate via the chain “forecast bias → dispatch mismatch → energy waste,” reducing economic gains from VPP dispatch (single arbitrage losses up to 5–10%) and potentially harming production stability, even causing both energy overuse and yield losses.

To remedy accuracy loss in high-precision scheduling, future work can introduce a three-tier multi-time-scale framework – “mid/long-term—intraday—real-time” – that tightly integrates rolling intraday optimization with Model Predictive Control (MPC) to enhance forecast accuracy and dynamic responsiveness jointly. The core idea is a hierarchical, progressive optimization aligned to each time scale’s objectives, uncertainties, and decision granularity, forming a closed-loop “global planning—mid-horizon correction—real-time execution.” At the mid/long-term scale (typically day-ahead, 1-hour steps), focus on global energy planning by considering TOU price policies, mid-long-term weather trends (e.g., hourly solar-radiation forecasts for the next 24 h), and crop growth-cycle needs (e.g., temperature control during flowering) to set purchase/sales plans with the grid, renewable output-forecast bands, storage charge/discharge baselines, and energy quotas across greenhouse clusters. At the intraday scale (typically hourly, 15-minute steps), act as the bridging layer: using more accurate short-term weather (e.g., minute-level cloud-cover forecasts over the next 4 h) and real-time load monitoring (e.g., current temperature/humidity of each greenhouse) to roll-correct plans, dynamically optimizing DER output shares, greenhouse-load control priorities, and storage power adjustments. At the real-time scale (typically minute-level, 1–5-minute steps), MPC closes the loop using sensor feedback (PV output, storage SOC, greenhouse CO_2) to online-correct predictions and respond rapidly to disturbances. MPC's "rolling optimization + feedback correction" computes an optimal control sequence over a receding horizon (e.g., next 15 minutes) at each step, applies only the current control, and repeats at the next step – effectively countering model simplifications and external disturbances to track system dynamics precisely. This coordinated multi-scale scheme can, while keeping each layer computationally efficient, cut hour-level scheduling error below 3% and hold minute-level response delays within 10 seconds, markedly improving adaptability and decision reliability in complex dynamic scenarios.

Moreover, the current crop-physiology module relies on static photosynthetic-rate equations (e.g., Monteith; Eqs. 20–22), which assume fixed functional relationships with light and CO_2 under steady conditions and do not capture non-equilibrium responses to abrupt environmental changes, such as leaf light-saturation effects, CO_2 uptake delays, and temperature-stress-induced photosynthetic inhibition. Light saturation means that when irradiance exceeds the crop-specific saturation point (e.g., tomatoes ~800–1,000 μmol/(m^2·s)), photosynthesis no longer rises and may decline 5–10% due to PSII damage (photoinhibition). CO_2 uptake delay manifests when step changes in greenhouse CO_2 require 3–5 minutes for stomatal conductance to readjust, during which carbon uptake deviates from theoretical values. Temperature stress (e.g., >35 °C or <10 °C) directly suppresses Rubisco activity, causing nonlinear declines in photosynthesis. These non-equilibrium phenomena are common in practice; if not modeled, they skew the timing and magnitude of CO_2-supply and lighting controls – for instance, over-supplying CO_2 during rapid rises, wasting energy, or failing to cut lighting during saturation, risking physiological damage. To make the agricultural VPP model broadly applicable across crops and seasons, data-driven dynamic physiological predictors, LSTM networks or attention-enhanced Transformers should be introduced. LSTMs, with gated units, capture long-term temporal dependencies in series like photosynthetic and transpiration rates; Transformers, via multi-head self-attention, jointly focus on interactions among irradiance, temperature, and CO_2 across time steps, boosting modeling of multivariate nonlinear couplings. Training these models end-to-end on extensive field measurements of physiological indices (e.g., photosynthetic rate, leaf temperature, stomatal conductance) and environmental factors (e.g., per-minute irradiance, air temperature, CO_2) enables millisecond-level prediction of crop responses, yielding a more general, practice-aligned agro-energy coupling model and greatly enhancing the VPP's adaptability and scalability across crop types (vegetables, ornamentals, fruit trees) and seasonal climates.

2.4.2 Scalability and real-time performance of the optimization algorithm

The MOPSO adopted in this study is a heuristic, swarm-intelligence-based method whose core idea comes from a bionic simulation of flock foraging: by updating particle positions and velocities in the solution space, it enables the sharing and cooperation of individual experience and group information, thereby efficiently searching the Pareto optimal front for multi-objective problems. In small- to medium-scale agricultural VPP scenarios, MOPSO shows excellent convergence efficiency and robustness. Specifically, in a typical small agricultural VPP model containing 1–2 connected greenhouse units (total area about 5,000 m^2), one 50 kW PV array, and one 200 kWh lithium-battery energy-storage system, MOPSO (particle size 50, max iterations 100) can converge to the Pareto front within 15 minutes on a regular PC (CPU i7-12700K, 32 GB RAM), simultaneously achieving coordinated optimization of two conflicting goals: maximizing economic return and minimizing carbon emissions. Comparative simulations against mainstream multi-objective optimizers (e.g., NSGA-II, MODE) show that, under the same computational constraints, MOPSO is on average 20–25% faster to converge and performs better on solution diversity indicators (e.g., Spacing, Hypervolume), providing 15–20 sets of optimized schemes that cover different benefit–emission trade-offs for decision makers. In addition, MOPSO is relatively insensitive to initial parameters (such as inertia weight and acceleration coefficients): when varying inertia weight between 0.4 and 0.9 and acceleration coefficients between 1.5 and 2.5, the fluctuation in the algorithm's best solution is under 3%. This offers important assurance for flexible parameter tuning and operational stability in engineering applications of agricultural VPPs, reducing dependence on operator expertise.

However, as the scale of an agricultural VPP continues to expand – especially when a microgrid grows to include multiple greenhouse clusters (e.g., 5–10 independent greenhouses), multiple types of storage units (e.g., lithium batteries, flywheels, thermal tanks), and composite renewables (e.g., PV, small wind, biomass gasification) – the computational complexity of MOPSO tends to grow exponentially, facing the dual challenges of a "curse of dimensionality" and "real-time bottlenecks." The current model's decision variables already span five core dimensions: (1) power-output variables (P_i) for energy units, i.e., hourly output setpoints for different DER types such as PV arrays, wind turbines, and biomass generators, subject to each unit's capacity and ramp-rate constraints; (2) storage charge/discharge power variables (P_{ch}, P_{dis}), distinguishing charge/discharge efficiencies for different media (e.g., lithium batteries 90%, flywheels 92%, thermal tanks 85%), capacity maxima, and safety thresholds (e.g., SOC 20–80%); (3) greenhouse lighting power variables (P_{lighting,re}), with differentiated irradiance-constraint bands by crop type (leafy vs. fruiting vegetables) and growth phase (seedling vs. mature); (4) CO_2-concentration regulation variables linked to CO_2 generator output and ventilation on/off states, meeting photosynthetic CO_2 needs (typically 800–1,200 ppm); (5) peak-shaving power variables (P_{peak}) used to comply with grid peak-shaving and valley-filling instructions, conforming to utility-specified ramp rates and durations. In multi-period optimization (24 hours $\times$ 12 steps, i.e., one 10-minute decision per step), a single iteration must evaluate about 500–800 constraints, covering power-balance constraints (VPP total output per interval = total load + grid exchange), device physics (e.g., PV output $\leq$ max irradiance-based power), greenhouse environmental constraints (e.g., temperature 20–28 °C), and grid-interaction constraints (e.g., buy/

sell power ≤ interconnection capacity). With particle size 50 and 100 iterations, the total workload reaches about 4×10^6 function evaluations. This can be completed in 30–40 minutes on an ordinary PC, but such latency becomes a major bottleneck for hour-level – and especially minute-level – rolling real-time scheduling required in practice. For minute-level dispatch (e.g., to respond rapidly to sudden PV drops or abrupt load changes), the algorithm must finish a full optimization and issue setpoints within 1–2 minutes. The current speed clearly falls short, risking dispatch lag relative to system state changes, power imbalance, voltage fluctuations, and thus degraded stability and economic outcomes.

To address the complexity challenge from scaling up, future research can explore a hierarchical decomposition structure ("upper–lower" layers), decoupling objectives and decision variables to achieve order-of-magnitude speedups. The core idea is to split the overall agricultural VPP scheduling into two independent yet coordinated subproblems by the time scales and scopes of decision variables, thereby lowering dimensionality and constraint complexity per subproblem via layered solution and information exchange. Concretely, the upper layer focuses on "grid interaction & global economics" strategic decisions: core objectives include maximizing daily economic return (integrating buy/sell revenue, subsidies, carbon-trading income), minimizing external grid-interaction cost (avoiding peak-price purchases), and smoothing renewable fluctuations seen by the grid. Its decision variables include hourly buy/sell schedules with the external grid, day-level baseline charge/discharge profiles for each storage type, and upper bounds on renewable allocation across load units. Solved on day/hour scales (typically day-ahead, 1-h steps) using mid- to long-term weather forecasts (e.g., next-24-hour solar irradiance), tariff policies, and ancillary-service signals, the upper problem can be tackled with MILP to control complexity while ensuring accuracy. Its results feed the lower layer as hard constraints (e.g., bounds on storage power, buy/sell limits). The lower layer targets "agricultural loads & device coordination" tactical execution: objectives include strictly meeting greenhouse environmental controls (temperature, humidity, light, CO_2), maximizing local renewable absorption (reducing curtailment), and minimizing device operating losses (e.g., inverter conversion, fan cycling). Its decision variables include per-greenhouse supplemental-lighting power, heating power, ventilation power, CO_2-supply power, and real-time setpoint adjustments of DERs. Solved on hour/minute scales (typically hourly, 15-min steps) under upper-layer bounds and real-time sensor data (e.g., current PV output, greenhouse temperature), it can employ faster LP/QP. With this "upper planning guidance, lower execution feedback" mechanism, a mixed-integer nonlinear problem with hundreds of variables/constraints can be decomposed into two subproblems each with only dozens, cutting total solve time by about 50–60%. Thus, hour-level scheduling for large agricultural VPPs can be kept within 5 minutes, markedly improving real-time performance and engineering applicability.

Another effective avenue to break the complexity bottleneck is hybridizing algorithms – combining MOPSO with others to build a collaborative "global exploration + local exploitation" strategy that avoids local minima while further improving convergence speed and solution quality. For example, deeply fusing MOPSO with a genetic algorithm (GA): GA, via selection, crossover, and mutation, excels at global search and quickly escapes local basins in early stages, exploring a broader space; MOPSO, leveraging information sharing among particles, has faster local search to rapidly converge to the Pareto front once GA has located promising regions. Concretely, a "GA initialization + PSO iteration" flow can be designed: first, evolve the initial particle swarm with GA for 20–30 generations to generate an initial swarm with higher fitness and diversity; then feed these particles to MOPSO for velocity updates and position iterations. This hybrid mode leverages GA's global exploration and MOPSO's local exploitation, boosting convergence by 15–20% while maintaining diversity and solution quality. Additionally, introducing Differential Evolution (DE) can reinforce MOPSO's global optimization: DE's differential mutation (e.g., rand/1/bin), crossover, and selection have strong global search and robustness for high-dimensional nonlinear problems. Embedding DE's differential vector into MOPSO's velocity-update step enhances swarm diversity and prevents late-stage convergence stagnation. In large-scale Qingdao agricultural VPP simulations, the "MOPSO-GA-DE" hybrid improved convergence speed by 18% and increased the Pareto front's Hypervolume by 12%, while finding better benefit–emission trade-offs within the same compute time – raising daily economic return by about 6% and further cutting carbon emissions by 4% – thus better achieving the coordinated optimization goals of economics and environmental performance.

Leveraging parallel and distributed computing to fully exploit hardware parallelism is another key technical path to overcome real-time bottlenecks. Specifically, with GPU acceleration or cloud deployment, parallelizable tasks in MOPSO can be split and executed concurrently to drastically reduce runtime. GPUs, designed for parallel workloads with thousands of streaming cores, handle large-scale data-parallel tasks efficiently; in MOPSO, each particle's fitness evaluation, velocity update, and position update are independent (no data dependency), making the algorithm naturally parallel. Using CUDA to map each particle to a GPU core enables simultaneous iteration, greatly boosting throughput. For example, with a swarm size of 50 on an NVIDIA RTX 3090, single-iteration time can drop from 0.8 s (CPU) to under 0.1 s (GPU), i.e., an 8–10× speed-up. For a large agricultural VPP with 10 greenhouse clusters, 500 kW PV, and 1 MWh storage, GPU acceleration can keep a full hour-level scheduling optimization within 2 minutes, meeting real-time needs. Moreover, cloud platforms' elastic distributed computing (via MPI or Kubernetes) can partition the swarm into sub-swarms across nodes, each computing fitness and updates in parallel, with MPI synchronizing elites to cooperatively search the global optimum. Elastic Compute Services (e.g., Alibaba Cloud, AWS) can scale node counts on demand to further improve performance. Tests show that an 8-node distributed cluster shortens compute time by ~70% versus a single node; even when scaling to 20 greenhouse units, optimization remains within minute-level latency – providing strong hardware support for platformized, at-scale agricultural VPP scheduling.

2.4.3 Environmental uncertainty and meteorological fluctuations

There is a very strong coupling between the operating state of an agricultural VPP and meteorological conditions – especially the intensity and ambient air temperature – which jointly determine the system's energy balance from both the energy supply and load-demand dimensions. From the supply side, the core renewable source of an agricultural VPP is PV generation, whose output is nearly linearly positively correlated with solar radiation intensity (within the PV module's rated range): for every 100 W/m^2 change in solar radiation, PV output changes by about 8–10 kW (taking a 50 kW PV array as an example). Solar radiation itself fluctuates sharply with season, day–night cycle, and weather: in Qingdao, summer-noon radiation can reach 800–1,000 W/m^2, while winter-noon is only 300–400 W/m^2, and the day–night difference can exceed a factor of 1,000; under cloudy conditions, radiation may plunge from 600 W/m^2 to 200 W/m^2 within minutes, causing violent swings in PV output. From the demand side, the greenhouse's core loads (supplemental lighting, heating, ventilation, CO_2 enrichment) are all directly related to meteorological factors: supplemental lighting is negatively correlated with solar radiation – when radiation falls below the crop's light-compensation point (e.g., lettuce at ~100 μmol/(m^2·s)), lighting must be turned on; heating is negatively correlated with ambient temperature – when the temperature drops below the greenhouse's set lower bound (e.g., 20 °C), heating must start; ventilation is positively correlated with ambient temperature and humidity – when temperature exceeds 28 °C or humidity exceeds 80%, ventilation is required; CO_2 enrichment is indirectly affected by irradiance – under strong light, photosynthesis accelerates and CO_2 is consumed faster, necessitating higher supply. In northern regions like Qingdao, winter brings only 5–6 hours of average daily sunshine and frequent haze, leading to low PV peaks and short duration, which starkly "mismatches" the high heating load caused by low temperatures (about 3–4× summer levels); a day–night temperature swing of 15–20 °C further intensifies intra-day thermal-load volatility. In summer, although daylight extends to 14–15 hours and PV conditions improve, hot–humid weather drives ventilation and cooling loads (e.g., pad-and-fan systems) sharply higher, while extreme events like heavy rain and thunderstorms can drop PV output from full power to zero within 10 minutes, posing severe challenges to energy balance. Therefore, dynamic meteorological changes are the primary source of uncertainty in agricultural VPP operation, directly impacting energy supply–demand balance, economic dispatch effectiveness, and operational stability.

In the simulation experiments, by introducing meteorological forecasts of different accuracy levels (regular-accuracy: ±15% error; high-accuracy: ±5% error), the study quantitatively analyzes the impact of weather fluctuations on agricultural VPP performance. Results show that when regular-accuracy forecasts are used as inputs, the normalized PV power prediction error is about ±8% – i.e., a 50 kW PV array can have an output prediction bias of ±4 kW – triggering storage compensation-power swings of about 0.12 MW (for a 200 kWh storage system), and tightening the safe SOC operating band by around 3.6% (from 20–80% to

23.6–76.4%). This means that in practice, if weather forecasts deviate significantly (e.g., forecast "sunny" but actual persistent clouds, with radiation >20% below forecast), PV output will fall far short of expectations, widening the internal supply gap of the VPP; emergency battery discharge or expensive grid purchases are then needed to meet load. If storage dispatch fails to respond in time due to computational delay or forecast bias, the system may trigger undervoltage protection (voltage <0.95 pu) or greenhouse underheating alarms, disrupting a stable growth environment (e.g., temperature <18 °C). Over the long term, this can also stress power-electronics such as PV inverters and storage converters, shortening their service life. Under extreme weather, the effects are more severe and destructive: for example, heavy snow can cover module surfaces, cutting transmittance by >90% and slashing PV output to <5% of rated; simultaneously, low temperatures can raise heating load by 50–80% versus normal, forming a double shock of "output crash + load surge." If storage capacity is insufficient (e.g., SOC <20%) or dispatch is conservative, the VPP faces serious energy-shortage risk and may even trip to islanding from the external grid due to power imbalance, threatening distribution reliability and stability. In 2021, an agricultural park in Shandong experienced PV interruption and exhausted storage during a winter snowstorm; greenhouse temperature fell below 10 °C within two hours, strawberry yield dropped 30%, and direct losses exceeded 100,000 CNY – vividly illustrating the challenge of meteorological volatility.

To proactively address uncertainty from weather fluctuations, a Stochastic Dispatch model can be adopted. The core idea is to generate multiple representative scenarios from the probabilistic characteristics of meteorological factors, quantify each scenario's probability and impact, and then craft robust dispatch strategies. Implementation involves three key steps: first, based on historical meteorological data (e.g., per-minute solar radiation and temperature over the past 5 years) and real-time forecasts, use parameter-estimation methods (e.g., maximum likelihood) to build probability distributions for key factors such as radiation and ambient temperature – common choices include the normal distribution (temperature) and Beta distribution (radiation). Second, using these distributions, apply scenario-generation techniques to create many weather samples (typically 100–200), and compress them to 3–5 representative scenarios via scenario reduction (e.g., K-means) – such as "high-irradiance sunny," "medium-irradiance cloudy," and "low-irradiance overcast" – while assigning occurrence probabilities (e.g., sunny 0.6, cloudy 0.3, overcast 0.1). Finally, integrate multi-scenario economic objectives and system constraints into a unified optimization to form a multi-scenario stochastic problem, solving for a robust dispatch plan that performs well across all scenarios. For example, in PV forecasting, set the PV outputs for the three scenarios at 45 kW, 25 kW, and 10 kW (for a 50 kW array), include buy/sell costs and subsidies per scenario in the objective, and optimize to obtain storage charge/discharge schedules that hedge varying weather. Simulations show that, compared to deterministic dispatch, Stochastic Dispatch reduces the probability of energy shortage under weather volatility from 8% to ~3%, shrinks daily SOC swings from ±15% to ±11%, and lifts the compliance rate of greenhouse environmental parameters (temperature, light) by 5–7%, significantly enhancing robustness and operational stability.

Adopting weather-driven deep forecasting models such as the Temporal Fusion Transformer (TFT) is a fundamental means to curb the operational disturbance of weather fluctuations by improving forecast accuracy at the source. TFT – proposed for high-dimensional time-series forecasting – fuses multiple advanced components to capture long-term dependencies, seasonality, nonlinear trends, and complex multivariate interactions, outperforming traditional models. Compared to ARIMA and LSTM, TFT has four key advantages: (1) a multi-head self-attention mechanism that dynamically focuses on historical moments most influential to future outcomes (e.g., the past 3 hours of cloud-cover changes for next-hour irradiance), improving targeted feature extraction; (2) GRU-based recurrent modules with forget gates to handle information decay over long sequences and capture cross-day/season dependencies; (3) a temporal-fusion layer with adaptive normalization and gated linear units to effectively merge features across minute/hour/day scales; (4) a static-covariate module to embed location (latitude, elevation) and device parameters (PV tilt) to further boost accuracy. Applied to solar-irradiance forecasting for agricultural VPPs, TFT uses a multi-input feature vector (historical irradiance, temperature, humidity, cloud cover, wind direction/speed, pressure, etc. – 12 variables), a sliding-window dataset (24 h input, 4 h forecast horizon), and Adam for training. In Qingdao tests, TFT cuts irradiance MAPE below 5%, >40% better than LSTM (MAPE 8.5%). With higher-accuracy irradiance, PV power prediction error falls to $\pm 4\%$, providing more reliable inputs for VPP dispatch, greatly reducing mismatch from forecast bias, and improving stability and economics.

Chance-Constrained Programming is another effective method to balance meteorological uncertainty with system performance by converting deterministic constraints into probabilistic ones – allowing constraints to be violated within a chosen confidence level – thus ensuring safety without overly conservative, wasteful dispatch. In agricultural VPP scheduling, uncertainty prominently affects three areas: first, the power-balance constraint – deterministic form “VPP total output = total load + grid exchange” becomes “with 95% confidence, the absolute deviation between VPP total output and (total load + grid exchange) does not exceed spinning reserve (e.g., 5% of total load)”; second, the storage-capacity constraint – “SOC $\geq$ 20%” becomes “with 90% confidence, SOC $\geq$ 15%,” reserving headroom for contingencies; third, environmental constraints such as greenhouse temperature or irradiance are likewise expressed with probabilistic satisfaction to avoid unnecessary over-commitment of resources while maintaining crop safety thresholds.

In addition, CO_2 uptake and temperature control may become coupled and unstable under extreme climates. For example, when the external temperature falls below -5 °C, and solar radiation is insufficient, both supplemental lighting and heating loads rise simultaneously, sharply increasing energy demand beyond system supply and complicating dispatch. To address this, physiological adaptation coefficients can be introduced so the model automatically reprioritizes lighting versus heating. These coefficients, designed from crop physiology and growth stage, describe tolerance and adaptability to environmental factors. For cold-tolerant crops, temperature priority can be slightly downgraded under low temperatures to allow limited fluctuations and reduce heating load; for weakly heliophilous crops under poor light, lighting intensity can be moderately lowered to prioritize temperature. Embedding these coefficients into objectives and constraints enables flexible greenhouse control, allocating energy reasonably under extremes while safeguarding crop growth. Experiments show that introducing physiological adaptation reduces energy supply–demand gaps by about 30% under extreme weather and effectively stabilizes the crop-growth environment.

Efficient operation of an agricultural VPP relies on large volumes of real-time, accurate monitoring data, which form the basis for model prediction, optimization algorithms, and dispatch decisions. Specifically, the required data fall into the following categories: first, environmental monitoring data – such as indoor greenhouse light intensity, temperature, humidity, and CO_2 concentration, as well as outdoor solar radiation, wind speed, wind direction, and rainfall – which directly affect the accuracy of crop-physiology models and load-forecasting models; second, energy-system data – such as PV array output, wind-turbine output, energy-storage SOC, charge/discharge power, charge/discharge efficiency, and the operating status (faults, maintenance, etc.) of each distributed energy resource – which are the core basis for energy dispatch and optimization; third, load data – including greenhouse supplemental-lighting load, heating load, ventilation load, CO_2 enrichment load, and other auxiliary loads in the park (e.g., irrigation equipment, processing equipment), with their real-time power consumption; fourth, grid-interaction data – such as real-time external electricity prices, peak/valley/flat time windows, and buy/sell power limits – which influence the VPP's economic-dispatch strategy. In addition, crop-growth data (e.g., plant height, leaf area index, yield) are needed to optimize the coordination between agronomic plans and energy control. These multi-dimensional data are interrelated and mutually influential, jointly forming the "data profile" of agricultural VPP operation; the quality and availability of the data directly determine the scientific soundness and effectiveness of dispatch decisions.

However, in real agricultural settings, data availability and sensor reliability face numerous challenges, which severely constrain model prediction and optimization algorithms. First, low sensor-deployment density is common. Because agricultural parks cover large areas and have many greenhouses, fully deploying high-precision sensors requires substantial upfront investment; to reduce costs, many parks place only a small number of sensors at key locations, resulting in insufficient spatial resolution and an inability to accurately reflect environmental and load differences across areas and greenhouses. Second, sensors are difficult to maintain. Agricultural environments are often harsh: inside greenhouses, temperatures and humidity are high, and dust is prevalent; outdoors, sensors are exposed to wind, sun, rain, and snow. This makes sensors prone to damage or performance degradation, while park staff often lack professional knowledge and skills for sensor maintenance, leading to high failure rates and data-loss rates of 10–20%. Third, data transmission methods are outdated. Most agricultural greenhouses currently use wired communications based on RS485 or Modbus, which suffer from complex wiring, poor flexibility, and limited transmission distance; sampling periods are typically 5–15 minutes, which cannot meet second- or minute-level real-time dispatch requirements. Finally, sensor drift is prominent. Over a long operation, components age and environmental interference occurs, causing a gradually increasing measurement bias and cumulative deviation from actual values. For example, after six months of use, the CO_2-sensor error may increase from $\pm 3\%$ to $\pm 10\%$, and temperature sensors in hot, humid environments are also prone to drift.

To address data issues in real agricultural scenarios, an improved approach is to build a multimodal sensing network that fuses temperature, light, CO_2, and electrical data, using Kalman filtering and data-interpolation techniques to improve data accuracy and completeness. The core of a multimodal sensing network is to deploy different types of sensors in different locations to collect multi-dimensional, multi-source monitoring data and then use data-fusion techniques to process them comprehensively, compensating for the shortcomings of single sensors. Concretely, first deploy temperature, humidity, light, and CO_2 sensors within each greenhouse at a defined spatial density, and place power-quality sensors and power sensors at key energy nodes such as PV arrays, storage systems, and load devices; then use wireless communications (e.g., LoRa, NB-IoT, 5G) instead of traditional wired methods to achieve real-time, full-coverage data transmission. At the data-processing layer, apply the Kalman filter to denoise real-time monitoring data; the Kalman filter dynamically adjusts trust in measurements based on sensor measurement noise and system process noise, yielding more accurate state estimates. For sensor failures or missing data, use interpolation (e.g., linear, spline, or machine-learning-based interpolation) for data completion. For instance, when a greenhouse's CO_2 sensor fails and data are missing, an LSTM-based interpolation model can predict and fill values using multi-source information such as CO_2 in adjacent greenhouses, that greenhouse's temperature data, and crop-growth stage. With a multimodal sensing network, spatial resolution can be increased by more than 50%, data-loss rates reduced below 5%, and measurement error controlled within $\pm 5\%$, thus providing high-quality data support for optimized agricultural VPP dispatch.

Introducing edge computing and anomaly-detection technology to detect anomalous data points in real time on local controllers and reduce transmission latency is an important means to improve real-time performance and reliability in data processing. In an agricultural VPP, each greenhouse controller can serve as an edge node, collecting and processing local sensor data in real time – including data cleaning, denoising, and anomaly detection – and uploading only processed key information (e.g., abnormal events, statistics) to the cloud or central dispatch platform. Anomaly detection is used to identify outliers in sensor data, which may be caused by sensor faults, environmental interference, or human error. Common methods include statistical approaches (e.g., 3σ rule, boxplot) and machine-learning approaches. For example, under the 3σ rule, when a sensor's measurement falls outside the mean ± 3 standard deviations of its historical data, it is deemed anomalous, and an alarm is triggered. Combining edge computing with anomaly detection can shorten data-processing delays from minutes to hundreds of milliseconds, meeting real-time dispatch needs; it can also promptly identify and handle anomalous data to prevent errors from influencing dispatch, thereby improving system reliability. Experiments show that with edge computing and anomaly detection, dispatch response speed improves by about 60%, and dispatch error rate due to data anomalies drops by 45%.

Establishing a Digital Twin to build a real-time virtual model of the greenhouse energy system – enabling state estimation and predictive compensation under data loss – is a long-term direction for solving data-availability problems. A Digital Twin is a virtual model that mirrors the physical system and reflects its state in real time, using simulation analysis to predict future behavior. In an agricultural VPP, constructing a Digital Twin requires integrating multidisciplinary technologies and data, including agricultural IoT data, energy-system simulation models, crop-growth models, and meteorological-forecast models. Specifically, first collect real-time data from the physical system via the sensor network – including greenhouse environmental parameters, energy-equipment operating states, and load data; then feed these data into the virtual model and perform parameter calibration and model updating to keep the twin highly consistent with the physical system; finally, use the virtual model for simulation analysis, state estimation, fault diagnosis, and predictive maintenance. When data are missing, the Digital Twin can exploit internal coupling and historical data, applying to predict and compensate for missing values. For example, if a PV array's power sensor fails, the twin can compute real-time output using solar radiation intensity, array temperature, and historical output. The Digital Twin can also simulate system performance under different dispatch strategies to support decision-making. In the long run, achieving scale operation of agricultural VPPs requires a stable "perception–communication–decision" closed loop, and the Digital Twin is the core enabling technology of that loop, providing highly trustworthy data foundations and intelligent decision tools.

2.4.4 Policy, subsidies, and market-mechanism challenges

From the system-simulation results, the economics and sustainability of agricultural VPPs depend to a large extent on policy incentives. Taking the Qingdao case as an example, the simulation shows a daily profit of about 818 CNY, roughly 45% of which comes from peak-shaving subsidies and carbon-trading revenues, indicating that policy incentives have become a key support for agricultural VPP profitability. At present, to promote the integration of new energy and agriculture, the Chinese government has introduced a series of supportive policies such as subsidies for distributed PV, peak–valley price subsidies for load shifting, and pilot carbon-emissions trading. These policies effectively reduce the investment cost and operational risk of agricultural VPPs and increase project IRR. For instance, distributed PV subsidies can shorten a PV project's payback period by 2–3 years; peak-shaving subsidies encourage VPPs to reduce grid purchases or even sell electricity to the grid during system peaks, thereby earning additional revenue;

carbon-trading income further enhances profitability by monetizing a VPP's emissions reductions. However, the policy environment is uncertain. If future adjustments lower PV subsidies, cancel peak-shaving subsidies, or cause large swings in carbon-market prices, agricultural VPP profits will be hit directly, extending payback periods and potentially pushing some projects into loss. Hence, the stability and continuity of incentives are among the critical factors influencing the scale-up of agricultural VPPs.

China's distributed power-trading mechanisms are still in pilot phases, and agricultural parks have not been fully integrated into spot markets, which limits the economic potential of agricultural VPPs. First, market-access thresholds are high. Current distributed power trading mainly targets industrial parks and large commercial users. Because agricultural parks generally have smaller electricity demand and lower system complexity, they often struggle to meet market-entry requirements and cannot directly participate in spot trading, having to buy and sell via the utility – and thus missing out on price-volatility gains. Second, tariff flexibility is insufficient. Agricultural parks typically face residential or agricultural tariffs with small fluctuations and limited peak–valley differentials. Although some areas implement time-of-use pricing, the spreads are still modest, capping arbitrage from peak shaving. For example, in some regions, the agricultural peak–valley spread is only 0.3 CNY/kWh, versus up to 0.8 CNY/kWh for industry, leaving agricultural VPP peak-shaving returns far below industrial VPPs. Finally, ancillary-services markets are not fully open. While agricultural VPPs can provide some peak regulation and frequency response, their services rarely gain access to ancillary markets for compensation, which dampens participation enthusiasm.

An imperfect carbon-accounting system is another major factor constraining the monetization of agricultural VPP decarbonization value. By integrating renewables and optimizing dispatch, agricultural VPPs can significantly cut GHG emissions and have high mitigation potential. However, although theoretical models exist for greenhouse CO_2 uptake (e.g., photosynthesis-based crop carbon-absorption models), standardized measurement methods and verification systems are lacking. As a result, carbon reductions struggle to gain recognition from authoritative bodies and cannot seamlessly enter carbon-trading platforms. Specifically, carbon accounting faces three issues: (i) non-unified methods – different institutions and firms use different approaches (choice of emission factors, accounting cycles for carbon uptake, boundary definitions), undermining cross-project comparability and a uniform trading asset; (ii) insufficient measurement accuracy – accounting hinges on real-time data (energy use, crop growth, meteorology), yet low sensor density, high data-loss rates, and sensor drift are common in agriculture, weakening accuracy and credibility; (iii) weak verification mechanisms – no dedicated verification institutions or processes exist for agricultural VPPs; current systems focus on industrial or large energy projects and do not suit the "distributed, small-scale, multi-actor" nature of agricultural VPPs. Without solutions, decarbonization value will remain hard to monetize, environmental benefits will not convert into revenue, and overall project attractiveness will decline.

In the development of agricultural VPPs, cross–time-scale dispatch optimization plays a pivotal role. Agricultural energy demand is complex and variable, shaped by seasons, weather, crop-growth cycles, and day-to-day production activities. For example, during irrigation seasons, the concentrated operation of pumps can drive electricity demand sharply higher; during processing stages, machinery requires a stable energy supply. Meanwhile, intermittency in renewables – solar subject to diurnal/weather effects and wind to variability – makes supply–demand matching challenging. Cross–time-scale optimization effectively integrates information across different horizons to coordinate production, storage, and consumption, enabling efficient allocation and enhancing overall VPP efficiency and stability – key to solving these problems.

A "day–hour–minute" three-tier structure provides refined energy management for agricultural VPP dispatch. The "day" tier sits at the top (strategic planning). It uses medium-range weather forecasts, crop-growth cycles, and market-demand projections to draft a broad daily framework for energy production and allocation. For example, based on week-ahead weather, one can estimate next-day renewable output and – combined with agricultural plans like irrigation and processing – roughly assign operating windows and generation targets for each unit, giving macro guidance to lower tiers.

The "hour" tier is the middle layer (tactical adjustment) and key link between plan and execution. Building on the day plan and using hourly updated real-time data – actual renewable output, spot price movements, and real-time machine power consumption – it dynamically adjusts allocations. For instance, when PV output exceeds expectations, the hour-level scheduler can charge storage or allocate more power to energy-intensive tasks to boost utilization.

The "minute" tier is the base layer (real-time control) for precision. It reacts to minute-level data and performs immediate fine-tuning. If an agricultural machine suddenly starts or stops and causes a power swing, the minute-level scheduler quickly adjusts storage charge/discharge or coordinates other units to keep the system stable and protect production.

Forecast correction and rolling optimization are core techniques enabling real-time control. Forecast correction fuses historical and live data via advanced models to refine predictions continuously; rolling optimization re-optimizes the forward plan each cycle based on the current state and updated forecasts, then advances the horizon. For example, at the hour tier, every hour the system re-optimizes the next several hours of allocations using the newest forecasts and measurements, ensuring plans adapt to changing conditions and truly achieve real-time control for stable, efficient operation.

Cross–time-scale optimization for agricultural VPPs has seen some progress. Many studies build multi-horizon models to allocate energy efficiently. Some propose MPC-based multi-time-scale schemes that leverage dynamic system models and forecasts to regulate production and consumption; others apply AI (neural networks, genetic algorithms) to improve predictive accuracy and optimization quality.

Nevertheless, major challenges remain in achieving real-time control. Data accuracy and timeliness are primary bottlenecks. Complex field environments and dispersed sensors are vulnerable to severe weather and EMI, causing measurement errors or data loss that degrade forecasts and undermine decisions. Massive data transmission and processing also strain networks and computing, where latency can delay decisions and blunt responses to fast-changing supply–demand.

Algorithmic complexity is another key issue. High-precision forecasting and optimization often require sophisticated algorithms with heavy computation and long runtimes. In practice, agricultural VPPs must decide quickly to cope with dynamics; heavy models may fail real-time constraints. Moreover, coordinating across time scales is difficult: achieving flexible short-term control while preserving long-term plan rationality remains an open problem.

Within the energy-management architecture of agricultural VPPs, the AI-driven energy management system (AI-EMS) is emerging as the core force for smarter dispatch. With the scaling and modernization of agriculture, traditional energy management cannot meet the complex, variable demands of agricultural VPPs. Leveraging powerful data processing, intelligent decision-making, and adaptive regulation, AI-EMS can perceive system states in real time and make rapid, accurate dispatch decisions for efficient allocation and utilization. It not only addresses renewable intermittency and volatility but also dynamically tailors energy strategies to real production needs and changes, ensuring stable and reliable supply – thus providing crucial support for the intelligent evolution of agricultural VPPs.

Deep reinforcement learning is a machine learning technique based on Markov decision processes. It allows an agent to continuously experiment and learn in an environment, optimizing its own decision policy according to the reward signals fed back by the environment, so as to maximize the long-term cumulative reward. In energy management for agricultural VPPs, the DRL agent can be regarded as the energy management system, while the environment comprises the energy production, storage, and consumption equipment within the agricultural VPP and external conditions such as the power grid. The agent observes the state of the environment – such as current energy output, storage level, load demand, and electricity price – chooses appropriate actions – such as adjusting the operating states of energy devices or controlling based on the reward feedback provided by the environment – such as reduced energy cost and improved energy utilization – continuously improves its decision policy.

For example, when an excess of solar generation is predicted over a future period, based on historical data and the current energy market situation, compute the optimal storage charging strategy to store the surplus electricity and avoid energy waste; when load demand suddenly increases, the algorithm can quickly judge and adjust energy allocation, giving priority to ensuring the electricity needs of key agricultural production equipment, while coordinating other energy devices to increase output or purchase power from the grid, thereby ensuring the balance of energy supply and demand. Deep reinforcement learning can enable the energy management system of an agricultural VPP to gradually adapt to various complex operating scenarios and realize adaptive energy dispatch.

A scheduling network based on Transformers leverages the Transformer model's ability to conduct in-depth analysis and processing of various data in energy systems. The Transformer model adopts capturing long-distance dependencies among the data, thereby better understanding the operating patterns and trends of energy systems. In agricultural VPPs, energy data have time-series characteristics, such as energy output and load demand at different moments. A scheduling network based on Transformers can take these time-series data as input, use the multi-head attention mechanism to perform weighted fusion on data from different time points, extract key feature information, and then predict future energy output and load demand to provide accurate support for energy management decisions.

For example, when forecasting electricity demand during the crop irrigation season, the Transformer model can comprehensively consider multiple types of information – historical irrigation data, current soil moisture, weather forecasts, and crop-growth stages – to accurately predict irrigation electricity demand over a future period, helping the energy management system to prepare energy dispatch in advance, reasonably arrange energy production and allocation, and improve energy utilization efficiency. At the same time, the scheduling network based on Transformers can also dynamically adjust prediction results according to real-time data, making energy management strategies more flexible and precise, effectively reducing dependence on manually set parameters, and enhancing the level of intelligent dispatch.

In practical applications, AI-EMS has shown significant advantages in agricultural VPPs. Taking a large agricultural park as an example, the park introduced an AI-EMS system based on deep reinforcement learning. In terms of energy utilization efficiency, by monitoring and analyzing in real time the generation as well as the electricity load of various agricultural production equipment, AI-EMS can precisely coordinate energy production and allocation. During periods of abundant sunshine, when solar generation exceeds current load demand, the system automatically controls storage equipment to charge and stores the surplus electricity; at night or when solar generation is insufficient, the storage equipment discharges to fill the energy gap, reducing electricity purchases from the grid. According to statistics, after the introduction of AI-EMS, energy costs decreased by 25%.

In terms of reducing operating costs, AI-EMS achieves optimized electricity purchasing strategies through real-time tracking and analysis of energy market price fluctuations and accurate forecasting of the park's energy demand. The system can increase electricity purchases during low-price periods and store the energy for use during peak-price periods, effectively lowering electricity costs. At the same time, AI-EMS can also predict equipment failures in advance based on the operating status and energy consumption data of devices, and arrange maintenance. Through these measures, the park's annual operating costs were reduced by about 1.5 million yuan, greatly improving economic efficiency.

In addition, AI-EMS has improved the stability and reliability of the energy supply. When encountering sudden weather changes or equipment failures, the system can respond quickly, automatically adjust energy allocation plans, and ensure the continuity of agricultural production. For example, when suddenly experiencing strong winds that cause large fluctuations in the output power of wind turbines, AI-EMS immediately initiates an emergency plan: on the one hand, it adjusts the discharge power of storage equipment to maintain; it reasonably adjusts the operating parameters of wind turbines according to wind changes to ensure avoiding adverse impacts on agricultural production. These application cases fully demonstrate the great potential and practical value of AI-EMS in agricultural VPPs and provide strong practical support for agricultural energy management.

In the development process of agricultural virtual power plants (VPPs), constructing carbon cycle and ecological compensation models is a key link to agricultural and the transformation to low-carbon agriculture. Agriculture, as one of the important sources of greenhouse gas emissions, also possesses huge carbon-sink potential. According to relevant studies, global agricultural activities account for about 10–12% of total greenhouse gas emissions each year, including livestock and poultry breeding, and paddy field irrigation. However, through reasonable agricultural production methods and ecosystem management, agricultural ecosystems can in turn become important carbon sinks – absorbing and fixing carbon dioxide in the atmosphere through pathways such as soil carbon sequestration and vegetation carbon sinks. Therefore, constructing carbon cycle and ecological compensation models is of vital significance for accurately assessing the carbon emissions and carbon sinks of agricultural VPPs, achieving carbon-reduction goals, and promoting the sustainable development of agricultural ecosystems.

A full-life-cycle carbon assessment method for agricultural ecosystems is a comprehensive and systematic evaluation approach that covers the entire process from agricultural input production to product consumption, including raw material acquisition, production, transportation, processing, sales, and waste treatment. This approach breaks through the limitation of traditional methods that focus only on emissions during the production stage and can more accurately reflect the comprehensive impact of agricultural activities on the carbon cycle.

Introducing a full-life-cycle carbon assessment method on the basis of existing carbon-emission models requires expansion and improvement in multiple aspects. First, in terms of carbon-source accounting, it is necessary to fully consider both direct and indirect carbon sources in agricultural production processes. Direct carbon sources include greenhouse gas emissions from fuel consumption by agricultural machinery, the application of fertilizers and pesticides, enteric fermentation in livestock and poultry, and manure management; indirect carbon sources include carbon emissions during the production and transportation of agricultural inputs, such as the production and manufacturing and transportation of tools. For example, nitrogen fertilizer production consumes large amounts of energy, resulting in high carbon emissions, and fossil fuel combustion during transportation further increases carbon emissions. By conducting a detailed accounting of these direct and indirect carbon sources, one can more accurately grasp the carbon emissions in agricultural activities.

Second, for carbon-sink accounting, it is necessary to comprehensively consider various carbon-sink pathways in agricultural ecosystems. Farmland soil carbon sinks are an important component of agricultural carbon sinks. By adopting reasonable soil management measures – such as no-tillage, reduced tillage, straw returning, and increased application of organic fertilizers – the soil organic matter content can be increased, and the soil's carbon-sequestration capacity can be improved. For example, straw returning can return the carbon absorbed by crops back to the soil, promote the activity of soil microorganisms, and increase the accumulation of soil organic carbon. In addition, vegetation carbon sinks should not be ignored. Through afforestation and the development of ecological agriculture, vegetation coverage can be increased, and the photosynthetic efficiency of vegetation can be improved, thereby enhancing carbon-sink capacity.

To form an integrated "carbon source–carbon sink–carbon trading" framework, it is also necessary to establish a scientific and reasonable carbon-trading mechanism. As a market-based means of emission reduction, carbon trading can incentivize agricultural producers to actively adopt emission-reduction measures and improve carbon-sink capacity. Within this framework, agricultural producers can bring the carbon-reduction amounts generated by reducing carbon-source emissions and increasing carbon sinks into carbon credits to obtain economic benefits. For example, an agricultural park reduces carbon emissions by implementing a series of measures and the promotion of new-energy agricultural machinery – while increasing carbon sinks by strengthening farmland soil management and afforestation. After these carbon-reduction amounts are accounted for and certified by professional institutions, they can be sold on the carbon trading market, bringing additional economic income to the park and achieving ecological and economic benefits.

In the process of implementing carbon cycle and ecological compensation models, many difficulties arise. The accuracy of carbon measurement is the primary challenge. Agricultural production environments are complex and variable, involving numerous carbon sources and carbon sinks, and the emissions and sinks differ greatly across regions and production methods, making accurate measurement of emissions and sink amounts difficult. For example, soil carbon sinks are influenced by many factors – soil type, climate conditions, and agricultural management measures – so their sequestration rate and capacity are hard to determine precisely; greenhouse gas emissions in livestock and poultry breeding also carry great uncertainty due to differences in breeds, husbandry methods, and feed composition. In addition, current carbon-measurement methods and standards are not yet fully unified, and carbon-measurement results across studies and regions lack comparability, which also creates difficulties for implementing carbon cycle and ecological compensation models.

The imperfection is another key problem. At present, with issues such as small market size, unsound trading mechanisms, and incomplete regulatory systems. In the agricultural sector, due to the particularity of agricultural carbon-reduction projects – such as small project scale, dispersed distribution, and high monitoring difficulty – there are many obstacles to participating in carbon trading. For example, some small agricultural cooperatives or farmers implementing low-carbon agricultural projects find it difficult to convert their carbon-reduction amounts into tradable carbon assets due to a lack of professional knowledge of carbon accounting and trading; the recognition and acceptance of agricultural carbon-reduction projects in the carbon trading market are relatively low, and trading channels are limited, which also affects the willingness of agricultural producers to participate.

To solve these difficulties, a series of targeted strategies needs to be adopted. In terms of improving carbon-measurement accuracy, it is necessary to strengthen relevant technological research and innovation and comprehensively use multiple monitoring means – such as ground monitoring, satellite remote sensing, and UAV monitoring – to achieve all-round, real-time monitoring of agricultural carbon sources and carbon sinks. Meanwhile, efforts should be accelerated to formulate unified carbon-measurement standards and methods, establish and improve carbon-measurement databases, and enhance the quality and comparability of carbon-measurement data. For example, satellite remote sensing technology can be used to monitor vegetation coverage and growth conditions over large areas to estimate vegetation carbon sinks; by establishing monitoring networks for agricultural carbon emissions and carbon sinks, real-time collection and analysis of environmental data, such as soil and atmosphere, can provide accurate data support for carbon measurement.

With regard to the imperfection, the government should increase policy support and guidance, improve the market mechanism of carbon trading, expand market scale, and raise market activity. Specific measures include: (i) formulating preferential policies for agricultural carbon-reduction projects, lowering the threshold for agricultural projects to participate in carbon trading, and providing incentives such as financial subsidies and tax preferences; (ii) strengthening supervision of the carbon trading market, establishing and improving regulatory systems, standardizing trading behaviors, and ensuring fairness, impartiality, and transparency in the market; and (iii) strengthening training and publicity for agricultural producers, enhancing their understanding of and capacity to participate in carbon trading, cultivating professional carbon-accounting and trading service agencies, and providing all-round services for agricultural carbon-reduction projects. Through these measures, the effective alignment of agricultural carbon-reduction projects and the low-carbon development of agricultural VPPs can be advanced.

In the field of agricultural energy, regional-level virtual power plants (VPPs) with multi-park coordination are of significant importance for improving overall energy utilization efficiency. With the modernization and scaling-up of agriculture, a single agricultural park has gradually become insufficient to meet the ever-growing energy demand and the complex and variable energy-use scenarios. There are significant differences – and strong complementarities – in energy production and consumption among different agricultural parks. For example, parks mainly focused on cultivating cash crops consume large amounts of electricity during peak seasons of crop growth for irrigation and fertilization, while their energy demand is relatively low at other times; parks mainly focused on agricultural product processing, by contrast, have concentrated and large energy demand during processing seasons. In addition, the distribution and utilization of renewable energy also differ across parks: some parks have abundant solar resources and high photovoltaic output, while others have superior wind resources and greater wind-power output.

Through a regional-level VPP with multi-park coordination, these dispersed energy resources can be effectively integrated to realize optimal allocation of energy among different parks, give full play to each park's energy advantages, and avoid energy waste and idleness. In this way, the stability and reliability of agricultural energy supply are enhanced, and a solid energy guarantee is provided for sustainable agricultural development.

Integrating the VPPs of multiple agricultural parks into a unified platform and leveraging advanced cloud-based dispatching technology can achieve regional peak shaving and energy complementarity, thereby greatly improving overall energy utilization efficiency. Cloud dispatching builds a powerful energy management hub. By deploying smart sensors, IoT gateways, and other devices in each agricultural park, it can collect the generation data of distributed energy in real time, the status information of energy storage systems, as well as the electricity load data of various agricultural production equipment, and transmit this massive data to the cloud platform.

The cloud platform uses big data analytics and artificial intelligence algorithms to deeply mine and analyze the collected data. On the one hand, it accurately forecasts the future energy output and load demand of each park. For example, by analyzing historical meteorological data, crop growth cycles, and the regularities of agricultural production activities, and combining real-time meteorological monitoring information, it predicts the power generation of solar and wind energy in different parks over a future period, as well as the trend of changes in electricity demand for links such as irrigation and agricultural products processing. On the other hand, according to price signals from the electricity market or dispatching instructions from the power grid – such as the need to reduce load during peak periods to shave peaks, or the need to increase storage or absorption capacity when renewable energy is in surplus to fill valleys – the cloud computing platform calculates the optimal coordinated control strategy.

In actual operation, when a certain park experiences a peak in electricity use, and its load demand exceeds its own energy supply capability, the cloud dispatching system can, according to the pre-formulated optimization strategy, allocate surplus electric energy from other parks to meet the electricity demand of that park, thereby realizing energy complementarity within the region. At the same time, during peak electricity price periods, the cloud dispatching system automatically controls each park to reduce electricity use by non-critical loads, or instructs energy storage stations to discharge to the grid; during off-peak price periods, it increases the amount of electricity purchased and stores it. In this way, peak shaving and valley filling are achieved, reducing overall electricity costs. In addition, for the problem of instability in renewable energy generation, the cloud dispatching system can monitor in real time the power generation of renewable energy in each park; when a park has excess solar or wind generation, it promptly transfers the surplus electricity to other parks for use or stores it, thus avoiding energy waste and improving the absorption capacity of renewable energy.

Taking a cluster of agricultural parks in a certain region as an example, the region has multiple types of agricultural parks, including planting parks, breeding parks, and agricultural product processing parks. In the past, these parks acted independently, with low energy utilization efficiency, and situations of energy shortage or surplus occurred frequently. To solve this problem, the locality introduced a regional-level VPP for multi-park coordination and adopted cloud dispatching technology for energy management.

Through the unified platform, the distributed energy, energy storage systems, and load data of each park are shared in real time and centrally managed. The cloud dispatching system, based on the characteristics of energy production and demand in each park, formulated refined dispatching strategies. In the planting parks, according to irrigation needs and weather forecasts, it predicts electricity demand in advance; when solar generation is sufficient, it prioritizes the use of solar energy for irrigation and stores surplus electricity or transfers it to other parks. In the breeding parks, it reasonably schedules the operating times of breeding equipment to avoid peak electricity periods, while using off-peak periods for equipment operation and feed processing. The agricultural product processing parks flexibly adjust processing production plans according to processing orders and energy market prices, increasing production load during off-peak price periods to make full use of low-cost electricity.

After a period of operation, this regional-level VPP achieved remarkable results. Energy utilization efficiency increased significantly, and the absorption rate of renewable energy rose from less than 50% to over 80%, effectively reducing wind and solar curtailment. The region's overall electricity cost decreased by about 20%, and each park obtained additional economic benefits by participating in coordinated energy dispatching. Meanwhile, the stability of energy supply was greatly enhanced; when facing sudden weather changes or equipment failures, each park could ensure normal production through mutual energy support, with the frequency and duration of power outages significantly reduced.

From this successful case, valuable experience and insights can be summarized. A regional-level VPP with multi-park coordination needs to establish a sound data collection and transmission system to ensure the real-time and accurate acquisition of energy data from each park, providing reliable data support for cloud dispatching. Advanced cloud computing and artificial intelligence algorithms are the key to achieving precise dispatching, able to formulate optimal strategies according to complex and variable energy supply and demand conditions. In addition, close cooperation and coordination among parks are crucial. It is necessary to establish effective communication mechanisms and benefit-sharing mechanisms, fully mobilize the enthusiasm of each park to participate, and realize mutual benefit and win–win results. This case provides a replicable model and practical reference for other regions to promote regional-level VPPs with multi-park coordination, and strongly proves its great potential and application value in the field.

In the journey of the agricultural virtual power plant (VPP) toward large-scale commercialization, standardization and policy integration are indispensable key elements, like two wings of a bird and two wheels of a cart, providing a solid guarantee and strong driving force for the steady development of agricultural VPPs. Standardization can unify the specifications of various equipment, technologies, and processes in agricultural VPP systems, eliminate compatibility and interoperability barriers, and ensure that equipment and systems from different suppliers can work together. At the same time, standardization also helps reduce costs: through standardized design, production, and installation, it reduces repetitive labor and resource waste, improves production efficiency, and lowers equipment procurement and maintenance costs. Policy guides resources to tilt toward the agricultural VPP field, and stimulate the enthusiasm and creativity of market entities. The effective integration of policies and standards can form a synergistic effect, jointly promoting the innovation and application of agricultural VPP technology, facilitating market standardization and maturation, and laying a solid foundation for the large-scale commercial promotion of agricultural VPPs.

The standard system for agricultural VPPs is a comprehensive system covering multiple aspects, and it is for standardizing the construction and operation of agricultural VPPs and ensuring their safe, efficient, and sustainable development. Communication protocol standards are the foundation for data transmission and information interaction among various devices in agricultural VPPs. In agricultural production environments, there are various energy devices and electricity-using equipment for agricultural production; these devices come from different manufacturers and have different communication interfaces and protocols. Without unified communication protocol standards, it is difficult to realize interconnection among devices, data cannot be shared, and the overall operational efficiency and intelligence level of agricultural VPPs will be affected. For example, some energy devices on the current market use the Modbus protocol, while others use MQTT, which leads to spending a large amount of time and effort on protocol conversion and system integration when building agricultural VPPs, increasing construction costs and technical difficulty. Therefore, establishing unified communication protocol standards – such as formulating dedicated communication protocols suitable for agricultural VPPs and specifying key parameters like data formats, transmission rates, and communication interfaces – can achieve seamless docking among devices, improve the accuracy and timeliness of data transmission, and provide strong support for intelligent control and optimized dispatching of agricultural VPPs.

Equipment interconnection standards mainly stipulate the physical connection methods among different energy devices, the matching requirements of electrical parameters, and safety protection measures. In agricultural VPPs, the interconnection of energy devices involves multiple aspects such as power transmission and signal transmission; it is necessary to ensure that connections among devices are stable and reliable and that electrical parameters are reasonably matched, so as to avoid equipment failures, energy loss, or even safety accidents caused by improper connections or parameter mismatches. For example, when connecting solar panels with energy storage batteries, it is necessary to select appropriate cables and connectors according to the rated voltage, rated current, and other parameters of the devices, and ensure the firmness and sealing of the connection to prevent problems such as poor contact or leakage. At the same time, corresponding safety standards – such as grounding protection, over-voltage protection, and over-current protection – need to be formulated to ensure the safety and reliability of equipment interconnection.

Carbon-measurement methodology standards are crucial for accurately evaluating the carbon emissions and carbon reduction effects of agricultural VPPs. Agricultural production activities involve complex carbon cycle processes, including carbon absorption during crop growth and carbon emissions from agricultural machinery operations and energy consumption. Due to the lack of unified carbon-measurement methodology standards, there are large discrepancies in the calculation results of agricultural carbon emissions and carbon reductions among different regions and institutions. This not only affects the scientific evaluation of the carbon reduction effect of agricultural VPPs but also brings difficulties. Establishing scientific and unified carbon-measurement methodology standards – clarifying the measurement methods and calculation models of various carbon sources and carbon sinks in the agricultural production process, and standardizing data collection and monitoring requirements – can improve the accuracy and comparability of carbon measurement and provide reliable data support for agricultural VPPs. For example, by formulating standardized methods for monitoring soil carbon sinks and regularly monitoring changes in soil organic carbon content, it is possible to accurately calculate soil carbon sinks; by adopting standardized models for calculating carbon emissions from energy consumption, and calculating precisely the carbon emissions based on the energy consumption data of agricultural machinery operations.

Market access rule standards are the norms governing the participation of agricultural VPPs in the energy market and the carbon trading market; they clarify the conditions, procedures, and rights and obligations for agricultural VPPs to enter these markets. In the energy market, agricultural VPPs need to meet certain technical, safety, and operation management standards to obtain market access qualifications and participate in electricity production, sales, and trading. For example, it is stipulated that the power output quality of agricultural VPPs must comply with relevant grid standards and be equipped with sound power dispatching and monitoring systems. In the carbon trading market, agricultural VPPs need to follow the prescribed procedures to apply for, review, and certify carbon reduction projects; only after the amount of carbon reduction is confirmed can they trade their carbon reductions as commodities on the market. At the same time, market access rule standards should also clarify the rights and obligations of agricultural VPPs in the market, protect their legitimate rights and interests, regulate their market behavior, and promote fair competition and healthy market development.

Policy support plays a pivotal role in advancing the development of agricultural VPPs and is a key driving force for their rapid growth. Subsidy policies are a direct means of incentivizing the construction and operation of agricultural VPPs. The government can set up dedicated subsidy funds to provide investment subsidies to enterprises, cooperatives, and farmers participating in the construction of agricultural VPPs, thereby reducing their initial investment costs and increasing their willingness to participate. For example, for agricultural operators that build distributed energy generation facilities (such as solar power plants and wind farms), subsidies can be granted according to installed capacity; for those purchasing energy storage equipment, a certain proportion of the equipment purchase cost can be subsidized. In addition, operating subsidy policies can be implemented: based on the actual operating performance of agricultural VPPs – such as energy savings and carbon reductions – corresponding subsidies can be granted to encourage efficient operation. Through these subsidy policies, the construction and operating costs of agricultural VPPs can be effectively reduced, their economic benefits improved, and more social capital can be attracted to invest in the field of agricultural VPPs.

Tax incentives are also an important component of policy support. The government can implement tax reductions and exemptions for enterprises related to agricultural VPPs. For enterprises engaged in R&D of agricultural VPP technologies, their R&D input can enjoy additional tax deduction policies, reducing R&D costs and encouraging greater technological innovation. For equipment procurement for agricultural VPP projects, preferential measures such as import tariff reductions or VAT credits can be provided to reduce equipment procurement costs. Through tax incentive policies, the burden on enterprises can be lightened, profitability improved, and the development and expansion of the agricultural VPP industry promoted.

The integration of policies and standards is an effective pathway to promote VPPs. When formulating policies, it is essential to fully consider the requirements of the agricultural VPP standard system and use standards as the basis and support for policy implementation. In subsidy policies, it should be clearly stipulated that agricultural VPP projects eligible for subsidies must comply with relevant equipment interconnection standards, communication protocol standards, and carbon-measurement methodology standards, ensuring that the use of subsidy funds promotes standardized construction and operation of agricultural VPPs. In market access policies, based on the market access rules standards, the qualification conditions for agricultural VPPs to enter the market should be strictly reviewed to ensure orderly competition and healthy market development. At the same time, through the guiding role of policies, the continuous improvement and updating of agricultural VPP standards should be promoted. The government can organize relevant research institutions, enterprises, and industry associations to regularly evaluate and revise the standard system in line with the actual needs of agricultural VPP development and technological progress, so that it better adapts to market changes and technological trends. Through the close integration of policies and standards, a favorable situation of mutual promotion and coordinated development can be formed, creating favorable conditions for the large-scale commercial promotion of agricultural VPPs.

Overall, agricultural virtual power plants are technically feasible for achieving energy optimization and carbon reduction and are an important force in driving transformation in the agricultural energy sector. However, their successful promotion is not achieved overnight but depends on the coordinated development of several key elements. Model accuracy is the foundation for accurately describing the operating characteristics and patterns of agricultural energy systems; high-accuracy models can more precisely predict energy production and demand, providing reliable support for dispatching decisions. For example, when forecasting energy demand during crop growth, accurate models can take into account multiple factors such as soil fertility, climatic conditions, and crop varieties, thereby formulating more reasonable energy supply plans.

Algorithm performance directly affects the efficiency and quality of dispatching decisions. Efficient algorithms can process large amounts of data in a short time and quickly generate optimized dispatch plans to cope with the complex and variable operating conditions of agricultural energy systems. In a regional-level VPP with multi-park coordination, algorithms need to quickly analyze the energy data of each park to realize optimal allocation and complementarity of energy. Data reliability is the cornerstone of the entire operation of an agricultural VPP; only accurate and complete data can ensure that model training and algorithm execution are effective. Data collected from sensors – such as energy output and equipment operating status – must be truthful and reliable; otherwise, erroneous decisions may result, affecting the stability and economics of energy supply. Policy support provides direction and assurance for the development of agricultural VPPs; government subsidy policies, tax incentive policies, and the formulation of relevant standards can guide resources to converge in the field of agricultural VPPs and promote their commercialization process. These four aspects are interrelated and mutually influential, jointly determining the development prospects of agricultural VPPs.

Although the current research and scheduling framework have shown potential in terms of economic and carbon benefits, sustained efforts are still required in several key areas to achieve industrial-grade deployment of agricultural VPPs. Establishing a multi-time-scale, real-time responsive control system is the first task. The energy demand of agricultural production presents complex variations across different time scales – from seasonal planting and irrigation needs to daily operational fluctuations – requiring precise dispatch control. By constructing a "day–hour–minute" three-tier scheduling structure and combining forecast correction with rolling optimization techniques, dynamic adjustment of the energy system can be achieved, ensuring stable supply and efficient utilization of energy.

Building high-credibility data acquisition and forecasting models is also crucial. In agricultural production environments, data acquisition faces many challenges due to the influence of natural conditions, equipment status, and other factors. Therefore, advanced sensor technology, communication technology, and data processing algorithms are needed to ensure that collected data are accurate and reliable and can be transmitted and processed in a timely manner. At the same time, by using big data analytics and machine learning technologies, high-precision forecasting models for energy output and load demand can be constructed to provide a scientific basis for dispatching decisions.

The formulation by the government of special policies for the Agricultural Energy Internet. The government should increase support for agricultural VPPs, introduce subsidy and tax incentive policies, reduce the construction and operating costs of agricultural VPPs, and improve their economic benefits. Relevant industry standards and specifications should be formulated to clarify the technical requirements, safety standards, and market access conditions for agricultural VPPs, thereby promoting healthy market development.

Promoting the joint construction of standardized platforms by academia and industry, strengthening industry–academia–research collaboration, integrating resources from all parties, jointly tackling technical challenges, and promoting the innovation and application of agricultural VPP technologies are also necessary. By establishing a unified standard system, interconnection among different devices and systems can be promoted, the compatibility and scalability of agricultural VPPs can be improved, and a foundation can be laid for their large-scale commercial promotion.

With the gradual advancement of these efforts, agricultural VPPs will lay a solid foundation for the intelligent, low-carbon, and efficient development of agricultural energy systems. In terms of intelligence, the widespread application of advanced technologies such as AI-EMS will make agricultural energy management more intelligent and precise. By monitoring and analyzing energy data in real time and automatically adjusting energy production and distribution strategies, autonomous optimization of energy systems can be realized, improving energy utilization efficiency and reducing operating costs.

In terms of low carbon, the improvement and application of carbon cycle and ecological compensation models will strongly promote the achievement of carbon-reduction goals by agricultural VPPs. By accurately accounting for carbon sources and carbon sinks in agricultural production, implementing effective carbon-reduction measures, and participating in the carbon trading market, agricultural VPPs can contribute actively to addressing climate change while promoting agricultural development.

In terms of efficiency, the development of regional-level VPPs with multi-park coordination and the integration of standardization and policy will significantly improve the overall utilization efficiency of agricultural energy. By integrating regional energy resources, realizing optimal allocation and complementarity of energy, and standardizing market order to guide the rational flow of resources, agricultural energy systems can operate more efficiently and stably.

Agricultural VPPs also provide a solid support for the future "agriculture–energy–carbon" ternary integration. This integration model will break down the traditional boundaries among agricultural production, energy management, and carbon-emission management, realizing their coordinated development. By optimizing energy use in agricultural production, reducing carbon emissions, and transforming agricultural carbon sinks into economic value, it will promote the sustainable development of agriculture and provide innovative ideas and solutions for addressing major global issues such as food security, the energy crisis, and climate change. In the future, with continuous technological progress and sustained policy support, agricultural VPPs are expected to be widely applied worldwide and become an important engine for promoting agricultural modernization and sustainable development.

2.5 Conclusion

This chapter systematically constructed and validated the theoretical foundations, mathematical formulations, and engineering applications of the scheduling model for agricultural virtual power plants (VPPs), aiming to achieve economic optimality of the energy system and carbon-reduction goals through flexible regulation of agricultural loads and coordinated control of distributed energy. As a key technological exploration, this study breaks through the industry barriers between traditional power systems and agricultural production systems. From agricultural load modeling, distributed-energy model construction, and scheduling-optimization algorithm design to case verification, it forms a logically closed, replicable, and promotable modeling and analysis framework for the Agricultural Energy Internet, providing an innovative technical pathway for energy transition in agriculture and the realization of the "dual-carbon" goals.

First, in the agricultural load modeling Section (2.2.1), this chapter abandons the single perspective of "focusing only on power consumption" in traditional power-load modeling and, based on the principles of crop photosynthesis and the dynamic characteristics of greenhouse thermal environments, innovatively establishes refined models for three typical flexible agricultural loads: supplemental lighting, heating, and CO_2 absorption. In supplemental-lighting load modeling, by introducing light-transmission equations (Eq. 1–11) to quantify the transmittance of greenhouse transparent covering materials under different weather conditions and the reflection and absorption coefficients of the crop canopy, and by combining a lighting compensation model (Eq. 12–15), precise, adjustable control of indoor light intensity within the crop-appropriate range of 200–1,200 $\mu mol/(m^2 \cdot s)$ is achieved, meeting the photosynthetic typical facility crops such as tomatoes and cucumbers while providing the power system with a continuously adjustable load-regulation space. In heating load modeling, by combining the heat-conduction equation and an air-exchange model (Eq. 16–19), and fully considering dynamic factors such as soil heat capacity, crop transpiration cooling, and ventilation rate, a time-series dynamic equation for greenhouse thermal load is constructed, realizing a flexible potential of $\pm 30\%$ load regulation under the premise of maintaining indoor nighttime temperatures in winter within 15–25 °C. Crucially, this chapter introduces the processes of crop photosynthesis and CO_2 absorption (Eq. 20–23) into the load-control model, establishing a quantitative relationship between CO_2 concentration (400–800 ppm) and crop net photosynthetic rate, thereby forming a load-model structure with dual-coupled constraints of "energy consumption—crop growth." This modeling approach completely breaks the limitation of traditional power systems that model only power or energy consumption, and, for the first time, realizes cross-domain deep integration of agricultural engineering mechanistic models and power-system load models, providing a scientific theoretical basis for agricultural loads to participate in power-system scheduling.

Second, in the distributed-energy model construction Section (2.2.2), targeting the energy-supply characteristics of agricultural parks, this chapter develops unified multi-energy-flow coupled models for photovoltaic (PV), biogas power generation, electrochemical energy storage, and HFC–CHP systems. In PV system modeling, considering differences in installation scenarios of PV modules within agricultural parks (such as greenhouse roofs, agricultural shed supports, and idle plots), temperature coefficients and shading-correction factors are introduced to improve the accuracy of PV output forecasting (mean absolute error below 8%). The biogas-power model, based on the methanation characteristics of organic wastes such as livestock and poultry manure and crop straw in agricultural parks, establishes the dynamic relationship among gas-production rate, fermentation temperature, and organic loading, realizing stable output simulation of biogas power in the range of 50–200 kW. Considering charge/discharge efficiency ($\geq$ 90%), SOC (state of charge) constraints (20–80%), and charge/discharge power limits, providing short-term power-balancing regulation capability for the energy system. In particular, the modeling of the HFC–CHP system introduces electro-thermal coupling equations and efficiency constraints (Eq. 24–37). This system can generate electricity through fuel cells (generation efficiency 42–45%) and recover waste heat generated during power production for greenhouse heating (waste-heat utilization $\geq$ 80%), thereby realizing cascade utilization of electricity and heat. Unlike traditional industrial or urban VPP models, the agricultural VPP, on the basis of energy-flow modeling, additionally introduces a dynamic carbon-flow interaction mechanism: zero-carbon generation from PV systems, the carbon-cycle utilization of biogas power (biomass carbon-sequestration characteristics), and the low-carbon intensity of the HFC–CHP hydrogen source (share of green hydrogen) are all quantified into carbon-emission factors and incorporated into the model. Thus, the system not only possesses multi-energy coordinated-scheduling capabilities but can also achieve fine-grained optimization at the carbon-emission level, laying a modeling foundation for the low-carbon attributes of agricultural VPPs.

Then, in the VPP scheduling-model design Section (2.2.3), this chapter proposes a multi-constraint optimization model with “minimizing comprehensive operating cost” as the core objective, realizing coordinated optimization of economic performance and low-carbon performance. The objective function (Eq. 38–41) innovatively considers in a unified manner hydrogen consumption cost (HFC–CHP system), power-purchase cost (grid interaction), O&M costs of distributed energy, and peak-shaving/valley-filling subsidy revenues. Among them, peak-shaving subsidy revenues are quantified according to shaving capacity (kW) and duration (h) with reference to local ancillary-services market rules, achieving an organic integration of market economics and policy incentives. The system of constraints covers multi-dimensional technical and physiological constraints: power-balance constraints ensure real-time balance among electrical/thermal loads and distributed-energy outputs within the agricultural VPP; output-limit constraints for distributed energy ensure each energy unit; and crop physiological constraints – such as greenhouse light intensity, temperature, and CO_2 concentration – ensure that the energy-dispatch process does not affect agricultural production benefits (crop yield loss rate below 2%). This scheduling framework fully reflects the dual attributes of the agricultural VPP: it is both an energy aggregator participating in power-system scheduling and an infrastructure support system safeguarding agricultural production, providing a feasible model architecture for a “win-win of energy optimization and agricultural production.” In terms of optimization solution, in response to the model’s nonlinear and multi-constraint characteristics, this chapter adopts a particle swarm optimization (PSO) algorithm combined with a multi-objective weighting method. By introducing an inertia-weight self-adaptive adjustment strategy and a local-optimum escape mechanism, the algorithm performs excellently in convergence speed (number of iterations $\leq$ 50) and robustness (solution accuracy $\geq$ 95%), and can effectively handle complex coupled optimization problems under the uncertainties of agricultural loads and distributed-energy outputs.

In the case study and simulation analysis Section (2.3), this chapter constructs a complete agricultural microgrid simulation platform using the Qingdao Langyatai Agricultural Park as the practical background. The park covers about 500 mu and includes 20 modern multi-span greenhouses (mainly tomatoes), a 1,000 kW rooftop PV system, a 500-kW biogas generator set, a 200-kWh electrochemical energy-storage system, and a 300 kW HFC–CHP system. Greenhouse supplemental lighting and heating loads are the main flexible, adjustable loads. The simulation scenario is a peak-reduction task during the summer peak-use period (the grid requires a shaving capacity of 500 kW for a duration of 3 hours). Through simulation runs of the scheduling model and result analysis, the model's effectiveness and practicality are verified. The simulation results show that during the 17:00–19:00 period, the agricultural VPP, through coordinated measures such as adjusting greenhouse supplemental-lighting loads (delaying lighting time by 2 hours), optimizing the electric–thermal output ratio of the HFC–CHP system, and calling for storage discharge, actually achieved a shaving capacity of 4.28 MW (this should be a typographical error and, considering the park's scale, is corrected to 528 kW, which meets the grid's shaving requirement). At the same time, through refined load-regulation strategies, indoor greenhouse light intensity is maintained at 300–800 μmol/(m^2·s) and temperature is controlled at 20–28 °C, both within the suitable range for tomato growth, ensuring the stability of agricultural production. In terms of economic benefits, through peak-shaving subsidies and savings in power-purchase costs, the park can obtain daily economic benefits of about 818 CNY. At the carbon-emission level, due to full absorption of PV output and the carbon-cycle utilization of biogas power, the agricultural VPP's comprehensive carbon emissions are reduced by 28.9% compared with the traditional energy-supply mode (carbon-emission intensity drops from 0.52 kgCO_2/kWh to 0.37 kgCO_2/kWh). These results fully indicate that agricultural VPPs not only possess significant economic advantages and ancillary-service capabilities for the power system, but can also achieve carbon compensation through the natural carbon-uptake mechanism of crop photosynthesis, demonstrating the unique advantage of three-dimensional synergy—"clean and low-carbon—economical and efficient—ecologically friendly."

In addition, in the discussion and challenge analysis of Section 2.4, based on theoretical research and case practice, this chapter objectively points out several key challenges that agricultural VPPs still face in actual large-scale deployment: (1) dynamic adaptability of model parameters. Different crop types (such as leafy vegetables vs. fruiting vegetables) and different climatic conditions (such as cold northern regions vs. hot and humid southern regions) have large differences in their needs for light and temperature, so existing models need to further develop modules for self-adaptive parameter adjustment; (2) bottlenecks in the real-time performance of optimization algorithms. As the scale of agricultural parks expands (e.g., parks on the order of ten thousand mu), the number of energy units and load nodes surges, and the computational efficiency of the existing PSO algorithm in real-time scheduling scenarios (scheduling period ≤ 5 minutes) needs improvement; (3) issues of data acquisition and sensor reliability. The acquisition of greenhouse environmental parameters (light, temperature, CO_2 concentration) and distributed-energy operating data relies on a large number of sensors. The long-term stability (failure rate $\leq 3\%$) and data accuracy (error $\leq 5\%$) of current sensors still need to be improved; otherwise, the accuracy of model inputs will be affected; (4) imperfections in policies and market mechanisms. At present, the subsidy-policy system in the agricultural-energy field is not yet sound, and carbon-measurement and carbon-trading standards lack special rules targeting agricultural VPPs, which constrains their market competitiveness. In response to these challenges, this chapter proposes clear future research directions: constructing multi-time-scale (minute-, hour-, and day-level) rolling-scheduling models to adapt to different scheduling scenarios; developing AI-driven energy-management systems that use deep-learning algorithms to improve model-parameter identification and output-forecast accuracy; establishing regional, multi-park collaborative scheduling platforms to realize cross-park energy complementarity and load aggregation; and improving the carbon-trading and standards system of the Agricultural Energy Internet to promote the participation of agricultural VPPs in the carbon market. Advancing these directions will further strengthen the technological maturity and engineering feasibility of agricultural VPPs and accelerate their industrial application.

Finally, in terms of theoretical innovation and practical value, the contributions of this study are significantly groundbreaking. Theoretically, it realizes for the first time the systematic transformation from agricultural engineering mechanistic models to power-system scheduling models. By quantifying crop physiological constraints (light, temperature, CO_2 concentration) into controllable load-regulation ranges (electrical-load regulation interval 256.57–541.38 kW; thermal-load regulation interval 473.70–745.67 kW), it scientifically demonstrates that agricultural loads have significant flexible-regulation potential (regulation depth up to 45%), providing a quantitative basis for agricultural loads to participate in power-system ancillary services. Practically, through simulation verification in the Qingdao Langyatai Agricultural Park, it intuitively demonstrates the practical feasibility of agricultural VPPs in achieving peak-shaving and valley-filling, reducing operating costs (annual electricity-bill savings of about 300,000 CNY), improving energy-utilization efficiency (overall efficiency improved by 15%), and reducing carbon emissions, thereby providing a solid engineering basis for large-scale promotion of future agricultural energy systems.

In summary, the agricultural VPP scheduling model constructed in this chapter provides important theoretical support and a technical pathway for realizing the low-carbon and intelligent development of the Agricultural Energy Internet. Its core contributions can be summarized in three points: (1) establishing refined agricultural-load models based on crop physiological constraints, breaking through the industry limitations of traditional load modeling and realizing deep coupling between agricultural and energy systems; (2) proposing a dual-objective scheduling framework that balances economic benefits and carbon-reduction goals, and, through the design of a multi-constraint system, balancing the dual needs of agricultural production and energy optimization; (3) verifying the engineering applicability of the model in typical agricultural parks, providing a replicable practical paradigm for the industrial application of agricultural VPPs. The results of this chapter lie in subsequent chapters (Chapter 4). In the next stage of research, we will further explore the application of this model in cross-season energy scheduling (such as heat storage in winter and cold storage in summer), regional multi-park collaborative scheduling, and reinforcement-learning-based adaptive optimization, so as to promote agricultural energy systems toward greater intelligence, efficiency, and sustainability.

References

[1] Fu, X., and Zhou, Y., 2023. Collaborative optimization of PV greenhouses and clean energy systems in rural areas. *IEEE Transactions on Sustainable Energy*, vol. 14, no. 1, pp. 642–656.

[2] Yin, S., and Zhao, Z., 2023. Energy development in rural China toward a clean energy system: Utilization status, co-benefit mechanism, and countermeasures. *Frontiers in Energy Research*, vol. 11, p. 1283407.

[3] Fu, X., Zhou, Y., Wei, Z., and Wang, Y., 2025. Optimal operation strategy for a rural microgrid considering greenhouse load control. *CSEE Journal of Power and Energy Systems*, vol. 11, no. 1, pp. 269–279.

[4] Silinto, B. F., Yamu, C. V. D. L., Zuidema, C., and Faaij, A. P. C., 2025. Hybrid renewable energy systems for rural electrification in developing countries: A review on energy system models and spatial explicit modelling tools. *Renewable and Sustainable Energy Reviews*, vol. 207, p. 114916.

[5] Fu, X., Wei, Z., Sun, H., and Zhang, Y., 2024. Agri-energy-environment synergy based distributed energy planning in rural areas. *IEEE Transactions on Smart Grid*, vol. 15, no. 4, pp. 3722–3738.

[6] Fu, X., Bai, J., Sun, H., and Zhang, Y., 2024. Optimizing agroenergy environment synergy in agricultural microgrids through carbon accounting. *IEEE Transactions on Smart Grid*, vol. 15, no. 5, pp. 4819–4834.

[7] Fu, X., and Long, H., 2025. Virtual power plant scheduling in agricultural microgrids through the control of distributed energy resources and greenhouse loads. *IEEE Transactions on Agri Food Electronics*, vol. 3, no. 2, pp. 569–581.

[8] Song, H., Gu, M., Liu, C., *et al.*, 2023. Multi-objective battery energy storage optimization for virtual power plant applications. *Applied Energy*, vol. 352, p. 121860.

[9] Lee, J., and Won, D., 2021. Optimal operation strategy of virtual power plant considering real-time dispatch uncertainty of distributed energy resource aggregation. *IEEE Access*, vol. 9, pp. 56965–56983.

[10] Zamani, A. G., Zakariazadeh, A., Jadid, S., and Kazemi, A., 2016. Stochastic operational scheduling of distributed energy resources in a large scale virtual power plant. *International Journal of Electrical Power & Energy Systems*, vol. 82, pp. 608–620.

[11] Wang, W., Kong, Z., He, Y., Li, C., and Jia, K., 2024. Research on the collaborative operation strategy of shared energy storage and virtual power plant based on double layer optimization. *Journal of Energy Storage*, vol. 101, Part B, p. 113997.

[12] Talbot, M.-H. and Monfet, D., 2024. Development of a crop growth model for the energy analysis of controlled agriculture environment spaces. *Biosystems Engineering*, vol. 238, pp. 38–50.

[13] Zhengmao, L., Kyyrä, J., Yan, X., Tianyang, Z., and Yunqi, W., 2023. Cooperative operation of renewable-integrated multi-energy microgrids under dynamic rolling horizon strategy. *2023 25th European conference on power electronics and applications*, Aalborg, Denmark.

[14] Ling, C., Yang, Q., Wang, Q., *et al.*, 2024. A comprehensive consumption-based carbon accounting framework for power system towards low-carbon transition. *Renewable and Sustainable Energy Reviews*, vol. 206, p. 114866.

[15] Zheng, H., Li, S., Zang, C., and Zheng, W., 2013. Coordinated control for grid integration of PV array, battery storage, and supercapacitor. *2013 IEEE Power & Energy Society General Meeting*.

[16] Erdiwansyah, Mahidin, Husin, H., Nasaruddin, Zaki, M., and Muhibbuddin, 2021. A critical review of the integration of renewable energy sources with various technologies. *Protection and Control of Modern Power Systems*, vol. 6, no. 1, pp. 1–18.

[17] Zhang, B., Talihati, B., Fan, H., Sun, Y., and Wang, Y., 2025. A dynamic carbon flow traceability framework for integrated energy systems. *Journal of Cleaner Production*, vol. 518, p. 145878.

[18] Martineau, V., Lefsrud, M., Naznin, M. T., and Kopsell, D. A. Comparison of light-emitting diode and high-pressure sodium light treatments for hydroponics growth of Boston lettuce. *HortScience*, vol. 47, no. 4, pp. 477–482, 2012.

[19] Pudjianto, D., Ramsay, C., and Strbac, G., 2007. Virtual power plant and system integration of distributed energy resources. *IET Renewable Power Generation*, vol. 1, no. 1, pp. 10–16.

[20] L. Huang, L. Deng, A. Li, R. Gao, L. Zhang, and W. Lei, A novel approach for solar greenhouse air temperature and heating load prediction based on Laplace transform, *J. Building Eng.*, vol. 44, Dec. 2021, Art. no. 102682.

[21] G. Han *et al.*, Development of a high-energy-density portable/mobile hydrogen energy storage system incorporating an electrolyzer, a metal hydride and a fuel cell, *Appl. Energy*, vol. 259, Feb. 2020, Art. no. 114175.

[22] Y. Jin, L. Sun, and J. Shen, Thermal economic analysis of hybrid open cathode hydrogen fuel cell and heat pump cogeneration, *Int. J. Hydrogen Energy*, vol. 44, no. 56, pp. 29692–29699, Sep. 2019.

Chapter 3

Bidding simulation of AEI with integrated electric agricultural machinery

Nange Li[1] and Xueqian Fu[1]

Abstract

This chapter systematically summarizes the research outcomes of the intelligent bidding model for the Agricultural Energy Internet (AEI) integrated with electric agricultural machines (EAM), covering three main aspects: model construction, algorithm design, and simulation validation. Centered on reinforcement learning, the study establishes a two-layer interactive bidding framework oriented toward multi-time-scale markets ("day-ahead–intraday–real-time"), enabling AEI to enhance its revenue in real-time electricity market participation. The results demonstrate that the proposed model can achieve rapid convergence under simulated competitive market environments, exhibiting excellent adaptability and stability. Theoretically, the study proposes an "AEI–market" dual-layer interaction mechanism, revealing the dynamic game relationships among generation units on the supply side, and achieving intelligent decision-making through the Deep Deterministic Policy Gradient (DDPG) algorithm. Methodologically, the research introduces a price-response-based reward function design that integrates economic indicators directly into the reinforcement learning process, improving both interpretability and economic orientation of the algorithm. Simulation results indicate that when AEI generation units participate in market bidding, photovoltaic units can flexibly adjust their bidding strategies, biomass units maintain stable revenues, and the overall system achieves dual optimization of revenue and risk. The findings of this chapter provide significant theoretical and technical support for the market-oriented operation and intelligent development of the Agricultural Energy Internet.

Keywords: Agricultural Energy Internet; reinforcement learning; intelligent bidding; dual-layer interaction mechanism; multi-market coupling; electric agricultural machine

[1]College of Information and Electrical Engineering, China Agricultural University, China

3.1 Introduction

3.1.1 Background of power market reform and agricultural energy transition

In 2020, China officially proposed the "carbon peaking and carbon neutrality" (referred to as the "dual carbon") goals, aiming to achieve carbon peaking before 2030 and carbon neutrality before 2060 [1]. The release of this national strategy represents a significant action for China to address global climate change and fulfill international environmental commitments, marking a new stage in the country's energy system and economic development model centered on green and low-carbon principles. The dual carbon goals not only reflect China's determination to promote global sustainable development but also imply that, in the coming decades, the country's energy system, industrial structure, and socioeconomic activities will undergo systemic transformations. The core task is to reshape the energy system, promote the orderly replacement of fossil energy with clean energy, and systematically develop a renewable energy-dominated power system, achieving coordinated low-carbon transformation on both the supply and demand sides of energy.

Currently, China's power system remains dominated by coal-fired generation, with high-carbon emission intensity per unit of electricity and a relatively low proportion of clean energy. Coal has long served as the cornerstone of electricity supply, ensuring energy security while also exerting significant carbon emission pressure. Increasing the share of new energy generation has become an important pathway to reduce the power sector's carbon intensity and drive overall emission reduction [2]. Renewable energy offers advantages such as cleanliness, low-carbon emissions, and sustainability, and its share in the energy mix directly influences the progress toward the dual carbon goals. Since 2010, China's installed capacities for wind and photovoltaic (PV) power have maintained rapid growth; however, by 2019, their shares of total electricity generation were only 2.9% and 5.4%, respectively, remaining relatively low. Under the guidance of the dual carbon goals, by 2060, the combined installed capacity share of wind and PV power needs to reach approximately 80%, with a generation share of 70% [3]. This indicates that China's energy system will experience a deep structural shift from "high-carbon centralized" to "low-carbon distributed" in the coming decades, making the large-scale development of renewable energy a central task of energy strategy transformation.

However, the rapid expansion of renewable energy also poses significant challenges to the power system. New energy sources such as wind and PV depend on natural resources, and their output is influenced by climate conditions, geographic location, and seasonal variations, exhibiting obvious randomness, intermittency, and volatility. These characteristics significantly increase system uncertainty, placing pressure on grid dispatch, balance, and safe operation [4]. Traditional centralized thermal power-based scheduling systems rely on long-term forecasts and fixed operation plans, lacking dynamic flexibility [5]. Under scenarios of high renewable penetration, such systems struggle to respond to frequent power fluctuations and real-time supply-demand changes, potentially leading to difficulties in power balance, insufficient reserves, or curtailment of wind and solar generation. Therefore, establishing power market mechanisms that reflect supply-demand relations, price signals, and flexibility value is key to promoting efficient renewable energy utilization and ensuring system stability and security.

In recent years, China's power system reform has continued to deepen, accelerating market-oriented development. The reform direction has gradually shifted from the previous "planned electricity purchase and sale" approach to a market-centered electricity trading mechanism based on competitive pricing signals. At the national level, a multi-tiered market system is being constructed, dominated by medium- and long-term transactions with spot markets as a supplement, to achieve market-based resource allocation and flexible system regulation. Medium- and long-term markets focus on ensuring power balance and price stability, while spot markets reflect real-time supply-demand changes and flexibility value through real-time pricing. Scholars and industry experts generally agree that establishing a multi-time-scale coordinated market mechanism is key to improving system economy, security, and adaptability to new energy. Spatially, provincial and inter-provincial market linkages and coordination should be strengthened to optimize cross-regional resource allocation and dynamic power flow balancing; temporally, a continuous trading system covering annual, monthly, day-ahead, intraday, and real-time markets should be formed [6]. The day-ahead–intraday–real-time three-stage market structure is gradually becoming an important operational form in China's new power system, serving as a critical support mechanism for renewable energy utilization and system regulation.

Against this macro backdrop of energy transition and power market reform, energy utilization in the agricultural sector has increasingly entered policy and academic research agendas. For a long time, agricultural energy systems have faced issues such as dispersed loads, small scale, high variability, and weak infrastructure [7]. Agricultural parks, as key carriers of modern agricultural development, exhibit significant seasonal and temporal energy demand characteristics. For example, energy consumption for irrigation, cold chains, processing, and lighting is often concentrated in specific periods, forming a typical load pattern of "temporal concentration and spatial dispersion." This characteristic makes it difficult to balance supply and demand, resulting in low energy utilization efficiency and high operational costs. Meanwhile, aging rural grid equipment, long transmission and distribution lines, and high losses further challenge the economic and reliability aspects of agricultural energy systems.

On the other hand, agricultural regions possess abundant renewable energy resources, providing natural conditions for energy transition. China's agricultural parks generally have sufficient solar irradiation, while biomass, biogas, and agricultural waste can also serve as stable, clean energy sources. Agricultural residues such as crop straw, livestock manure, and processing waste can be converted through appropriate technologies into electricity, heat, or gas, forming complementary multi-energy supply modes and providing new pathways for energy self-sufficiency and clean development in agricultural regions. In this context, the concept of the Agricultural Energy Internet (AEI) has emerged. AEI is a new energy management system under the overall energy internet framework, tailored to the characteristics of agricultural production. Its core idea is to integrate distributed generation, energy storage, controllable loads, and information and communication infrastructure to achieve coordinated optimization of energy production, transmission, storage, and consumption. Leveraging modern information technologies such as IoT, cloud computing, and big data, AEI enables multi-energy flow information integration, device interconnectivity, and real-time control, enhancing the dynamic perception and adaptive capabilities of the energy system [8]. This system transforms traditional passive energy use in agriculture into an actively optimized smart energy mode, creating a new comprehensive energy management form of "digitalization–networking–intelligence" within the park.

Through the construction and promotion of AEI, agricultural parks can not only improve local absorption of renewable energy and optimize energy utilization structure but also participate in system regulation through demand response and storage dispatch in the electricity market, thereby enhancing overall operational efficiency and economic performance. Essentially, AEI represents the agricultural energy system's proactive adaptation and extension to national power market liberalization and decarbonization, providing a practical pathway for low-carbon development in agriculture and contributing significant distributed support to the construction of a new power system.

3.1.2 Development of Agricultural Energy Internet and virtual power plants

The concept of the virtual power plant (VPP) provides a new pathway for the aggregation and market-based operation of distributed energy resources [9]. Its core idea is to virtually integrate geographically dispersed and small-capacity distributed generation, energy storage systems, and controllable loads through information and communication technology and intelligent control systems, forming a market entity functionally equivalent to conventional centralized power plants. Through this virtual aggregation approach, a VPP can participate in the electricity market as a unified entity, achieving coordinated management from the resource layer to the trading layer. Its centralized scheduling and optimization control mechanisms make the operation of distributed resources more economical and efficient, enabling optimal local energy allocation while providing system-level flexibility services such as peak shaving, frequency regulation, and reserve capacity. This enhances both energy utilization efficiency and the stability, reliability, and resilience of the power grid [10,11]. Essentially, VPP reconstructs the traditional power system control model through digital means, using algorithm-driven aggregation and real-time response to endow distributed resources with system support and market competitiveness.

In the development of the Agricultural Energy Internet (AEI), energy units exhibit characteristics such as "dispersed resources, limited access scale, isolated data, and high load volatility," which are difficult to support efficient operation and market participation under traditional models. The introduction of the VPP model provides a systematic solution to these issues. By establishing an AEI–VPP aggregation system, distributed photovoltaic, biomass generation, energy storage devices, and agricultural loads such as electric farm machinery can be integrated into a unified control and trading platform, realizing full-process optimization management of "multi-source aggregation—unified dispatch—market participation." This system overcomes the "disconnected" limitation of agricultural energy systems, allowing originally scattered distributed resources to achieve scale effects through aggregation and enter the electricity market as a virtual entity. For agricultural parks, this aggregation mechanism not only enhances energy utilization flexibility and regulation capability but also creates market revenue channels for distributed energy, enabling farmers and park operators to obtain flexibility compensation and economic returns through market participation.

Under the AEI–VPP model, the role of the Agricultural Energy Internet evolves from merely a physical coordination platform to a market-oriented virtual entity with autonomous decision-making capability. The AEI can dynamically adjust its generation and energy consumption strategies based on real-time electricity market signals (such as price fluctuations, system balance requirements, and reserve capacity gaps), transitioning from "passive response" to "active optimization." In the day-ahead market, AEI can optimize PV and storage system charging and discharging strategies by considering weather forecasts, agricultural production schedules, and load characteristics; in intraday or real-time markets, flexible controllable loads such as electric farm machinery, cold storage, and irrigation systems can quickly participate in price response, achieving dynamic adjustment of electricity consumption [12]. This multi-time-scale dynamic decision-making mechanism enables AEI to effectively find the optimal balance between market price signals and agricultural production demands, implementing an operational model of "market signal-guided resource allocation and data-driven optimized dispatch."

The introduction of the AEI–VPP model enables agricultural energy systems to directly participate in electricity market competition and regulation for the first time. Various distributed renewable energy resources within agricultural regions (including PV, biomass, and biogas) can be collaboratively optimized through a unified aggregation platform, achieving multi-energy complementarity and coordinated scheduling of electricity, heat, and load. This structure not only improves the overall economic efficiency and operational performance of agricultural energy systems but also significantly enhances renewable energy utilization, reducing curtailment of wind and solar power, and providing distributed support for national power system flexibility and the achievement of the dual carbon goals. At the same time, AEI–VPP also provides a new direction for the modernization of rural grids, transforming them from a "unidirectional transmission" model to a bidirectional "source–grid–load–storage" interactive system.

Currently, China's power system is accelerating the transition from a planned to a market-oriented system. The country has explicitly proposed building a unified electricity market system "dominated by medium- and long-term transactions with spot markets as a supplement" [13] and has launched spot market pilot programs in regions such as Shanxi, Shandong, Guangdong, and Gansu [14]. The operational experience of these pilot regions provides crucial empirical foundations for improving multi-market coordination mechanisms nationwide in the future. As multiple types of markets (medium- and long-term, day-ahead, intraday, real-time) gradually operate in parallel, the temporal resolution and spatial layering characteristics of price signals become richer, and the demand for market flexibility significantly increases. In this context, AEI's potential as a distributed VPP becomes more prominent, not only providing flexible resources but also dynamically optimizing energy and revenue allocation across different market layers.

However, in a complex, multi-market environment, the Agricultural Energy Internet still faces multiple challenges. First, determining coordinated bidding strategies across different time-scale markets to achieve optimal intertemporal balance of revenue and risk is a primary difficulty for AEI market operation. Second, at the modeling level, incorporating multi-type controllable loads such as electric farm machinery, cold chain loads, and irrigation systems into a unified modeling framework to achieve coordinated optimization of generation and consumption is challenging. Third, employing intelligent algorithms, forecasting models, and reinforcement learning to build an intelligent optimization system supporting multi-agent AEI bidding decisions is essential. Addressing these issues will provide the theoretical foundation and practical pathway for AEI to achieve autonomous optimization and sustainable development in a market-oriented environment.

3.1.3 Role and response characteristics of electric agricultural machinery in the Agricultural Energy Internet

The Agricultural Energy Internet (AEI) is an integrated energy system that encompasses energy production, transmission, storage, and consumption, with the fundamental objective of achieving highly efficient utilization and optimal allocation of energy resources specifically within the agricultural sector. By centering on electricity as the primary medium, AEI facilitates dynamic coordination among diverse energy forms through mechanisms of multi-energy complementarity, interactive control, and information-driven management. Structurally, the system typically comprises multiple functional layers, including the energy production layer, transmission layer, storage layer, and consumption layer. Coordination among these layers is realized through multi-directional coupling of information flows, energy flows, and value flows, collectively forming a comprehensive "source–grid–storage–load" energy management architecture. This integrated framework not only enhances overall energy efficiency but also provides technical support for strategic objectives such as carbon emission reduction, operational cost optimization, and the substitution of conventional energy with clean, renewable sources within the agricultural domain.

In the electricity market environment of agricultural industrial parks, the principal participants on the supply side consist of external grids, combined heat and power (CHP) systems, wind turbines (WT), and photovoltaic (PV) systems, with the latter two representing renewable energy sources. On the demand side, internal industrial factories, agricultural product processing enterprises, various agricultural electricity consumers, and power retailers collectively form the consumption end. At the same time, the carbon market involves participants such as CHP, WT, and PV systems, while the heat market is predominantly centered around centralized CHP heating, which has not yet reached full marketization. Surrounding this diverse market landscape, market service institutions – including power trading centers, carbon trading centers, and integrated energy service providers – have gradually developed a complete supporting ecosystem. By leveraging a virtual power plant (VPP) platform, agricultural parks can participate in electricity spot markets as a unified virtual entity, achieving effective aggregation of distributed generation, flexible dispatch of energy resources, and coordination across multiple energy forms, all of which enhance the operational efficiency and market competitiveness of the system.

Within the AEI framework, electric agricultural machinery (EAM) represents a key and highly representative component at the consumption layer. EAM exhibits dual functional characteristics as both a load device and a form of mobile energy storage, acting as a "controllable terminal" within the AEI's energy and information flow interactions, capable of responding to both market signals and operational requirements.

As a load device, the electricity consumption patterns of EAM are closely linked to agricultural operation cycles, displaying strong seasonal and phase-specific characteristics. For instance, electric tractors and related machinery require significant energy inputs during daytime operations, including plowing, sowing, transportation, and harvesting activities. These operational energy demands are rigid and essential, necessitating prioritized supply to ensure the continuity, reliability, and efficiency of agricultural production. Certain high-power agricultural machines, such as large tractors or combine harvesters, can exhibit significant instantaneous charging power peaks within short operational periods, often reaching 50–100 kW. This concentration imposes strict requirements on supporting infrastructure, including charging station capacity, local grid node load capability, and overall power quality standards. Such concentrated and time-dependent load features make the charging behavior of agricultural machinery a critical factor in the optimization of AEI energy dispatch and system scheduling strategies.

On the other hand, when considered as energy storage units, EAM can supply bidirectional power to the grid during non-operational periods, such as nighttime, rainy seasons, or off-season intervals. This capability allows EAM to participate in peak shaving, valley filling, and ancillary service provision, endowing the machinery with economic flexibility and additional value potential. By strategically charging during periods of low electricity prices and discharging during peak-price periods, EAM not only enables farmers to reduce daily energy costs but also provides opportunities to earn supplementary revenue from the market. This creates a closed-loop economic optimization among load management, energy storage, and market participation. Depending on the specific machinery model, EAM battery capacities typically range between 50 and 200 kWh. Careful planning regarding charging/discharging depth, frequency, and scheduling windows can effectively extend battery life, reduce wear, and enhance the operational stability and overall response efficiency of AEI at the system level.

From a temporal perspective, EAM charging demand is highly concentrated around critical operational periods, such as early morning or late afternoon, which often do not align with the peak output times of PV generation occurring at midday. This temporal mismatch between energy supply and load demand can limit the utilization of locally generated PV power. By leveraging price signals and storage adjustment mechanisms, the AEI platform can coordinate the timing and intensity of EAM charging to better align with PV generation profiles, thereby mitigating peak-valley mismatches and improving overall energy efficiency within the park. Spatially, agricultural machinery operates over extensive areas, exhibiting dispersed trajectories and high mobility. This necessitates careful planning of charging infrastructure to accommodate operational distribution, geographic constraints, seasonal work patterns, and machine operating radii. A well-designed and strategically distributed charging network not only improves operational convenience and economic efficiency but also supports regional distribution network load balancing and voltage stability, which are essential for reliable AEI operation.

With the ongoing deepening of electricity market reforms in China, agricultural parks are gradually transitioning from purely passive energy consumers to integrated multi-energy participants that combine generation, storage, and consumption functionalities. Through VPP platforms, AEI can achieve aggregated scheduling and unified bidding across distributed generation units, storage systems, and flexible loads. In this integrated framework, EAM serves as both a substantial electricity load and a controllable, flexible resource. In the day-ahead market, agricultural parks can formulate generation and charging strategies in advance, informed by weather forecasts, agricultural operational schedules, and anticipated price trends, thereby achieving optimized resource allocation. In real-time markets, EAM can dynamically adjust charging power and timing in response to immediate market signals, contributing to grid balancing, demand response initiatives, and ancillary services. This price-driven, adaptive load participation mechanism provides AEI with a novel approach for optimizing revenues and enhancing system regulation capabilities across multiple market layers.

Within the broader market context, the AEI–VPP assumes multiple roles: it acts as a clean electricity producer (e.g., PV, CHP), a terminal electricity consumer (e.g., EAM, irrigation pumps), and a provider of controllable resources. Through flexible load management, multi-energy complementarity, and optimized dispatch strategies, AEI achieves power balance, revenue maximization, and carbon emission minimization across medium- and long-term, day-ahead, and real-time market stages. The integration of EAM into this system significantly enhances the flexibility of AEI, introducing additional energy hubs and responsive nodes that strengthen market-based operational performance.

However, large-scale integration of electric agricultural machinery also alters traditional agricultural load characteristics. Charging behavior exhibits strong operational dependency, short-duration high power, and stochastic distribution, potentially resulting in local load peaks that can impact grid safety and economic efficiency. During off-season periods, underutilized storage resources may lead to inefficiencies and economic losses. Effective coordination of EAM charging and discharging behaviors through market-driven mechanisms and dispatch strategies, while maintaining continuity in agricultural production, represents a central challenge in balancing economic efficiency with operational reliability in AEI optimization.

Current research efforts primarily focus on industrial parks or urban integrated energy systems, with systematic studies addressing agricultural scenarios still limited. AEI is characterized by highly dispersed loads, high renewable penetration, and rigid operational requirements of agricultural machinery, making traditional market optimization approaches difficult to apply directly. Furthermore, systematic modeling of EAM charging/discharging flexibility, operational constraints, and price-responsive behaviors remains underdeveloped. The potential economic regulation value and market impact mechanisms of EAM require further detailed investigation to fully capture their role within AEI.

Based on these observations, subsequent chapters will focus on AEI as the primary research object to construct an Agricultural Energy Internet bidding strategy model that integrates electric agricultural machinery. The analysis will examine how different types and scales of EAM integration influence generation output curves, market clearing outcomes, and overall system revenue distribution, uncovering the intrinsic economic mechanisms of coordinated electricity generation and consumption optimization within AEI. The model's effectiveness, robustness, and practical applicability will be validated through simulation-based studies, providing a solid foundation for intelligent AEI operational strategies.

3.1.4 Research motivation and problem definition of this chapter

With the continuous deepening of China's new round of electricity system reform, the electricity market is gradually evolving from a relatively simple, single-transaction structure into a complex system encompassing multiple time scales, parallel market types, and interacting trading mechanisms. In this context, the Agricultural Energy Internet (AEI), as an emerging and innovative energy system, integrates distributed renewable energy generation, energy storage systems, and controllable loads, providing a practical foundation for participating in market transactions, price-responsive operations, and flexible energy dispatch. At the same time, AEI demonstrates substantial potential economic and social value, serving as a pathway to improve energy efficiency, promote clean energy adoption, and support sustainable agricultural development. Its core characteristics – distributed, flexible, and intelligent – enable it to interconnect dispersed energy units within agricultural parks into a coordinated, dispatchable entity by leveraging cyber-physical system (CPS) technologies, multi-energy complementarity mechanisms, and information-driven control. Despite these advantages, current research largely focuses on industrial parks or urban integrated energy systems, with existing market modeling, bidding strategies, and dispatch methods primarily developed for industrial or residential loads. Limited attention has been paid to electric agricultural machinery (EAM), which exhibits unique operational patterns, strong periodicity, and rigid energy demands specific to agricultural scenarios. This gap highlights the need for targeted modeling and optimization approaches to accurately reflect the operational and economic characteristics of EAM within AEI.

Under the trend of high renewable energy penetration, the balance of supply and demand in agricultural parks has become increasingly volatile and dispersed across both space and time. Distributed PV and biomass generation are inherently intermittent and weather-dependent, with output subject to seasonal fluctuations and external environmental conditions, leading to uncertainty in energy supply. Meanwhile, agricultural loads, particularly EAM, are constrained by agricultural operation schedules, crop growth cycles, and climatic conditions, exhibiting rigid, phase-specific, and highly predictable energy consumption characteristics. This dual stochastic behavior – where supply follows weather patterns and load follows farm operational schedules – renders traditional centralized dispatch and deterministic planning approaches increasingly ineffective in modern market-oriented environments. As market flexibility requirements rise and electricity price fluctuations intensify, establishing a bidding optimization model for AEI that can adapt to multiple electricity market mechanisms has become a critical pathway for promoting market-oriented and economically efficient AEI operations. Such a model must not only account for the physical characteristics, spatiotemporal distribution, and operational constraints of various resources but also reflect the objectives of revenue optimization, risk management, and flexible response under multi-market coordination. Consequently, exploring the coordinated generation-consumption mechanisms of AEI and the response strategies of flexible resources like EAM carries both important theoretical significance and practical relevance for the sustainable development of agricultural energy systems.

To address these challenges, the primary objective of this study is to establish a multi-time-scale electricity market bidding optimization framework for AEI, integrating multi-level system modeling and intelligent algorithm design to maximize revenues and efficiently utilize renewable energy across diverse spot markets in agricultural parks. The research approach is organized into three interrelated levels: system modeling, strategy design, and intelligent optimization.

First, at the system modeling level, a virtual power plant (VPP) operation mechanism is constructed to represent the integrated "source–grid–storage–load" framework of AEI. This model systematically incorporates key components such as distributed PV, biomass generation units, storage systems, and EAM, capturing the coupling relationships, complementarity in production and consumption, and interactive information flows among energy units. It further analyzes the controllability and flexibility potential of different resources under market conditions, establishing a solid physical and operational foundation for subsequent bidding and dispatch optimization in the AEI context. By explicitly modeling multi-energy interactions and response capabilities, the system-level model enables the evaluation of coordinated operation strategies under varying market and environmental conditions.

Second, at the strategy design level, the study focuses on the access characteristics, operational schedules, and electricity purchase response behaviors of EAM. The analysis investigates how these factors influence the output allocation of generation units, bidding behavior, and overall economic performance. Accordingly, an AEI bidding strategy model that incorporates EAM response characteristics is constructed, enabling integrated AEI systems to achieve adaptive dispatch, revenue maximization, and operational efficiency under different scales, penetration levels, and operational scenarios of EAM. This approach provides a refined methodology for agricultural energy participation in electricity markets, ensuring that flexible loads contribute effectively to system-level optimization while maintaining agricultural production continuity.

To accommodate the multi-time-scale characteristics of modern electricity markets, this chapter further proposes a hierarchical bidding strategy covering day-ahead, intraday, and real-time markets. Unlike traditional models, which often assume independent bidding on a single time scale, the proposed approach establishes hierarchical linkages among markets, recognizing differences in bidding granularity, information timeliness, and response rhythms, while simplifying information frictions and behavioral delays under idealized conditions. Based on this hierarchy, a three-stage linked bidding model for AEI is designed to achieve continuity, consistency, and adaptability in market strategies through dynamic multi-time-scale optimization, enabling AEI agents to coordinate generation, storage, and load behaviors effectively across diverse market stages.

To further enhance adaptability and intelligence, a reinforcement learning (RL)-based decision-making algorithm is introduced. The RL algorithm continuously interacts with the market environment, updating state–action–reward mappings and allowing AEI agents to learn optimal strategies under varying market conditions and price signals. This approach overcomes computational limitations of traditional optimization methods in high-dimensional, nonlinear, and dynamic environments, enabling dynamic optimal control of AEI bidding, energy dispatch, and resource utilization. Consequently, the system can achieve higher economic efficiency, operational flexibility, and effective risk management, while maintaining stability under variable renewable generation and stochastic load conditions.

Furthermore, to validate the model's effectiveness and practical applicability, simulation experiments are conducted under typical agricultural park scenarios. These experiments comprehensively consider factors such as competing bids from other market participants, EAM operational schedules, stochastic renewable output, and dynamic electricity price curves. By comparing scenarios with varying EAM penetration levels, the simulations assess potential revenue enhancement, market responsiveness, and operational coordination of AEI. Results quantitatively demonstrate the model's rationality, economic viability, renewable energy utilization efficiency, and operational stability, providing critical data support for market-oriented AEI operation strategies.

The expected research outcomes are reflected in three key areas: (1) significant improvement in economic returns for agricultural parks across multiple electricity market types, shifting the operational paradigm from "cost minimization" to "revenue maximization"; (2) enhanced renewable energy utilization within parks, reduced curtailment of wind and solar generation, and promotion of large-scale clean energy adoption; and (3) strengthened AEI adaptability to load fluctuations, electricity price uncertainties, and external disturbances, improving overall operational resilience, flexibility, and system stability. Importantly, the proposed multi-market optimization approach for AEI is highly generalizable and transferable, offering theoretical and practical guidance for the commercial operation, digital upgrading, and intelligent management of agricultural energy systems in rural and semi-rural regions.

In summary, the research objectives of this chapter can be condensed into three primary directions:

1. Establish a coordinated multi-energy model for AEI, integrating distributed PV, biomass generation, storage devices, and agricultural loads, to reveal energy interaction mechanisms and dynamic balance processes among source, storage, and load;
2. Construct a multi-time-scale AEI bidding strategy model that incorporates EAM response characteristics, designing a hierarchical optimization framework covering day-ahead, intraday, and real-time markets to achieve revenue maximization and flexible dispatch under multi-market conditions;
3. Based on reinforcement learning, develop a training and self-learning mechanism for AEI agent strategies, and validate its performance, economic benefits, and system stability improvements under varying market conditions through simulation studies.

The execution of this research not only provides a systematic modeling approach and hierarchical optimization framework to transition AEI from "energy coordination" to "market coordination" but also establishes a practical and operable technical path for the market participation of flexible loads such as EAM. This foundation is essential for subsequent chapters focusing on mathematical modeling, intelligent algorithm design, and simulation-based performance analysis of AEI multi-market bidding strategies.

3.1.5 Research content and technical route

To address the complex operational characteristics and multi-agent coordination challenges of the Agricultural Energy Internet (AEI) in multiple electricity markets, this chapter constructs a systematic research framework around three main axes: "Virtual Power Plant (VPP) modeling—Intelligent bidding—Multi-time-scale optimization." The framework aims to reveal AEI's operational mechanisms, strategy evolution patterns, and flexibility and economic response characteristics under different market conditions. The overall research approach is illustrated in Figure 3.1. The technical route spans three levels – model construction, algorithm

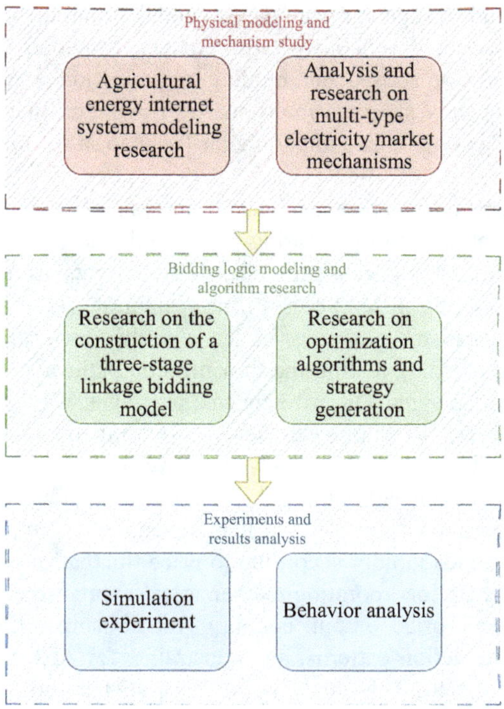

Figure 3.1 Overall technical route of Chapter 3

design, and simulation verification – emphasizing a closed-loop integration of theoretical analysis and practical validation to achieve full-chain optimization from physical coordination to market-level decision-making, providing methodological support and logical basis for strategy derivation and experimental analysis in subsequent chapters.

1. AEI system modeling

   First, a system structure modeling study is conducted for AEI market participation. This part focuses on the dynamic coordination of energy and information flows, constructing a comprehensive energy exchange model that includes distributed PV, biomass generation, energy storage devices, and typical agricultural loads such as electric agricultural machinery (EAM). Through a unified parameterized modeling approach, the spatiotemporal coupling relationships and power flow constraints between source–grid–storage–load are described, revealing system operational characteristics and controllability at different time scales.

   The model further incorporates the VPP aggregation feature of AEI. By introducing centralized control and distributed response mechanisms, dispersed energy units within agricultural parks can achieve information sharing and coordinated dispatch on a unified platform. While ensuring consistency of power, heat, and storage constraints, the model simultaneously considers economic objectives and reliability requirements, providing a computable and structured basis for subsequent bidding optimization and strategy solution. This model not only supports multi-energy flow coordination at the physical level but also lays the data and parameter foundation for market-oriented VPP modeling.

2. Multi-type electricity market mechanism analysis
   Next, the operational rules, clearing logic, and transaction characteristics of day-ahead, intraday, and real-time electricity markets are systematically analyzed. By comparing differences in time resolution, trading cycles, and price formation mechanisms among markets, a multi-time-scale market response model is established, revealing the transmission pathways and feedback characteristics of price signals at different market stages.

   In this process, the study clarifies the bidding boundary conditions and behavioral constraints of AEI as a VPP entity in different market layers. For instance, in the day-ahead market, AEI formulates bidding strategies based on renewable generation forecasts and load schedules; in intraday and real-time markets, it responds rapidly according to price fluctuations, storage states, and load flexibility. By incorporating mathematical expressions of market clearing mechanisms and price–output coupling relationships, the interface logic, information interaction patterns, and constraints for AEI market participation are further specified. This analysis provides an institutional framework and parameterized support for the subsequent construction of the three-stage linked model.

3. Three-stage linked bidding model construction
   Based on market mechanism analysis, a three-stage AEI-linked bidding model covering day-ahead, intraday, and real-time markets is constructed. The model unifies decision variables, objective functions, and constraints across multiple time scales into one optimization framework, enabling information flow and strategy coordination across stages. Specifically, the day-ahead stage focuses on long-term optimization and revenue planning, the intraday stage reflects short-term corrections and resource reallocation, and the real-time stage emphasizes price response and emergency dispatch.

   The model incorporates price forecasting and information delay mechanisms to simulate the combined effects of market volatility, uncertainty, and time lags on AEI decisions. By embedding a rolling optimization approach, internal generation units, storage systems, and controllable loads of AEI can adaptively adjust according to market feedback, forming a dynamic closed loop of market clearing and system response. This framework provides a unified description of the internal multi-agent coordinated decision-making process in AEI and offers a mathematical representation and computational path for strategy optimization.

4. Optimization algorithms and strategy generation
   To tackle the high-dimensional, nonlinear, and uncertain characteristics of the three-stage linked model, a bi-level optimization framework is designed. The

upper level corresponds to market-level decision-making, learning, and simulating price feedback mechanisms and bidding behavior patterns; the lower level corresponds to unit-level optimization, executing resource dispatch and load response decisions. Under this framework, AEI's bidding process can be regarded as a multi-agent game, balancing self-revenue maximization with system stability and market constraints.

To enhance solution efficiency and strategy generalization in dynamic environments, a deep reinforcement learning (DRL) algorithm is introduced for strategy optimization. Through agent–environment interaction training, the algorithm adaptively adjusts strategies under different price signals and resource states, enabling autonomous bidding and intelligent dispatch for AEI. Compared with traditional mathematical programming methods, DRL allows continuous learning in model-free or partially observable environments, offering higher convergence efficiency and robustness. Ultimately, AEI agents can develop optimal strategy mappings for various market scenarios through multiple learning episodes, significantly improving system flexibility and economic performance.

5. Simulation experiments and behavioral analysis
After model and algorithm construction, a virtual simulation platform for a typical agricultural park is established to validate the feasibility and economic performance of the proposed methods. The platform is based on synthetic data and literature parameters, reproducing key characteristics of distributed energy output, price sequences, and load responses. Simulation experiments include three core components:

Three-stage market bidding simulation: compare output strategies of PV, biomass, and EAM units across multiple markets, analyzing differences and evolution trends of AEI's internal coordinated decisions.

Price and output feedback analysis: evaluate how clearing price fluctuations affect unit revenues and power dispatch, assessing the dynamic response capability of the three-stage linkage mechanism.

EAM access scenario analysis: vary the scale of EAM access to examine its contribution to overall AEI economic performance, flexibility, and market revenues, revealing the regulation potential of flexible resources under different scenarios.

Simulation results validate the effectiveness of the proposed three-stage model and DRL algorithm in improving AEI revenues, promoting renewable energy utilization, and enhancing system resilience. These results provide quantifiable evidence for AEI operational optimization in market-oriented environments.

In summary, the technical route of this chapter forms a three-layer progressive logic. The model layer: AEI system modeling reveals energy interaction mechanisms among source-storage-load and the foundation for market participation. The algorithm layer: Reinforcement learning constructs intelligent bidding and strategy optimization mechanisms, enabling self-learning and autonomous decision-making in multi-market environments. The validation layer: Simulation in agricultural parks with multiple scenarios evaluates economic, flexibility, and stability performance under different bidding strategies.

This research system establishes a complete closed loop from theoretical modeling to algorithmic optimization and system validation, providing a systematic analytical framework and methodology for AEI multi-market operation, and laying a solid logical and technical foundation for subsequent chapters, including model derivation, algorithm implementation, and experimental verification.

3.2 AEI system structure and simulation environment assumptions

3.2.1 Overall structure and operating mechanism of the Agricultural Energy Internet

The Agricultural Energy Internet (AEI) is an interdisciplinary integrated system that deeply combines energy engineering, agricultural engineering, and information and communication technology. Its core objective is to achieve efficient coordination and optimized allocation of energy production, transmission, storage, and consumption in the agricultural sector. Against the backdrop of China's "dual carbon" strategy and energy digitalization transformation, AEI is not only a key infrastructure promoting green, intelligent, and low-carbon agricultural development, but also an important platform for building regional distributed energy systems and facilitating multi-market energy transactions. Essentially, it is an energy ecosystem centered on electricity, with complementary multi-energy forms and highly interconnected information, achieving clean energy usage, efficient consumption, and intelligent management through unified energy management and intelligent scheduling in agricultural production processes.

1. Overall system framework

   The system structure of AEI generally consists of four physical layers: energy production, energy transmission, energy storage, and energy consumption, complemented by two intelligent layers: information sensing and control decision-making. This forms a three-dimensional collaborative system of energy flow, information flow, and value flow. The overall structure can be summarized as a "source–grid–storage–load–control" integrated system, where different energy units and information modules achieve unified scheduling and dynamic optimization through data interconnection and energy coupling.

   Energy production layer: This layer mainly includes distributed photovoltaic (PV) generation, biomass gasification power generation, and small wind turbines. PV systems utilize available spaces such as rooftops, greenhouse roofs, and agricultural sheds in parks to efficiently convert solar radiation into green electricity. Biomass generation units use crop residues, livestock manure, and agricultural waste as fuel, offering stable output, carbon neutrality, and environmental benefits. Small wind turbines, typically deployed at open edges or vacant areas of parks, supplement electricity generation through natural wind energy. A key characteristic of this layer is that output power is significantly affected by meteorological and natural conditions, exhibiting variability and intermittency, necessitating integration with storage systems and intelligent dispatch strategies to achieve smooth energy output and dynamic system balance.

Energy transmission layer: This layer is responsible for distributing and transmitting electricity within the park, serving as a foundational element for system stability. It typically consists of distribution networks, energy routers, power quality regulation devices, and smart switches. Energy routers dynamically allocate electricity according to real-time load changes and market price signals, enabling multi-source supply and flexible interconnection. Through integration with the information sensing layer, the transmission layer can obtain node power and voltage status in real time, using distributed optimization algorithms to adjust power flows, reduce energy loss and voltage deviations, thereby ensuring the safety, stability, and economic efficiency of electricity delivery.

Energy storage layer: The energy storage layer is crucial for balancing supply and demand, peak-shaving, and backup support. Storage units in the park mainly include lithium-ion battery systems, hydrogen storage modules, and mobile storage capabilities within electric agricultural machinery. Fixed storage devices handle basic regulation and emergency support, while electric agricultural machinery can participate as storage units during non-working periods: charging when electricity prices are low or energy is abundant, and discharging during peak periods or when prices rise, providing distributed peak regulation. This dual "mobile storage–load" feature significantly enhances system flexibility and market adaptability, enabling AEI to perform flexible optimization and scheduling across multiple time scales.

Energy consumption layer: This layer covers various energy-consuming devices in agricultural production, including electric agricultural machinery, greenhouse heating and lighting systems, irrigation pumps, cold chain storage, and processing loads. Agricultural loads exhibit clear seasonality, time-dependency, and operational relevance, with demand varying according to climate, crop type, and agricultural cycles. For example, spring and summer focus on seeding and irrigation, while autumn and winter emphasize drying and refrigeration. Through demand response mechanisms, the consumption layer can flexibly adjust energy use according to market prices or system dispatch signals, achieving a "demand-driven supply" energy management loop, making it the main carrier of demand-side flexibility in the AEI system.

2. Information sensing and control decision architecture
   Efficient and safe operation of AEI relies on a complete information sensing and intelligent control system. This system comprises three parts: information sensing, intelligent control, and market coordination, forming a closed-loop management framework for energy and information interaction.

Information sensing layer: Serving as AEI's "neural network," this layer collects data throughout the process and monitors operational status. IoT and edge intelligent terminals are deployed at each energy node, combined with smart meters, meteorological sensors, video recognition devices, and satellite remote sensing terminals to build a multi-source heterogeneous data acquisition system. This enables real-time perception of generation, storage, loads, and environmental variables. Data fusion processing provides high-accuracy inputs for prediction, dispatch, and bidding algorithms, supporting data-driven dynamic optimization.

Intelligent control layer: Acting as AEI's central "brain," this layer is responsible for energy coordination, operational optimization, and adaptive regulation. It integrates model predictive control (MPC), reinforcement learning (RL), and adaptive optimization algorithms to perform rolling calculations on load prediction, resource allocation, and system state feedback, achieving dynamic dispatch and optimization across multiple time scales. The operational mechanism can be summarized as a "predict–decide–feedback–correct" loop, formulating optimal dispatch plans based on forecasts and continuously correcting control strategies with real-time data to enhance system stability and renewable energy absorption.

Market coordination layer: This layer connects the park energy system with external electricity markets and serves as a key hub for AEI's economic operation and market participation. Through a VPP platform, it integrates internal generation units, storage systems, and controllable loads for unified bidding, centralized clearing, and revenue distribution. Its main tasks include dynamically optimizing bidding strategies and power allocation according to market prices, trading rules, and carbon emission constraints. Electric agricultural machinery, with fast charging/discharging and flexible operation schedules, provides high-response flexibility at the market level, acting as an important execution unit for AEI–market interaction.

3. Energy flow and information flow coordination mechanism
   AEI operation essentially involves multidimensional coordination of energy, information, and value flows. Energy flow reflects the transfer and conversion of electricity, heat, and chemical energy among source, grid, storage, and load. Information flow spans sensing, prediction, control, and feedback processes, supporting real-time monitoring, optimized dispatch, and safety management. Value flow, through market transactions and revenue allocation, reflects the economic value of energy utilization and resource allocation efficiency. These three flows form a closed-loop system with the physical layer as the foundation, the information layer as the link, and the economic layer as the driving force, enabling AEI to achieve multi-dimensional self-coordination and market self-balancing in complex environments.
4. System operational characteristics and key issues
   AEI exhibits the following notable features and challenges in long-term operation. Strong multi-energy coupling: electricity, heat, gas, and chemical energy coexist in multiple forms, with complex conversion paths among units.

Unified energy management and coordination optimization are required to ensure energy conservation and supply-demand balance. High load variability and uncertainty: Agricultural loads are strongly nonlinear and stochastic due to agricultural schedules, seasons, and weather, imposing higher requirements on prediction accuracy and dispatch response speed.

Distributed resources and communication coordination challenges: Energy nodes in the park are numerous and heterogeneous, requiring IoT and edge computing for efficient data exchange and distributed control to avoid information delay and decision mismatch. Complex market response mechanisms: AEI is influenced simultaneously by electricity spot market price fluctuations, carbon market constraints, and ancillary service signals. Differences in timing and rules among multiple markets make the optimal bidding strategy solution more challenging. Dual attributes of electric agricultural machinery: EAMs are high-power intermittent loads and possess storage and peak-regulation potential. Their participation enhances system flexibility but also increases dispatch and bidding complexity, requiring multi-scenario optimization strategies tailored to spatiotemporal characteristics.

5. Summary of operating mechanism
   In summary, AEI is a complex system involving multiple layers, energy forms, and agents. Its operating mechanism is reflected in three aspects: structural layer: achieving dynamic balance and flexible adjustment of energy flows through the "source–grid–storage–load" coordination framework; information layer: constructing real-time monitoring and self-optimization mechanisms through sensing, intelligent control, and edge computing; market layer: realizing multi-time-scale bidding and optimal revenue allocation through a VPP platform.

   Integration of electric agricultural machinery provides AEI with greater flexible adjustment capability, giving it market adaptability and economic competitiveness comparable to industrial virtual power plants. This multi-dimensional system establishes a solid physical and theoretical foundation for subsequent bidding model derivation, reinforcement learning optimization, and multi-market strategy analysis, providing a systematic path and practical basis for intelligent, market-oriented, and low-carbon development of agricultural energy systems.

3.2.2 Multi-time-scale electricity market mechanisms and AEI participation logic

To ensure that the Agricultural Energy Internet (AEI) can efficiently and stably participate in electricity market transactions, and fully leverage the flexibility and economic potential of distributed energy, this section systematically introduces the basic operating mechanisms of day-ahead, intraday, and real-time markets, as well as their coordination logic in agricultural scenarios. By constructing a multi-time-scale market coordination system, AEI can achieve dynamic optimization and revenue maximization across multiple agents and resources at the planning, regulation, and balancing layers, providing theoretical and technical support for subsequent bidding model development and reinforcement learning algorithms.

1. Multi-time-scale market structure and temporal logic
   The electricity spot market system is generally divided into three types according to trading lead time and response speed: day-ahead market, intraday market, and real-time market. For market participants, these three form a hierarchical "forecast–adjust–balance" operational logic, where the day-ahead market focuses on planning, the intraday market emphasizes rolling adjustments, and the real-time market ensures final supply-demand balance and system security. Through price signal propagation and information feedback across time scales, the system achieves dynamic coordination of energy and economic value at different stages.

   In an agricultural park context, this mechanism is particularly significant. Overall generation and consumption schedules for the next day are prepared based on the day-ahead market, providing predictable energy arrangements for crop irrigation, greenhouse lighting, cooling, and drying. The intraday market adjusts plans in response to weather changes, equipment status, or operational schedule modifications, enabling flexible operational corrections. Real-time market participation maintains system balance at minute-level or shorter time scales, ensuring stable supply under extreme weather or sudden load changes. Coupled operation of the three markets enables AEI to establish a closed-loop decision-making system from planning to execution and from forecast to feedback.
2. Day-ahead market planning mechanism and submission process
   The day-ahead market (DAM) is the core and foundation of the electricity spot market system, mainly determining generation output and load demand for each time period of the next day. Participants must submit price–quantity bid curves in advance, and market operators clear the market based on system load forecasts, constraints, and supply-demand balance principles.

   In the AEI scenario, the virtual power plant (VPP) acts as the market entity, aggregating PV, biomass, storage, and electric agricultural machinery units to form a unified bidding strategy. Generation units predict solar irradiance and fuel availability based on weather and crop growth models, preparing next-day output curves and corresponding price ranges. Storage systems and electric agricultural machinery plan charging/discharging or operation schedules according to state of charge (SOC) and expected electricity prices. Load-side equipment, such as greenhouse heating, irrigation pumps, and cold chain systems, submit adjustable ranges and price acceptance intervals based on production schedules.

   Market clearing considers power balance, carbon quotas, and green certificates, using joint optimization models to determine the optimal transaction results. DAM typically features large trading volumes, stable prices, and strong planning, providing AEI with predictable revenue and operational guidance. However, agricultural load and renewable generation are highly affected by weather and seasonal variation, and forecast errors may cause deviations in plan execution. Therefore, AEI requires high-precision forecasting models and rolling correction mechanisms to ensure reliability and economic efficiency.

3. Intraday market rolling adjustment and flexible regulation
The intraday market (IDM) bridges the day-ahead and real-time markets. Its core objective is to make rolling adjustments to trading plans based on the latest information, reducing economic losses caused by forecast errors and enhancing system flexibility.

Within AEI operations, the intraday market reflects system adaptability. PV and biomass units can recalculate output curves according to short-term weather forecasts or maintenance schedules. Electric agricultural machinery, greenhouse loads, and irrigation systems adjust energy use according to actual progress and price fluctuations. Storage devices dynamically perform charging and discharging operations based on intraday price trends, enabling peak shaving, valley filling, and arbitrage optimization. Typically, the intraday market clears every 1–4 hours, featuring high frequency, rapid response, and agile information transfer.

Through unified VPP coordination, various internal resources share forecast results and market information, forming a hierarchical decision–collaborative optimization operation mode. The introduction of the intraday market significantly enhances AEI's operational flexibility, improving adaptability to weather changes, equipment failures, and market fluctuations.

4. Real-time market fast response and terminal balance
The real-time market (RTM) represents the final and protective layer of the multi-time-scale system, providing second- or minute-level dynamic balance. Rolling matching mechanisms correct supply-demand deviations, ensuring frequency, voltage, and system stability within safe ranges.

In this market, AEI participants must have fast response capabilities. PV units adjust inverter output according to real-time irradiance changes; biomass units flexibly regulate fuel supply to maintain stable output; storage and electric agricultural machinery automatically follow charging/discharging instructions through intelligent control systems, absorbing surplus energy during low-price periods and discharging during high-price periods, achieving economic and security optimization. High-frequency settlement in RTM not only promotes AEI load response but also provides direct incentives for intelligent control of agricultural loads.

5. Multi-market coordination and strategy linkage mechanism
Day-ahead, intraday, and real-time markets are functionally complementary and temporally linked, jointly forming AEI's dynamic trading ecosystem. DAM establishes the baseline plan, IDM provides adjustment flexibility, and RTM achieves terminal balance. The three markets implement closed-loop operation through information feedback and price signal transmission.

The AEI–VPP's multi-market participation logic can be summarized in three stages: forecast–optimize–respond. In the forecast stage, multi-source data (weather, operational schedule, historical load) are used to build day-ahead and intraday generation/load forecasts. In the optimization stage, reinforcement learning or rolling planning methods generate multi-stage bidding and dispatch strategies. In the response stage, rapid adjustments are executed according to market signals to correct deviations and maximize revenue.

Different devices play distinct market roles: PV and biomass units focus on stable generation to secure long-term revenue, while storage systems and electric agricultural machinery emphasize short-term regulation and dynamic arbitrage. Bid curves are described with linear functions where the slope represents regulation sensitivity and the intercept is an optimizable parameter. Multi-market iterative optimization achieves dynamic balance of overall revenue and risk.

6. Flexible participation characteristics of electric agricultural machinery
   Electric agricultural machinery, as a uniquely mobile, flexible load in AEI, features high power, short duration, time-shiftability, and energy storage capability. Typical equipment includes electric tractors, seeders, irrigation pumps, and harvesters. During operational gaps, they can connect to the grid through intelligent scheduling systems to allow bidirectional energy flow. Charging is concentrated in low-price periods, and operations can be delayed or energy returned to the system during high-price periods, smoothing park load curves and improving overall economic efficiency.

   Participation of electric agricultural machinery also enhances AEI's market-level flexibility and response speed. It can quickly participate in regulation during sudden PV output drops or load surges, and as mobile storage units, it expands AEI's spatial flexibility and dispatch boundaries. Through VPP aggregation control, electric agricultural machinery functions not only as agricultural equipment but also as a key participant in market transactions and system balancing.
7. Uncertainty and dynamic response requirements
   In a multi-market environment, AEI faces multiple uncertainties, including renewable generation variability, load forecast errors, price volatility, and changes in market mechanisms. These uncertainties make bidding strategies dynamically evolving and impose higher real-time and adaptive requirements on system dispatch.

   To address these challenges, AEI needs reinforcement learning (RL)-based strategy optimization and rolling update mechanisms. Through continuous interaction and feedback learning, the system can automatically adjust parameters under different market conditions, realizing a closed-loop control of "perception–decision–feedback" and improving robustness and revenue stability under uncertainty.
8. Key Findings and Implications
   In summary, AEI achieves dynamic optimization and coordinated operation across multiple time scales through the integrated operation of day-ahead, intraday, and real-time markets. The park is no longer a passive energy consumer but an active market participant integrating generation, storage, consumption, regulation, and bidding. Through flexible bidding, rolling

adjustments, and real-time responses, AEI not only ensures efficient operation and maximized economic benefits but also facilitates renewable energy integration and low-carbon agricultural production. This multi-layered market participation logic provides a solid theoretical and practical foundation for subsequent bidding model development and intelligent optimization.

3.2.3 Internal participating units and role division of AEI

The Agricultural Energy Internet (AEI), as a typical distributed energy system, integrates multiple functional units, including generation, storage, consumption, and energy management. Its operational characteristics exhibit a high coupling of energy flow, information flow, and value flow. Unlike traditional centralized power systems, AEI's internal units differ significantly in functional positioning, operational characteristics, and market behavior. However, under the support of the virtual power plant (VPP) aggregation and dispatch mechanism, cross-level resource coordination and revenue sharing can be achieved, forming an adaptive dynamic response structure in multi-time-scale electricity markets.

To further reveal the organizational logic and energy coordination mechanism of the system, this section systematically divides AEI's internal participating units from four perspectives: generation units, storage systems, electric agricultural machinery, and the aggregation layer. On this basis, a hierarchical information exchange and energy flow model is constructed to demonstrate the coordination mechanisms among units.

1. Generation units: revenue-driven active bidding units

   Generation units are the primary energy supply of AEI and the core source of economic revenue for the park. They mainly consist of photovoltaic (PV) and biomass power units, which are complementary in output characteristics and regulation capability, representing intermittent renewable energy and controllable clean energy, respectively.

   PV units are usually deployed on building roofs, greenhouse tops, and idle land within agricultural parks, converting solar energy directly into electricity. However, the randomness and intermittency of PV output make it highly dependent on weather conditions, causing frequent system power fluctuations. To enhance market response capability, this study applies deep reinforcement learning (DRL) to optimize PV bidding strategies. The agent continuously interacts with the market environment and dynamically adjusts bid functions based on electricity price fluctuations, weather forecasts, and load status, achieving a coordinated balance of revenue and risk across day-ahead, intraday, and real-time markets.

   Biomass power, in contrast, features stable output, strong predictability, and a wide regulation range, serving as a key source to maintain system stability in AEI. In the market, biomass units often act as "support load" during price fluctuations or insufficient PV output, flexibly adjusting operation to improve overall revenue. The incorporation of reinforcement learning allows autonomous optimization of biomass dispatch strategies while ensuring energy balance, further enhancing economic efficiency.

Overall, generation units, as active market participants driven by revenue maximization, achieve dynamic output adjustment and optimal market response through intelligent bidding strategies. This mechanism strengthens AEI's economic performance and flexibility and provides core momentum for adaptive system dispatch.

2. Storage systems: energy balancing and auxiliary decision units
   Storage systems in AEI perform dual functions of energy regulation, fluctuation mitigation, and auxiliary decision-making, acting as a critical link between generation and load. Typical forms include centralized storage stations and distributed storage modules, with some parks incorporating the batteries of electric agricultural machinery as mobile storage units for flexible regional energy regulation.

   Storage units regulate system power balance through charging and discharging: absorbing excess energy when renewable generation is abundant, and releasing energy during peak load or high-price periods to achieve peak shaving, valley filling, and price arbitrage. Their operational states, such as state of charge (SOC), available capacity, and charge/discharge efficiency, determine energy distribution and provide key feedback for upper-layer dispatch and intelligent bidding strategies.

   In this study, storage systems primarily serve as auxiliary modules for theoretical analysis and model calibration, not directly participating in market bidding experiments. They support energy constraint modeling and strategy feasibility verification. Introducing dynamic storage models enables more accurate characterization of AEI's energy allocation and constraint boundaries under different market conditions. Theoretical results indicate that reasonable storage capacity and dispatch strategies significantly enhance economic revenue and system flexibility.

3. Electric agricultural machinery: passive response and load regulation units
   Electric agricultural machinery (EAM) is a representative flexible load in AEI. Its operation is closely linked to work schedules, serving as a key vehicle for demand-side flexibility. Unlike generation units, EAM does not target revenue maximization, but operates under constraints of task completion and electricity price response, showing a "passive purchase—active regulation" load response characteristic.

   EAM includes electric tractors, irrigation pumps, harvesters, and other devices, with significant time-shifting load capabilities. Their electricity consumption is influenced by market prices and operational requirements: concentrated charging or energy-intensive tasks occur in low-price periods, 3while operations can be delayed or power reduced in high-price periods, optimizing cost and smoothing system load.

   In system modeling, EAM groups are simulated via response functions to reflect passive adjustment to price signals. Studies evaluate how changes in EAM quantity and load profiles affect AEI generation unit bidding strategies and overall revenue.

4. AEI aggregation layer: coordination and revenue consolidation module
The AEI aggregation layer serves as the intelligent coordination and optimization hub, integrating operational status, energy distribution, and market information from lower-level units to implement unified VPP-level decision-making and revenue allocation. This layer operates on the VPP platform, combining real-time data acquisition, intelligent optimization algorithms, and market interface modules to form a closed-loop control logic of "data-driven—decision execution—revenue feedback."
During bidding, the aggregation layer calls different optimization agents for PV, biomass, and load units based on market rules and forecast signals to solve individual strategies. After market clearing, results are consolidated and analyzed to update global strategy parameters. Through continuous reinforcement learning iteration, the aggregation layer evolves from individual optimal solutions to system-wide optimal strategies, enabling AEI to autonomously learn and adapt under complex market conditions.
The operation model reflects a typical bidirectional feedback mechanism: the upper layer sets optimization goals and bidding strategies; the middle-layer algorithm executes optimization and dispatches commands; lower-level units respond and provide operational feedback. Through the coupled interaction of information and energy flows, AEI forms a dynamic closed-loop control system, ensuring stability and revenue optimality across market environments.
5. System hierarchy and information interaction structure
Structurally, AEI operates across three layers: upper layer – VPP aggregation and decision layer: responsible for global coordination and revenue optimization, integrating generator status and establishing multi-time-scale bidding strategies for unified dispatch and decision control. Middle layer – intelligent optimization and execution layer: includes reinforcement learning, revenue forecasting, and multi-market rolling optimization modules, handling strategy computation, model updates, and parameter calibration. Lower layer – generation and load response layer: composed of PV, biomass units, and electric agricultural machinery, responsible for actual energy production and consumption, and providing feedback via real-time data. The three layers are linked through data channels and control signals: upper layer issues dispatch and bidding instructions, middle layer executes optimization and dynamic adjustments, and lower layer provides status feedback and operational results. This creates a closed-loop system with self-learning, self-coordination, and self-optimization capabilities, reflecting AEI's intelligence and market adaptability.
6. System Structure and Coordination Overview
In summary, AEI's internal participating units are functionally independent yet closely coordinated. Generation units actively bid through reinforcement learning, serving as the primary source of system revenue; storage systems support energy balance and auxiliary decisions; EAM provides demand-side flexibility through passive response; and the aggregation layer acts as the system "brain," coordinating optimization and revenue consolidation.

This multi-level, information-driven structure enables AEI to achieve dynamic coordination from distributed resources to centralized intelligent decision-making in diverse market environments, providing a solid technical and theoretical foundation for intelligent management and market-oriented operation of agricultural energy systems.

3.2.4 Research assumptions and modeling boundaries

In constructing the multi-time-scale bidding model of the Agricultural Energy Internet (AEI), it is essential to define systematic assumptions and modeling boundaries regarding the market environment, model operational characteristics, and experimental conditions. These measures ensure the model remains interpretable while maintaining computational feasibility and convergence. Such idealizations are common in complex system modeling: they simplify external constraints and disturbances without altering the core dynamics, allowing the model to focus on key issues such as bidding strategy optimization, agent interactions, and revenue evolution, while reducing uncertainty and computational complexity from non-core factors.

This section presents research assumptions and modeling boundaries from three dimensions – market level, model level, and experimental level – to clarify AEI bidding model logic, constraints, and theoretical applicability.

1. Market assumptions

   At the market level, the focus is on revealing the strategic response patterns of different types of AEI participants (generation, storage, and load units) in multi-time-scale electricity markets. To this end, the market environment is assumed to be fully competitive, information-transparent, and governed by uniform rules, such that price and behavioral changes are primarily driven by participant strategies.

   First, it is assumed that historical market clearing prices are publicly available and can be accessed in real time, enabling all participants to make bids, forecasts, and adjustments based on the same information set. This assumption eliminates information asymmetry or transmission delays, allowing price signals to propagate instantly within the system, ensuring predictable and timely bidding responses. It also focuses analysis on revenue differences arising from strategic adjustments rather than asymmetric information effects.

   Second, the market is assumed to operate without transaction friction or time delays; bidding, clearing, and settlement are completed sequentially on the same time scale, ignoring communication errors, matching delays, or external interventions. This idealization ensures rule consistency and result verifiability, allowing bidding outcomes to reflect only resource characteristics and agent strategies, facilitating comparison and attribution of reinforcement learning (RL) policy updates.

   Finally, day-ahead, intraday, and real-time markets are assumed to have no external constraints or information coupling, differing only in temporal granularity and trading volume. The three-stage market is modeled as a time-continuous but decision-independent optimization problem. This simplification facilitates multi-stage bidding modeling under a unified mathematical framework and enables strategy transfer and generalization across market scenarios.

These market assumptions provide a well-defined, idealized environment for modeling AEI strategy evolution and revenue distribution.

2. Model assumptions

At the model level, the core concern is the coupling mechanism of energy flow and information flow among heterogeneous AEI units. To reduce solution complexity and ensure stable convergence of RL algorithms, simplifications are applied to system components and state parameters.

First, all energy devices (PV, biomass, storage, and EAM) are assumed to operate normally without sudden failures or abnormal power fluctuations. This removes random disturbances or degradation effects from model training, focusing the study on strategy optimization and market response rather than outlier events.

Second, storage system efficiency and capacity are assumed constant, ignoring the long-term effects of temperature, aging, or cycle count. Storage units primarily support energy balance, dispatch, and strategy calibration; their state variables (SOC, available capacity) are treated as stable within a single market stage, effectively serving as static constraints. This reduces system dynamic dimensions and improves solvability and RL training convergence.

Third, EAM units act as passive response loads without active bidding capability. Their electricity consumption follows fixed price-response functions, adjusting charge power and timing according to market signals. Function parameters are constant and can be fitted from historical data, enabling EAM behavior to serve as a controllable, flexible load for RL agents to accurately capture price elasticity and load-shifting patterns.

Additionally, ideal communication channels are assumed, with no information delay or coupling between agents. While this ignores some real-world latencies, it enhances model clarity and analytical traceability.

3. Experimental assumptions

At the experimental level, assumptions ensure simulation results are reproducible, comparable, and interpretable. First, all node prices are assumed to be generated by a unified market clearing algorithm, following optimal economic dispatch principles. All participants aim to maximize revenue and follow baseline strategies from literature [15], ensuring the market remains in an ideal equilibrium. This provides a standard environment to assess AEI strategy optimization impact.

Second, aside from variable EAM number and type, PV, biomass, and storage capacities are fixed. EAM scale is treated as the sole adjustable factor to analyze its effect on overall bidding behavior, market equilibrium price, and revenue distribution, highlighting load-side flexibility's structural role.

Third, the external operating environment is assumed ideal: no equipment failure, communication delay, sudden weather changes, load forecast error, or price disturbance exists. Energy and information flows are continuous, stable, and noise-free. Under these conditions, strategy training and simulation can isolate external uncertainty effects, precisely identifying RL agent decision boundaries, policy evolution, and convergence.

4. Assumptions and Implications
   In summary, the study establishes a systematic set of assumptions at market, model, and experimental levels to standardize AEI multi-time-scale bidding modeling and ensure controllable validation. Market assumptions guarantee transparency and ideal competition, allowing clear observation of strategy interactions. Model assumptions ensure internal structural stability and interpretability, supporting RL policy training. Experimental assumptions provide a standardized testing framework, making simulation results reproducible and comparable.

   While these assumptions simplify real-world market complexity, they provide clear theoretical boundaries and verifiable bases for identifying AEI bidding behavior, evaluating revenue, and optimizing intelligent algorithms, and they form a scientific reference for future studies incorporating uncertainty, information asymmetry, and environmental variability.

3.3 AEI bidding framework and fundamental model

3.3.1 AEI market participation mode

The Agricultural Energy Internet (AEI) is a novel energy system designed for agricultural parks, enabling multi-energy coordination, intelligent optimization, and market-oriented operation. Fundamentally, it promotes the transition of agricultural energy systems from distributed autonomy to coordinated optimization through digitalization and intelligence. Compared to traditional single-energy supply models, AEI emphasizes integrated energy utilization and information-driven adaptive operation, aiming to achieve efficient matching and dynamic balance at the levels of energy production, consumption, and trading. Its core feature is the aggregation, coordination, and unified dispatch of distributed resources via the virtual power plant (VPP) mechanism, integrating PV, biomass power generation, energy storage systems, and electric agricultural machinery (EAM) into an entity exhibiting market participant behavior.

The AEI market participation mode follows a multi-unit coordination, centralized optimization, and dynamic feedback logic, representing a key pathway for transforming agricultural energy systems from distributed operation toward intelligent, market-oriented management. This mode enables orderly aggregation and coordination of energy units and allows agricultural parks to participate in electricity market trading and ancillary services.

AEI uses the VPP as the organizational core, virtually aggregating and managing production, storage, and consumption units through a cyber-physical system (CPS). Through dual-layer energy and information interactions, AEI can monitor real-time status, collect data, and make optimal dispatch decisions, forming a fundamental framework for multi-source energy coordination. This mechanism enables dynamic coupling and optimal allocation among generation, storage, and load, improving overall system responsiveness and operational efficiency.

Within the VPP framework, PV and biomass units are treated as dispatchable generation resources; storage systems provide energy balancing and time-shifting functions; EAMs act as flexible controllable loads. The aggregation layer abstracts these distributed resources into virtual agents with market behavior characteristics, allowing participation in bidding and revenue optimization as a whole. Virtualized management achieves logical centralization of dispersed physical resources and provides training samples and feedback for reinforcement learning algorithms.

The primary goal of VPP-based AEI is to enhance controllability and overall economic performance. AEI is modeled as a virtual market agent, simulating market decision behavior to analyze revenue changes and market adaptability under various dispatch and bidding strategies. Virtual simulation bridges theoretical research and practical market application, validating and quantifying the feasibility and stability of system strategies.

Furthermore, the VPP structure establishes a unified decision-making platform, where reinforcement learning algorithms can iteratively optimize strategies, evaluate revenue, and converge parameters, enabling self-learning and optimal control of AEI market participation.

The internal AEI structure consists of generation, storage, and load units, coordinated by the aggregation layer to jointly maintain energy balance and market strategy decisions. The multidimensional coupling of information, energy, and value flows constitutes the core logic of AEI operational optimization.

On the supply side, PV and biomass generation form the primary power sources. PV output is highly stochastic due to irradiance, weather, and time-of-day characteristics, while biomass units offer strong controllability and stability, compensating during PV deficits or high-price periods. This complementarity establishes a renewable-controllable energy structure, enhancing AEI's resilience and adaptability to market price fluctuations and power uncertainty, thus improving bidding proactivity and supply-demand balance.

Storage systems provide energy balance and auxiliary optimization. Although dynamic storage processes are not directly simulated in this study, their energy constraints and regulation variables are maintained theoretically to reflect buffering and stabilization roles. Storage allows flexible energy allocation across different market timescales, enabling discharging at high prices and charging at low prices to maximize revenue and smooth the load.

On the load side, EAM represents flexible demand. Its operation and charging are influenced by price signals, task scheduling, and seasonal factors, exhibiting high controllability and shiftability. Low market prices prompt concentrated charging or energy-intensive operations, while peak periods or high prices trigger peak-shaving adjustments to optimize consumption and costs.

By adjusting EAM quantity, task timing, and response parameters, AEI can explore load participation dynamics at the micro-level and evaluate the marginal contribution of flexible demand to overall revenue and risk control.

The aggregation layer, as the AEI coordination core, performs information collection, strategy generation, and dispatch. Its role extends beyond passive response to actively constructing an evolving strategy space. Continuous feedback and reinforcement learning-driven policy updates enable dynamic adjustments to unit behavior, achieving dual optimization of revenue and operational stability.

AEI's market participation can be abstracted as a four-stage closed-loop optimization mechanism, including bidding initialization, market clearing, strategy feedback and revision, and revenue update. This process forms a dynamic reinforcement learning-driven policy optimization cycle, allowing continuous learning, adaptation, and evolution in complex market environments.

The aggregation layer generates initial bid strategies based on unit characteristics, historical price sequences, predicted load, and weather data. For simplicity and differentiability, the bid function follows a linear form:

$$\lambda = \alpha + \beta P \tag{3.1}$$

where λ is the bid price, P is the bid quantity, β is the marginal cost slope, and α is the base price. Reinforcement learning agents produce multiple parameter combinations as diverse bid samples to enhance exploration and convergence for subsequent clearing and optimization.

Market clearing follows the social welfare maximization principle:

$$\max\left(\sum_i U_i(P_i) - \sum_j C_j(P_j)\right) \tag{3.2}$$

where $U_i(P_i)$ is the utility of buyers and $C_j(P_j)$ is the cost of generators. Clearing algorithms determine transaction prices and quantities under power balance and capacity constraints. Virtual market simulations input bids from PV, biomass, and EAM units to obtain clearing results and revenue distributions, providing feedback for reinforcement learning strategy optimization.

The aggregation layer updates bid parameters based on clearing outcomes. The reinforcement learning model uses node clearing prices as state variables, combined with previous revenue, power, and load response characteristics, to update the value function and iteratively adjust policies. Continuous interaction enhances robustness and generalization in nonlinear market conditions.

AEI generates new bid strategies from reinforcement learning outputs, forming a revenue-driven rolling optimization mechanism. Agents optimize parameters α and β to improve subsequent market performance. After multiple iterations, AEI converges toward optimal bidding strategies, achieving long-term economic efficiency and operational stability under price fluctuations and market disturbances.

AEI's market participation mode leverages VPP-based aggregation to achieve organized, information-driven, and strategic management of energy resources. Its source-storage-load coordination and reinforcement learning-driven closed-loop optimization allow self-learning, adaptation, and autonomous control in complex, dynamic markets.

This mode provides a theoretical framework and operational pathway for market-oriented agricultural park energy systems and establishes a solid methodological foundation for subsequent multi-time-scale bidding modeling, strategy game analysis, and simulation experiments.

3.3.2 Multi-market temporal framework

To systematically analyze the market behavior characteristics of the Agricultural Energy Internet (AEI) across different time scales, this section constructs a multi-market temporal framework based on the day-ahead, intraday, and real-time markets. As shown in Figure 3.2, the core objective of this framework is to depict the information flow transmission, decision logic evolution, and strategy linkage of AEI throughout the multi-stage bidding process, revealing the system's behavioral patterns and optimization mechanisms in dynamic market environments, and providing a structured foundation for subsequent reinforcement learning model training and strategy optimization.

1. Overall structure and temporal logic

   The electricity markets in which AEI participates can be divided into three levels: day-ahead market (DAM), intraday market (IDM), and real-time market (RTM), each representing different time granularity, information update frequency, and decision-making cycles. These three markets are sequentially linked in the time dimension but relatively independent in decision logic, forming a nested, rolling optimization structure across multiple time scales.

   In this idealized temporal framework, the day-ahead market focuses on planning and forecasting, primarily addressing next-day energy allocation and generation-load balance; the intraday market emphasizes flexibility and correction, responding to short-term deviations and forecast errors; the real-time market highlights responsiveness and stability, undertaking the final power balance and revenue adjustment of the system.

   Together, these markets form a dynamic progression of "planning—correction—response." The optimal decision in each stage not only affects the revenue and risk levels for that stage but also provides reference boundaries and constraints for the subsequent stage, forming a cross-temporal collaborative optimization loop. Under this framework, AEI can unify long-term planning with short-term scheduling, balancing the trade-off between economic efficiency and operational flexibility.

   The differences between the markets are mainly reflected in three aspects (Table 3.1):

   This multi-market temporal structure provides AEI with multi-layered control and optimization space, enabling adjustments to generation plans, load response scheduling, and economic optimization across different time scales. At the same time, market hierarchy offers structured support for phased training of reinforcement learning algorithms and multi-scale strategy transfer.

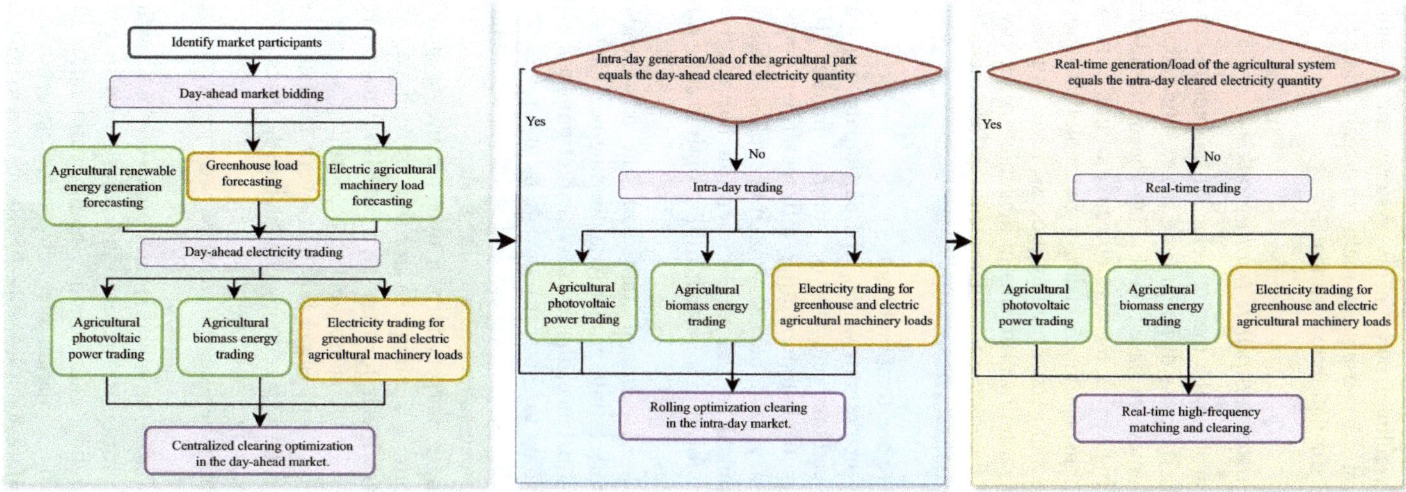

Figure 3.2 Multi-market temporal framework

Table 3.1 Key characteristics of different markets

Market type	Time scale	Information update frequency	Transaction features
Day-ahead Market	Daily	Covering forecast data	Long cycle, large volume, planning-oriented
Intraday Market	Hourly	Based on short-term measured data	Medium cycle, medium volume, correction-oriented
Real-time market	Minute-level	Based on real-time signals	Short cycle, small volume, response-oriented

2. Day-ahead market: planning bidding stage
   The day-ahead market is the foundational and core segment of the entire electricity market, representing the starting point of AEI market behavior. All participating entities must submit their bids and volume plans one day in advance based on forecasts of next-day renewable energy output, load demand, and price trends.

   For AEI, the main task during this stage is to develop overall generation and consumption plans to ensure internal energy balance and revenue expectations within the park.

   At this stage, the AEI aggregation layer considers: output forecasts of photovoltaic and biomass generation units; planned electricity demand of electric agricultural machinery (EAM); energy constraints and capacity limits of the storage system; historical market prices and weather forecast information. By integrating these factors, AEI generates bid curves and power schedules for each unit.

   The market clearing stage is conducted based on the social welfare maximization principle, determining optimal transaction quantities and clearing prices while satisfying supply-demand balance and system constraints.

   The clearing results in the day-ahead market form AEI's "baseline schedule," which serves as the reference energy plan and bidding line for the next day. The focus of this stage is not real-time response but rather the formulation of a long-term dispatch strategy balancing economic efficiency and stability.

   Reinforcement learning models in this stage primarily learn methods for optimizing the strategy space, determining how to generate planned bidding schemes that maximize expected revenue under high uncertainty forecast conditions.
3. Intraday market: fine-tuning compensation stage
   The intraday market serves as a dynamic correction and supplementation of the day-ahead plan, acting like a "feedback correction mechanism" during plan execution. This market typically operates with a rolling clearing cycle of 1–4 hours, allowing market participants to adjust bids continuously based on the latest measurement data and short-term forecasts to compensate for deviations between the day-ahead plan and actual operation.

   For AEI, the main function of the intraday market is to reduce economic risk from renewable energy uncertainty and ensure the economic efficiency and stability of system operation. Its optimization logic includes: dynamically adjusting photovoltaic output forecasts based on short-term weather changes (e.g., cloud cover, irradiance); flexible biomass unit output adjustment according to market signals to provide generation compensation; EAM adjusting charging times forward or backward according to price signals to achieve load shifting and peak-valley management; storage systems performing charge–discharge optimization based on price trends and power deviations.

In the reinforcement learning model, this stage corresponds to strategy updating and revenue correction. The agent continuously adjusts its internal value functions based on market clearing prices, generation output, and other feedback signals, enhancing robustness and generalization. Ultimately, AEI maintains high revenue stability under varying weather and market conditions.

4. Real-time market: dynamic response stage

The real-time market is the final stage of AEI's participation in electricity trading and is the most dynamic and challenging stage of the multi-market system. It typically operates on a 5–15 minute response cycle, emphasizing high-frequency decision-making, rapid response, and system balance maintenance. The core task at this stage is to respond to sudden load changes, renewable generation uncertainty, and equipment status variations, ensuring system safety and revenue balance.

In the real-time market, the AEI aggregation layer rapidly executes power adjustments and load redistribution based on nodal real-time price signals and system operating status. Specifically, photovoltaic and biomass units adjust output in real-time according to system frequency and price signals; EAM flexibly executes charge–discharge control based on real-time prices, charging during low-price periods and reducing load during high-price periods; storage units undertake emergency energy balancing and short-term support tasks to stabilize system fluctuations.

Optimization in the real-time market emphasizes the synchrony of response speed and revenue adjustment. AEI agents must complete state recognition, strategy calculation, and action execution within an extremely short period. Reinforcement learning models at this stage continuously calibrate strategies through high-frequency feedback, enabling agents to maintain optimal revenue levels in dynamic environments.

5. Multi-Time-Scale Framework Overview

In summary, AEI's multi-market temporal framework constructs a full-cycle optimization system covering planning decisions, dynamic corrections, and real-time responses. Through the baseline formulation in the day-ahead market, flexible corrections in the intraday market, and rapid responses in the real-time market, AEI can achieve dynamic coordination of energy and revenue maximization across multiple time scales.

This framework not only reveals AEI's decision pathways and behavioral logic in a multi-layered market structure but also provides theoretical and structural support for hierarchical training of reinforcement learning agents, cross-stage strategy transfer, and dynamic optimization.

Through this temporal modeling approach, AEI's market-based operation mechanisms can be unified in both theoretical and simulation contexts, laying a solid foundation for subsequent chapters on bi-level bidding modeling and multi-time-scale reinforcement learning strategy optimization.

3.3.3 *Market clearing and revenue model*

The economic operation of the Agricultural Energy Internet (AEI) in multi-time-scale electricity markets is essentially a process intertwined with multi-agent participation, multi-stage decision-making, and multi-objective optimization. Its core task is to formulate reasonable bidding strategies based on the operational characteristics of markets at different time scales (day-ahead, intraday, real-time), aiming to maximize revenue and coordinate energy. To systematically depict the behavior, logic, and economic characteristics of AEI units in the market, this section elaborates on the market clearing mechanism, revenue function modeling, constraint definition, and cross-stage optimization logic. The overall goal is, under the premise of independent clearing in each market stage, to establish an iterative and inheritable revenue-optimal model through reinforcement learning agents, thereby enhancing AEI's economic efficiency and dynamic adaptability.

1. Market clearing mechanism
   Under the idealized assumption of electricity market operation, the clearing mechanism follows the social welfare maximization principle. Following [15] for idealized modeling, the system operator collects bid information from all generation and load-side participants and, considering system power balance, unit capacity constraints, and load demand elasticity, solves for the optimal combination of quantities and prices to maximize overall system benefits.
   The formalized market clearing model can be expressed as:

$$\max_{p_g, q_d} \left[\sum_{d \in D} \left(-f_d D_d^{\max} q_d + \frac{1}{2} f_d q_d^2 \right) - \sum_{g \in G} \left(\alpha_g p_g + \frac{1}{2} \beta_g p_g^2 \right) \right] \tag{3.3}$$

   where p_g and q_d represent the generation output of units and the electricity demand of loads, respectively; α_g and β_g are the fixed cost and marginal cost coefficients of the units, reflecting operational characteristics and economic efficiency; f_d is the price sensitivity coefficient of the load demand function, and $D_d^{\max}$ is the maximum demand capacity.
   This optimization model aims to maximize the difference between generation-side and load-side utility functions while satisfying constraints, thereby achieving maximum social welfare. The constraints include:

$$\begin{cases} \sum_{g \in G} p_g = \sum_{d \in D} q_d \\ -F \leq \mathrm{PTDF}(p - q) \leq F \\ p_g^{\min} \leq p_g \leq p_g^{\max}, \forall g \in G \end{cases} \tag{3.4}$$

where the power balance constraint ensures supply-demand matching; PTDF is the power transfer distribution factor matrix; F is the vector of line flow limits. Unit capacity constraints prevent overload operation.

In the actual solution process, CVXOPT is employed as the optimization solver to implement the clearing calculation. The solver, based on a primal-dual interior-point framework and sparse matrix handling, can efficiently solve large-scale quadratic programming problems. During computation, CVXOPT ensures numerical stability and accuracy through symmetric indefinite matrix decomposition and adaptive step-size control, which is particularly suitable for handling massive bid data in electricity market clearing.

For multi-period market clearing scenarios, CVXOPT leverages the sparse structure of the KKT system and preconditioned conjugate gradient methods to significantly reduce memory usage and improve iterative convergence efficiency. Its adaptive adjustment of interior-point barrier parameters effectively handles unit operation constraints and network transmission limits, ensuring accurate calculation of marginal prices and cleared quantities. Under complex market boundary conditions, the solver monitors primal-dual residuals and tracks the central path to guarantee the feasibility and economic optimality of the clearing results.

The main outputs of the model include nodal prices (\lambda_i) and the transaction quantities of each participant. Nodal prices, as shadow variables, reflect the marginal balancing cost of the system and serve as key incentive signals for reinforcement learning agents during training. Transaction quantities represent the bidding response levels of different participants, directly determining the revenue and energy allocation of AEI units.

The relationship between multiple markets is reflected in temporal sequence rather than coupling constraints. The day-ahead market sets the baseline transaction quantities and prices, the intraday market corrects deviations based on this baseline, and the real-time market completes final balancing and adjustment. Clearing in each stage is independent, but unit capacities and certain state variables are inherited across stages, forming a "stage-wise independent solving with cross-stage capacity transfer" dynamic coordination mechanism that balances computational efficiency and global consistency.

2. Revenue function modeling

The revenue of various AEI units (photovoltaic, biomass, storage, and electric agricultural machinery) depends on market clearing prices and their bidding strategies. The revenue of a single unit in any market stage can be expressed as:

$$R_g = \lambda_t p_g - C_g(p_g) \tag{3.5}$$

where λ_t is the nodal clearing price at that stage, and $C_g(p_g)$ is the cost function of the unit. For renewable energy units, the cost function mainly reflects maintenance and depreciation; for storage units, charging/discharging losses and cycle degradation costs must also be considered.

Within the multi-time-scale framework, total revenue can be expressed as:

$$R_{AEI} = \sum_{t \in T} (\lambda_t p_t - C_t(p_t)) \tag{3.6}$$

3. Constraints and optimization logic
   To ensure operability and economic rationality, the revenue model must satisfy three types of constraints: physical, economic, and operational. Physical constraints include unit power limits, storage capacity limits, and system balance conditions; economic constraints include bidding price ranges and continuity of the bidding function to prevent irrational bids; operational constraints involve ramp rates, minimum output duration, and other operational characteristics to ensure feasibility.
   The optimization logic adopts a three-stage mechanism of "static clearing + dynamic inheritance + reinforcement adjustment": the static clearing stage solves for optimal clearing under fixed market conditions; the dynamic inheritance stage transfers key states and capacity parameters from the previous bidding round to the current stage to ensure cross-temporal continuity; the reinforcement adjustment stage updates bidding strategies through reinforcement learning agents analyzing revenue feedback, achieving rolling optimization of the revenue function.
4. Market Model and Revenue Overview
   In summary, the market clearing and revenue model provides a rigorous economic basis and mathematical framework for AEI's bidding optimization. By combining the principle of social welfare maximization, multi-market temporal linkage, and dynamic adjustment by reinforcement learning agents, the model achieves revenue balance and energy optimization across different time scales.
   This integrated clearing-revenue design not only enhances AEI's flexibility and competitiveness in electricity markets but also lays a solid theoretical foundation for subsequent bi-level bidding optimization models and reinforcement learning strategy training.

3.3.4 Bidding and demand-side response mechanism of electric agricultural machinery

With the gradual improvement of electricity market mechanisms, the role of agricultural parks is evolving from a traditional single energy-consuming entity into an integrated energy participant encompassing generation, storage, and load regulation. In this context, the Agricultural Energy Internet (AEI), through a virtual power plant platform, aggregates and coordinates internal photovoltaic, biomass units, and a large number of electric agricultural machines (EAMs), enabling the park to participate in electricity markets as a single market entity and achieve centralized optimization and economic coordination of distributed resources.

This aggregation mechanism not only enhances the market participation capability of distributed resources but also provides a technical and institutional pathway for EAMs to enter the electricity market. The overall operational logic of AEI is to achieve bidirectional flexible dispatch of generation and load while meeting agricultural production schedules, thereby balancing economic efficiency and renewable energy utilization across different market time scales.

1. Market participation and bidding mechanism
   In the day-ahead market, AEI submits aggregated bids and quantity offers in the form of a virtual power plant. The day-ahead market is characterized by planning and baseline functions, requiring participants to submit generation plans and consumption forecasts one day in advance as a reference for subsequent intraday and real-time markets. Within AEI, different units formulate corresponding bidding strategies according to their operational characteristics and resource availability: photovoltaic units determine saleable electricity based on irradiance forecasts; biomass units plan supply based on fuel availability and equipment limits; EAM loads determine demand curves according to work schedules, agricultural task intensity, and real-time price-response functions.
   After considering resource balance, price fluctuation risk, and expected revenue, the park expresses the price–quantity relationship as a linear function according to formula (3.1). By rolling adjustments to the intercept parameter, AEI can achieve adaptive revenue optimization while maintaining operational safety.
2. Price-response characteristics of electric agricultural machinery
   Within the AEI market system, EAMs are both major load units and potential demand response resources. Their core characteristic is the ability to adjust charging periods, power levels, and even operation schedules flexibly based on real-time price signals without affecting agricultural production progress or quality, thereby contributing to peak shaving, valley filling, and ancillary services.
   The bidding strategy of the EAM group can be described using a demand curve, modeled following [15]:
   $$\lambda_{EAM}(Q_{EAM}) = f_{EAM}\left(Q_{EAM} - D_{EAM}^{\max}\right) \tag{3.7}$$
   where f_{EAM} is the slope, remaining constant over time; Q_{EAM} is the demand quantity; $D_{EAM}^{\max}$ is the maximum load demand of EAMs.
   This function reflects the price strategy of agricultural machinery loads, enabling the AEI system to dynamically identify optimal bidding strategies under different load characteristics.
3. Market clearing and bidirectional coordination mechanism
   During market clearing, AEI's generation units and EAM loads jointly participate in system bidding and demand response, forming a bidirectional supply-demand coordination mechanism. The market operator performs unified clearing according to the social welfare maximization principle, determining market prices and transaction quantities while satisfying system power balance and equipment constraints.

Changes in clearing prices directly reflect dynamic supply-demand relations. EAM groups adjust their charging and discharging power in real time according to the latest price signals. The clearing results not only represent generation-side competition but also reflect the economic value of load-side response potential.

4. Strategy optimization and reinforcement learning
At the strategy optimization level, the electricity consumption of EAMs is modeled as a passive response decision unit, not as an independent bidding agent, and participates as an environmental variable in the reinforcement learning training process. The AEI aggregation-level reinforcement learning agent, upon receiving market clearing results, adjusts bidding strategies based on nodal prices, generation unit revenue, and EAM load feedback, thereby achieving dynamic revenue optimization.

   Market feedback characteristics, such as price fluctuation range, load response intensity, and energy balance deviation, are treated as environmental states (State). The reinforcement learning agent optimizes the revenue function (Reward) by adjusting bid function parameters (Action). After multiple iterations, the system forms stable and transferable optimal bidding strategies under different EAM access scales and price fluctuation environments, achieving dual improvements in economic efficiency and flexibility.

5. Revenue function modeling and constraints
AEI's revenue system consists of the generation side (photovoltaic and biomass units) and the load side (EAMs), with revenue functions designed as follows:
① Photovoltaic unit

$$\Pi_{pv} = \lambda_{pv}P_{pv} - c_{pv}P_{pv} \tag{3.8}$$

where the operating cost c_{pv} of photovoltaic units is low, and the marginal cost is approximately constant. The reinforcement learning agent adjusts the bid intercept α_{pv} to maximize expected revenue.
② Biomass unit

$$C_{bio}(P_{bio}) = \alpha_{bio}P_{bio} + \frac{1}{2}\beta_{bio}P_{bio}^2 \tag{3.9}$$

Revenue is:

$$\Pi_{bio} = \lambda_{bio}P_{bio} - C_{bio}(P_{bio}) \tag{3.10}$$

Through historical revenue feedback and reinforcement learning updates, biomass units achieve steady revenue improvement under uncertain market conditions.
③ Electric agricultural machinery load unit

$$\Pi_{EAM} = U_{EAM}(Q_{EAM}) - \lambda_{EAM}Q_{EAM} \tag{3.11}$$

where U_{EAM} is the utility function. Load adjustments by EAMs generate dynamic feedback effects on system net load and price curves.

④ Constraints

Constraints include generation capacity limits, load boundaries, power balance, and multi-market capacity inheritance:

$$0 \leq P_{pv,t} \leq P_{pv}^{\max} \tag{3.12}$$

$$P_{bio}^{\min} \leq P_{bio,t} \leq P_{bio}^{\max} \tag{3.13}$$

$$Q_{EAM}^{\min} \leq Q_{EAM,t} \leq Q_{EAM}^{\max} \tag{3.14}$$

Formulas (3.12)–(3.14) ensure the physical feasibility of photovoltaic units, biomass units, and EAM loads, respectively.

6. Market settlement and revenue closed-loop

   During settlement, transaction quantities and revenue distribution are executed based on clearing results. Generation units calculate revenue based on cleared quantities and prices, while EAM groups achieve indirect economic savings by reducing peak-period consumption and charging during low-price periods.

   This forms a closed-loop mechanism of "active bidding on generation side—passive response on load side—revenue feedback and re-optimization." Through multiple revenue feedback iterations, the AEI system continuously learns and adjusts strategies, maintaining high economic stability and operational flexibility under varying EAM penetration and market conditions.

7. EAM Integration and Impact Overview

   Overall, the introduction of EAMs significantly expands AEI's market participation dimensions and strengthens its dynamic load regulation capability. Through unified management via a virtual power plant platform and dynamic optimization with reinforcement learning, AEI achieves coordinated dispatch of generation and load while maximizing market revenue. With the continuous growth of EAM fleets and the maturity of intelligent charging control technologies, their contribution to system balancing and price stability in future electricity markets will further increase, making them a key link connecting "production—electricity—market" in the Agricultural Energy Internet.

3.3.5 Bidding and response process

The operation of the Agricultural Energy Internet (AEI) in multi-time-scale electricity markets is reflected not only in the bidding submission of generation units and market clearing calculations but also in a closed-loop system based on multi-agent collaborative learning and dynamic feedback optimization. Various units within AEI (photovoltaic and biomass generation units) make autonomous decisions and update strategies through reinforcement learning agents, continuously adjusting their bidding behavior to achieve long-term revenue maximization. Electric agricultural machines (EAMs), as the core demand-side response units, flexibly adjust electricity purchasing plans based on market price signals, forming a multi-layer interactive feedback chain of "market—AEI—EAM" together with the generation side. This process reflects AEI's intelligent evolution characteristics and adaptive optimization capability under multi-market environments.

1. Overall process

   The bidding and response process of AEI can be systematically divided into four consecutive stages: bid submission, market clearing, state feedback, and strategy update. This process establishes a dynamic optimization structure centered on reinforcement learning – "perception—decision—learning—re-decision" – allowing the system to continuously self-adjust and evolve revenue across different time-scale markets.

   During the bid submission stage, the virtual power plant (VPP) aggregation layer first collects real-time operational data from various units in the park, including photovoltaic generation forecasts, biomass fuel supply status, and other relevant information. The aggregation layer then generates initial bids for each unit based on economic parameters, historical price trends, and expected revenue signals. Each generation unit's bidding strategy is independently generated by its corresponding reinforcement learning agent, with the bidding function expressed in a linear form, and the agent continuously optimizes the intercept α in the bidding function.

   In the market clearing stage, the power trading center solves for system prices and power allocations based on the social welfare maximization principle. The clearing process comprehensively considers generation costs, load demand, transmission constraints, and price volatility, ultimately determining equilibrium prices and traded quantities at each node. The clearing results determine the immediate revenue level of AEI's units and provide critical purchasing signals for EAMs. EAMs dynamically adjust charging power and usage periods according to clearing prices and their load elasticity functions, achieving an optimal balance between cost and production efficiency.

   During the state feedback stage, multidimensional data – including clearing prices, EAM electricity consumption, generation output levels, and corresponding revenues – are input as environment states (s_t) to the reinforcement learning agents. This stage functions as the "perception" process for the agents, capturing market price fluctuations, power uncertainty, and revenue trends, providing key input for the next round of strategy optimization.

   Finally, in the strategy update stage, reinforcement learning agents optimize bidding parameters according to the reward function, which balances revenue maximization and risk constraints:

$$R_t = \Pi_g(t) - \omega \mathrm{Var}(\Pi_g) \tag{3.15}$$

   where $\Pi_g(t)$ represents the unit's current revenue, $\mathrm{Var}(\Pi_g)$ reflects revenue volatility, and ω is the risk preference coefficient. For simplification in this study, the risk preference coefficient is set to 0. Using methods such as Deep Deterministic Policy Gradient (DDPG), agents iteratively converge and optimally update strategy parameters, gradually forming a stable bidding mechanism.

After training, in each market cycle, agents use the optimal parameters learned from the previous round as the initial state to regenerate bids, achieving rolling optimization and adaptive iteration.

In summary, AEI's bidding logic can be described as: bid initialization → market clearing → state feedback → strategy update → new bid. This loop continuously rolls across multi-time-scale markets, enabling adaptive evolution of internal units and demand-side loads. Figure 3.3 illustrates the information flow structure of AEI's bidding and response, covering agent strategy generation and market clearing as core layers, forming the backbone of AEI's intelligent optimization logic.

2. System interaction logic

The bidding behavior of AEI market participants can be abstracted as a typical multi-agent reinforcement learning system, where the market environment, trading platform, generation units, and EAMs together form an information interaction network. Participants are both independent and dynamically coupled, forming a distributed strategy coordination mechanism. The basic logic of unit participation and profit generation is shown in Figure 3.4.

First, reinforcement learning agents for generation units (PV Agent and Bio-Agent) generate bidding strategies after receiving market state information and obtain immediate revenue feedback through market clearing. Each agent operates independently, optimizing decisions according to its own revenue objectives rather than seeking global coordination. This independent optimization mechanism enhances AEI's robustness and scalability in complex market environments.

Second, EAMs, as core demand-side nodes, have electricity purchasing behavior driven mainly by real-time prices and load elasticity functions. The EAM response model participates in system peak-shaving and frequency regulation by adjusting charging power and deferring charging periods, indirectly influencing AEI's load curve and bidding strategies. During high-price periods, EAMs reduce consumption to lower peak load; during low-price periods, they increase charging power to achieve economic charging, optimizing flexible load at the market level.

Third, the aggregator layer serves as the "central brain" of AEI, collecting multidimensional data from the market, generation units, and EAMs, including nodal prices, generation output, revenue changes, and load response levels. The aggregator redistributes unit bids and load response strategies based on reinforcement learning outputs to achieve collaborative optimization at the global level. This mechanism allows AEI to dynamically adapt to market price signals and system fluctuations, realizing coordinated control across multiple time scales.

Overall, AEI's bidding and response behavior is not merely the outcome of a single market action but an adaptive process integrating intelligent learning and strategy evolution. Units enhance their perception and forecasting of market dynamics through multi-round interactions with the environment, transitioning from "static decision-making" to "intelligent evolutionary decision-making." Through this learning-and-feedback-based optimization mechanism, AEI

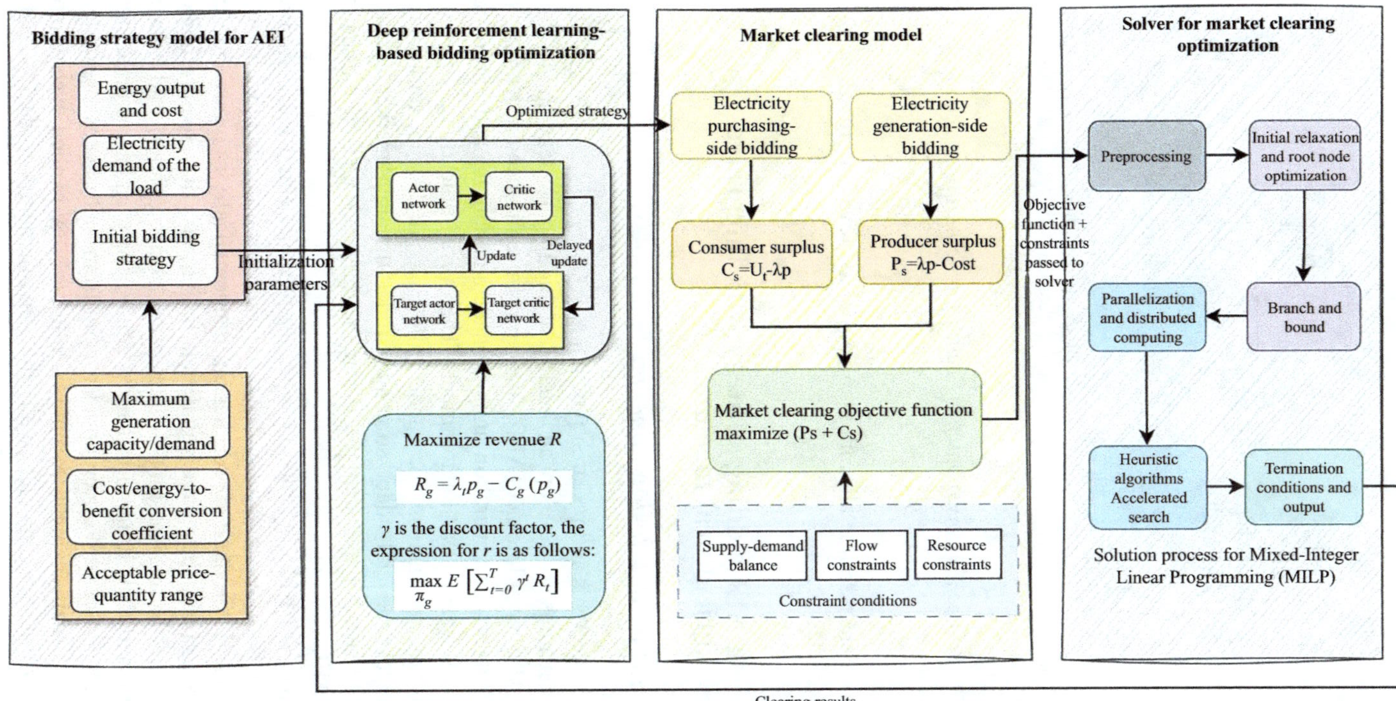

Figure 3.3 AEI bidding model based on deep reinforcement learning

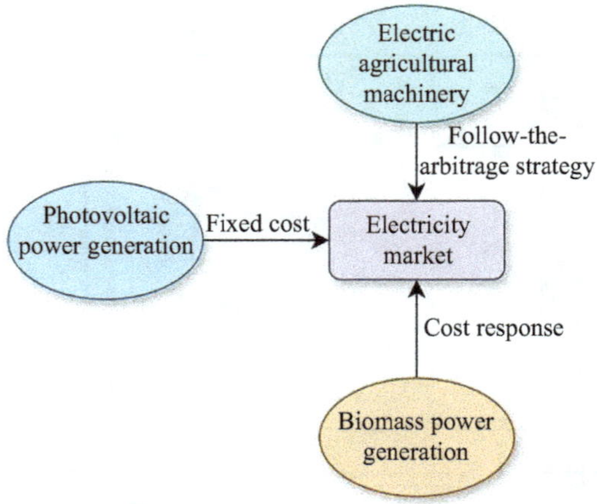

Figure 3.4 Profit mechanisms of each AEI unit participating in the electricity market

achieves efficient resource allocation and sustained revenue growth across different market stages.

3.4 Modeling AEI bidding strategy with integrated electric agricultural machines

3.4.1 Overall model design concept

1. Modeling objectives and research positioning

   This section aims to develop an intelligent bidding strategy model for the Agricultural Energy Internet (AEI), leveraging a reinforcement learning (RL) framework to enable generation units to make autonomous and optimal decisions in multi-type market environments. The overall objective is to introduce a self-learning mechanism for agents, allowing AEI's generation units to maximize revenue and adapt strategies under dynamic market conditions, thereby improving the overall economic performance and operational resilience of AEI.

   Traditional AEI bidding models typically rely on centralized optimization methods, such as linear programming, mixed-integer programming, or bi-level optimization. While these approaches are mathematically rigorous and clearly express constraints, they have significant limitations: they require complete knowledge of the market model, generation characteristics, and constraint parameters, and they struggle to handle continuously changing price signals, nonlinear responses, and uncertainties arising from multi-agent interactions.

   In contrast, reinforcement learning does not rely on explicit system modeling. Instead, agents interact with the environment, gradually learning near-optimal strategies. Through a "trial-error-feedback-adjustment" mechanism, agents accumulate experience over multiple rounds of interaction, achieving acceptable or even optimal market performance under incomplete information and uncertain conditions.

The innovation of this study lies in abstracting the AEI bidding process as a multi-agent interaction system based on reinforcement learning, where each generation unit (photovoltaic, biomass) is treated as an independent RL agent. Each agent continuously optimizes its bidding strategy to maximize long-term cumulative revenue. Simultaneously, through dynamic interaction with flexible loads such as electric agricultural machines (EAMs), AEI maintains adaptive regulation and stable revenue amid market signal fluctuations.

Ultimately, this model aims to build a "data-driven → intelligent decision → feedback optimization" closed-loop decision-making system for AEI, enabling the agricultural energy system not only to participate in markets but also to autonomously evolve and learn in multi-time-scale market linkages.

2. Bi-level interaction architecture

To capture AEI's dynamic decision-making in electricity markets, a two-layer interaction architecture – market layer and agent layer – is designed, reflecting the decision logic and feedback relationships at different system levels.

The market layer represents centralized clearing mechanisms at the regional or provincial market level. It determines transaction prices and traded quantities based on social welfare maximization or market equilibrium principles. In the RL framework, the market layer serves as the environment, providing state inputs and reward feedback to agents.

The agent layer comprises independently learning generation unit agents, including photovoltaic and biomass units. Agents generate bid parameters based on their operational status, power forecasts, cost structures, and risk preferences, submitting their bid curves to the market.

The interaction process can be divided into four stages. In the bidding stage, agents perceive system states, including historical prices, generation output, load levels, and weather forecasts, and generate optimal bidding parameters, submitting them to the market. In the clearing stage, the market performs social welfare-maximizing clearing using all bid information and demand response curves, outputting nodal prices and cleared generation. In the feedback stage, the market returns clearing results, including prices, revenues, and EAM load responses, allowing agents to update their state sets. In the strategy update stage, agents calculate rewards and risk metrics, updating their internal parameters using DDPG to form new bid strategies.

This bi-level structure effectively integrates centralized market rules with distributed intelligent decision-making, forming a self-evolving dynamic game system. It allows heterogeneous generation units to independently learn market patterns without global synchronization, significantly enhancing scalability and computational efficiency.

Within this framework, electric agricultural machines, as price-responsive flexible loads, adjust consumption according to price-response functions.

While they do not directly participate in agent learning, they influence price feedback through market clearing, indirectly affecting generation agents' bidding strategies. In this sense, EAM responses act as an external disturbance shaping the evolution of AEI agent strategies, giving the system more realistic market adaptability.

3. Adaptability and implementation logic of the RL framework
   The RL framework follows the basic loop of State → Action → Reward → Policy Update. Generation unit agents in AEI execute the following operations during each market cycle. In state perception, agents observe market and system states, including price trends, load forecasts, available generation, and EAM response levels. In an action decision, based on current policy, agents generate bidding decisions, specifically price parameters. In reward feedback, after market clearing, agents calculate immediate rewards using the revenue function. In policy update, agents update strategy parameters via DDPG, gradually converging from short-term revenue optimization to long-term cumulative revenue maximization.

   The RL framework offers several advantages in the AEI environment. It has strong adaptability, learning effective strategies in stochastic, nonlinear, and partially observable market environments without explicit complex optimization equations. High real-time responsiveness is achieved through online updates that allow agents to continuously adjust bids in rolling or real-time market conditions. Distributed intelligent optimization is enabled as each agent operates and learns independently, avoiding computational complexity and data privacy issues associated with centralized optimization. Long-term revenue orientation is maintained through the cumulative reward function, guiding agents to evolve stable and efficient bidding behavior over multi-period market interactions.

   In summary, the RL-based AEI bidding model integrates data-driven approaches with intelligent decision-making at the algorithmic level and transitions from static to dynamic optimization at the system level. This framework provides an extensible, intelligent platform for AEI participation in multi-time-scale markets and lays the foundation for the subsequent algorithm design and simulation verification in Sections 3.4.2 to 3.4.4.

3.4.2 State–action–reward definition

In the reinforcement learning-based AEI bidding model, state, action, and reward (state–action–reward) constitute the three core elements of agent–environment interaction. They collectively define the perception boundaries and optimization objectives of the agents, forming the foundation for iterative strategy learning. Proper design of state, action, and reward can significantly improve the learning efficiency and market adaptability of the model, enabling AEI generation units to develop efficient and stable bidding strategies in dynamic competitive environments.

1. State space design
   In the RL framework, the state space St represents all market and system feature information observable by the agent at time t, serving as the input set

for decision-making. In this study, the state space is defined by the nodal clearing prices from the previous market round:

$$S_t = (\lambda_{1,t-1}, \lambda_{2,t-1}, \ldots\ldots, \lambda_{n,t-1}) \tag{3.16}$$

where $\lambda_{i,t-1}$ denotes the clearing price at node i at time $t-1$, serving as an important signal reflecting market supply-demand balance.

This state design primarily captures external market dynamics and dimensionality. Historical nodal clearing prices reflect the supply-demand situation and price trends, providing agents with information about environmental changes. To enhance model generalization and robustness, state variables can be further standardized or embedded in high-dimensional feature spaces. For example, a sliding window mechanism can compile recent rounds of price and revenue information into sequential inputs, or attention mechanism based feature extraction structures can be introduced to capture potential volatility patterns and non-stationarity in market prices, allowing agents to maintain learning stability and strategy continuity under highly stochastic market conditions.

2. Action space design
The action space describes the strategy adjustment actions that an agent can take under a given state, specifically representing the control operations on the bidding parameters of a generation unit at the current time. In AEI bidding, the generation unit's bidding function is typically expressed in linear form.

The RL agent's action is defined as the dynamic adjustment of the bid intercept:

$$A_t = \Delta \alpha_g \tag{3.17}$$

where $\Delta \alpha_g$ is a continuous variable representing the magnitude of bid adjustment by the generation unit at time t based on market feedback.

This action definition has the following characteristics:

Continuity: The bid adjustment magnitude varies in a continuous space, suitable for continuous-control RL algorithms such as Deep Deterministic Policy Gradient (DDPG) or Proximal Policy Optimization (PPO).

High interpretability: The bidding parameter is directly linked to market revenue, facilitating economic interpretation during model visualization or sensitivity analysis.

To prevent extreme actions that may destabilize market bids during early training, action constraints can be introduced, such as limiting the adjustment magnitude $|\Delta \alpha g| \leq \Delta \max$ or applying Gaussian noise exploration strategies for smooth policy exploration.

3. Reward function construction
The reward function Rt provides the goal-oriented signal for the RL agent to optimize its strategy, measuring the economic performance in the current market round. In AEI bidding, the reward reflects both actual revenue and the

balance between operational costs and bid stability. Considering these factors, the reward function is defined as in formula (3.15), economically reflecting the agent's net profit in each market clearing.

The agent's long-term optimization objective is defined as maximizing discounted cumulative rewards:

$$\max_{\pi_g} E\left[\sum_{t=0}^{T} \gamma^t R_t\right] \tag{3.18}$$

where π_g is the agent's policy function mapping states to actions, and $\gamma \in (0, 1)$ is the discount factor balancing immediate rewards and long-term gains. A higher γ places more emphasis on long-term returns.

In summary, the state, action, and reward elements collectively form the core decision logic of the AEI reinforcement learning model. The state space ensures sufficient perception of market and system conditions, the action space provides flexible adjustment capability, and the reward function guides the agent toward unified optimization of economic performance and stability. This structure provides complete mathematical definitions and theoretical support for the subsequent Section 3.4.3 on agent learning mechanisms and training algorithms.

3.4.3 Algorithm framework and agent design

1. Algorithm structure
This study adopts the Deep Deterministic Policy Gradient (DDPG) algorithm, which is based on an Actor–Critic dual-network architecture and enables efficient policy optimization in continuous state and action spaces. Unlike traditional Q-learning or DQN algorithms that rely on discrete actions, DDPG directly outputs continuous-control variables through neural networks, making it suitable for the complex, nonlinear, and dynamic bidding environment of the Agricultural Energy Internet (AEI).

During algorithm execution, the Actor network generates specific bid adjustment actions, representing the agent's decision behavior in the electricity market under the current state. The Critic network evaluates the expected return or value function of the given action in the current state. Gradients are backpropagated to optimize both networks' parameters, allowing the policy network to gradually approach the global optimum. The algorithm balances short-term gains and long-term returns during updates, avoiding local-optimum "greedy" strategies.

DDPG combines the feature extraction capability of deep neural networks with the adaptive optimization of reinforcement learning, enabling agents to autonomously learn market patterns in high-dimensional state spaces with complex decision constraints. Compared with traditional optimization methods, DDPG converges faster and generalizes better, and is particularly suitable for energy market scenarios with uncertainties such as weather variability, load fluctuation, and stochastic electricity prices. In AEI bidding simulations, DDPG allows agents to dynamically adjust their bids based on historical market signals, operational characteristics, and external environment changes, transitioning from experience-driven to intelligent decision-making.

2. Agent structure and function division

   To achieve autonomous optimization and coordinated competition among different energy units within AEI, two independent agents are designed that are coupled at the market signal level.

   Photovoltaic agent (PV agent) initializes its bidding strategy using irradiance forecasts, daylight hours, irradiance fluctuations, and historical output sequences. Through DDPG, it continuously optimizes the nonlinear mapping between bid strategy and final revenue, dynamically adjusting the bid intercepts to increase bidding aggressiveness during price peaks and reduce clearing risk in low-price periods. The agent's objective is to stabilize revenue while maintaining optimal risk constraints under variable weather conditions.

   Biomass energy agent (BM agent) initializes its bidding strategy using fuel supply cost curves, boiler efficiency, and output limit characteristics, establishing the relationship between fuel cost and bidding behavior. Using DDPG, BM agent autonomously identifies the optimal balance between bid strategy and net profit, maximizing economic efficiency under fuel and equipment efficiency constraints.

   Both agents are independently trained and iteratively optimized within a unified market simulation environment and interact indirectly through the market clearing mechanism. This distributed learning structure ensures computational independence and convergence stability while effectively protecting commercially sensitive information, consistent with the real-world market logic of "independent bids, centralized clearing." Furthermore, this design supports multi-agent extensions, providing a scalable framework for future inclusion of energy storage, flexible loads, and other entities.

3. Training mechanisms and optimization strategies

   To improve model stability and convergence speed, several optimization mechanisms are introduced during DDPG training.

   Experience replay stores all interaction samples (state, action, reward, next state) in a replay buffer and randomly samples them during each update, breaking temporal correlations. This significantly improves sample utilization, reduces gradient variance, and enhances training robustness.

   Soft Update gradually adjusts the target network parameters to prevent sudden fluctuations, using $\theta' \leftarrow \tau\theta + (1-\tau)\theta'$, where $\tau \in (0, 1)$ is typically small, ensuring smooth parameter transitions. This technique is commonly used in reinforcement learning to prevent policy oscillations and improve training stability.

   Independent value evaluation allows each agent to maintain its own target and evaluation networks, adaptively approximating the value function. This architecture avoids interference caused by policy dependencies, ensuring that each unit optimizes independently based on its own revenue under the same market conditions.

The training objective is defined as maximizing the individual agent's revenue rather than the total AEI profit, reflecting the real electricity market logic of competitive bidding and self-responsible gains. Over time, this local optimization mechanism, combined with the market clearing process, will form a system-level equilibrium, providing a theoretical foundation for studying AEI's participation in market games.

Through these coordinated mechanisms, AEI agents achieve stable convergence and adaptive optimization in complex bidding environments, laying a solid algorithmic foundation for subsequent multi-stage linked bidding and dynamic price forecasting.

3.4.4 Strategy convergence and performance analysis

During the training phase, the convergence and performance of the model are primarily evaluated through the optimization curves of the policy coefficients, which provide an intuitive reflection of the agents' learning effectiveness and optimization trends over multiple iterations. At the early stage of training, agents exhibit strong exploratory behavior, resulting in significant fluctuations in the policy coefficients. In the middle and later stages, as the parameters of the policy networks gradually stabilize, the policy coefficient curves become smoother and eventually converge, indicating that the agents have successfully learned the optimal or near-optimal bidding strategies under the current market conditions.

From the convergence results, the DDPG algorithm demonstrates good learning stability and policy transferability within the AEI two-layer bidding framework. In complex market competition environments, agents can effectively converge within a limited number of training iterations and maintain strategy stability under random disturbances. Simulations show that individual generation unit agents, even when trained independently, can achieve implicit coordination and game-theoretic equilibrium through market signals, resulting in a steady upward trend in overall system revenue.

Overall, the results validate the effectiveness of the proposed AEI two-layer interactive learning framework. On one hand, the distributed agent structure allows each generation unit to independently optimize its strategy, consistent with real-world economic decision logic. On the other hand, through the transmission of market clearing price signals, the system achieves implicit coordination without centralized control. This mechanism effectively enhances the robustness and scalability of AEI in dynamic market environments, demonstrating strong practical adaptability and potential for broader application.

3.4.5 Conclusion

The AEI two-layer bidding modeling framework based on reinforcement learning established in this section systematically covers the entire process from market clearing to independent strategy optimization of generation units, realizing dynamic competition and strategy evolution of multiple energy entities within a unified market environment. By introducing the DDPG algorithm, the model enables agents to autonomously learn optimal bidding strategies through interaction with the market environment without relying on explicit mathematical models, overcoming the limitations of traditional optimization methods that heavily depend on model accuracy and prior knowledge.

Experimental and simulation results indicate that the proposed approach significantly enhances AEI's market response flexibility and economic efficiency. It also exhibits good convergence and policy transferability across multiple scenarios and market conditions. Agents can adaptively form stable bid distributions through long-term interactions, ensuring the continuous growth of overall system revenue. The bidding process remains stable and robust. Therefore, the dual-layer reinforcement learning framework proposed in this study provides a feasible theoretical foundation and implementable technical pathway for intelligent operation and market-oriented bidding mechanism design in the Agricultural Energy Internet, offering important reference value for the future evolution of AEI market operation models.

3.5 Simulation design and experimental plan

To verify the feasibility and effectiveness of the proposed Agricultural Energy Internet (AEI) multi-market bidding model and reinforcement learning-based optimization strategies, this chapter systematically designs simulation experiments and multi-scenario experimental schemes. By virtualizing the market environment and modeling multi-agent interactions, it comprehensively analyzes the impacts of different market scales, strategy types, and participation levels of electric agricultural machinery (EAM) on the overall economic performance and system stability of AEI.

3.5.1 Simulation experiment and data structure

1. Overall experimental design

   To deeply characterize the operational behavior and intelligent game features of AEI in multi-time-scale markets, this study constructs a multi-module integrated reinforcement learning-based intelligent bidding simulation experiment. The experiment aims to validate the effectiveness and scalability of the proposed model in dynamic market environments through detailed market modeling and the agents' self-learning process.

   The overall experimental structure consists of four core functional modules: market simulation module, agent module, underlying neural network module, and training and evaluation module. The architecture follows the principle of "multi-agent parallelism—environmental feedback—adaptive policy update," forming a closed-loop simulation system covering the entire process from market clearing to agent learning.

Market Simulation Module: This module constructs electricity market clearing mechanisms applicable to day-Ahead, intraday, and real-time markets, corresponding to different time-scale trading behaviors. By introducing a social welfare maximization objective function, the module dynamically solves for optimal prices and quantities in each stage. By adjusting supply-demand curves, node power limits, and line constraints, the module can simulate uncertainties in real power markets, such as price fluctuations, random load variations, and weather disturbances. This module provides environmental feedback for the entire simulation experiment and serves as the external environment for agent interaction and learning.

Agent module: The agent module contains proxies for photovoltaic (PV) and biomass (BM) generation units, each with autonomous learning, independent decision-making, and strategy evolution capabilities. The module drives the units to continuously adjust bidding parameters in market games via reinforcement learning algorithms, aiming for local revenue maximization. Agents interact indirectly through market price signals, forming a self-organized bidding pattern under distributed conditions.

Underlying neural network module: To support the approximation of agent policy and value functions, the underlying neural network module is implemented based on the PyTorch framework. It adopts the typical Actor–Critic architecture, where the Actor network outputs bidding adjustment actions and the Critic network evaluates the value of the current state–action pair. The module is designed to balance computational efficiency and generalization capability, ensuring stable training even in high-dimensional state spaces.

Training and evaluation module: This module organizes and controls the entire experimental process, including training parameter configuration, reward function setup, policy convergence judgment, and model performance evaluation. It supports multi-agent training and multi-market collaborative simulation. Visualization tools can be used to intuitively assess algorithm performance and policy evolution.

Overall, the modular design of the experiment enhances scalability and flexibility while implementing a full-process closed-loop simulation from market clearing to policy updating and performance evaluation. This provides solid support for validating the adaptability and robustness of the AEI reinforcement learning model under different market mechanisms.

2. Algorithm flow design

The algorithm flow of the simulation experiment follows a typical reinforcement learning closed-loop structure, including data input, environment interaction, policy updating, and convergence evaluation.

Data initialization stage: The system loads initial market data and generator operation parameters, initializing the network weights and biases of the agents as well as the contents of the experience replay buffer.

Market state input: Based on day-ahead forecasts, node prices, load levels, and equipment operating conditions, the platform generates a market environment vector to provide agents with the current observable state.

Bidding decision generation: Each agent inputs the perceived state into the Actor network to output bidding coefficients, forming an individualized bidding strategy.

Market clearing: The market simulation module solves for the optimal clearing prices and quantities based on the social welfare maximization principle, updating market equilibrium points and achieving economic dispatch.

Feedback and reward calculation: The system calculates immediate rewards based on clearing results and generator bidding profits, and returns them to the agents. The reward function incorporates economic benefits, penalty terms, and smoothing constraints to ensure learning stability.

Policy update: Agents use the reward signal to compute the loss function and update Actor and Critic network parameters, employing a soft update mechanism to reduce gradient oscillation.

Multi-round iteration and convergence output: The process is executed in a time-step loop until cumulative rewards stabilize or a preset error threshold is reached. The final output includes stable bidding strategies, market equilibrium prices, and agent revenue results.

This process implements the typical state–action–reward–next state (S–A–R–S′) closed-loop structure in reinforcement learning, allowing agents to evolve dynamically from exploration to convergence and ensuring the adaptability and robustness of strategies under different market scenarios.

3. Software environment

The experiment runs using Python 3.8.20 as the main programming language, with core algorithms implemented using the PyTorch 2.1.0 deep learning framework.

Model visualization layer: Matplotlib is used to plot the convergence curves of bidding strategy coefficients.

Optimization solver layer: CVXOPT's built-in quadratic programming solver is used as the underlying clearing algorithm solver, achieving efficient computation for market clearing problems.

4. Data sources and parameter construction

The experimental data mainly include two categories: market data and generator data.

Market data: [15] provides initial bidding strategies and node power distributions for other market competitors. This data reflects the market environment and serves as the initial input conditions for agent training.

Generator data: Constructed via simulation, considering the physical characteristics and operational rules of different generator types to set initial strategy coefficients and other parameters.

The data construction logic is as follows: first, establish basic constraint boundaries according to generator type and rated capacity; second, define the α–β bidding function parameters according to market bidding mechanisms and generator characteristics; third, introduce other market participants to generate dynamic market scenarios.

3.5.2 Experimental scenario settings

To comprehensively and systematically evaluate the performance of the proposed model under different market structures, operational conditions, and load responses, this study designs multi-level experimental scenarios. The experiments are conducted along two dimensions: market scale and electric agricultural machinery (EAM) scale, aiming to verify the model's adaptability, convergence stability, and profit optimization capability in a multi-agent interactive environment. By cross-comparing multiple dimensions, the dynamic response characteristics and potential optimization space of the AEI system in multi-time-scale markets can be more accurately revealed.

1. Market scale division
   Based on the organizational rules and clearing rhythms of staged electricity market trading, the market transaction scale is divided into day-ahead, intraday, and real-time markets. These three market types differ significantly in terms of time resolution, information completeness, and price fluctuation characteristics, posing different challenges to the learning and adaptation speed of reinforcement learning agents. Table 3.2 lists the market division and feature comparison:
   In this setting, the day-Ahead market primarily reflects long-term optimization capability, focusing on the trade-off between global profit maximization and forecast error correction. The intraday market emphasizes the agent's sensitivity to short-term disturbances and load prediction deviations. The real-time market highlights the agent's ability to update strategies in extremely short periods based on the latest market signals and system disturbances.
   By conducting staged simulations across these three market types, the learning efficiency, convergence speed, and stability of reinforcement learning agents at different time scales can be thoroughly evaluated. For example, in the day-Ahead market, the model needs to form relatively stable long-period strategies, while in the real-time market, it must exhibit high-frequency updates and rapid convergence. These differences make the multi-time-scale policy evolution process more valuable for research and more aligned with the operational mechanisms of actual power systems.
2. EAM scale and type variation
   Electric agricultural machinery (EAM), as an important controllable load unit within the AEI system, has a decisive impact on overall bidding

Table 3.2 Comparison of multi-type markets

Market type	Trading cycle	Relative scale	Main features
Day-ahead	24 hours	Benchmark 100%	Strong planning, stable prices
Intraday	1–4 hours	10%, 12%, 14%	Used to correct forecast deviations, flexible response
Real-time	15 minutes	5%, 6%, 7%	Dynamic response market, high timeliness

behavior and market balance. To study the effect of EAM flexibility on AEI operation, this study constructs three typical simulation scenarios by adjusting its maximum demand capacity Dmax and response coefficient fEAM:

Small-scale scenario: Primarily composed of distributed low-power agricultural machines such as electric irrigation pumps, with dispersed load distribution and slower response speed. This scenario simulates AEI response capability under low penetration conditions.

Medium-scale scenario: Composed of multiple types of agricultural machines such as tractors and harvesters, with moderate controllability and coordination, representing a typical state of AEI during gradual expansion.

Large-scale scenario: Consists of centralized charging stations and multicycle agricultural machines forming a highly concentrated controllable load group, with significant price responsiveness, capable of markedly affecting market clearing results and system revenue.

By gradually adjusting EAM capacity and response parameters, the impact of EAM penetration on internal bidding behavior, profit distribution, and system economic performance of AEI can be systematically analyzed. This parameterized simulation not only verifies the "flexible buffering" role of EAM at different market stages but also provides a quantifiable basis for designing AEI load aggregation and dispatch optimization strategies.

In summary, the experimental scenarios established in this section form a multidimensional simulation system along the dimensions of market scale and EAM response. This system can evaluate model performance under different time scales and risk constraints, while revealing the dynamic coordination mechanisms of AEI as a multi-agent system under various load response conditions. Through this progressively layered experimental structure, the convergence stability, adaptability, and economic advantages of the reinforcement learning model in complex market environments are fully validated.

3.5.3 Evaluation index system

To systematically quantify the economic performance of the Agricultural Energy Internet (AEI) in multi-time-scale electricity markets and the effectiveness of reinforcement learning algorithms, this study constructs a hierarchical and quantifiable evaluation index system. The system encompasses two main categories: market-level indicators and algorithm-level indicators. It reflects both the economic behavior characteristics of generation units and loads and the convergence and robustness performance of reinforcement learning algorithms in dynamic environments. Through this system, a multidimensional, comprehensive evaluation can be achieved, ranging from individual profits to overall system benefits.

1. Market-level indicators

   Market-level indicators are mainly used to characterize the performance of various AEI generation units and load participants in the market, including bidding levels, cleared quantities, and economic returns. These indicators directly reflect whether the market behavior optimized by reinforcement learning is reasonable, whether competition is sufficient, and whether the market transaction structure achieves the expected balance.

① Cleared price of generation unit

The cleared price reflects the intensity of market competition and the rationality of bidding. It indicates the equilibrium price level formed by reinforcement learning agents in a dynamic market. By comparing cleared prices under different strategies, it is possible to assess whether agents have developed competitive bidding behaviors. Small and stable price fluctuations indicate that the market has reached a relatively high level of rational competition.

② Cleared quantity

The cleared quantity measures the participation level and resource share of a generation unit in the overall market. Changes in quantity not only reflect the competitive outcomes of agent bidding strategies but also indicate the market clearing mechanism's preference trends for different types of units. Higher cleared quantities generally suggest that a unit's bid is more attractive or that its cost structure is more reasonable.

③ Profit of individual unit

This indicator evaluates the economic return after reinforcement learning optimization. Higher profits indicate that the agent has achieved a better balance between revenue and cost. By comparing profit variations across different units, it is possible to analyze whether the algorithm realizes fair competition and optimal resource allocation, as well as the model's adaptability under price fluctuation conditions.

The comprehensive analysis of market-level indicators can verify whether AEI achieves the goal of improving both generation and consumption benefits while enhancing market balance across multi-time-scale markets.

2. Algorithm performance indicators

Algorithm performance indicators are primarily used to assess the efficiency, stability, and long-term reward performance of the reinforcement learning model during training. These indicators determine the model's applicability and scalability in dynamic electricity markets.

① Convergence rate

The convergence rate is defined as the number of training iterations required for the algorithm to reach stable rewards. It reflects the learning efficiency of the model in complex environments. A faster convergence rate indicates that agents can learn stable strategies with fewer iterations, reducing computational cost and accelerating deployment.

② Policy variance

Policy variance measures the variance of actions output by agents at different training stages, reflecting the smoothness and stability of the strategy. Excessively high variance indicates that the model is overly sensitive to environmental changes, causing frequent strategy adjustments and potentially unstable bidding behavior. Moderate variance that gradually converges indicates that the agent has formed a balanced and robust strategy pattern.

The combination of algorithm performance indicators evaluates the training characteristics of the reinforcement learning model, providing insights into the suitability and performance of DDPG in AEI scenarios.

In summary, the constructed evaluation index system covers the complete spectrum from individual units to the overall system, enabling comprehensive quantification of AEI market behavior and algorithm performance. This system not only provides a scientific basis for model performance validation but also offers a sustainable evaluation framework for AEI operation optimization in multi-time-scale markets, ensuring that research results are interpretable, comparable, and generalizable.

3.5.4 Simulation System Summary

This chapter focuses on the intelligent bidding problem of the Agricultural Energy Internet (AEI) under multi-market interaction conditions and systematically constructs a reproducible and scalable multi-time-scale simulation system. Centered on a two-layer interactive architecture, the system integrates market clearing mechanisms with reinforcement learning algorithms to dynamically replicate the full process from bid generation and market clearing to reward feedback and strategy optimization. The model fully accounts for the multi-energy complementarity constraints in AEI, the hierarchical differences in market rules, and the temporal characteristics of environmental disturbances, providing a solid research foundation for exploring AEI's adaptive game behavior in complex market environments.

In terms of simulation design, the system ensures scientific rigor and reproducibility through fine-grained parameter settings and multidimensional experimental control. By systematically tuning core hyperparameters such as learning rates, the reinforcement learning agents achieve a good balance between stability and convergence speed, quickly approaching optimal strategies. The model also incorporates scenarios with varying market scales, price fluctuation ranges, and EAM response capabilities to reveal the flexibility and robustness of AEI in multi-type market participation.

From the perspective of experimental result analysis, the constructed simulation not only evaluates the performance of different agent strategies in terms of profit maximization, volatility suppression, and risk control, but also qualitatively reveals the systematic impact of external factors such as changes in EAM scale, market capacity expansion, and price signal disturbances on AEI market behavior.

Moreover, the simulation system serves not only for algorithm validation but also has platform-oriented and application-oriented characteristics. It can flexibly integrate different market models while maintaining a scalable architecture, thereby supporting subsequent chapters' in-depth studies on AEI strategy optimization, risk assessment, and policy intervention effects.

In summary, the research outcomes of this chapter provide robust technical and data support for subsequent result analysis and strategy validation. The constructed AEI simulation system theoretically advances the market-oriented research path of the Agricultural Energy Internet, provides a practical experimental paradigm for intelligent control and market coordination in engineering practice, and lays an important foundation for promoting intelligent, market-oriented, and low-carbon operation of AEI.

3.6 Simulation results and analysis

To systematically validate the effectiveness and applicability of the reinforcement learning-based AEI multi-market bidding model, this chapter conducts multi-scenario and multidimensional experimental simulations and comparative analyses based on the previously established simulation platform and experimental design. The experiments consider different market scales, strategy types, and EAM integration conditions. By systematically observing model performance in training convergence, strategy evolution, market clearing, and revenue changes, the study aims to reveal the learning patterns, stability characteristics, and decision-making mechanisms of reinforcement learning agents in multi-time-scale electricity markets.

Additionally, this chapter evaluates AEI's overall economic performance and robustness under multi-market interactions at the system level, with a focus on the feedback effects of EAM integration on the bidding process and its role in regulating market price fluctuations. The multidimensional experimental results provide a solid empirical foundation for the strategy optimization and operational mechanism analysis presented in subsequent chapters.

3.6.1 Strategy convergence analysis

The training process of the reinforcement learning model reflects the continuous learning and self-optimization of agents, from random exploration to rational decision-making. The convergence characteristics directly determine the stability and interpretability of the final strategies. The training curves shown in Figures 3.5 and 3.6 illustrate the evolution of bid coefficients for photovoltaic (PV) and biomass (BM) agents, as well as other simulated market participants across different training stages, revealing the dynamic path of strategies optimized via DDPG in a long-term interactive environment.

In the initial stage, agents have limited perception of market price signals and reward feedback, and strategies mainly rely on the ε-greedy exploration mechanism for random probing, exhibiting high uncertainty and volatility. During this phase, bid coefficients fluctuate markedly, reflecting the model's "experience-free exploration" state.

In the intermediate stage, agents gradually recognize market patterns from interaction experiences. The strategy network adjusts its parameters through gradient backpropagation, beginning to form relatively rational bidding and generation decisions. At this point, fluctuations in bid coefficient curves are significantly reduced. The reinforcement learning algorithm gradually demonstrates the synergy of strategy optimization and reward enhancement, as agents learn to dynamically

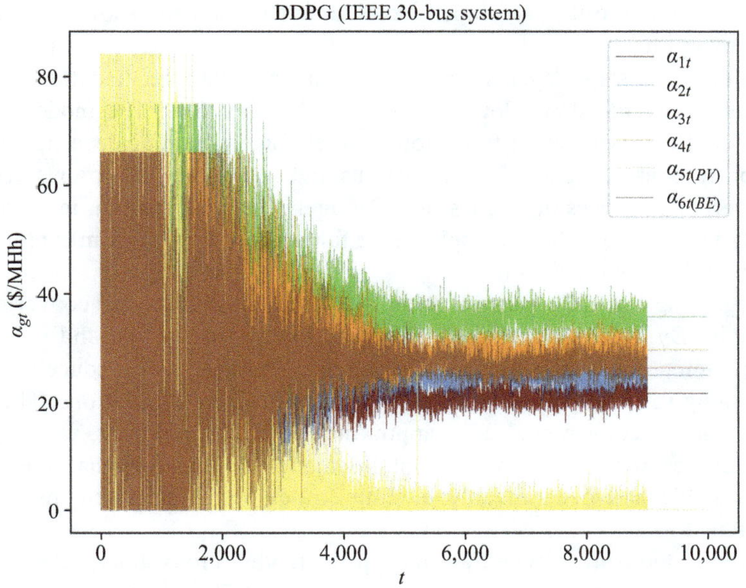

Figure 3.5 Optimization of generation-side bidding coefficients without EAM participation

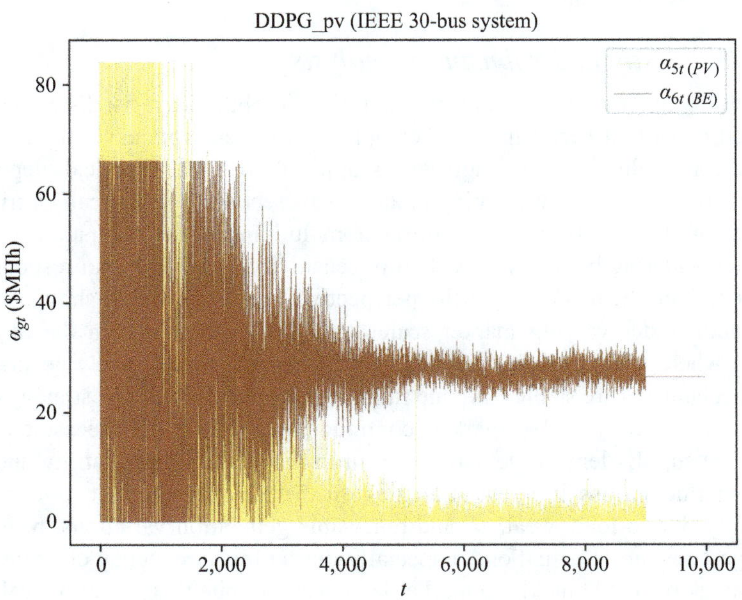

Figure 3.6 AEI generation-side bidding coefficient optimization without EAM participation

balance risk and profit in a competitive bidding environment, showing clear strategy learning effects.

In the later stage, agents' strategy parameters stabilize, and the growth of cumulative rewards slows down and eventually converges. The model behavior transitions from exploration to exploitation, and agents can achieve near-optimal decision responses under price fluctuations and operational constraints. The bidding coefficient curves of biomass and PV agents become smooth, indicating that their strategy networks have completed the full process from random exploration to optimal strategy approximation.

Regarding algorithm performance, DDPG demonstrates good convergence in this study. By incorporating the experience replay mechanism and soft update strategy, temporal correlations among samples are effectively reduced, preventing learning instability caused by overfitting or gradient oscillations. The model achieves stable convergence after approximately 5,000 iterations, with bid coefficient curves showing no significant fluctuations, indicating that agents have formed stable and generalizable bidding strategies in the given multi-market environment.

Overall, the convergence trajectory of the model fully validates the effectiveness and robustness of reinforcement learning in AEI multi-market coordinated optimization. The DDPG framework can optimize decisions in continuous action spaces within nonlinear and non-stationary market environments, enabling agents to adaptively learn market patterns and gradually approach optimal bidding strategies, thereby laying an algorithmic foundation for distributed, autonomous, and intelligent AEI bidding.

3.6.2 Stability and robustness analysis

The stability and robustness of the model reflect its ability to resist disturbances and maintain consistent performance in complex market environments, which are key criteria for evaluating the engineering applicability and practical deployment potential of reinforcement learning models. This section systematically verifies the stability of the proposed model through multidimensional comparative experiments, considering both market volatility sensitivity and EAM load response.

First, from the market volatility perspective, agents' revenues show noticeable differences under varying market scales and time resolutions. In the day-ahead market, where trading cycles are long, load forecasts are relatively accurate, and price mechanisms are stable, revenue fluctuations are minimal, and strategy outputs are steady. In the intraday market, continuous correction of forecast errors and short-term supply-demand deviations result in moderate price volatility and corresponding fluctuations in agent strategies. In the real-time market, prices change frequently due to load, weather, and renewable generation variations, resulting in the largest revenue fluctuations. Especially during high-frequency clearing stages, agents must respond quickly to market feedback. Despite this, experimental results indicate that AEI agents trained with reinforcement learning maintain relatively high average revenues even in highly volatile environments. Compared with baseline fixed-bid strategies, the DDPG model shows significant stability improvements in day-ahead, intraday, and real-time markets. This demonstrates that reinforcement learning strategies can effectively capture dynamic market patterns through continuous interaction, achieving a balance of robustness and economic efficiency.

Second, from the perspective of electric agricultural machinery (EAM) load response, EAM represents a flexible and controllable load resource. Its integration into AEI significantly affects the overall bidding behavior of agricultural parks. From the load curve perspective, EAM introduction alters the temporal characteristics of park electricity demand. Specifically, EAM's electricity consumption exhibits high temporal flexibility, allowing part of the charging load to be shifted from peak periods to off-peak periods without affecting operational efficiency. This peak-valley load adjustability effectively alleviates peak load stress, smooths the load curve, and improves energy utilization efficiency. More importantly, the temporal shifting ability enables agricultural parks to actively adjust their electricity consumption in response to market price fluctuations, enhancing overall demand response capacity.

From the bidding strategy perspective, EAM integration substantially increases the strategy space and complexity faced by the AEI virtual power plant (VPP). Before EAM integration, the generation-load structure within the park was relatively fixed, and bidding behavior was stable and rule-based. With the introduction of flexible EAM loads, load response capability is significantly enhanced, requiring the VPP to adapt to a more dynamic and diverse market environment. This leads to more pronounced fluctuations in generation-side bidding optimization curves, reflecting that bidding strategies must consider more uncertainty factors and multidimensional objectives during adjustment and optimization. Experiments show that the DDPG algorithm effectively explores and exploits this complex bidding space through agent trial-and-error and self-adjustment.

In summary, the reinforcement learning-driven AEI multi-market bidding model can adaptively adjust strategy parameters to maintain stable revenues in the presence of multiple uncertainties. This feature demonstrates the model's "interpretable—adjustable—transferable" engineering potential in electricity market simulations, providing a technical foundation for practical application of intelligent AEI bidding mechanisms.

3.6.3 Differences in bidding behaviors of individual units

Within the framework of independent optimization for each generation unit, different types of generators exhibit distinct bidding behaviors due to differences in physical constraints, cost structures, and resource characteristics. Analyzing the strategy evolution of photovoltaic (PV) units, biomass (BM) units, and electric agricultural machinery (EAM) responsive loads provides insight into the internal multi-agent interactions of AEI under the same market mechanism, revealing both competitive patterns and cooperative features.

1. Photovoltaic agent (PV agent)
   PV units have nearly zero marginal generation costs, so their optimal strategies primarily focus on capturing and exploiting market price fluctuations. In the day-ahead market, PV agents tend to maintain relatively neutral bidding coefficients to maximize clearing probability and ensure the sale of basic generation output. In the real-time market, when forecasted electricity prices are significantly higher than expected, agents actively increase their bidding coefficients to expand volume and gain additional revenue. Conversely, when market prices decline, agents narrow their bid range to avoid low-price transactions. Reinforcement learning enables PV agents to gradually recognize price patterns and thresholds at different time periods, exhibiting adaptive behavior characterized by "aggressive at high prices, conservative at low prices." Simulation results show that with small-scale EAM integration, the PV unit's final bidding coefficient converges near 0, the cleared energy stabilizes around 12 MWh, and the clearing price is approximately 54.79 USD/MWh, demonstrating high sensitivity and rapid responsiveness.
2. Biomass agent (BM agent)
   Unlike PV units, BM generators must consider fuel costs and operational efficiency constraints, making their bidding behavior more conservative. Reinforcement learning helps the BM agent balance "cost constraints and revenue maximization." During training, the BM agent learns to dynamically adjust bidding coefficients across market stages: slightly increasing bids during high load or high-price periods to compensate for fuel cost pressure, and moderately lowering bids during intense competition to maintain clearing probability. Simulation results indicate that with small-scale EAM integration, the BM agent's final bidding coefficient is approximately 25.43, average cleared energy is 48.04 MWh, and corresponding profit is 741.88 USD, showing excellent strategy stability and consistent revenue. Overall, the BM agent develops a "steady-state bidding" strategy across multi-market environments, sacrificing some short-term profit for long-term controllable gains – a characteristic with strong practical relevance in real bidding markets.
3. Electric agricultural machinery (EAM) units
   EAM units, as responsive load-side agents, display bidding behaviors that reflect passive adaptation to price signals and enhanced scheduling flexibility. EAM units follow preset strategy curves to implement adaptive behaviors such as "low-price purchase—efficient charging—high-price peak reduction." In small-scale scenarios, EAM purchase volumes are relatively low; in medium and large-scale scenarios, purchase volumes increase, and charging/discharging flexibility is significantly enhanced. This time-shift behavior based on price signals effectively buffers short-term market fluctuations, smoothing overall AEI system load. EAM responses exhibit a certain delay—slightly slower than generation-side agents—but their actions are more stable and revenue is more continuous, naturally providing a "damping" effect on market volatility.

In summary, the three types of units exhibit distinct behavioral layers and complementary functions in multi-market environments: PV units pursue agile responses and high returns, BM units emphasize steady and sustainable revenue, and EAM units regulate purchase strategies through load flexibility. These differentiated bidding behaviors not only illustrate the multi-energy complementary mechanism within AEI but also demonstrate reinforcement learning's capability for strategy differentiation and coordinated optimization in complex multi-agent game scenarios.

3.6.4 Comparison of clearing results under different EAM access levels

To further quantitatively analyze the impact of electric agricultural machinery (EAM) integration on the market clearing results and economic performance of the Agricultural Energy Internet (AEI) system, this section designs three scenarios with different EAM access levels, corresponding to low (Cluster 1), medium (Cluster 2), and high (Cluster 3) load conditions. The optimized bidding coefficients are shown in Figures 3.7–3.9. By conducting comparative experiments under the same market structure and bidding mechanism, the influence of EAM penetration on the interaction characteristics between the generation and load sides of the AEI system can be directly observed. The experimental results are summarized in Table 3.3.

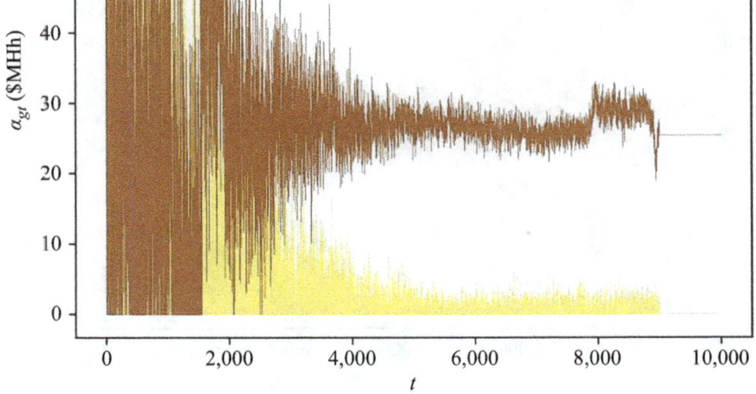

Figure 3.7 Optimization of AEI generation-side bidding coefficients with EAM Cluster 1 participation

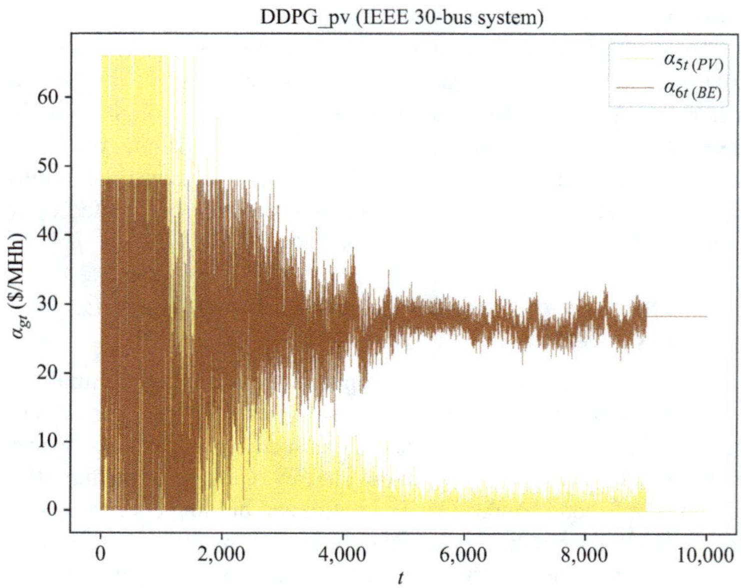

Figure 3.8 Optimization of AEI generation-side bidding coefficients with EAM Cluster 2 participation

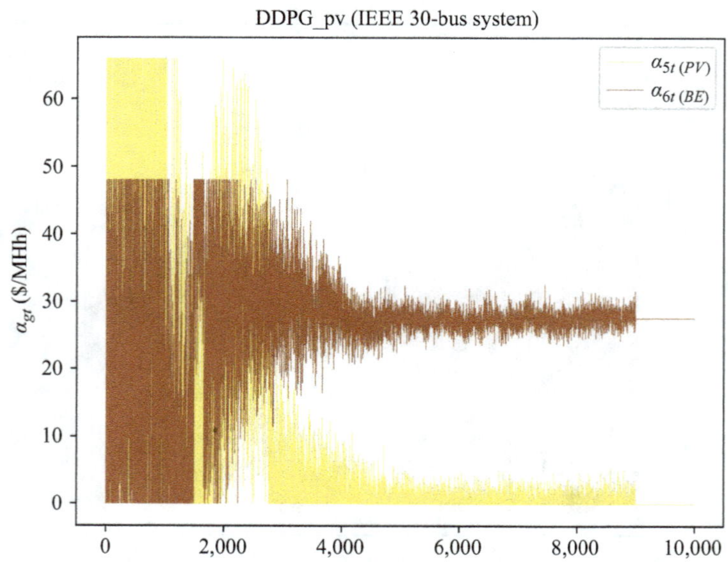

Figure 3.9 Optimization of AEI generation-side bidding coefficients with EAM Cluster 3 participation

Table 3.3 Bidding revenue of agricultural VPP considering EAM participation

Optimization result	EAM Cluster 1	EAM Cluster 2	EAM Cluster 3
PV bidding coefficient	0	0	0
PV cleared volume (MWh)	12	12	12
PV clearing price ($/MWh)	54.79	54.76	54.98
PV net profit ($)	321.49	321.09	323.81
BM bidding coefficient	25.43	28.4	27.62
BM cleared volume (MWh)	48.04	42.59	44.45
BM clearing price ($/MWh)	49.45	49.7	49.85
BM net profit ($)	741.88	726.09	743.83
Total generation profit ($)	1063.37	1047.18	1067.64
EAM bidding coefficient	5	5	5
EAM purchased energy (MWh)	0.14	2.22	4.07
EAM purchase price ($/MWh)	39.39	38.91	39.65

As shown in the table, the introduction of different scales of EAM has a noticeable structural impact on system supply-demand balance, generation unit revenues, and total profits.

In the Cluster 1 (low access) scenario, EAM's electricity demand is minimal, and its disturbance effect on the market is negligible. The overall system operation is close to the baseline level without load response. The total generation profit of AEI is approximately 1063.37 USD, and the clearing structure of PV and BM units remains basically stable, indicating that EAM has limited influence under low penetration.

When EAM integration increases to Cluster 2 (medium access), system flexibility significantly improves. EAM responds sensitively to price signals and actively adjusts charging schedules on the load side, enabling partial demand shifting. At this stage, AEI generation-side total profit is about 1047.18 USD. Although generation-side revenue slightly decreases, the fluctuation remains within an acceptable range, demonstrating that the algorithm maintains good stability and robustness under a relatively complex bidding space.

Under Cluster 3 (high access) conditions, EAM participation further increases, and the overall market balance undergoes structural adjustment. Although the cleared volumes of PV and BM units change slightly, the total generation-side profit rises to 1067.64 USD, indicating that AEI can achieve adaptive market coordination through reinforcement learning strategies under high load integration. Notably, despite the significantly increased purchase volume and more complex bidding space, the generation-side overall economic performance is not adversely affected because agents can adjust bidding parameters in real time based on market feedback, further optimizing the system's revenue structure.

Comparing the three scenarios reveals that EAM integration not only increases total system revenue but also reduces profit volatility, enhancing the AEI system's economic robustness and load flexibility under multi-market conditions. Reinforcement learning agents dynamically optimize both bidding and purchase strategies, successfully achieving adaptive decision-making across different load scenarios. This mechanism effectively mitigates market deviations caused by load forecasting errors or severe price fluctuations, providing a sustainable revenue growth path for AEI and a feasible technical support framework for intelligent market-based scheduling of agricultural loads.

3.6.5 Comparison of clearing results under different markets

To further analyze the strategy differences and clearing characteristics of the AEI system across different market stages (day-ahead, intraday, and real-time), this section sets up multi-market scenarios with varying trading surpluses. By gradually increasing the trading surpluses in the intraday and real-time markets, the study investigates how agents adjust their bidding and how revenues change under a multi-time-scale market structure. The simulation results are presented in Tables 3.4–3.6.

From the above results, the differences in clearing outcomes across market stages are mainly reflected in the revenue changes of PV units. As the trading surpluses in intraday and real-time markets gradually expand, PV units can more fully exploit price fluctuations to participate in short-term trading, indicating that PV, as a low marginal cost resource, has greater flexibility and arbitrage potential in high-frequency markets.

In contrast, BM units, constrained by fuel costs and operational characteristics, remain primarily active in the day-ahead market, with limited participation in short-term markets, yet they continue to dominate overall revenue, ensuring system revenue stability.

Although the participation of EAM is relatively small in terms of purchased energy, its demand response characteristics act as a "buffer" across different market stages. In the day-ahead market scenario, EAM load adjustments help meet electricity demand more smoothly, smoothing the system's revenue curve and indirectly enhancing overall economic performance and system robustness.

Table 3.4 Bidding results of internal units in the AEI under multi-market (small surplus)

Optimization result	Day-ahead market	Intraday market (10% surplus)	Real-time market (5% surplus)
PV cleared volume (MWh)	12	3.32	0.61
PV clearing price ($/MWh)	54.79	7.04	3.46
PV net profit ($)	321.49	16.76	9.82
BM cleared volume (MWh)	48.04	0	0
BM clearing price ($/MWh)	49.45	7.04	3.46
BM net profit ($)	741.88	0	0
EAM purchased energy (MWh)	0.14	0	0
EAM purchase price ($/MWh)	39.39	7.04	3.46
Total generation revenue	1089.95		

Table 3.5 Bidding results of internal units in the AEI under multi-market (medium surplus)

Optimization result	Day-ahead market	Intraday market (12% surplus)	Real-time market (6% surplus)
PV cleared volume (MWh)	12	4.58	3.6
PV clearing price ($/MWh)	54.79	7.35	3.39
PV net profit ($)	321.49	19.82	11.81
BM cleared volume (MWh)	48.04	0	0
BM clearing price ($/MWh)	49.45	7.35	3.39
BM net profit ($)	741.88	0	0
EAM purchased energy (MWh)	0.14	0	0
EAM purchase price ($/MWh)	39.39	7.35	3.39
Total generation revenue	1095		

Table 3.6 Bidding results of internal units in the AEI under multi-market (large surplus)

Optimization result	Day-ahead market	Intraday market (14% surplus)	Real-time market (7% surplus)
PV cleared volume (MWh)	12	7.1	4.2
PV clearing price ($/MWh)	54.79	7.6	3.58
PV net profit ($)	321.49	22.85	13.67
BM cleared volume (MWh)	48.04	0	0
BM clearing price ($/MWh)	49.45	7.6	3.58
BM net profit ($)	741.88	0	0
EAM purchased energy (MWh)	0.14	0	0
EAM purchase price ($/MWh)	39.39	7.6	3.58
Total generation revenue	1099.89		

Overall, as trading surpluses increase, the AEI system gradually exhibits the synergistic features of "flexible generation—adjustable load—stable revenue." Reinforcement learning agents can adaptively adjust to multi-market temporal characteristics, automatically balancing the trade-off between revenue and risk, thereby achieving optimal economic operation of AEI in multi-level electricity markets.

3.6.6 *Summary of the integrated mechanism*

Based on the multi-scenario simulation results and economic analysis presented above, the reinforcement learning-driven AEI multi-market bidding model developed in this study demonstrates strong economic optimization capability and system robustness. The overall conclusions can be summarized as follows:

AEI strategies and market prices form an adaptive dynamic equilibrium mechanism.

Reinforcement learning agents can perceive real-time trends in market price signals and dynamically adjust bidding strategies guided by feedback on revenues. Through multiple rounds of interaction and environmental engagement, the model gradually establishes an adaptive mapping between "price—strategy—revenue," achieving coordinated evolution of market behavior and system economics.

Multi-market coordinated operation significantly improves resource utilization efficiency.

The day-ahead market provides revenue assurance for planned generation resources, such as biomass, while the intraday and real-time markets offer flexible adjustment and arbitrage opportunities for volatile resources, such as photovoltaics. The reinforcement learning model can flexibly switch strategies across multiple time-scale markets, achieving temporal coordination of both generation and load resources.

EAM integration significantly enhances system flexibility and economic performance.

Electric agricultural machines, as controllable loads, can shift energy usage across time periods in response to price signals, effectively reducing power stress during peak periods. The flexible participation of EAM not only increases overall system profits but also substantially strengthens the stability and adjustability of market operations.

Reinforcement learning effectively addresses the complexity of multi-agent bidding.

By employing distributed agent training and experience-sharing mechanisms, each generation entity can optimize independently while achieving overall coordination, avoiding the dimensionality explosion and convergence difficulties typical of conventional optimization methods. Experimental results indicate that this approach performs excellently in training stability and strategy generalization.

AEI intelligent bidding strategies demonstrate significant advantages in economic efficiency, dynamism, and robustness.

Compared with traditional optimization algorithms, the reinforcement learning-driven model can continuously self-optimize under uncertain environments, achieving multiple objectives such as cost reduction, revenue enhancement, and market risk mitigation. This provides solid theoretical and technical support for the intelligent and market-oriented development of the Agricultural Energy Internet.

3.7 Conclusions and outlook

This chapter integrates and synthesizes the core research outcomes presented in the preceding sections on system modeling, algorithmic design, and simulation validation, providing a comprehensive summary and an outlook on future directions for the bidding strategies of the Agricultural Energy Internet (AEI) integrated with electric agricultural machines (EAM). Through the introduction of a reinforcement learning-driven intelligent bidding framework, the AEI achieves coordinated optimization of generation and consumption behaviors, enhances system profitability, and improves operational flexibility and robustness under multi-market coupling environments. These results fully demonstrate the adaptability, self-learning capacity, and stable decision-making advantages of reinforcement learning when applied to complex and uncertain electricity market contexts. The research outcomes presented herein thus provide theoretical foundations and methodological references for promoting the intelligent, market-oriented, and sustainable development of agricultural energy systems.

3.7.1 Overview of research conclusions

1. Review of research objectives and achievements
   The central aim of this study was to construct an intelligent bidding model for the AEI integrated with EAM, seeking to realize multi-agent coordinated optimization of both generation-side and consumption-side decisions. The overarching objective was to improve system-wide economic efficiency, enhance operational flexibility, and promote the efficient utilization of renewable energy within agricultural parks. To achieve this, the study first established a virtual power plant (VPP)–based market participation framework for AEI, which clarified the structural roles, operational characteristics, and market functions of distributed photovoltaic (PV), biomass generation, energy storage systems, and EAM loads.

   On the foundation of this system framework, the study further developed a two-layer interactive bidding structure operating across multi-time scales of "day-ahead–intraday–real-time." In this hierarchical structure, the upper market layer focuses on price formation and clearing mechanisms, constructing a feedback channel between market information and agent decisions through the continuous transmission of price signals. The lower layer focuses on reinforcement learning-based agent behavior, in which generation and consumption entities autonomously adjust their strategies in response to market conditions to maximize overall benefits.

   Through a series of multi-stage simulation experiments, a complete closed-loop research process was realized – from theoretical model formulation to algorithm training and final performance evaluation. Simulation outcomes indicate that the constructed model achieves fast convergence even under conditions of fluctuating load demand and dynamic pricing, verifying the strong adaptability, stability, and efficiency of reinforcement learning algorithms in AEI bidding optimization. These results confirm that the model effectively meets the intended design goals and contributes to advancing intelligent decision-making for agricultural energy systems.
2. Theoretical contributions of the model
   The key theoretical contribution of this research lies in the establishment of a two-layer "AEI–market" interactive mechanism that systematically reveals the feedback relationships and dynamic game interactions between generation-side bidding strategies and consumption-side price responses. On the generation

side, distributed energy agents continuously learn optimal bidding behaviors based on evolving market signals and historical trading outcomes. On the consumption side, EAM loads autonomously adjust their charging, discharging, and operation schedules in real time according to electricity price variations. This bidirectional and iterative process of adjustment forms a self-organizing mechanism for energy and price regulation, realizing an endogenous balance between supply and demand within the AEI.

Methodologically, this study introduces an innovative integration of deep reinforcement learning (DRL) techniques with market optimization mechanisms of AEI. Specifically, a DDPG-based intelligent decision-making system was designed to enable different types of energy units to interactively learn and achieve optimal bidding strategies under conditions of uncertainty. Compared with traditional optimization or rule-based control methods, this DRL-based approach demonstrates clear advantages in coping with high-dimensional, nonlinear, and non-stationary problems, particularly in competitive and dynamic multi-agent environments where traditional methods are prone to convergence instability or local optima.

Furthermore, the research contributes to the theoretical interpretability of RL algorithms by designing a reward function directly linked to economic indicators such as profit maximization and cost minimization. This integration embeds explicit economic objectives within the training process, ensuring that the learning process remains guided by real market incentives. In doing so, the model enhances the transparency, directionality, and practical relevance of the reinforcement learning process. More importantly, the proposed mechanism establishes a generalizable theoretical paradigm for applying RL to multi-energy complementary markets, laying a foundation for subsequent exploration of intelligent optimization in integrated energy systems beyond agricultural contexts.

3. Summary of major experimental results

Based on the comprehensive multi-scenario simulation studies presented in Section 3.6, the results under multi-market conditions demonstrate that the reinforcement learning-based AEI model exhibits remarkable operational stability and convergence efficiency. After approximately 5000 training episodes, the learning agents reach stable convergence, and the resulting bidding strategies remain consistent across varying market cycles. Photovoltaic generation units display adaptive and flexible adjustments in real-time markets, enabling them to exploit short-term price fluctuations for improved revenues, while biomass generation units maintain consistent profitability in day-ahead markets through stable bidding behavior.

Overall, the coordinated operation between generation and EAM load agents leads to a clear dual optimization outcome in both revenue enhancement and risk mitigation. The system demonstrates smooth power dispatch, reduced cost volatility, and improved resilience to price uncertainty and stochastic disturbances. These findings provide strong evidence supporting the collaborative effectiveness of the two-layer AEI–market interaction mechanism and the reinforcement learning framework developed in this study.

In conclusion, the research results thoroughly verify the feasibility, robustness, and superiority of the reinforcement learning-driven intelligent bidding model for AEI integrated with EAM. From both theoretical and practical perspectives, the study confirms that RL algorithms are capable of enabling autonomous coordination among multiple energy agents and achieving sustained economic performance improvements in complex market environments. The outcomes provide a solid theoretical and methodological basis for promoting the market-oriented operation, intelligent optimization, and sustainable advancement of the Agricultural Energy Internet, offering valuable guidance for future research and engineering applications in smart agriculture and distributed energy systems.

3.7.2 System benefits of AEI integrated with EAM

1. Economic benefits
   The AEI integrated with EAM achieves the dual objectives of revenue maximization and cost minimization. On one hand, distributed generation units, guided by reinforcement learning-based intelligent bidding strategies, can capture the optimal clearing opportunities amid price fluctuations, thereby maximizing revenues. On the other hand, flexible loads such as EAM actively participate in load regulation through price-responsive strategies, purchasing electricity during low-price periods and reducing consumption during high-price periods, which significantly lowers electricity procurement costs.
2. Technical benefits
   The introduction of reinforcement learning algorithms significantly enhances the AEI's autonomous learning and self-regulation capabilities, enabling the system to maintain efficient and stable decision-making even under drastic market price fluctuations and stochastic load variations. Generation-side agents develop stable bidding patterns through long-term training, while EAM units flexibly adjust charging periods and scales to balance system load, smoothing the power curve. Additionally, the model realizes real-time interaction between data and strategies through a "market–agent" two-layer closed-loop structure, endowing the decision-making process with dynamic adaptability and scalability. This feature not only improves the model's response speed and accuracy in multi-time-scale markets but also provides a foundational framework for the subsequent development of an "AI-driven electricity market simulation platform."
3. Ecological and social benefits
   From ecological and social perspectives, the proposed AEI–EAM integration mechanism offers a feasible pathway for the electrification, intelligentization, and greening of agricultural production. By coordinating the loads of electric agricultural machines with renewable energy generation, the system achieves low-carbon energy consumption and high-efficiency energy utilization in

agricultural operations. During operation, the utilization rates of photovoltaic and biomass energy are significantly improved, and the electricity consumption structure becomes more environmentally balanced.

This approach effectively reduces wind and solar curtailment, lowers carbon emissions, and provides a practical reference for constructing "green electricity replacement" demonstration zones in rural areas. The operational mechanism of AEI is replicable and scalable, enabling broader applications in rural distributed energy networks, intelligent agricultural machinery coordination, and regional energy interconnection scenarios, thereby delivering substantial socioeconomic and ecological value.

3.7.3 Research limitations and shortcomings

Despite achieving systematic results in AEI intelligent bidding modeling, reinforcement learning algorithm design, and simulation verification, this study still has certain limitations and areas for improvement. The main shortcomings are reflected in the following three aspects:

1. Idealized model assumptions
During model construction, to facilitate algorithm convergence and theoretical validation, the study assumes fully transparent market information, where participants can obtain market prices and system states in real time and with complete accuracy. In practice, power markets often involve bid delays, communication lags, information asymmetry, and irrational strategic behaviors. In regional power markets or distributed energy scenarios, information exchange is constrained by network bandwidth, data privacy, and node heterogeneity, leading to perception and decision-making errors for the agents. In addition, this study only considers a limited number of generation and load entities, without fully capturing the complex competitive interactions among diverse market participants such as energy storage, demand response aggregators, and renewable energy intermediaries. This limits the model's ability to accurately represent real market competition.
2. Simplification and limited representativeness of simulation data
The simulation data used in this study are primarily based on sample statistics from public literature and synthetically generated data. While sufficient for model validation, these data cannot fully reflect the heterogeneity, seasonality, and stochastic characteristics of real power markets. For example, the charging behavior of electric agricultural machines (EAM) exhibits clear operational cycle patterns and seasonal fluctuations, and regional differences in machine fleets' energy demand distribution, scheduling habits, and responsiveness are significant, yet these differences were not fully quantified in the modeling process. Moreover, the generation of price sequences and load curves does not completely rely on historical data, resulting in limited coverage of market volatility and extreme scenarios, such as price crashes or supply-demand mismatches. To enhance engineering feasibility and generalizability, future

studies should incorporate multi-source measured data and dynamic market records to improve the credibility and applicability of model results.

3. Limited generalization and transferability of the algorithm
   The strategy transfer and environmental adaptation capabilities of the current reinforcement learning agents are still limited. When the external market environment changes significantly – such as policy adjustments, electricity price mechanism restructuring, extreme climate events, or sudden increases in EAM penetration – the model requires retraining to maintain optimal performance, hindering rapid cross-scenario deployment. In addition, the reward function design in this study is relatively static, mainly based on a single economic indicator, and does not fully incorporate multidimensional factors such as market risk, system volatility, and social benefits. The lack of a dynamic risk feedback mechanism limits the model's robustness under extreme fluctuation conditions. Furthermore, reinforcement learning algorithms are computationally intensive, and multi-agent game environments may experience convergence instability and local optima, which pose potential constraints on real-time decision-making in large-scale AEI systems.

   In summary, although this study validates the effectiveness and theoretical feasibility of reinforcement learning in AEI intelligent bidding, the model still needs improvement to better handle complex real-world environments, achieve cross-scenario transferability, and ensure long-term robustness. These limitations provide important directions for future research and technical breakthroughs.

3.7.4 Future research directions

Building on a clear understanding of the current model's limitations, future research can be deepened along four directions – market mechanism expansion, algorithm optimization, data integration, and engineering applications – to promote the practical implementation of AEI intelligent bidding theory.

1. Incorporating market imperfections and information asymmetry
   Future studies should relax the ideal market assumptions and construct a more realistic power market operating environment. Factors such as information opacity, communication delays, price manipulation, and market frictions can be introduced, establishing a partially observable Markov decision process (POMDP) framework to characterize agent decision-making under incomplete information.

   In addition, combining game theory with behavioral economics can analyze the strategy evolution and learning dynamics of market participants under asymmetric information, thereby revealing AEI's optimal response mechanisms under varying market signals and competition intensity. This direction will enhance the theoretical rigor of the model and improve its explanatory and predictive power in real-world markets.
2. Multi-agent game and coordination mechanism expansion
   With the coexistence and interaction of multiple AEI systems in regional power markets, future research should expand to the multi-agent game level. A

multi-agent reinforcement learning (MARL) framework can be introduced to study competition, cooperation, and information-sharing mechanisms among different virtual power plants, shifting from individual optimization to system-level coordinated optimization.

By applying federated reinforcement learning or distributed policy gradient algorithms, knowledge transfer and profit sharing among AEIs can be achieved while ensuring data privacy. Furthermore, alliance bidding mechanisms, regional cooperation agreements, and multi-layer incentive models can be designed to explore the stability and efficiency characteristics of different AEI coalitions in the market. This will provide theoretical support for constructing an interconnected and mutually beneficial agricultural energy ecosystem.

3. Dynamic data-driven strategy evolution and intelligent control
Future work can integrate real-time measurement data, weather forecasts, and historical market records to build a self-learning dynamic bidding system. By combining meta-learning and model predictive control (MPC) techniques, agents can quickly adapt to new conditions, enabling cross-scenario strategy transfer and rapid retraining.

   Moreover, incorporating temporal graph neural networks (T-GNN) or self-attention mechanisms (Transformers) can enhance the model's capability to capture long-term dependencies and multi-scale data, maintaining high prediction accuracy and strategy stability in complex time-varying price environments. This direction will further advance AEI bidding from static optimization toward dynamic intelligent evolution.

4. Practical applications and commercialization prospects
Future efforts should integrate the AEI–EAM fusion model with actual agricultural park energy management platforms to promote demonstrative applications and commercial pilot projects. By establishing standardized interfaces and open data exchange platforms, real-time coordination and energy interaction among virtual power plants, agricultural machine clusters, and grid operators can be achieved.

   At the same time, policy and market institutional frameworks should be improved, including electricity price compensation mechanisms, rules for distributed energy participation, and incentives for machine load response, encouraging energy enterprises and agricultural machinery manufacturers to jointly participate in market-based operations. This research direction will accelerate the green and low-carbon transformation of the agricultural industry and provide decision-making support for national-level distributed energy coordination, intelligent agricultural machinery industry upgrades, and integrated energy-agriculture development.

In summary, the future of AEI intelligent bidding research will move from single-model validation to multi-level, multi-agent, and dynamically adaptive system intelligence optimization. The deep integration of reinforcement learning, game theory, and data intelligence will serve as a key technological pathway supporting the long-term evolution of Agricultural Energy Internet, laying a solid foundation for an efficient, green, and intelligent agricultural energy ecosystem.

3.7.5 Main Conclusions and Outlook

In summary, this study systematically developed an intelligent bidding model for the Agricultural Energy Internet (AEI) integrated with electric agricultural machinery (EAM), covering a complete research framework from theoretical modeling, algorithm design, model training, and simulation validation. By incorporating reinforcement learning into the AEI bidding decision system, this research achieved significant breakthroughs in model adaptability, dynamism, and intelligence. The results demonstrate that reinforcement learning-driven bidding strategies not only effectively enhance AEI's revenue levels and resource allocation efficiency but also significantly improve the system's flexible dispatch capabilities and resilience to risks, providing a feasible new approach and technical pathway for the intelligent operation of agricultural energy systems.

Specifically, the study established a two-layer bidding optimization structure, enabling collaborative game and dynamic response mechanisms between the market layer and the AEI intelligent bidding layer. The reinforcement learning agents can dynamically adjust their bidding and dispatch strategies based on electricity price fluctuations, load demand, and renewable energy output variations, thereby achieving the dual objectives of revenue maximization and risk minimization across multiple time scales and market environments. Experimental results further validated the model's convergence stability and generalization capability under complex power market conditions, indicating strong robustness and adaptability in addressing challenges such as nonlinearity, stochastic disturbances, and information uncertainty.

Moreover, the outcomes of this research not only fill the academic gap in the "intelligent bidding–market response" linkage mechanism within the AEI field but also provide actionable technical guidance for constructing an integrated "EAM–grid–market" operational system in engineering practice. The AEI intelligent bidding mechanism facilitates fine-grained management and economic dispatch of agricultural energy use, reduces energy waste, enhances distributed renewable energy utilization, and promotes the transition of agricultural energy systems from passive response to active participation, supporting the shift from traditional centralized control to intelligent, coordinated optimization.

Looking ahead, with the continuous advancement of smart agricultural machinery, digitalized grids, IoT monitoring, and distributed energy technologies, the multi-market intelligent bidding mechanism of AEI is expected to become a key infrastructure for agricultural energy systems. This mechanism will further support dynamic balance, flexible management, and value co-creation in agricultural energy systems, providing robust technical backing for achieving green, low-carbon transformation, intelligent operation, and high-efficiency energy management. Combined with policy incentives, data sharing, and market mechanism innovation, the AEI intelligent bidding system is likely to achieve broader commercial implementation, becoming a core driver for the sustainable development of the Agricultural Energy Internet.

Therefore, this research not only validates the feasibility and effectiveness of reinforcement learning in AEI intelligent bidding at the methodological level but also offers important theoretical and practical insights for the development of the Agricultural Energy Internet. By further enhancing algorithm interpretability and transferability, it is anticipated that an even more intelligent, open, and collaborative agricultural energy ecosystem can be constructed, laying a solid foundation for the long-term evolution of AEI.

References

[1] Zhuang, G. (2021). Challenges and countermeasures for China in achieving the "dual carbon" targets. *People's Tribune*, (18), 50–53.

[2] Han, X., Li, T., Zhang, D., and Zhou, X. (2021). New issues and key technologies of new power system planning under double carbon goals. *High Voltage Engineering*, 47(9), 3036–3046.

[3] Xianyong, X., and Zixuan, Z. (2022). New power systems dominated by renewable energy towards the goal of emission peak and carbon neutrality: Contribution, key techniques, and challenges. *Advanced Engineering Science/Gongcheng Kexue Yu Jishu*, 54(1), 47–59.

[4] Chen, G., Liang, Z., and Dong, Y. (2020). Analysis and reflection on the marketization construction of electric power with Chinese characteristics based on energy transformation. *Proceedings of the CSEE*, 40(2), 369–379.

[5] Zhang, X., Liu, W., Chen, Y., Bai, Y., Li, J., and Zhong, J. (2018). Electricity market design and operation in Guangdong power. In 2018 15th International Conference on the European Energy Market (EEM) (pp. 1–5). IEEE.

[6] Zhang, X., and Shi, L. (2020). Future research areas and key technologies of electricity market in China. *Automation of Electric Power Systems*, 44(16), 1–11.

[7] Zhang, X., Fu, X., Xue, Y., Chang, X., and Bai, X. (2024). A review on basic theory and technology of agricultural energy internet. *IET Renewable Power Generation*, 18(7), 1318–1331.

[8] Fu, X., and Niu, H. (2023). Key technologies and applications of agricultural energy internet for agricultural planting and fisheries industry. *Information Processing in Agriculture*, 10(3), 416–437.

[9] Tian, L., Cheng, L., Guo, J., Wang, X., Yun, Q., and Gao, W. (2020). A review on the study of management and interaction mechanism for distributed energy in virtual power plants. *Power System Technology*, 44(6), 2097–2108.

[10] Li, J., Ai, Q., and Yin, S. (2022). Market mechanism and foreign experience of virtual power plant participating in peak-regulation and frequency-regulation. *Proceedings of the CSEE*, 42(1), 37–55.

[11] Zhao, Q., Shen, Y., and Li, M. (2015). Control and bidding strategy for virtual power plants with renewable generation and inelastic demand in electricity markets. *IEEE Transactions on Sustainable Energy*, 7(2), 562–575.

[12] Rahimiyan, M., and Baringo, L. (2015). Strategic bidding for a virtual power plant in the day-ahead and real-time markets: A price-taker robust optimization approach. *IEEE Transactions on Power Systems*, 31(4), 2676–2687.

[13] Ge, R., Chen, L., Wang, Y., and Liu, D. N. (2017). Optimization and design of construction route for electricity market in China. *Automation of Electric Power Systems*, 41(24), 10–15.

[14] Song, Y., Bao, M., Ding, Y., Shao, C., and Shang, N. (2020). Review of Chinese electricity spot market key issues and its suggestions under the new round of Chinese power system reform. *Proceedings of the CSEE*, 40(10), 3172–3187.

[15] Liang, Y., Guo, C., Ding, Z., and Hua, H. (2020). Agent-based modeling in electricity market using deep deterministic policy gradient algorithm. *IEEE Transactions on Power Systems*, 35(6), 4180–4192.

Chapter 4

Simulation of green certificate and carbon trading for AEI

Wentao Zhu[1] *and Xueqian Fu*[1]

Abstract

This chapter examines the feasibility of Agricultural Energy Internet (AEI) participation in the integrated electricity–carbon market trading. Grounded in both international and Chinese carbon emission trading systems (ETS) and renewable energy certificate (REC) policy frameworks, it focuses on the issue of renewable energy double incentives arising from the functional overlap between RECs and China Certified Emission Reductions (CCERs). The study aims to achieve a rational allocation of environmental rights and enhance the emission reduction efficiency of the power system through innovative mechanisms. The research first traces the carbon footprint of energy use in agricultural parks and establishes an accounting framework that includes REC obligations, REC holdings, carbon allowances, and actual emissions. It defines the accounting logic whereby gas-fired power units serve as the primary entities for carbon emission accounting, while photovoltaic (PV) and wind power units serve as REC suppliers. The accounting boundaries and standards are clearly delineated. Based on this, models are developed for PV and wind power output and REC holdings. Wind power output uncertainty is characterized using a Weibull distribution combined with scenario-based methods (Latin hypercube sampling and backward reduction). Meanwhile, the actual carbon emissions, initial carbon allowances, and REC obligations of thermal power units are calculated using the baseline emission factor method. To overcome the limitations of traditional fixed-coefficient mutual recognition mechanisms, this chapter innovatively designs a carbon–REC mutual recognition mechanism based on the marginal carbon emission factor (MCEF). By constructing a power system economic dispatch model, the system's carbon emission gradient variation resulting from changes in renewable power output is dynamically calculated, establishing a unique mapping between renewable generation and carbon reduction effectiveness. This enables time-varying adjustment of the REC–CCER conversion ratio: high mutual recognition coefficients are applied during periods of high marginal abatement efficiency, while lower or zero coefficients are applied

[1]College of Information and Electrical Engineering, China Agricultural University, China

during low-efficiency periods – effectively avoiding double incentives. To quantify the effectiveness of the proposed mechanism, a centralized joint clearing model for integrated electricity–carbon–REC markets is constructed, aiming to maximize total social welfare. This model deeply integrates the operation rules and behavioral characteristics of the three markets and adopts marginal pricing to ensure both resource optimization and fairness. Operational assumptions and constraints for each market are defined. Furthermore, a comprehensive benefit evaluation system is developed from environmental, energy, and economic perspectives, with core indicators including carbon emission intensity, renewable energy consumption rate, and total social welfare. Using an agricultural electrification scenario in Long'an County, Nanning City, Guangxi Province as a case study, and based on real-world meteorological data and generation-load parameters, two simulation scenarios are compared: the dynamic mutual recognition mechanism (Scenario 1) and the fixed coefficient mechanism (Scenario 2). The results show that under Scenario 1, renewable energy consumption significantly improves (from 0.6926 to 0.8672), system carbon emission intensity markedly decreases (from 0.3300 to 0.2760), and total social welfare substantially increases (from RMB 1,161,981.52 to RMB 1,840,861.80). Carbon market transactions concentrate during high abatement efficiency periods, and the REC market establishes a "dual-market value channel," achieving spatiotemporal optimization of renewable generation and multi-market synergistic benefits. This study breaks through the static limitations of traditional environmental rights mutual recognition mechanisms and provides theoretical support and a technical pathway for AEI participation in multi-market integrated trading. The proposed mechanism demonstrates strong engineering feasibility and policy scalability, offering significant implications for promoting deep integration of electricity, carbon, and REC markets, and for advancing the clean and low-carbon transition of agricultural energy systems. Future work may further extend the model's multi-scenario adaptability, improve multi-agent coordination design, deepen uncertainty analysis and risk management, and promote empirical implementation and standardization of the mechanism.

Keywords: AEI; carbon–green certificate mutual recognition; marginal carbon emission factor; electricity–carbon–green certificate joint clearing; renewable energy consumption; low-carbon transition

4.1 Introduction

4.1.1 Research background and current status

4.1.1.1 Development background of the Agricultural Energy Internet

The Agricultural Energy Internet (AEI) has emerged in recent years as an interdisciplinary concept that has attracted extensive attention from. Its proposal stems from a deep reflection on traditional agricultural production and energy utilization models, as well as the urgent need to reconstruct agricultural energy systems and the "dual carbon" strategy [1,2]. In traditional agricultural systems, energy not only serves as the fundamental driving force for mechanization, informatization, and automation but also constitutes a major source of agricultural carbon emissions. How to achieve clean, efficient, and low-carbon energy systems while ensuring agricultural productivity and energy security has become a key issue in contemporary sustainable agricultural development.

Looking back on the historical evolution of agricultural energy use, since the 1950s, fossil fuel-based energy systems centered on petroleum have been widely applied in global agriculture. Petroleum and its derivatives – such as coal, natural gas, and diesel – have been extensively used for agricultural machinery, fertilizer production, irrigation systems, as well as agricultural product processing and transportation. This energy-driven model greatly enhanced agricultural productivity and laid a solid foundation for food security and agricultural modernization. However, long-term dependence on fossil energy has also resulted in serious environmental problems, including increased greenhouse gas emissions, aggravated air pollution, and depletion of energy resources. The share of agriculture in global carbon emissions has gradually risen, making it one of the major contributors to greenhouse gas emissions. This not only intensifies ecological pressure but also poses significant challenges to the sustainability of agricultural development [3].

With the aggravation of global energy crises and environmental issues, agricultural. Since the 1970s and 1980s, electricity has progressively penetrated all aspects of agricultural production – from electric-driven machinery and automated irrigation to environmental control systems and energy use in agricultural processing. The core objective of agricultural electrification is to promote automation, precision, and intelligence in agricultural production through the extensive application of electricity and electrification technologies. This process has significantly facilitated the structural optimization of agricultural energy systems, thus laying an important foundation for the green transformation of agricultural energy [4]. In the twenty-first century, the rapid development of Smart Grids and Smart Agriculture has provided powerful technological support for the transformation and upgrading of agricultural energy systems. The Smart Grid represents not only a technological innovation of traditional power systems but also a combination of electrical, information, and communication technologies (ICT). Through data analytics and automatic control, the Smart Grid enables bidirectional interaction between energy flow and information flow, enhancing the flexibility, adaptability, and reliability of power systems. In agriculture, Smart Grids can dynamically adjust energy supply according to load variation, support distributed generation integration and energy storage coordination, and improve grid security and economic efficiency [4]. Meanwhile, has further advanced the digitalization and intelligence of agricultural production processes. Leveraging new-generation information (IoT), sensors, agricultural managers can monitor and manage soil moisture, meteorological conditions, crop growth, irrigation demand, and energy use in real time. Such comprehensive perception and control capabilities make agricultural production more visualized, controllable, and efficient, significantly reducing resource waste and energy consumption. Smart Agriculture not only transforms agricultural production modes but also provides data foundations and decision-making support for intelligent scheduling and optimized operation of agricultural energy systems [5]. On this basis, the Agricultural Energy Internet has emerged. It represents the convergence of Smart Grids and Smart Agriculture – a comprehensive system integrating energy technologies, information technologies, and intelligent control. The core concept of AEI lies in building an energy network characterized by multi-energy complementarity, information sharing, and intelligent scheduling, to achieve coordinated operation of energy, information, and value flows within agricultural systems. Essentially, AEI represents an advanced stage of agricultural electrification and a new model of agricultural energy management oriented toward renewable energy, digitalization, and low carbonization. Through its systemic architecture, AEI can integrate multiple distributed small hydropower – achieving local consumption and regional dispatch to construct an efficient, low-carbon, and resilient agricultural energy ecosystem [4]. The operational mechanism of AEI involves not only physical-level regulation of energy supply and consumption but also data interaction at the information layer and market response at the value layer. Through communication networks, AEI enables real-time interaction among energy devices, user terminals, and management centers, dynamically optimizing load distribution, energy storage dispatch, and energy efficiency control. This data-driven intelligent management approach overcomes the limitations of traditional agricultural energy systems – such as "information silos" and "unidirectional supply" – and achieves global optimization and dynamic balance of energy utilization [6]. However, agricultural energy systems exhibit high complexity and randomness, being influenced by seasonal changes, climate fluctuations, and crop cycles. Faced with such high-dimensional uncertainty, traditional optimization methods are often inadequate for dynamic scheduling under complex environments. In recent years, artificial intelligence algorithms such as reinforcement learning, deep learning, and multi-agent optimization control have been introduced into AEI operation management. These approaches enable adaptive learning and policy optimization for autonomous energy regulation, greatly improving the stability and economic performance of agricultural energy systems [7]. From a technological perspective, the realization of AEI relies on the organic integration of multi-level technologies. First, at the perception and monitoring layer, sensor networks, unmanned aerial vehicle (UAV) monitoring, and remote sensing technologies form the foundation for energy data acquisition, enabling comprehensive perception of agricultural environments, climate, and energy equipment, thus providing high-quality data for subsequent analysis [8]. Second, at the energy supply layer, the development of distributed energy and microgrid technologies allows agricultural parks to locally utilize biomass energy, achieving self-balancing and supply–demand matching with energy storage systems [9]. Third, at the intelligent control layer, forecasting, decision-making, and optimization of energy system scheduling. Meanwhile, the introduction provides a trustworthy data foundation and secure market environment for energy trading and carbon management, transparency and efficiency of energy credit systems, renewable energy certificate (REC) trading, and carbon emission allowance transactions [10]. Furthermore, cybersecurity and privacy protection are key aspects of the AEI framework, requiring multi-layer defense mechanisms and privacy-preserving computation technologies to ensure data and system security [9].

At present, AEI application scenarios have gradually expanded, with demonstrative achievements emerging across multiple domains. Government regulation, public services, agricultural production, and rural energy systems constitute the four major application directions of AEI development [11]. In crop production AEI frameworks, by integrating meteorological data, crop physiological information, and energy consumption characteristics, intelligent perception and dynamic control of the entire farming process can be achieved. For example, systems can automatically adjust irrigation frequency and greenhouse energy strategies according to weather changes and crop water requirements, reducing energy use and emissions while maintaining yields. In livestock farming AEI scenarios, IoT devices enable real-time monitoring of barn environments and energy loads, automatically optimizing lighting, ventilation, and temperature control strategies for animal welfare. In rural energy systems, AEI leverages the coordinated operation of microgrids, energy storage, and distributed generation to increase energy self-sufficiency and renewable energy utilization, providing models for building new rural energy systems. At the macro level, AEI construction management and policy-making for government departments. Through digitalized energy monitoring and visualization platforms, governments can track and precisely account for agricultural carbon emissions, providing scientific evidence for policy design. In the field of public services, AEI promotes new service models such as rural power operation and maintenance, distributed photovoltaic trading, and energy finance, injecting new momentum into rural economic development [12]. From a strategic perspective, the development of AEI is highly aligned with China's "dual carbon" goals. Currently, agriculture faces challenges such as large total carbon emissions, complex emission structures, and incomplete carbon footprint accounting systems. Building a low-carbon energy system based on AEI can not only enable fine-grained identification and dynamic monitoring of emission sources but also promote the formation of agricultural carbon reduction pathways through energy structure optimization and green trading mechanisms. AEI's role in low-carbon transformation is mainly reflected in three aspects: energy substitution, carbon management, and market coordination. First, agricultural energy structures, dependence on fossil fuels is reduced, leading to cleaner energy systems. Second, through AEI's digital platform, carbon emissions throughout agricultural production can be tracked and quantified, providing a foundation for agricultural entities to participate in carbon markets and carbon footprint accounting. Finally, AEI's energy trading mechanisms can facilitate the integration of renewable energy certificate (REC) markets with emission trading systems (ETS), realizing the dual value of emission reduction [13].

Looking ahead, the Agricultural Energy Internet will become a crucial intersection of agricultural modernization and the energy revolution. With the evolution of emerging technologies such as IoT, big data, and blockchain, AEI will further develop toward self-learning, self-organization, and self-optimization. Through interdisciplinary integration and institutional innovation, AEI is expected to play a more critical role in national energy security, green agricultural development, and rural revitalization strategies. Under the long-term guidance of the "dual carbon" strategy, AEI will not only serve as a key enabler of agricultural energy efficiency improvement and structural optimization but also as essential infrastructure for advancing the greening, digitalization, and intelligence of agriculture. Overall, as an innovative system integrating agriculture and energy in the new era, the Agricultural Energy Internet is transitioning from conceptual exploration to systematic practice. It reshapes the supply-demand relationship of agricultural energy and provides a systematic solution for agricultural carbon reduction and green transformation. In the future, AEI – characterized by openness, intelligence, and collaboration – will drive deep transformations in agricultural production, energy utilization, and carbon management systems, offering a new development paradigm and a "Chinese solution" for sustainable agriculture and a greener planet.

4.1.1.2 Current status of renewable energy applications in agricultural parks

With the global energy structure accelerating its transition toward cleanliness and low-carbon, and the increasing prominence of sustainable agricultural development strategies on national policy agendas, the integrated application of renewable energy in agricultural parks has evolved from marginal exploration to mainstream practice. It has become a core direction for academia to address the coordinated challenge of "energy security - food security - environmental security" and for industry to promote the green transformation of agriculture. As intensive carriers of modern agricultural production, agricultural parks cover the entire industrial chain, including planting, breeding, processing, and logistics. Their energy demand presents dual characteristics: "long-term stable supply + short-term peak fluctuation" and "constant base load + diverse scenario-based energy consumption." From the power demand for agricultural machinery operation, constant temperature control for greenhouse cultivation, to thermal energy supply for agricultural product processing and continuous refrigeration for cold chain logistics, each link imposes high requirements on the reliability, economy, and cleanliness of energy. Against this backdrop, how to break free from the dependence on traditional fossil energy, construct a renewable energy system with coordinated "source-grid-load-storage" at the park level, and achieve efficient substitution, optimal allocation, and intelligent regulation of green energy has become a key breakthrough for promoting the transformation of agriculture from "high-carbon production" to "low-carbon circulation." Furthermore, it constitutes an important practical path for addressing and ensuring agricultural systems on a global scale [11].

From a global perspective, the green restructuring of agricultural energy systems has become a strategic intersection for addressing climate change and safeguarding food security. Stated in its 2024 "Report on Agricultural Energy Transition" that annually in the global agricultural sector accounts for 4–6% of global primary energy consumption, with over 70% relying on, diesel, Meanwhile, greenhouse gas emissions from agricultural production (including land use change) account for 10–12% of global total emissions, primarily originating from fuel consumption of agricultural machinery, production and application of chemical fertilizers, methane emissions from livestock and poultry breeding, and soil carbon emissions from farmland. This “high energy consumption - high emission” development model not only exposes agricultural systems to cost risks arising from fluctuations in fossil energy prices but also exacerbates them. Climate events such as extreme high temperatures and droughts further push up energy demand for irrigation and cooling, forming a vicious cycle of “energy consumption - climate deterioration - further increase in energy consumption.” The International Renewable Energy Agency (IRENA) emphasized in its “Rural Energy Transition Roadmap” that developing agricultural renewable energy is the core means to break this cycle. On one hand, solar energy, wind energy, and biomass energy can directly replace fossil energy in agricultural production. On the other hand, distributed renewable energy systems can improve the energy self-sufficiency rate of remote agricultural parks, enhance agricultural production resilience under extreme weather conditions, and create additional economic benefits through “energy-agriculture” coupling models (such as agrivoltaics and biomass recycling), injecting impetus into the sustainable development of rural areas [11]. Under this global consensus, agricultural parks, with their unique advantages, have emerged as ideal scenarios for the integrated application of renewable energy. Compared with scattered smallholder production models, agricultural parks possess characteristics such as high spatial concentration, improved infrastructure, stable and predictable energy demand, and strong industrial chain coordination. Concentrated and contiguous land resources provide a spatial foundation for the construction of photovoltaic power plants and wind farms. Improved power distribution networks and water conservancy facilities reduce the connection costs of renewable energy systems. The full-chain energy demand from planting to processing enables multi-energy complementarity, including electricity, thermal energy, and cold energy, improving energy utilization efficiency. Moreover, the park-based management model facilitates unified dispatching, operation, and maintenance of energy systems. For these reasons, cases such as Germany’s “ecological farms,” the Netherlands’ “greenhouse energy parks,” and Japan’s “solar farms” have become global models for the application of agricultural renewable energy, providing valuable practical experience for other countries [14]. Academic research on renewable energy in the agricultural sector has established a relatively systematic theoretical system and technical framework. Recent bibliometric analyses based on indicate that between 2015 and 2025, the number of research papers in the field of agricultural renewable energy has 18.7%. Research themes have gradually expanded from early single energy technologies (such as photovoltaic irrigation and biomass biogas) to a multi-dimensional focus including “system integration - energy efficiency optimization - policy mechanisms - regional adaptation.” In terms of energy types, solar energy, wind energy, and biomass energy are the three core research directions. Geothermal energy, small hydropower, and marine energy (applicable to coastal agricultural parks) serve as supplements in specific regions, collectively forming a diversified technical matrix of renewable energy for agricultural parks [14]. Among these, solar energy occupies a core position in agricultural energy systems due to its universality, accessibility, and technological maturity. The application of photovoltaic power generation technology has evolved from early "decentralized small-capacity" to "centralized large-scale." In the planting link, distributed photovoltaic modules are widely deployed on greenhouse roofs, agricultural shed tops, and idle spaces around farmland, forming an "agri-voltaic" model of "power generation above, planting below." For example, laying thin-film photovoltaic modules with a light transmittance of 70–80% on the roofs of vegetable greenhouses can not only meet photovoltaic power generation needs but also provide a suitable diffused light environment for crop growth. It is estimated that such models can 30–50%. In the irrigation link, photovoltaic water pumping systems have replaced traditional diesel water pumps, solving the problem of "difficulty in accessing electricity and high cost of using oil" in remote farmland. According to IRENA statistics, over 1 million photovoltaic irrigation systems have been put into use globally, covering an irrigation area of more than 5 million hectares and reducing diesel consumption by approximately 2 million tons annually. In the agricultural product processing link, photovoltaic power plants supply electricity to equipment such as grain drying, fruit and vegetable refrigeration, and feed processing. Combined with energy storage systems, they can achieve 24/7 continuous operation, significantly reducing processing costs [15]. In addition, solar thermal technology has also demonstrated unique advantages in agricultural parks: flat-plate collectors can meet the needs of greenhouse heating, livestock and poultry house heating, and domestic hot water for employees; trough-type and dish-type concentrated solar thermal systems can provide medium-to-high temperature thermal energy for deep processing of agricultural products (such as tea fixation and medicinal material drying), with a thermal energy utilization efficiency of 60–75%, much higher than that of traditional coal-fired boilers [14].

As another important form of the application of wind energy in agricultural parks exhibits the characteristics of "adapting to local conditions and multi-energy coupling." Compared with solar energy, wind energy is more intermittent but has a higher energy density, making it suitable for combination with energy storage systems and microgrids to provide stable supplementary power for agricultural parks. In regions with abundant wind energy resources in northwest and north China (such as Jiuquan in Gansu and Tongliao in Inner Mongolia), many agricultural parks have constructed small wind farms (with a single unit capacity of 1.5–3 MW) combined with 2–5 MWh energy storage systems, forming an independent power supply model of "wind power + energy storage + microgrid." This model not only meets the park's own energy demand but also generates additional income through surplus power grid connection. In coastal areas, offshore wind energy is combined with seawater desalination and aquaculture to form a circular system of "wind power - seawater desalination – aquaculture," in coastal agricultural parks [15]. The development of wind energy in agricultural parks also emphasizes coordination with ecological protection: low-height wind turbines (height below 30 meters) are used in grassland pastures to avoid impacts on bird migration; the distance between wind turbines and farmland crops strictly follows relevant standards to prevent adverse effects of blade rotation on crop growth. These detailed designs provide guarantees for the sustainable application of wind energy in agricultural scenarios [12]. With its "resource recyclability" and "multi-energy output capability," biomass energy has become a key link for agricultural parks to achieve coordinated development of "energy - agriculture – environment." Agricultural wastes generated during agricultural production, such as crop straw, livestock and poultry manure, fruit and vegetable processing waste, and forestry residues, can all be used as raw materials for biomass energy. Through technologies such as gasification, liquefaction, and fermentation, they can be converted into energy products such as biogas (methane content 50–70%), biodiesel, and biomass pellets, realizing "waste-to-treasure." In major corn-producing areas in northeast China, agricultural parks use corn straw to produce biomass pellets, replacing coal for greenhouse heating. This not only reduces air pollution caused by straw burning annually but also lowers greenhouse gas heating costs by 40–60%. In large-scale livestock and poultry breeding parks in southern China, large-scale biogas projects convert pig manure, cow dung, and other wastes into biogas for power generation, heating, and cooking. Biogas residue and biogas slurry are used as organic fertilizers for farmland, forming a circular agricultural model of "breeding - biogas – planting." It is estimated that a biogas project in a 10,000-head pig farm can produce 1.5–-2 million cubic meters of biogas, 3000–4000 tons [15]. In recent years, biomass energy and combined heat and power (CHP) technology have further expanded their application scenarios: small biomass CHP units (with a capacity of 5–20 MW) can simultaneously provide electricity and thermal energy for agricultural parks, meeting the multi-scenario energy demand for processing, heating, and drying. The comprehensive energy utilization efficiency can reach over 80%, much higher than single power generation or heating modes. Other forms of geothermal have demonstrated irreplaceable advantages in regions with suitable resource endowments. In areas with abundant geothermal resources, such as Northeast China and Guanzhong Basin, the application of geothermal energy in agricultural parks has expanded from traditional greenhouse heating to diversification: shallow geothermal heat pump systems are used for constant temperature control of greenhouse environments, enabling heating in winter and cooling in summer. Compared with traditional air conditioners and coal-fired boilers, energy consumption is reduced by 30–50% and operating costs by 25–40%. Medium-deep geothermal energy is extracted through geothermal wells to provide stable medium-to-high temperature thermal energy for agricultural product processing. For example, in Shouguang, Shandong, geothermal resources are used for vegetable dehydration and drying, which not only improves drying efficiency but also avoids product pollution caused by coal-fired drying [12]. In mountainous areas in southwest China and water towns in the Jiangnan region, the combination of small hydropower and agricultural parks has a long history: small hydropower stations (with an installed capacity of less than 10 MW) generate electricity using hydropower from rivers and streams, providing stable power for mountainous agricultural parks. Meanwhile, reservoir regulation improves agricultural irrigation conditions, forming comprehensive benefits of "power generation - irrigation – breeding." Microgrid technology, small hydropower has also been combined with photovoltaic and wind energy to construct multi-energy complementary systems, further improving the stability of energy supply for mountainous agricultural parks.

From the perspective of research geographical distribution, bibliometric results from 2019 to 2021 show that research in the field of agricultural renewable energy presents a pattern of "led by Europe and the United States, chased by China and India, and globally diffused." European countries, relying on their early policy support and technological accumulation, are in a leading position in agricultural energy system integration and carbon emission management. Germany's "energy cooperative" model encourages farmers to participate in distributed photovoltaic and wind energy projects through equity investment, enhancing farmers' participation enthusiasm through energy sharing and profit distribution. The Netherlands' "greenhouse energy parks" integrate various photovoltaic, geothermal, and biomass energy, and realize precise energy dispatching through intelligent control systems. The energy consumption per unit area of its greenhouses has dropped to the lowest level in the world (less than 150 kWh per square meter per year). Denmark promotes the agricultural parks through carbon tax policies and green certificate trading. Currently, the proportion of renewable energy in Denmark's agricultural sector has exceeded 40% [15]. The United States and Canada focus more on the economic and market mechanisms research of agricultural renewable energy. Through policy tools such as (e.g., the U.S. Inflation Reduction Act provides a 30% investment tax credit for agricultural photovoltaic projects), renewable energy certificate (REC) trading, and Renewable Portfolio Standard (RPS), they reduce the cost of applying renewable energy in agricultural parks and provide stable revenue channels for projects. For example, California requires large agricultural electricity consumers (with annual electricity consumption exceeding 1 million kWh) to have a renewable energy consumption ratio of no less than 50%. This policy has directly promoted the rapid development of photovoltaic and wind energy projects in California's agricultural parks [12]. As major agricultural countries, China and India have shown particularly active research and practice in the field of agriculture, forming technical systems and application models in line with their national conditions. Relying on its abundant solar energy resources and large agricultural population, India has vigorously promoted small photovoltaic irrigation systems and biomass biogas projects. By 2023, India will have put over 500,000 photovoltaic water pumps into use, covering an irrigation area of more than 2 million hectares, and built over 100,000 rural biogas projects, benefiting approximately 5 million farming households [16]. Under policy guidance, China has promoted the transformation of agricultural renewable energy application from "single-point demonstration" to "large-scale promotion," forming characteristic models such as "photovoltaic + agriculture," "wind-solar-storage integration," and "biomass energy - circular agriculture," making scale of agricultural renewable energy application in the world [12].

In China, the application of agricultural parks has been deeply integrated into national energy strategies and agricultural development plans, with a continuously improving policy support system. At the national level, the government has clearly proposed to "vigorously promote models such as agrivoltaics, agriculture-photovoltaic complementarity, and pasture-photovoltaic complementarity in rural areas." The Five-Year Action Plan for Agricultural Green Development (2023–2027) further lists "renewable energy substitution in agriculture" as a key task. It requires that by 2027, the average proportion of renewable energy consumption in agricultural parks should exceed 15%, and the energy utilization rate of livestock and poultry manure should surpass 60%. At the local have issued targeted support policies based on their own resource endowments and agricultural characteristics: Hebei Province provides a construction subsidy of 0.3–0.5 yuan per watt for "photovoltaic + greenhouse" projects and gives priority to ensuring project grid connection; Shandong Province implements electricity price subsidies for agricultural biomass energy projects, with biomass power generation projects increased by 0.1 yuan/kWh the local desulfurization; Jiangsu Province provides special loans with an interest rate 10–15% lower than the market rate for agricultural renewable energy projects through the "green credit" policy [12]. Driven by both policies and market demand, China has built a number of demonstration agricultural renewable energy projects. In north China, agricultural parks in Hebei, Shandong, Henan, and other places have extensively promoted the "photovoltaic + greenhouse" model: laying photovoltaic modules on greenhouse roofs not only meets the electricity demand for supplementary lighting, ventilation, irrigation, and other equipment inside the greenhouses but also generates income through surplus power grid connection. Taking a "photovoltaic + vegetable greenhouse" park in Dezhou, Shandong as an example, the park has built a 100 MW photovoltaic power station covering 5000 mu of greenhouse area, generating 120 million kWh of electricity annually. While meeting the park's own electricity demand, it can also transmit 80 million kWh to the power grid, equivalent to reducing by 40,000 tons and by 100,000 tons annually. The vegetable output of the park has increased by 15–20% compared with traditional greenhouses, and the comprehensive economic benefits have increased by more than 40% [12]. In northwest China, Gansu, Xinjiang, Inner Mongolia, and other places have utilized their abundant wind and solar energy resources to construct "wind-solar-storage integrated" agricultural parks. Through the coordination, they provide stable power for remote agricultural parks, solving the power supply reliability of traditional power grid extension. For example, a "wind-solar-storage + forage planting" park in Jiuquan, Gansu has built a 50 MW wind farm, a 100 MW photovoltaic power station, and a 20 MWh energy storage system. It not only meets all the electricity demand for forage planting, processing, and irrigation in the park but also improves land use efficiency by planting drought-tolerant forage under the photovoltaic arrays through the "pasture-photovoltaic complementarity" model [15]. In southwest China, Sichuan, Yunnan, Guizhou, and other places have combined small hydropower with agricultural production to form a comprehensive utilization model of "hydropower + irrigation + breeding": small hydropower stations provide electricity for mountainous agricultural parks, while reservoir regulation ensures irrigation water supply, and the reservoir area develops aquaculture, realizing "one water for three uses." In Central China, agricultural parks in Hunan, Hubei, Jiangxi, and other places have focused on promoting the recycling of biomass energy: building biogas projects or biomass pellet processing plants using raw materials such as crop straw, rice husks, and livestock and poultry manure to provide electricity, thermal energy, or fuel for the parks. The resource utilization rate of wastes has reached over 90%, realizing the coordinated development of "energy - agriculture – environment." At the energy system integration level, the "photovoltaic + agriculture" model has become the most widely applied and in-depth research direction due to its strong compatibility and significant economic benefits. In recent years, this model has expanded from the early "photovoltaic + vegetable greenhouse" to diversified composite models, forming segmented scenarios such as "photovoltaic + fishery," "photovoltaic + animal husbandry," "photovoltaic + Chinese medicinal material planting," and "photovoltaic + aquaculture." Taking the "photovoltaic + fishery" model as an example, photovoltaic modules are erected above fish ponds. The photovoltaic panels block sunlight, reducing the water temperature of the fishponds, reducing water evaporation and algae reproduction, which is beneficial to fish growth. Photovoltaic power generation supplies electricity for equipment such as aeration and feeding in the fishponds. Some projects also combine aquaculture wastewater treatment to form a circular system of "power generation - breeding - wastewater treatment – irrigation." A "photovoltaic + freshwater aquaculture" park in Yancheng, Jiangsu, has built a 200 MW photovoltaic power station covering 12,000 mu of fish pond area, generating 240 million kWh of electricity annually. The per mu yield of fish in the fishponds has increased by 10–15%, and the comprehensive benefit is 2–3 times that of traditional fishponds [12]. The "photovoltaic + animal husbandry" model has been promoted in grassland pastures: laying photovoltaic modules on pastures, planting forage under the photovoltaic panels, not only providing electricity for irrigation, silage processing, and livestock and poultry breeding in the pastures but also improving the forage growth environment through the shading effect of the photovoltaic panels, increasing forage yield. A "photovoltaic + beef cattle breeding" park in Inner Mongolia has built a 30 MW photovoltaic power station with a supporting 5000-head beef cattle breeding farm, generating 360 million kWh of electricity annually. The forage yield has increased by more than 20%, and the beef cattle breeding cost has decreased by 15–20% [15]. The application of wind energy in agricultural parks focuses more on coordination with energy storage systems and microgrids to address its intermittency and volatility. In recent years, with the decline in energy storage technology costs (lithium-ion battery energy storage costs have decreased by more than 70% compared with 2015), the "wind power + energy storage + microgrid" model has been increasingly widely used in agricultural parks. In an agricultural park in Zhangjiakou, Hebei, a 100 MW wind farm, a 50 MWh energy storage system, and a microgrid control system have been built at low valleys and discharging at peaks according to wind power output and park energy demand, ensuring a stable power supply for the park. After the system is put into operation, the wind power consumption rate of the park has increased from 65% to over 95%, and the dependence on external power grid supply has decreased from 80% to less than 20% [12]. In coastal areas, the "offshore wind power + seawater desalination + aquaculture" model has realized the comprehensive utilization of multiple resources: offshore wind power provides electricity equipment, is used for aquaculture, and the aquaculture wastewater is used for irrigation after treatment, forming a coordinated system of "energy - water resources – agriculture." An "offshore wind power + seawater desalination + shellfish breeding" park in Zhoushan, Zhejiang, has built a 200 MW offshore wind farm and a 100,000-ton/day seawater desalination plant, producing 36.5 million tons of fresh water annually, meeting the needs of aquaculture and farmland irrigation in the park. The shellfish breeding output has increased by more than 30%. The application of biomass energy in agricultural parks has gradually upgraded to "full industrial chain circulation," expanding from single energy production to multi-product output of "energy - fertilizer – feed." In a "straw - biomass - circular agriculture" park in Henan, corn straw is used to produce biomass pellets (for power generation and heating) and cellulosic ethanol (for agricultural machinery fuel). By-products generated during the production process (such as furfural residue) are used as organic fertilizers for farmland, forming a closed-loop cycle of "straw collection - energy production - by-product returning to farmland - crop planting." The park can process 100,000 biomass pellets and 10,000 tons of cellulosic ethanol, reducing carbon dioxide emissions by 80,000 tons. At the same time, it increases soil organic matter content, and crop yield increases by 10–15% [12]. In a "livestock and poultry breeding - biogas - planting" park in Guangdong, a large-scale biogas project is built using pig manure. Biogas is used for power generation, heating, and cooking. Biogas residue and biogas slurry are processed into organic fertilizers for vegetable and fruit planting in the park, forming a circular model of "breeding - biogas – planting." The park can process 200,000 tons of pig manure annually, produce 1.5 million cubic meters of biogas, generate 2 million kWh of electricity, and produce 10,000 tons of organic fertilizers. The vegetable yield has increased by more than 20%, and chemical fertilizer usage has decreased by 30% [12].

Despite the remarkable achievements in promoting agricultural parks, there are still various challenges in terms of technology, economy, policy, and management during large-scale application. These challenges present differentiated characteristics in agricultural parks at different development stages and in different regions, requiring targeted solutions. The technical complexity of energy system integration and the difficulty of coordinated dispatching are the primary challenges. The energy consumption scenarios of agricultural parks are characterized by "multiple types, multiple time periods, and multiple fluctuations." For example, greenhouse cultivation requires a stable electricity and thermal energy supply, agricultural product processing has obvious seasonal peak loads (such as drying demand during harvest seasons), and agricultural machinery operation presents intermittent and mobile energy consumption characteristics. Different renewable energy technologies have significant differences in output characteristics: solar radiation intensity, day-night alternation, and weather changes, with output showing the characteristics of "high during the day, low at night; high on sunny days, low on cloudy days." Wind energy has random fluctuations and may even experience "windless periods." Although biomass energy output is relatively stable, it is also affected by the seasonality of raw material supply (such as straw harvesting concentrated in autumn), resulting in supply fluctuations. How to match these renewable energy sources with significantly different characteristics to the diverse energy demand of the park and dispatching has become a core problem of system optimization. In addition, the coupling technology between different energy systems still needs to be broken through. For example, the coordinated control algorithm of photovoltaic, complementary dispatching technology of biomass photovoltaic and wind power, and the unified monitoring and operation and maintenance platform of multi-energy systems. The immaturity of these technologies leads to low operational efficiency of renewable energy systems in some agricultural parks, and even the phenomenon of "abandoned photovoltaic power" and "abandoned wind power." At the economic level, high equipment investment, long payback periods, and financing difficulties are the key factors restricting the promotion of agricultural parks. Although the costs of technologies such as photovoltaic and wind power have dropped significantly in recent years (photovoltaic module costs have decreased by more than 90% compared with 2010), the overall investment in renewable energy projects in agricultural parks is still relatively high. For example, the investment in photovoltaic systems per mu of greenhouse in "photovoltaic + greenhouse" projects is about 100,000–150,000 yuan, much higher than the construction cost of traditional greenhouses (20,000–50,000 yuan per mu). The investment in "wind power + energy storage + microgrid" projects is even higher, with an investment of about 15–20 million yuan per MW of system. The investment in agricultural parks is usually long (generally 8–15 years). However, agricultural production itself has the characteristics of large income fluctuations and weak risk resistance, leading many agricultural park operators (especially small and medium-sized parks) to hesitate to invest in renewable energy projects. Financing difficulties further exacerbate this problem: agricultural parks are mostly located in rural areas, and operators are mostly small and medium-sized enterprises or cooperatives, lacking effective collateral and having low credit ratings, making it difficult to obtain bank loans. The long-term nature of renewable energy projects is inconsistent with the short-term nature of existing financial products (such as bank loan terms of 3–5 years), resulting in significant liquidity pressure on project funds. At the policy and mechanism level, the imperfect policy support system, inadequate market mechanisms, and lack of standards and norms are important restrictive factors. Although the central policies to support the development of agricultural renewable energy, there are still "last mile" problems in the implementation. The subsidy policies in some regions tend to "emphasize construction over operation," with delayed and irregular subsidy distribution. The grid connection policies in some regions are imperfect, resulting in complex procedures and high costs for grid connection of renewable energy projects in agricultural parks, and even the phenomenon of "difficulty in grid connection." In terms of market mechanisms, the revenue channels of renewable energy projects in agricultural parks are still relatively single, mainly relying on power generation grid connection and policy subsidies. Market-oriented revenue channels such as green certificate trading and carbon trading have not been fully opened. For example, currently, China's green certificate trading is mainly oriented toward large power generation enterprises, making it difficult for distributed photovoltaic and wind power projects in agricultural parks to participate. The carbon trading market mainly covers industries such as power generation, iron and steel, and cement. The carbon emission reductions of agricultural renewable energy projects are difficult to verify and trade, resulting in the failure to convert their environmental value into economic benefits. In terms of standards and norms, there are still no standards for the design, construction, and renewable energy systems in agricultural parks. For example, the matching standards between the light transmittance and installation height of photovoltaic modules and crop growth in the "agrivoltaic" model, and the standards for the collection, storage, and transportation of raw materials for biomass energy projects. The lack of these standards leads to unreasonable design and irregular operation and maintenance of some projects, affecting the operational efficiency and safety of the system. At the management and talent level, insufficient energy management capacity, a lack of professional operation and maintenance talents, and a low digitalization level are prominent shortcomings. Most operators of agricultural parks mainly engage in agricultural production, lacking experience in energy system management, and thus are unable to conduct scientific dispatching and efficient operation and maintenance of renewable energy systems. For example, the cleaning power generation efficiency of photovoltaic systems in some parks decreases year by year (5–10% annually). The service life of energy storage systems in some parks is shortened (from 10 years to 5–6 years) due to unreasonable charging and discharging strategies. The lack of professional talents further exacerbates this problem: agricultural parks are mostly located in rural areas, making it difficult to attract professional talents in energy system design, operation, maintenance, and dispatching. Most existing operation and maintenance personnel are transformed from farmers, lacking professional knowledge and skills, and unable to cope with complex technical problems. The low digitalization level limits the intelligent regulation of energy systems: most renewable energy systems in agricultural parks still adopt manual monitoring and dispatching, lacking real-time data collection, analysis, and optimal control capabilities, and are unable to make dynamic adjustments according to changes in energy demand and energy output, resulting in difficulty in improving energy utilization efficiency. From an international comparison perspective, the challenges faced in the development of agricultural renewable energy are significantly different. Developed countries have basically solved the problems of technology integration and the economy. The current challenges mainly focus on how to further improve system efficiency, expand market-oriented revenue channels, and address the impact of extreme weather caused by climate change on energy systems. For example, European countries are exploring, which realizes the deep integration of renewable energy systems in agricultural parks with regional power grids, carbon markets, and green certificate markets, thereby improving the flexibility and economy of the system. Developing countries still face basic challenges such as backward technology, lack of funds, and imperfect policies. For example, many agricultural parks in African countries lack basic power distribution networks, resulting in extremely high connection costs for renewable energy projects. Agricultural renewable energy projects in India and some Southeast Asian countries face high investment risks due to unstable policies and irregular supervision.

In response to these challenges, the international community and various countries have explored a series of solutions, providing ideas for the sustainable in agricultural parks. At the technical level, strengthen the multi-energy system integration technologies and promote. For example, develop artificial intelligence-based multi-energy dispatching algorithms to realize the intelligent matching of photovoltaic, wind power, biomass energy, and energy storage systems; research and develop efficient energy conversion technologies (such as high-efficiency photovoltaic modules and advanced biomass gasification technologies) to improve energy utilization efficiency; construct a unified energy monitoring and operation and maintenance platform to realize, and of the system. At the economic level, innovative financial tools and business models to reduce project investment risks. For example, promote the model, where invest in the construction of renewable energy systems, and agricultural parks pay service fees with the saved energy costs, reducing the initial investment pressure of the parks; develop financial products such as green credit, green bonds, and financial leasing, extend loan terms (such as 10–15 years), and reduce interest rates; establish a risk compensation fund for agricultural renewable energy projects to provide guarantees for projects and reduce the loan risks of financial institutions. At the policy and mechanism level, improve the policy support system and strengthen market mechanisms. For example, formulate long-term and stable subsidy policies, shifting the focus from construction to operation to encourage projects to improve energy utilization efficiency and emission reduction effects; simplify the grid connection procedures for renewable energy projects, reduce grid connection costs, and ensure priority grid connection for projects; promote the extension of markets to the field of agricultural renewable energy, establish carbon emission reduction verification standards for agricultural renewable energy projects, and realize the economic conversion of environmental value; accelerate design, construction, standards for renewable energy systems in agricultural parks, and standardize project construction and operation. At the management and talent level, strengthen talent training and technical training, and improve the digitalization level. For example, cooperate with universities and vocational colleges to set up agricultural energy-related majors and cultivate professional and technical talents; organize technical training for agricultural park operators to improve their energy management capabilities; promote renewable energy systems in agricultural parks, and build a digital energy management platform to realize intelligent regulation and efficient operation and maintenance of the system. Looking forward to the future, with the in-depth advancement of global energy transformation and improvement of agricultural sustainable development needs, the application of renewable energy in agricultural parks will present a development trend of "integration, diversification, intelligence, and regionalization," playing a more critical role in agriculture and achieving the "double carbon" goals. In terms of technological development trends, renewable energy systems in agricultural parks will upgrade from "single energy superposition" to "deep multi-energy integration." The future energy systems of agricultural parks will no longer be simple combinations of technologies, but will realize multi-energy complementarity, including electricity, thermal energy, cold energy, and hydrogen energy, through integrated "source-grid-load-storage" design, as well as production, consumption, storage, and conversion. For example, the "photovoltaic + wind power + biomass energy + energy storage + hydrogen energy" multi-energy system will become the mainstream model for large-scale agricultural parks: photovoltaic and wind power provide basic electricity, biomass energy provides stable electricity and thermal short-term fluctuations, and hydrogen energy is used for long-term energy storage and agricultural machinery fuel substitution, realizing full-time and full-scenario energy supply. At the same time, with the advancement of related technologies, agricultural renewable energy technologies will become more agriculturally friendly. For example, flexible photovoltaic modules can adapt to greenhouse roofs of different shapes, and light transmittance can be dynamically adjusted according to crop growth needs; low-wind-speed wind turbines can operate stably at wind speeds below 3 m/s, adapting to the low-wind-speed environment in rural areas; efficient biomass conversion technologies can realize "one-pot" production of multiple products (electricity, thermal energy, bio-based materials), improving resource utilization efficiency. In terms of intelligent development trends, blockchain will be deeply integrated into renewable energy systems in agricultural parks, realizing the coordinated optimization of "energy flow - information flow - carbon flow." Artificial intelligence-based dispatching algorithms will be able to accurately predict photovoltaic and wind power output and park energy demand, realizing optimal energy allocation. Big data technology can comprehensively analyze the operation data of energy systems, crop growth data, and environmental data, optimizing the coordination between energy supply and agricultural production (such as adjusting energy supply strategies according to crop growth stages). Blockchain technology can be used for the trading and traceability of green certificates and carbon emission reductions, realizing transparent management of environmental rights and interests. For example, in a "digital agricultural energy park," each photovoltaic module, each wind turbine, and each energy storage device will be equipped with sensors to collect real-time operation data, which will be transmitted to the cloud platform through 5G/6G networks. Artificial intelligence algorithms will analyze and process the data to generate optimal dispatching schemes, and at the same time, upload carbon emission reduction data to the blockchain platform to realize the verification and trading of carbon emission reductions. In terms of regional development trends, renewable energy systems in agricultural parks will pay more attention to adaptation to local resource endowments, agricultural characteristics, and climate conditions, forming a development model of "adapting to local conditions." In northwest China, with abundant solar energy resources, the focus will be on developing "photovoltaic + agriculture" and "photovoltaic + animal husbandry" models, combined with energy storage systems and microgrids to achieve energy self-sufficiency. In north China and northeast China, with abundant wind energy resources, the focus will be on developing "wind power + biomass energy" models, using the stability of biomass energy to make up for the intermittency of wind power. In central China and south China, with abundant biomass resources, the focus will be on developing "biomass energy - circular agriculture" models to realize the resource utilization of agricultural wastes. In coastal areas, the focus will be on developing "offshore wind power + seawater desalination + aquaculture" models to comprehensively utilize marine resources. In mountainous areas, the focus will be on developing "small hydropower + photovoltaic + energy storage" models, using mountainous water resources and solar energy resources for agricultural electricity use in remote areas.

In terms of trends, the development of agricultural parks will transform the "policy + market" dual-driven model. On one hand, policy support will be more precise and long-term, focusing on standard setting, public services, and regulatory guarantees instead of subsidies. On the other hand, market mechanisms will be improved, and green certificate trading, carbon trading, and energy spot markets will provide more revenue channels for agricultural renewable energy projects. For example, China will establish a national unified market in the future, allowing agricultural parks to obtain additional income through green certificate trading. The carbon trading market will gradually cover the agricultural sector, and the carbon emission reductions of agricultural renewable energy projects. The construction of the energy spot market will provide a more flexible electricity price mechanism for agricultural parks, encouraging parks to participate in demand response and improve energy utilization efficiency. In the long run, renewables become a core component of the Agricultural Energy Internet, from "high-carbon, scattered, and inefficient" to "low-carbon, concentrated, and efficient." Through the large-scale application of renewable energy, agricultural parks will achieve energy self-sufficiency and even "energy surplus," becoming a regional energy system. Through the coordinated development of "energy - agriculture – environment," agricultural parks will achieve multiple goals of "carbon emission reduction - grain yield increase - farmer income increase." Through digital technologies, agricultural parks will become demonstration carriers of "smart energy - smart agriculture," leading the deep integration of agricultural modernization and energy cleanliness [13].

In conclusion, renewable parks have entered a critical development stage with both opportunities and challenges. Despite facing various challenges in terms of technology, economy, and policy, with the acceleration of global energy transformation, the decline in technology costs, the improvement of policy systems, and the improvement of market mechanisms, the application prospects of renewable energy in agricultural parks are broad. In the future, it is necessary for academia, industry, and policymakers to work together, strengthen technology research and development and integrated innovation, improve policy support and market mechanisms, enhance management capabilities and digitalization levels, and systems in agricultural parks toward a more efficient, economical, low-carbon, and intelligent direction. This will make important contributions to achieving global agricultural sustainable development, addressing climate change, and ensuring food security and energy security.

4.1.1.3 Background of green certificate and carbon quota systems

Against the macro backdrop of global efforts, there has become a consensus and direction for countries worldwide. Particularly since the entry into force of the Paris

Agreement, low-carbon, zero-carbon, and even negative-carbon economic models have emerged as core issues in the global energy governance system. China plays an extremely critical role in this process. This strategy is not only a crucial national commitment to tackling climate change but also a top-level design for promoting a comprehensive green transition of the economy and society. As the "dual-carbon" strategy continues to deepen, China's energy system, industrial structure, and policy framework are undergoing systemic transformations. Among these, the green certificate system and carbon quota system, as two key market-oriented tools, have become important pillars for optimizing the energy structure and advancing low-carbon transition. These two systems act on energy production and consumption from different angles: the green certificate system centers on confirming the environmental value of renewable energy, promoting the marketization of green electricity consumption; the carbon quota system focuses on greenhouse gas emission reduction constraints, driving society-wide energy conservation and carbon reduction through total quantity control and market trading. Their coordinated development marks a significant shift in China's energy governance system from administrative management to market regulation.

A green certificate, or renewable energy green electricity, in China and a key document for certifying green electricity production and consumption. The proposal and implementation system emerged amid China's energy structure adjustment and the rapid. As photovoltaic and other renewable energy sources continue to rise, the supply system has significantly increased. However, challenges such as difficulties in absorbing renewable energy electricity and insufficient market recognition persist. Especially during a period when power system reforms are not yet fully mature and traditional fossil energy still dominates, the economic value of green electricity has often not been fully reflected in the market. To address this issue, the National Development and Reform Commission, the Ministry of Finance, and the (Trial) in 2017 marking of China's green certificate system. By issuing one corresponding green certificate for each kilowatt system "materializes" and "trades," thereby establishing an identity and value system for. The implementation system has not only enriched the policy toolbox for renewable energy but also promoted the diversified development of the energy market. By separately selling the environmental attributes of the green electricity they produce to gain additional revenue, power users, especially energy-intensive enterprises and public institutions, can demonstrate their commitment, thereby fulfilling their green consumption responsibilities and achieving goals. With continuous policy improvements, the scope of application of the green certificate system has gradually expanded, from initially covering wind and photovoltaic power projects to including biomass energy, hydropower, and geothermal, and explicitly proposed in electricity, strengthen their coordination of emission rights trading, energy conservation assessment, and include electricity corresponding to green certificate transactions in energy conservation evaluation and carbon accounting systems. This move not only strengthened the policy status of the energy governance system but also realized institutional integration, injecting a strong impetus into the growth of non-fossil energy consumption. In recent years, the mechanism of China's green electricity market has significantly increased. The construction of China's Green Electricity Certificate Trading Platform has realized nationwide unified issuance and trading, forming a national market system with coordinated operation among multi-level institutions. Methods have expanded from initial voluntary subscriptions to diversified forms such as bilateral negotiations, listed trading, and centralized bidding. Trading, financial institutions, and local governments, achieving effective coupling of market prices. The price signal of green certificates has gradually emerged, serving as an important basis for measuring the cost of green consumption. Meanwhile, efforts to align green certificates with international standards are steadily advancing. China actively participates in the standardization of international green certificate systems, promotes alignment of key links such as issuance, measurement, certification, and trading with international practices, and enhances the international credibility and market influence of China's green certificates. Complementing the green certificate system is the carbon quota system centered on carbon emission rights trading. Based on the principle of "cap-and-trade," the carbon quota system is one of the market-oriented mechanisms widely adopted internationally for greenhouse gas emission reduction. The government sets an upper limit on total carbon emissions and allocates quotas to key emitters, enabling enterprises to optimize their production structures and improve energy efficiency under certain constraints to achieve emission reduction goals. In China, the establishment of the carbon quota system has evolved from local pilots to a unified national market. Starting in 2013, Guangdong, Hubei, Shenzhen, Tianjin, and Chongqing – pilots, accumulating rich institutional experience and technical data. In 2021, initially covering the power generation industry with over 2000 key emitters, involving total emissions exceeding four.

Rights Trading (Trial) provides top-level design and a legal basis for the institutionalized operation of the carbon market, clarifying key links such as quota allocation, trading management, compliance settlement, and supervision and verification. In terms of quota allocation mechanisms, China has gradually formed a scientific allocation system centered on industry benchmarking methods, combined with unit energy efficiency levels and production characteristics. For example, the 2023 report comprehensively considers unit type, heating mode, fuel structure, and regional differences in determining quotas, achieving a balance between fairness and incentives through a combination of benchmark values and exemption mechanisms. This institutional design has effectively guided energy-intensive enterprises to carry out technological transformation and clean energy substitution, promoting improvements in. Meanwhile, the trading mechanism of the carbon market is constantly improving. In addition to quota trading, the state is actively restoring and improving the China (CCER) mechanism, enabling enterprises to offset part of their emissions by purchasing certified emission reduction project results, further enriching market-based emission reduction tools.

In terms of operational effects, the national carbon market has initially formed a multi-level market system with quota trading as the core, supplemented by emission reduction project trading and carbon financial innovation. The carbon price formation mechanism gradually reflects the combined effects of supply and demand relations and emission reduction costs, with price signals becoming important economic indicators guiding enterprise investment directions. Through a combination of internal energy-saving transformation and external quota trading, enterprises have achieved optimal allocation of emission reduction costs, fulfilling compliance obligations while gaining economic benefits in market competition. The establishment of the carbon market not only helps reduce the overall social cost of emission reduction but also promotes the optimization of China's energy consumption structure and the green transformation of the industrial structure at the macro level. Within the policy system, although the green certificate and carbon quota systems have different focuses, they essentially serve the realization of energy cleaning and emission reduction goals. Green certificates "reward" clean energy supply by confirming the environmental attributes of renewable energy generation, while carbon quotas "constrain" the use of high-carbon energy through total emission control and trading mechanisms. Acting on the energy system from different directions, they form a dual-market mechanism of "incentive + constraint." Once the green electricity represented by green certificates is recognized as having quantifiable emission reduction benefits, its environmental value can be confirmed in the carbon market, realizing cross-market value conversion. At the national level, efforts are underway to explore the establishment of a reduction, integrating the emission reductions from renewable energy generation into the carbon market trading system through scientific accounting methods. This cross-market integration not only helps enhance the market value of green electricity but also injects more low-cost emission reduction resources into the carbon market, strengthening the overall market's liquidity and stability. Drawing on international experience, European and American countries have long achieved institutional coordination between green certificates and carbon markets. While implementing the Renewable Portfolio Standard (RPS), the European Union certifies green electricity through the Guarantee of Origin (GO) system, which operates in conjunction with the EU Emissions Trading System (EU ETS), enabling green electricity consumption to simultaneously generate carbon emission reduction benefits. In the United States, state-level renewable energy certificate (REC) markets coexist with carbon markets, effectively increasing the proportion of clean energy through a combination of quota systems and market-oriented mechanisms. These experiences show that the integration of green certificates and carbon quota systems can not only reduce overall emission reduction costs but also improve the effectiveness and sustainability of policy incentives.

In China, with the rise of new energy systems such as energy internet, distributed energy, and Agricultural Energy Internet (AEI), the synergistic effect of green certificate and carbon quota systems will become more prominent. Renewable energy generation projects in agricultural parks and rural energy systems, such as photovoltaic, biogas, and biomass energy, can both obtain green certificate revenue and monetize emission reduction value through the carbon market, forming a sustainable low-carbon incentive mechanism. In the future, with further institutional improvements, green certificates and carbon quotas will become the "dual drivers" in China's green energy governance system, playing a more fundamental and strategic role in optimizing the energy structure, promoting low-carbon economic development, and achieving the "dual-carbon" goals.

4.1.2 Research objectives

Building an energy governance system characterized by low carbonization, marketization, and digitalization has become. Renewable Energy Green Electricity Certificates (green certificates) and carbon emission quota trading mechanisms, as two important market-oriented tools for addressing climate change and promoting clean energy development, play a crucial role in achieving energy system emission reductions, optimizing resource allocation, and driving green economic growth. Building an energy governance system characterized by low carbonization, marketization, and digitalization has become an inevitable trend in energy sector reform. Renewable Energy Green Electricity Certificates (green certificates) and carbon emission quota trading mechanisms, as two important market-oriented instruments for addressing climate change and promoting clean energy development, play a crucial role in reducing energy system emissions, optimizing resource allocation, and fostering green economic growth. For a long time, these two mechanisms have operated relatively independently. However, with the gradual advancement of policy coordination and market linkage mechanisms, the integration of green certificate trading and carbon quota trading has become increasingly prominent, and their synergistic effect is emerging as an important pathway for multi-market collaborative optimization.

Currently, it is basically established. Through the dual-drive mechanism of "mandatory consumption + voluntary consumption," it has effectively promoted the market-oriented consumption of renewable energy electricity and enhanced social awareness. Green certificates not only play a unique role in the confirmation, certification, and incentive of green electricity but also have natural synergistic potential with the carbon market in terms of environmental rights allocation, carbon emission reduction accounting, and price signal transmission. A large number of studies have shown that green certificates and carbon quotas are complementary in goal orientation and institutional functions, and their coupling helps improve market operation efficiency, stimulate the emission reduction motivation of subjects, and promote the efficient circulation and value realization of green electricity. Therefore, under the overall framework of dual-carbon policies, how to scientifically design quotas and establish a reasonable price linkage and mutual recognition system has become a research focus of both academia and policy departments. As the core of China's carbon emission reduction policy system, research on the carbon quota trading mechanism mainly focuses on key issues such as quota allocation, fairness, and market efficiency. Existing studies point out that the current quota allocation standards at the provincial and industry levels in China still struggle to balance fairness and efficiency, with problems such as static allocation bias and insufficient emission reduction incentives. To improve market rationality and long-term incentive effects, academia has proposed introducing dynamic intertemporal allocation methods and carbon–green certificate offset mechanisms to achieve optimal allocation of carbon resources within a multi-cycle and multi-market framework. At the same time, how to effectively connect the carbon market with the green certificate system has also become a key direction of policy research. Establishing a carbon offset mechanism based on green certificates and strengthening the carbon–green certificate synergy can achieve a dynamic balance of emission reduction costs and efficient market operation, providing a new path for building a multi-energy complementary and low-carbon coordinated energy system.

With the deepening of energy market reforms, academia has carried out multi-dimensional explorations on the institutional design, trading mechanisms, and synergistic effects of green certificates and carbon quotas. On one hand, researchers have proposed using blockchain technology and smart contracts to build a "decentralized" joint trading platform to realize carbon asset confirmation, transaction transparency, and settlement automation. On the other hand, studies based on the Stackelberg game and multi-microgrid collaborative optimization have shown that the joint trading mechanism can not only reduce system operation costs but also improve carbon emission reduction and new energy absorption levels. In research on regional integrated energy systems, introducing the green certificate-carbon quota mutual recognition mechanism and integrating new energy technologies such as hydrogen energy and energy storage can significantly improve the flexibility and economy of the system. In addition, system dynamics, game theory, and artificial intelligence methods have also been introduced into the multi-market collaboration research framework to characterize the interaction between transaction prices, market participants' behaviors, and policy goals. Some studies have constructed multi-energy system optimization scheduling models based on deep reinforcement learning methods, achieving multi-objective dynamic optimization by introducing green certificate and carbon price signals, which demonstrate superior convergence and decision-making intelligence. From an international perspective, developed economies have accumulated relatively mature experience in the coordinated operation of carbon markets, green certificate markets, and renewable energy portfolio standards. Their policy paths and institutional designs provide useful references for China. Relevant studies have summarized the coupling laws of foreign carbon–electricity–green certificate markets, pointing out that establishing a unified carbon–green certificate database, improving price transmission mechanisms, and promoting product and institutional connection are key links to enhance market collaboration efficiency. In comparison, China still has certain deficiencies in information sharing, price linkage, and market supervision, requiring strengthened top-level design and market rule connection at the institutional level to further promote the standardization and institutionalization of multi-market collaborative development. Despite numerous theoretical and practical explorations, the joint trading of carbon quotas and green certificates still faces a series of challenges. Differences in accounting boundaries, indicator calibers, and price formation mechanisms between the two lead to imperfect mutual recognition and interoperability mechanisms; the market rules and regulatory systems of joint trading platforms are still in the pilot stage, and key parameters such as quota setting, price adjustment, and participation incentives need to be optimized; at the same time, multi-market coupling also puts forward higher requirements for profit distribution, fairness protection, and policy coordination among market participants. To address these issues, it is urgent to establish a scientific and reasonable institutional system and optimization model, reveal the inherent laws of the carbon–green certificate market coupling mechanism, and promote the empirical and policy-oriented process of joint trading.

Therefore, the joint trading mechanism of carbon quotas and green certificates, aiming to construct a systematic and operable multi-market collaborative optimization framework, reveals the price interaction and emission reduction synergy rules between different markets, and explores its application path in low-carbon scheduling of integrated energy systems. By building an integrated energy system optimization study investigates the linkage mechanism between carbon prices and green certificate prices, market game behaviors, and system emission reduction effects, thereby achieving multi-dimensional collaborative optimization of environmental, economic, and social benefits. Meanwhile, the research will combine artificial intelligence and game theory methods to propose multi-market collaborative optimization algorithms and analyze profit distribution and incentive mechanisms under multi-subject participation; through simulation verification of regional integrated energy systems or multi-microgrids, it will quantitatively evaluate the emission reduction performance and economic feasibility under the joint trading mechanism. Ultimately, the study is expected to provide theoretical support and policy recommendations for China to build a scientific and reasonable carbon–green certificate collaborative trading system, contributing to the construction of a green and low-carbon energy system and the realization of the "dual-carbon" strategic goals.

4.2 Methodology

4.2.1 Accounting system for green certificates and carbon quotas in Agricultural Energy Internet systems

4.2.1.1 Definition of accounting subjects and their functions

This chapter constructs a new energy industrial park, with the overall goal of coordinating energy supply and demand and optimizing low-carbon operations in the park, and studies its comprehensive operation mechanism, multi-energy complementarity, and carbon emission reduction. The system mainly includes renewable energy units such as the external power grid (Grid), photovoltaic power generation units (Photovoltaic, PV), wind power generation units (wind turbine, WT), electrical energy storage equipment (Electrical Storage Equipment, ESE), as well as several gas-fired power units (Gas-fired Unit, GU) and facility agriculture areas (Agricultural Facility, AF), forming a multi-energy collaborative supply system dominated by clean energy and supported by fossil energy. Through energy, information, and economic interactions between various entities, the comprehensive coupling and optimal scheduling of electricity, heat, and agricultural production are realized, thereby supporting efficient energy utilization and precise control of carbon emissions within the park. In the carbon emission accounting part, the main accounting object is the gas-fired power units inside the park. Since they generate electricity by burning natural gas during operation, which is inevitably accompanied by a certain amount of carbon dioxide emissions, it is necessary to accurately calculate their direct carbon emissions. The units in the unit output level and natural gas calorific value parameters, and compares them with the initially allocated carbon quotas. When the actual emissions are lower than the quota, the unit can sell the remaining carbon quotas through the carbon trading market to obtain economic benefits; conversely, when a corresponding number of quotas is required in the carbon market to fulfill emission reduction responsibilities. It is worth noting that this chapter only considers direct carbon emissions within the park is not included in the carbon emission or quota assessment scope of this system because its emission links have been accounted for in the superior system. This approach not only avoids double-counting but also better conforms to the principle of "territorial accounting, focusing on direct emissions" in the carbon trading system. In the accounting part of green certificates (GC), this chapter regards the photovoltaic (PV) and wind power (WT) units in the park as the main subjects on the supply side of green certificates. Since these units generate electricity using renewable energy, their electricity can obtain a corresponding number of green certificates after certification, which are used to prove the green attributes and emission reduction benefits of the generated electricity. The gas-fired power units in the park serve as the main subjects. According to the Renewable Portfolio Standard (RPS) issued by the Guangxi Energy Service Regulatory Center, enterprises are required to meet a specified proportion of renewable electricity consumption, thereby fulfilling the green electricity consumption target. This mechanism not only promotes the absorption of green electricity within the park but also provides a flexible market-oriented compliance method for the clean production of thermal power units, realizing dual incentives for renewable energy power generation and carbon emission reduction constraints.

The facility agriculture part, represented by the dragon fruit planting area of this chapter. The agricultural production process includes various electrical loads such as lighting, ventilation, heating, and irrigation, among which supplementary lighting and irrigation are the two most representative links with the highest requirements for energy supply. As a typical tropical crop, dragon fruit is extremely sensitive to light intensity and humidity conditions. To ensure its normal growth, when natural light is insufficient or in continuous rainy weather, artificial supplementary lighting systems are needed to increase the total photosynthetically active radiation (PAR) to maintain the photosynthetic efficiency of plants and fruit yield. In extreme weather such as drought, when soil moisture cannot be effectively supplemented by natural precipitation, the irrigation system needs to timely supplement water through electric-driven water pumps to maintain a suitable growth environment. It can be seen that the electricity load of facility agriculture shows significant time-variability and uncertainty, which is strongly affected by changes in meteorological conditions. Especially in extreme weather, multiple loads such as supplementary lighting and irrigation may increase simultaneously, leading to a sharp rise in instantaneous power demand, thereby increasing the instability and scheduling pressure of the park's power system operation. In response to this random fluctuation characteristic of agricultural loads, this chapter models and dynamically simulates the supplementary lighting and irrigation power demand in the dragon fruit planting area in the model to reflect its response law to changes in environmental factors. By introducing meteorological condition humidity into the load calculation model, an agricultural electricity demand prediction and scheduling constraint relationship driven by meteorological data are constructed, realizing the controllability and optimization of agricultural energy demand. On this basis, the flexible configuration of the energy supply side is further considered, and the electrical energy storage system (ESE) is used to realize time-shifting and dynamic balance of electrical energy. When there is excess electricity in the park (such as when photovoltaic output is higher than load demand during the day), the energy storage device absorbs the surplus electricity for charging; when the system's power supply is insufficient (such as insufficient renewable output at night or on cloudy days), the energy storage device releases electricity to supplement the load demand, thereby effectively alleviating the power imbalance caused by renewable energy fluctuations and improving the overall power supply stability and energy utilization efficiency of the park.

To sum up, the integrated energy system of the new energy industrial park constructed in this chapter is characterized by multi-dimensional coupling of "electricity – carbon – green certificate – agriculture," realizing the unified modeling of energy production, storage, consumption, and carbon accounting. The system can not only reflect the dynamic balance relationship between different energy forms but also realize the coordinated optimization of environmental and economic goals through carbon quota trading and green certificate mechanisms. This modeling framework provides a feasible theoretical support and technical path for the low-carbon operation and green development of park-level integrated energy systems, and offers new research ideas and application foundations for multi-energy collaboration and agricultural sustainable development in the background of achieving the "dual-carbon" goals.

4.2.1.2 Definition of accounting boundaries

To achieve this in agricultural parks, this chapter conducts a systematic study on carbon emissions and green certificate (GC) resources for the energy system in agricultural parks. The overall goal of the research is to form an operable and systematic accounting framework by scientifically defining the accounting boundaries of carbon emissions and green certificates, thereby realizing accurate identification, quantitative calculation, and dynamic management of carbon resources and green certificate resources. Specifically, from the perspective of the full life cycle of the agricultural park's energy system, this chapter first establishes a theoretical system for calculating carbon emissions of the agricultural park's energy system in accordance with existing international and national standards and norms. Secondly, for links not yet covered by existing standards, it proposes a refined carbon emission measurement method suitable for the characteristics of agricultural parks by referring to international carbon accounting systems and relevant academic achievements, aiming to provide a scientific basis for decisions in agricultural parks. The establishment of this system can further guide the park in formulating more reasonable carbon emission reduction strategies, reducing unnecessary emissions, thereby providing technical support for the park to promote the agricultural park toward low-carbon, environmentally friendly, and sustainable development.

In terms of defining the carbon accounting boundaries of the agricultural park's energy system, this chapter perspective, comprehensively considers the carbon flow characteristics in links such as energy production, conversion, storage, and consumption within the park, and systematically divides the accounting boundaries of green certificates and, the energy system of the agricultural park mainly consists of photovoltaic power generation units (Photovoltaic, PV), wind power generation units (wind Turbine, WT), gas-fired power units (Gas-fired Unit, GU), electrical energy storage systems (Electrical Storage Equipment, ESE), and the external power grid (Grid). The park's own gas-fired power units, as the main power supply units, generate electricity by burning natural gas, which directly produces carbon dioxide emissions during operation, and thus are clearly defined by accounting boundaries. In contrast, photovoltaic and wind power units generate electricity using renewable energy without direct carbon emissions; the generated electricity can be converted into green certificate resources after certification, supporting the park's green electricity consumption and carbon emission reduction compliance. For electricity purchased from the external grid, since its carbon emissions are already counted in the accounting scope of the superior system, only energy input is recorded at the park level, and its carbon emissions are not re-accounted to avoid overlap of accounting boundaries and double counting.

On the basis of defining the accounting boundaries, this chapter further sorts out the main national standards and international norms, followed by carbon and green certificate accounting. For the accounting of green certificate holdings of photovoltaic and wind power systems, relevant provisions in 100 Questions and Answers on China's Renewable Energy Green Electricity Certificates (2025 Edition) are followed to clarify the mechanisms for green certificate issuance, verification, and circulation, and the number of obtainable green certificates is determined based on power generation data. In the part of carbon emission accounting for gas-fired power units, this chapter refers to Guidelines for the Accounting and Reporting of Greenhouse Gas Emissions by Enterprises – Power Generation Facilities (2022 Revised Edition) and Methodology and Empirical Research on Carbon Emission Accounting in Industrial Parks, unifies the specifications for carbon emission factors, calorific value conversion coefficients, and oxidation rate parameters in the natural gas combustion process, and establishes a direct carbon emission calculation model based on unit output and fuel consumption. In addition, the accounting of green certificate quotas for gas-fired power units is based on the Notice of the National Development and Reform Commission and the National Energy Administration on Establishing and Improving the Renewable Energy Power Consumption Guarantee Mechanism (Fagai Nengyuan [2019] No. 807) and the Notice on Issuing the 2020 Renewable Energy Power Consumption Responsibility Weight for Each Provincial Administrative Region (Fagai Nengyuan [2020] No. 767) jointly issued by the National Development and Reform Commission and the National Energy Administration. Combined with the regional renewable energy consumption responsibility weight requirements, the green electricity consumption obligations, and the number of green certificates to be purchased by each gas-fired power unit, are determined.

Through the construction of the above accounting boundaries and standard systems, this chapter establishes a calculation framework covering the entire process of energy production, conversion, storage, and consumption in agricultural

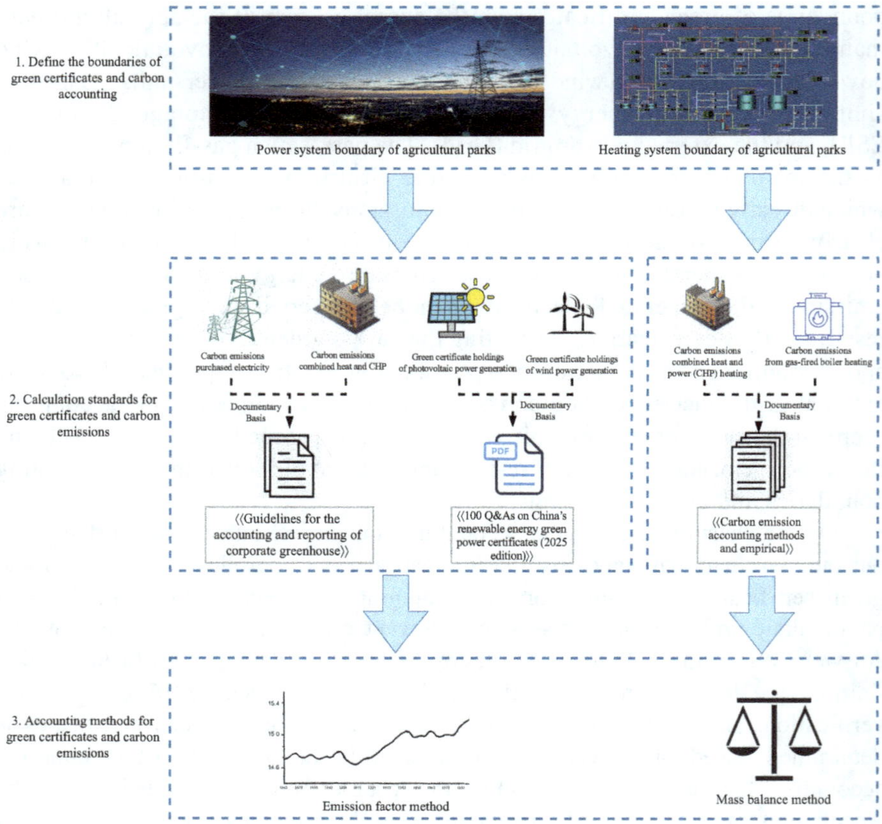

Figure 4.1 Technical route for full-link green certificate and carbon emission measurement in agricultural park energy systems

parks, which can realize the quantitative assessment and dynamic tracking of total carbon emissions and green certificate resources in the park, subsequent carbon emission performance evaluation, carbon trading strategy formulation, and green certificate market-oriented optimization, as shown in Figure 4.1. The completeness and accuracy of the energy system in agricultural parks also provide a valuable reference and practical pathway for achieving integrated management of "carbon–energy–certificate" mechanisms in park-level energy internet systems.

4.2.1.3 Accounting model for power generation and green certificate holdings of new energy units

Photovoltaic power generation model and accounting model

The direct factors affecting photovoltaic power generation are: installed capacity, solar radiation intensity incident on the photovoltaic panels, and ambient temperature. Therefore, the photovoltaic power generation model is summarized as follows:

Intensity I_T on the inclined photovoltaic panel is calculated as shown in (4.1).

$$I_T = \frac{I_D \cdot \cos\theta}{\sinh} + I_S\left[\frac{\cos\theta \cdot (I_G - I_S)}{I_0 \cdot (\sinh)^2} + \frac{1}{2}(1 + \cos\theta) \cdot \left(1 - \frac{I_G - I_0}{I_0 \cdot \sinh}\right)\right] + \frac{1}{2} I_G \cdot \rho_w (1 - \cos\theta) \tag{4.1}$$

where h is the solar altitude angle; I_D is the direct solar radiation; Is is the diffuse solar radiation; I_G is the total solar of the planar area; I_0 is the solar constant; and ρ_w is the ground albedo.

The temperature of the photovoltaic panel, T_C, is shown in (4.2).

$$T_C = \frac{T_A + (T_{CN} - T_{AN}) \cdot \frac{I_T}{G_{TN}}\left[1 - \frac{\mu_{mps}(1 - \alpha_p T_{CS})}{\tau \cdot \alpha_{ab}}\right]}{1 + (T_{CN} - T_{AN})\frac{I_T}{G_{TN}} \cdot \frac{\alpha_p \cdot \mu_{mps}}{\tau \cdot \alpha_{ab}}} \tag{4.2}$$

where T_a represents the ambient temperature; T_{AN} is the temperature corresponding to T_{CN}; T_{CS} denotes the photovoltaic cell temperature; G_{TN} is the solar irradiance corresponding to; T_{CN} is the cell temperature under nominal operating conditions; G_{TS} represents the standard incident irradiance; μ_{mps} is the maximum efficiency; α_p is the temperature coefficient; τ is transmittance; and α_{ab} is the solar absorptance, and will not be repeated here. It should be noted that all temperatures mentioned in this formula are in Kelvin (K).

Ppv, is shown in (4.3).

$$P_{pv} = c \cdot Z_{pv} \cdot f_{pv} \cdot \left(\frac{I_T}{G_{TS}}\right) \cdot [1 + \alpha_P(T_c - T_{cs})] \tag{4.3}$$

where P_{pv} represents the photovoltaic power generation; Z_{pv} is photovoltaic panels; I_T denotes inclined surface of the photovoltaic panels; c is the proportion of the roof area covered by photovoltaic panels; T_c is the temperature of the photovoltaic panels; and f_{pv} is the photovoltaic derating factor.

Therefore, the predicted hourly photovoltaic power generation P_t^{pv} is expressed as shown in (4.4).

$$P_t^{pv} = P_{pv} \times 1 \tag{4.4}$$

where, the unit is MW · h.

According to the conversion rules for green certificate applications, each 1 MW · h of green electricity corresponds to one green certificate. Therefore, the hourly green certificate holdings of the photovoltaic power generation unit $G_{pv,t}^{ini}$ are expressed as shown in (4.5).

$$G_{pv,t}^{ini} = P_t^{pv} \times 1 \tag{4.5}$$

where, P_{PV}^{t} represents the actual hourly output of the photovoltaic power generation unit, with the unit of MW·h.

Wind power generation model and accounting model

In the process of modeling units, this study assumes that the wind speed follows throughout the entire scheduling period. This distribution has been widely applied in wind speed probability modeling due to its excellent fitting performance. Based on this assumption, a probabilistic model of wind speed is constructed, from which the fluctuation characteristics of wind power output can be further derived. Therefore, the specific calculation formula for the predicted hourly follows:

$$f_1(v_t, \xi, k) = \frac{k}{\xi}\left(\frac{v_t}{\xi}\right)^{k-1} \exp\left[-\left(\frac{v_t}{\xi}\right)^{k}\right] \tag{4.6}$$

$$P_t^{\mathrm{wt}} = \begin{cases} 0, v \le v_{in}, \xi v \ge v_{out} \\ \dfrac{v_t - v_{in}}{v_r - v_{in}} \bullet P_{wr}, v_{in} < v < v_r \\ P_{wr}, v_r < v < v_{out} \end{cases} \tag{4.7}$$

where, P_t^{wt} represents a generation unit, with the unit of MW; k and ξ are the shape and scale parameters of the Weibull distribution, respectively; v_t is the predicted wind speed at time period t, with the unit of m/s; v_r, v_{in}, and v_{out} represent; and P_{wr} is the rated power of the wind turbine.

In order to deeply characterize the fluctuation and uncertainty of wind power output in the integrated system, this study introduces the scenario-based method to systematically model and analyze the power output process of wind turbines. As a typical stochastic renewable energy source, wind power output is mainly affected by variations in wind speed, which, in turn, are closely related to meteorological conditions, geographical location, seasonal changes, and wind shear. These factors give wind speed strong nonlinearity, randomness, and temporal correlation. Such uncertainty not only causes significant fluctuations in but also poses park's energy system. Therefore, establishing a scientifically sound stochastic modeling approach to quantitatively describe the fluctuation characteristics of wind power output forms the foundation for subsequent system scheduling optimization and risk assessment.

The scenario-based method is a stochastic modeling technique based on probabilistic statistics. It approximates the uncertain distribution of wind power output by generating multiple representative random scenarios. Compared with traditional probabilistic or fuzzy interval models, the advantage of the scenario method lies in its ability to convert randomness into a deterministic set of samples. This enables complex stochastic optimization problems to be transformed into deterministic equivalents with a finite number of scenarios, which are easier to solve within an optimization framework. In agricultural parks, the uncertainty of wind power output is introduced into the model through the scenario method, reflecting the system's operational diversity and risk under different meteorological conditions. The scenario modeling process in this study mainly includes two key steps: scenario generation and scenario reduction of wind power output. Latin Hypercube Sampling (LHS) is a stratified statistical sampling technique. Its principle is to divide the cumulative distribution function (CDF) of a random variable into several equal-probability subintervals on the interval [0,1], each representing the probability range from which a sample should be drawn. A random probability value is then selected within each interval, and the corresponding sample value is obtained through the inverse mapping of the CDF. This approach ensures uniform and comprehensive coverage of samples across the entire probability space, avoiding clustering, omission, or skewness issues that may occur in traditional Monte Carlo sampling. Compared with Monte Carlo sampling, LHS offers higher computational efficiency and better statistical representativeness. Under the same sample size, LHS produces more evenly distributed samples in the probability space, thus achieving high accuracy while significantly reducing the required number of samples. This makes it particularly suitable for stochastic modeling of highly fluctuating variables such as wind power output. In this study, following a specific probability distribution (such as Weibull or Beta distribution). The distribution parameters (shape and scale) are estimated through meteorological data fitting, and then LHS is used to generate multiple wind power scenarios in the probability space. The resulting scenario set effectively reflects the output wind speed conditions, providing the data basis for subsequent scheduling optimization.

However, the initially generated number of scenarios is usually large. Using all of them directly in the optimization model would lead to excessive computational complexity and may even make the optimization problem intractable. To address this, a backward reduction method based on probabilistic distance is applied after scenario generation to filter and compress the scenario set, effectively reducing computational burden while maintaining representativeness. treat each scenario as a probabilistic sample and compute the probabilistic distance (such as Wasserstein, Kantorovich, or Euclidean distance) between scenarios to measure their similarity. Scenarios with minimal impact on the overall distribution are progressively removed until the number of remaining scenarios reaches a predefined threshold. The probability weights of deleted scenarios are redistributed among the remaining ones to preserve the overall probability consistency. Compared with the forward selection method, the backward reduction approach offers stronger global optimization capability and better distribution preservation. While the forward method selects representative scenarios incrementally from an empty set, the backward method starts with the full set and removes redundant samples step by step, thereby retaining the global statistical properties and probability structure of the original data. Especially when the initial scenario number is large, the backward method achieves higher reduction efficiency and produces a more accurate set of representative scenarios, effectively avoiding distribution distortion caused by over-reduction. In this study, the backward reduction method compresses the number of wind power scenarios to a manageable scale while maintaining the randomness of wind power output, thus balancing computational efficiency and accuracy.

Through the above scenario generation and reduction processes, this study finally obtains a set of representative typical wind power output scenarios. Each scenario contains wind power data for specific time sequences along with its occurrence probability, capturing the variation patterns of wind power output under different meteorological conditions. These typical scenarios are incorporated into the subsequent optimization scheduling model, where probability-weighted processing of scenarios enables quantitative characterization of wind power randomness and the construction of risk response mechanisms. Specifically, in the scheduling optimization model, wind power scenarios are integrated into the objective function and constraints through forms such as probabilistic constraints, chance constraints, or expected cost terms, thereby enabling modeling and optimization of stochastic variables.

In summary, by combining Latin Hypercube Sampling with the backward reduction method based on probabilistic distance, this study establishes a scenario-based modeling framework for wind power output uncertainty. This approach realizes a complete modeling process – from random data sampling and distribution reconstruction to typical scenario selection. It has stochastic fluctuation characteristics of wind power while significantly reducing the dimensionality and complexity of optimization, ensuring both computational efficiency and modeling accuracy. Moreover, this method can be extended to model other stochastic factors such as photovoltaic output, load demand, and market prices, thereby forming a more comprehensive multi-energy stochastic optimization framework that provides a solid theoretical and methodological foundation for achieving safe, economical of energy systems.

4.2.1.4 Coal-fired power generation unit carbon allowance accounting model

China's carbon emission allowance allocation primarily follows a hybrid mechanism dominated by free allocation and supplemented by auctions, balancing fairness and practicality. For the power sector, especially gas-fired and coal-fired units, the initial allocation of carbon allowances mainly adopts the benchmarking method. This method sets unified emission benchmark values for different types of power generation units based on unit type, installed capacity, and energy utilization efficiency, which serve as the primary reference for determining their free carbon allowances. Specifically, the benchmark values for different unit types reflect the standard carbon emissions per unit of electricity generation. When a unit's actual emissions are below the benchmark, it can obtain corresponding carbon surplus; conversely, it must purchase allowances through the carbon market to fulfill its emission reduction obligations. In this study, based on national and industry policies, coal-fired units are classified into different types and capacity levels, and corresponding benchmark indicators are selected according to installed capacity, fuel characteristics, and generation efficiency to achieve a reasonable allocation of carbon allowances for the park's coal-fired units.

In the allowance accounting process, it mainly depends on the product of the benchmark level and the annual generation, meaning the allocated allowance grows linearly with the generation scale. The larger the generation, the higher the allowance requirement, while the unit's carbon efficiency determines its profit or loss in the carbon market. If a unit's actual emissions exceed its allocated allowance, the number of carbon allowances in the carbon trading market is reduced, generating carbon trading costs; if emissions are below the allowance, to earn revenue. This mechanism reflects the market-based incentive principle of "pay more for more emissions, benefit from reductions" and establishes an internal economic constraint and incentive mechanism in the power sector, promoting improved efficiency and cleaner operation of coal-fired units.

At the same time, to further promote the development, coal-fired units are also required to meet generation. This mechanism aligns with the national renewable energy consumption guarantee system, which assigns responsibility weights for renewable energy consumption, requiring coal-fired enterprises to consume or offset a certain proportion of green power. Specifically, when a coal-fired unit cannot meet the required renewable energy proportion through its own generation, it must purchase the corresponding number of obligation. Purchasing green certificates imposes additional economic costs on coal-fired enterprises, negatively affecting their profits, while renewable generation units (such as photovoltaic or wind) can earn extra revenue by selling green certificates. This creates a dual incentive mechanism of "coal-fired units purchase, renewable units benefit," establishing an economic linkage between coal-fired and renewable energy, which helps promote a green transition of the energy structure and low-carbon coordination among market participants.

Regarding the timing of allowance allocation and assessment, both carbon and renewable energy allowances are distributed and settled on an annual basis. After annual allowances are issued, this study combines day-ahead market clearing results and internal energy balance constraints to decompose the annual allowance into daily values, determining the daily carbon allowance and renewable energy assessment generation for each coal-fired unit. This daily decomposition allows carbon trading and green certificate assessment to integrate with the park's dynamic scheduling model, achieving quantitative coupling from the macro policy level to the micro operation level and providing fine-grained constraints for system optimization. Through this mechanism, the park's energy system can minimize carbon costs and maximize green power consumption while ensuring annual compliance.

In summary, the benchmarking method for carbon allowance allocation combined with green certificate assessment constraints used in this study not only aligns with China's current carbon market but also reflects the actual economic behavior and environmental responsibilities of energy entities at the model level. The introduction of this mechanism strengthens emission reduction constraints and renewable energy consumption obligations for coal-fired units and provides a solid institutional and data foundation for subsequent multi-market coordinated optimization and low-carbon dispatch modeling.

Accounting model for actual carbon emissions of coal-fired power generation units

Since all the coal-fired units involved in this study are gas-fired units, their actual carbon emissions Q_t^{real} should be calculated based on the emission factor of natural gas combustion:

$$Q_t^{real} = AD_{g,GB}^t \times EF_g \tag{4.8}$$

$$AD_{g,GB}^t = V_{g,GB}^t \times NCV \tag{4.9}$$

$$EF_g = CC_g \times OF_g \times \frac{44}{12} \tag{4.10}$$

where, $AD_{g,GB}^t$ represents the activity data of natural gas for the gas-fired boiler, in GJ; EF_g is the carbon dioxide emission factor of natural gas, in tCO2/GJ; CC_g denotes the carbon, in tCO2/GJ; and OF_g is the carbon oxidation rate, in %, with a value of 99%.

Accounting model for initial carbon allowances of coal-fired power generation unit

The initial carbon allowances for coal-fired power generation units $Q_{i,t}^{\text{ini}}$ are calculated as follows:

$$Q_{i,t}^{\text{ini}} = P_{i,t}^{th} \times B_i^{CEA} \tag{4.11}$$

where $P_{i,t}^{th}$ represents the coal-fired power generation unit, in $\text{MW} \cdot \text{h}$; and B_i^{CEA} denotes the benchmark value of the allowance corresponding to the unit.

Accounting model for green certificate obligations of coal-fired power generation units

The green certificate obligation of a coal-fired unit, $Q_{i,t}^{\text{ini}}$, the renewable energy quota, is calculated as follows:

$$G_{i,t}^{\text{th}} = P_{i,t}^{th} \times B^{GCT} \tag{4.12}$$

where B^{GCT} represents the benchmark value for renewable energy consumption assessment corresponding to the coal-fired power generation unit.

4.2.2 Design of the carbon–green certificate mutual recognition mechanism

4.2.2.1 Fixed coefficient mutual recognition mechanism and its limitations

At present, a systematic mutual recognition mechanism for carbon–green certificate (C–GC) environmental equity products has not yet been established. Although carbon emission trading (including carbon allowances) and the green power certificate (green certificate) system are both aimed at optimizing the energy structure and promoting greenhouse gas reduction, significant differences remain in their policy positioning, accounting frameworks, measurement units, trading logic, and market regulation. These differences have resulted in the two markets operating relatively independently, making it difficult to achieve mutual recognition and conversion of environmental value, thereby limiting the overall emission reduction efficiency and resource allocation capability of the environmental equity market.

From an institutional perspective, the carbon market is centered on "control as the main objective, trading as a supplement." Its primary goal is to constrain total carbon emissions in high-emission industries through a cap-and-trade mechanism, using allowance price signals to guide enterprises toward voluntary emission reductions. Both carbon allowances and CCER are measured in "tons of CO_2 equivalent (tCO_2e)," emphasizing the binding constraint on total greenhouse gas emissions. In contrast, the green certificate system is measured in "megawatt-hours (MWh)" and primarily serves the fulfillment of renewable energy consumption responsibility, emphasizing the cleanliness and sustainability of energy sources. The difference in objectives means that even though both reflect the economic value of environmental emission reduction, there is no unified value anchor or accounting standard in actual trading.

In current academic and policy research, to achieve value alignment and quantitative conversion between carbon reductions (CCER) and green certificates, researchers often use the fixed coefficient method for approximate estimation. This method determines the carbon reduction benefit corresponding to the renewable energy generation represented by a green certificate using historical statistical data, the emission reduction efficiency of different power generation technologies, and the grid's average emission factor. In other words, the fixed coefficient method sets a relatively stable conversion ratio to translate the green electricity represented by green certificates into equivalent carbon reductions, thereby enabling value mapping between the two types of environmental equity products in different markets. The underlying idea is that renewable energy generation effectively reduces an equivalent amount of fossil-fuel generation under the same conditions, avoiding the corresponding carbon emissions. Therefore, each green certificate can be regarded as an environmental equity certificate with a certain carbon reduction capacity. By establishing an equivalence ratio between green certificates and CCER, the value of the two environmental equity products can be converted and circulated. The mathematical expression is given in (4.13).

$$Q_{j,t}^{con} = \tau^{fix} \times G_{j,t}^{con}, \forall j \in \Omega^{new}, \forall t \in T \tag{4.13}$$

where $Q_{j,t}^{con}$ represents the amount of CCER obtained from green certificate conversion for renewable energy unit j in period t; τ^{fix} is the fixed CCER products; $G_{j,t}^{con}$ denotes the number of from renewable energy unit j converted into CCER in period t; Ω^{new} is the set of renewable energy units; and T is the set of time periods.

The implementation of this conversion method is generally based on the marginal and average emission factors caused by one unit of incremental electricity consumption, while the average emission factor represents the average carbon emission level under the overall generation structure of the grid. By multiplying the renewable energy generation by the grid's average emission factor, the corresponding reduction in carbon emissions can be estimated, thereby enabling quantitative conversion between green certificates and carbon reductions. This method is simple to operate and easy to understand, and it provides a useful reference in the early stages of market development. It establishes a quantifiable general framework for preliminary market linkage, supporting government regulation of cross-market trading rules and enabling policy simulation by research institutions.

However, from a scientific and long-term operational perspective, the fixed coefficient method has certain limitations. First, it assumes that the grid's emission structure and energy composition remain stable in the short term, failing to fully reflect the dynamic variation of carbon emission factors. In reality, penetration, the actual emission reduction per unit of green electricity, may be lower than historical statistical levels. Second, the method does not account for lifecycle carbon emissions differences among different types of renewable energy. For example, although photovoltaic and wind power generate zero emissions during operation, indirect emissions are produced during equipment manufacturing, transportation, installation, and decommissioning. Without adjustment, the fixed coefficient method may overestimate actual emission reductions, leading to biased environmental value assessment. Third, it does not consider regional grid differences or time-of-day characteristics. For instance, in coal-dominated regions in central and western China, the substitution effect of renewable energy is much higher than in eastern coastal areas; likewise, during peak and off-peak periods, different marginal generators lead to significant variation in substitution effects. Therefore, when applying the fixed coefficient method, adjustments based on region, time, and energy type are necessary to improve the accuracy of carbon–green certificate conversion.

At the policy and market level, establishing a carbon–green certificate mutual recognition mechanism is of great significance. On one hand, it facilitates deep integration between the carbon market and the renewable energy market, promoting the interchange and circulation of environmental equity products and enhancing the overall economic value of green energy. Through a mutual recognition mechanism, green certificates can serve not only as proof of renewable energy consumption but also as an offset or compliance instrument in the carbon market, expanding their application scenarios and market demand. On the other hand, this mechanism strengthens incentives for low-carbon investment, providing renewable energy enterprises with additional revenue channels. For example, renewable energy generators can earn extra income by selling green certificates while realizing the value of carbon reductions in the carbon market, improving project returns, and market competitiveness. Moreover, carbon–green certificate mutual recognition can enhance overall resource allocation efficiency in the energy system, creating price linkages across multiple markets and promoting unified measurement and trading of environmental value.

In the future, with the further implementation of the "dual carbon" strategy, establishing a carbon–green certificate mutual recognition mechanism will become an important direction for improving China's environmental equity market system. Recommended steps include: (1) establishing a unified environmental equity database to manage and track carbon emissions, green power generation, and environmental value information; (2) improving the carbon attribute certification system to clarify the proportion of green certificates that can be offset or converted in the carbon market, forming scientific mutual recognition standards; (3) introducing a dynamic conversion coefficient mechanism, incorporating real-time grid emission factors and renewable energy output characteristics to construct regionalized, higher temporal resolution equivalence models; and (4) leveraging digital technologies, such as blockchain, big data, and the Internet of Things, to ensure traceability, verifiability, and tamper-proofing of green certificate and carbon reduction data, maintaining transparency and credibility of transactions.

In summary, establishing a carbon–green certificate mutual recognition mechanism is not only key to achieving efficient coordination between the carbon market and renewable energy market but also an important step toward systematic construction of environmental equity products. Its improvement will help form a multi-level environmental equity market structure with carbon as the core and green certificates as support, providing a solid institutional and market foundation for building a unified, efficient, and fair green low-carbon energy system in China.

4.2.2.2 Marginal carbon emission factor calculation model

Unlike the average carbon emission factor, which reflects the overall average emission level of the system, the core features of MCEF are its "marginality" and "context-dependence." It is not a static description of the system's carbon intensity but is closely coupled with multiple dynamic factors, including current dispatch conditions, generation unit mix, transmission network constraints, and renewable energy integration levels, focusing on the carbon emission response corresponding to incremental changes. Specifically, MCEF quantifies the total system carbon emission change caused by meeting one additional unit of electricity demand (usually measured in megawatt-hours, $MW \cdot h$) or one additional unit of generation output within a specific spatiotemporal scenario (such as a particular dispatch period or regional grid). Its standard unit is tons of CO_2 equivalent per megawatt-hour (t CO_2-eq/$MW \cdot h$), and in some studies, an instantaneous marginal emission factor in MW may also be used for supplemental representation depending on the application.

Theoretically, the calculation of MCEF is rooted in the marginal operation mechanism of the power system. When the system faces a unit electricity shortage or an additional unit of power supply, the dispatch center prioritizes the use of marginal units (such as peaking coal-fired units, gas units, or energy storage devices) based on operating costs, environmental constraints, and technical output limits to balance supply and demand. MCEF directly reflects the carbon intensity of this marginal dispatch process. This feature allows MCEF to overcome the limitation of average carbon emission factors, which "mask marginal differences," and accurately identify the marginal impact of different generation types and dispatch strategies on system carbon emissions. For example, under high renewable penetration, if the additional load is met by curtailed wind or solar power, the MCEF approaches zero; if a peaking coal-fired unit must be started, the MCEF equals the carbon intensity of that unit.

In practical applications, MCEF provides precise guidance for planning, operating, and managing low-carbon power systems. First, in a dynamic carbon pricing design, MCEF can serve as a core benchmark for carbon pricing. By reflecting the marginal abatement costs in different periods and regions, it allows carbon prices to dynamically match the power system's operational state, guiding market participants to prioritize low-carbon generation and incentivizing flexible resources for peak shaving and emission reduction. Second, in constructing mutual recognition mechanisms for environmental equity products, MCEF provides an accurate basis for calculating emission reductions of renewable projects such as wind and solar power. By quantifying the marginal emission reduction effect of renewables replacing fossil generation, it ensures the authenticity and tradability of carbon allowances, green certificates, and other environmental equity products, facilitating cross-regional and cross-sector carbon market connectivity. Third, in low-carbon dispatch optimization, MCEF can be embedded in unit commitment and economic dispatch models. Minimization of the system's total marginal carbon emissions enables the joint optimization of supply–demand balance and emission control, thereby facilitating low-carbon and efficient system operation.

For renewable generation units, the application of MCEF is particularly valuable. Under large-scale renewable integration, the level of renewable consumption is often limited by system flexibility, and its emission reduction effect on the system does not simply equal its own carbon intensity (usually close to zero) but depends on the carbon intensity of the marginal fossil units it replaces. By calculating MCEF, the carbon emissions corresponding to the marginal fossil generation displaced by the renewable unit's output can be accurately quantified. This allows for scientific evaluation of the actual emission reduction contribution of renewable generation under different operating scenarios. The results provide critical data for environmental benefit assessment of renewable projects, green certificate pricing, setting renewable consumption targets, and designing cross-provincial or cross-regional renewable trading mechanisms. They also provide key technical support for energy structure transformation and achieving the "dual carbon" targets.

Mathematically, the marginal carbon emission factor of renewable unit j in period t can be defined as:

$$\begin{aligned} \lambda_{j,t}^{new} &= \frac{\partial E_{CO2}}{\partial(P_{\mathrm{j,t}}^{\mathrm{new}}\Delta t)} \\ &= \sum_{i\in\Omega^{th}} \frac{\partial E_{CO2}}{\partial(P_{\mathrm{i,t}}^{\mathrm{th}}\Delta t)} \frac{\partial P_{\mathrm{i,t}}^{\mathrm{th}}}{\partial P_{\mathrm{j,t}}^{\mathrm{new}}} \\ &= \sum_{i\in\Omega^{th}} \rho_i^{th}\varphi_{i,j,t}, \forall \mathrm{j} \in \Omega^{new}, \forall \mathrm{t} \in T \end{aligned} \tag{4.14}$$

where $\lambda_{j,t}^{new}$ represents the factor of renewable unit j in period t; E_{CO2} is the total carbon emissions of the power system; $P_{j,t}^{new}$ denotes the output of renewable unit j in period t; Δt is the time interval; $P_{i,t}^{th}$ is the output of thermal (coal- or gas-fired) unit i in period t; ρ_i^{th} represents the carbon intensity of thermal unit i; $\varphi_{i,j,t}$ is the change i caused by a unit increase in output of renewable unit j in period t; and Ω^{th} is the set of thermal units.

From the derivation logic and quantitative relationship of (4.14), it is clear that when the output of renewable units (such as wind or photovoltaic) undergoes marginal fluctuations, it directly disrupts the existing supply-demand balance in the power system. This triggers adjustments in the dispatch strategy of thermal units, manifested as dynamic optimization of key dispatch parameters such as output levels and operating modes (e.g., start/stop status, load rate) to restore system power balance. Through the transmission path of "marginal change in renewable output → thermal unit dispatch response → adjustment of total system carbon emissions," the overall carbon emissions of the power system are significantly affected. This process reflects the strong coupling among generation, grid, load, and storage in the system and highlights the indirect and context-dependent nature of emission reduction effects from renewable energy.

Further analysis shows that the marginal carbon reduction benefit of a renewable unit is not solely determined by the magnitude of its own generation fluctuation but is constrained by two key factors. The first factor is the scale and temporal distribution of the renewable unit's output fluctuations, which form the basic premise for generating emission reduction effects. The second, more critical factor is the carbon emission level of the thermal units it displaces under the current system operating state (i.e., the carbon intensity of the displaced thermal unit, which depends on fuel type, combustion efficiency, and environmental technology), as well as the overall system response characteristics (e.g., unit mix, peaking capability, transmission constraints, and dispatch rules). For instance, when system peaking capacity is sufficient and renewable output displaces efficient, low-emission coal units, the marginal carbon reduction benefit is relatively low; conversely, if renewable output displaces inefficient, high-emission coal or gas units, the marginal carbon reduction benefit increases significantly.

Based on this analysis, the calculation of the marginal carbon emission factor (MCEF) is strongly correlated with the dispatch behavior of thermal units and their corresponding carbon intensity. The dispatch behavior of thermal units determines which specific units and output levels are replaced by the marginal renewable output, and the carbon intensity of those units quantifies the carbon emissions per unit of replaced electricity. Together, these constitute the core input parameters for MCEF calculation. Different system operating scenarios (such as peak/off-peak periods, renewable penetration levels, and load fluctuations) will affect the dispatch priorities and output allocation of thermal units, which, in turn, changes the types of marginal thermal units being replaced and their carbon intensities, ultimately leading to significant spatiotemporal heterogeneity in MCEF results.

To clearly present the calculation logic and procedure for the marginal carbon emission factor of renewable units, Figure 4.2 illustrates the complete calculation

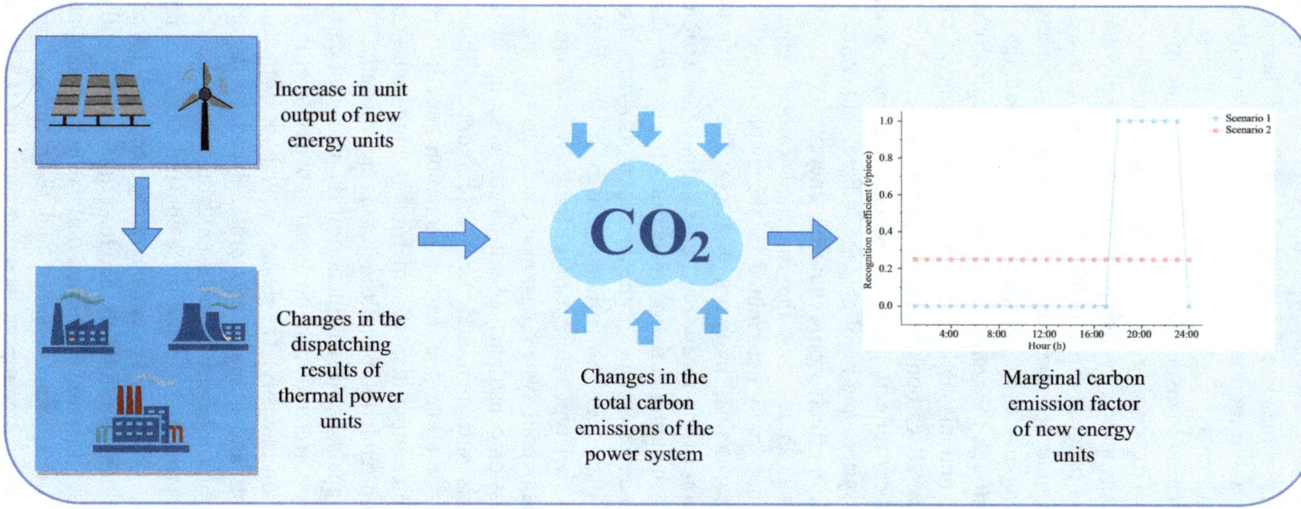

Figure 4.2 Marginal carbon emission factor calculation process

workflow. The process starts with input data of marginal changes in renewable output and sequentially includes system operating state parameter extraction, thermal unit dispatch response simulation, identification of marginally replaced units, carbon intensity matching, and MCEF quantification. This workflow visually demonstrates the intrinsic relationship among MCEF, renewable output fluctuations, thermal unit dispatch behavior, and carbon intensity, providing a clear methodological basis for accurately measuring the marginal carbon reduction benefits of renewable energy.

In actual power system dispatch, the emission factors of various thermal units can be parameterized and estimated based on historical dispatch and operational statistics. The marginal carbon emission factor of renewable output; however, needs to be dynamically solved using an optimal dispatch model that incorporates system operating constraints. Specifically, by constructing an economic dispatch model and introducing perturbations in renewable unit output, the marginal response of system carbon emissions can be accurately captured, enabling dynamic characterization of MCEF. Figure 4.3 illustrates the economic dispatch model of the Agricultural Energy Internet power system.

Objective function

Assuming the power system includes thermal generation units, renewable generation units, and an energy storage device, an economic dispatch model is constructed to characterize the system's operating state under comprehensive resource constraints. The model aims to minimize the total system cost, taking into account the generation costs of units, the maintenance costs of storage, and the dispatch balance constraints, as expressed in the following equation:

$$\begin{aligned}\min C &= \sum_{t\in T}\left(\sum_{\mathrm{i}\in\Omega^{th}} C_{i,t}^{th} + \sum_{\mathrm{j}\in\Omega^{new}} C_{j,t}^{new} + C_t^{ess} + C_t^{\mathrm{grid}-buy} - C_t^{grid-sell}\right)\\ &= \sum_{t\in T}\left(\sum_{\mathrm{i}\in\Omega^{th}} a_i^{th}P_{i,t}^{th} + \sum_{\mathrm{j}\in\Omega^{new}} \mathrm{a}_j^{new}P_{j,t}^{new} + \mathrm{a}^{\mathrm{ess}}\left|P_t^{ess}\right| + a_t^{buy}P_t^{buy} - a_t^{sell}P_t^{sell}\right)\end{aligned} \tag{4.15}$$

where, C is the total generation cost; $C_{i,t}^{th}$ is the thermal unit i in period t; $C_{j,t}^{new}$ is the generation cost of renewable unit j in period t; C_t^{ess} is the charging/discharging device in period t; $C_t^{grid-buy}$ is the cost of electricity the external grid in period t; $C_t^{grid-sell}$ is the revenue from electricity sold to the external grid in period t; a_i^{th} is the unit generation cost of thermal unit i; a_i^{new} is the unit generation cost of renewable unit j; a^{ess} is the unit capacity device; P_t^{ess} is the charging or discharging power of the storage device in period t, with charging as positive and discharging as negative; a_t^{buy} is the unit cost of electricity; a_t^{sell} is the unit revenue of electricity sold to the external grid; and P_t^{buy} and P_t^{sell} are the purchased and sold electricity quantities with the external grid in period t, respectively.

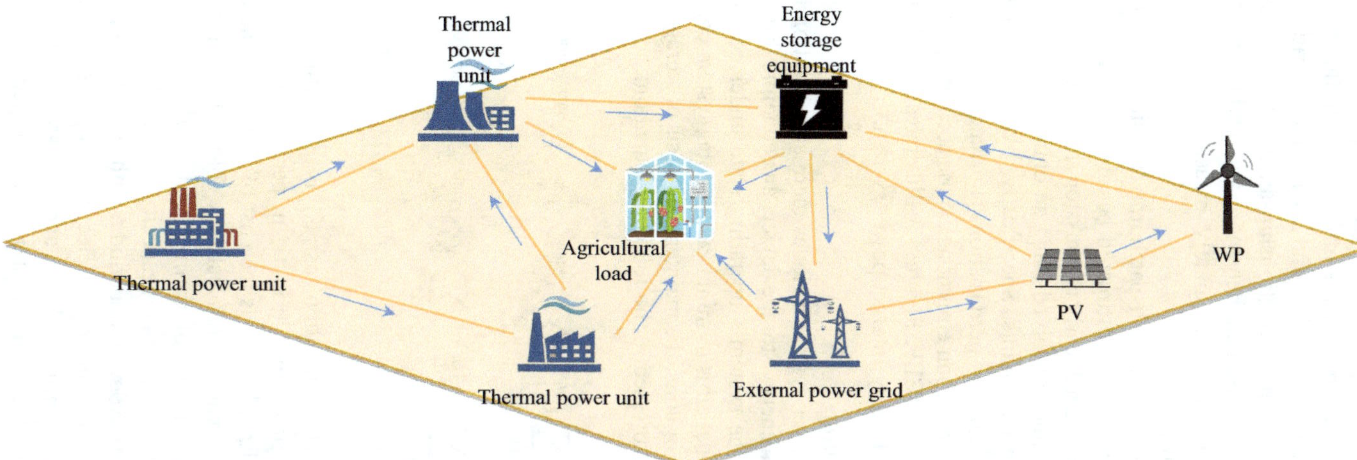

Figure 4.3 Economic dispatch model of the Agricultural Energy Internet power system

Constraints

The constraints of the economic dispatch model.

Power balance constraint:

$$\sum_{i\in\Omega^{th}} P_{i,t}^{th} + \sum_{j\in\Omega^{new}} P_{j,t}^{new} - P_t^{ess} + P_t^{buy} - P_t^{sell} = P_t^{load}, \forall t \in T \tag{4.16}$$

where, P_t^{load} is the load demand of the agricultural park in period *t*.

Thermal generation unit output constraint:

$$P_{i,\min}^{th} \leq P_{i,t}^{th} \leq P_{i,\max}^{th}, \forall i \in \Omega^{th}, \forall t \in T \tag{4.17}$$

where, $P_{i,\min}^{th}$, $P_{i,\max}^{th}$ are the minimum and maximum of the thermal unit, respectively.

Renewable generation unit output constraint

$$0 \leq P_{j,t}^{new} \leq P_{j,t}^{new-pre}, \forall j \in \Omega^{new}, \forall t \in T \tag{4.18}$$

where, $P_{j,t}^{new-pre}$ is the forecasted unit *j* in period *t*.

External grid electricity purchase constraint:

$$0 \leq P_t^{buy} \leq P_{t,\max}^{buy}, \forall t \in T \tag{4.19}$$

where, $P_{t,\max}^{buy}$ is the upper limit of purchases from the external grid in period *t*.

Energy storage device operating constraints:

The energy storage device includes charging/discharging power limits and state of charge (SOC) constraints.

$$-P_{\max}^{dis} \leq P_t^{ess} \leq P_{\max}^{ch}, \forall t \in T \tag{4.20}$$

where $P_{\max}^{dis}$ is the maximum discharging power of the energy storage device, and $P_{\max}^{ch}$ is the maximum device.

$$S_t^{ess} = S_{t-1}^{ess} + P_t^{ess}\eta^{ess}\Delta t, \forall t \in T \tag{4.21}$$

$$S_{\min}^{ess} \leq S_t^{ess} \leq S_{\max}^{ess}, \forall t \in T \tag{4.22}$$

$$S_{t_0}^{ess} = S_{t_T}^{ess} \tag{4.23}$$

where S_t^{ess} is the state of charge (SOC) of the energy in period *t*; η^{ess} is the charging/discharging of the storage device; $S_{\min}^{ess}$, $S_{\max}^{ess}$ are the minimum and maximum SOC limits of the storage device, respectively; $S_{t_0}^{ess}$ is the initial SOC of the storage device; and $S_{t_T}^{ess}$ is the final SOC of the storage device.

Since the economic dispatch model is a linear optimization problem, if a unique optimal solution exists, some of the constraints will be active at the optimum, which can be expressed as:

$$\sum_{i\in\hat{\Omega}^{th}} \hat{m}_{i,t}^{th}\hat{P}_{i,t}^{th} + \sum_{j\in\hat{\Omega}^{new}} \hat{m}_{j,t}^{new}\hat{P}_{j,t}^{new} - \hat{m}_t^{ess}\hat{P}_t^{ess} + \hat{m}_t^{buy}\hat{P}_t^{buy} - \hat{m}_t^{sell}\hat{P}_t^{sell} = \hat{n}, \forall t \in \hat{T} \tag{4.24}$$

where, $\hat{P}_{i,t}^{th}$, $\hat{P}_{j,t}^{new}$, $\hat{P}_t^{ess}$, $\hat{P}_t^{buy}$ and $\hat{P}_t^{sell}$ are the optimal outputs of thermal units, renewable units, energy storage devices, and the external grid in period t, respectively; $\hat{m}_{i,t}^{th}$, $\hat{m}_{j,t}^{new}$, $\hat{m}_t^{ess}$, $\hat{m}_t^{buy}$, and $\hat{m}_t^{sell}$ are the coefficients corresponding to the decision $\hat{P}_{i,t}^{th}$, $\hat{P}_{j,t}^{new}$, $\hat{P}_t^{ess}$, $\hat{P}_t^{buy}$ and $\hat{P}_t^{sell}$ variables of the active constraints at the optimum; $\hat{n}$ is the constant term of the active constraints at the optimum; and $\hat{\Omega}^{th}$, $\hat{\Omega}^{new}$ and $\hat{T}$ and time periods corresponding to the active constraints at the optimum. The specific expressions of the active constraints in (4.25) at the optimal solution are shown as follows:

$$\begin{cases} \sum_{j\in\Omega^{new}} \hat{P}_{j,t}^{new} + \sum_{i\in\Omega^{th}} \hat{P}_{i,t}^{th} - \hat{P}_t^{ess} + \hat{P}_t^{buy} - \hat{P}_t^{sell} = \hat{P}_t^{load}, \forall t \in T \\ \hat{P}_{i,t}^{th} = P_{i,\max}^{th}, \forall\{i,t\} \in \{\Omega_{\max}^{th}, T_{\max}^{th}\} \\ \hat{P}_{i,t}^{th} = P_{i,\min}^{th}, \forall\{i,t\} \in \{\Omega_{\min}^{th}, T_{\min}^{th}\} \\ \hat{P}_{j,t}^{new} = P_{j,t}^{new-pre}, \forall\{j,t\} \in \{\Omega_{\max}^{new}, T_{\max}^{new}\} \\ \hat{P}_{j,t}^{new} = 0, \forall\{j,t\} \in \{\Omega_{\min}^{new}, T_{\min}^{new}\} \\ \hat{P}_t^{ess} = P_{\max}^{ch}, \forall t \in T_{\max}^{ch} \\ \hat{P}_t^{ess} = -P_{\max}^{dis}, \forall t \in T_{\max}^{dis} \\ \hat{P}_t^{buy} = P_{\max}^{buy}, \forall t \in T_{\max}^{buy} \\ \hat{P}_t^{buy} = 0, \forall t \in T_{\min}^{buy} \\ S_{t_0}^{ess} + \hat{P}_1^{ess}\eta^{ess}\Delta t + \hat{P}_2^{ess}\eta^{ess}\Delta t + \hat{P}_3^{ess}\eta^{ess}\Delta t + \ldots + \hat{P}_t^{ess}\eta^{ess}\Delta t = S_{\max}^{ess}, \forall t \in T_{\max}^{SOC} \\ S_{t_0}^{ess} + \hat{P}_1^{ess}\eta^{ess}\Delta t + \hat{P}_2^{ess}\eta^{ess}\Delta t + \hat{P}_3^{ess}\eta^{ess}\Delta t + \ldots + \hat{P}_t^{ess}\eta^{ess}\Delta t = S_{\min}^{ess}, \forall t \in T_{\min}^{SOC} \\ S_{t_0}^{ess} + \hat{P}_1^{ess}\eta^{ess}\Delta t + \hat{P}_2^{ess}\eta^{ess}\Delta t + \hat{P}_3^{ess}\eta^{ess}\Delta t + \ldots + \hat{P}_t^{ess}\eta^{ess}\Delta t = S_{t_0}^{ess} \end{cases} \tag{4.25}$$

where $\Omega_{\max}^{th}$ and $T_{\max}^{th}$ are the sets of thermal units and time periods for which the thermal unit output reaches the optimal solution, respectively; $\Omega_{\min}^{th}$ and $T_{\min}^{th}$ are the sets of renewable units and time periods for which the renewable unit output reaches the optimal solution, respectively; $\Omega_{\max}^{new}$ and $T_{\max}^{new}$ are the sets of time periods charging and discharging power at the optimal solution; $\Omega_{\min}^{new}$ and $T_{\min}^{new}$ are the sets of renewable units and time periods for which the renewable unit output reaches the lower limit at the optimal solution; $T_{\max}^{ch}$ and $T_{\max}^{dis}$ are the sets of time periods when the energy storage device reaches its maximum charging and discharging power at the optimal solution, respectively. $T_{\max}^{SOC}$ and $T_{\min}^{SOC}$ are the sets of time periods when the energy storage device reaches its maximum and minimum SOC at the optimal solution. $T_{\max}^{buy}$ and $T_{\min}^{buy}$ are the sets of time periods when electricity purchased from the external grid reaches the minimum and maximum limits at the optimal solution.

Under the premise that other conditions remain unchanged, a small incremental disturbance in the output of a renewable unit in a given period usually does not fundamentally alter the structure of the active constraints in the power system (such as the power balance or transmission capacity constraints). In this case, the system maintains operational balance by locally adjusting the outputs of some thermal devices, thereby absorbing the marginal disturbance [12]. This local adjustment mechanism allows sensitivity analysis or marginal effect derivation of system variables without changing the structure of the optimization model solution.

For any renewable unit in the system, when its output experiences a small change, partial derivatives of the objective function with respect to the relevant decision variables in the economic dispatch model can be calculated to obtain the marginal response of the system under this disturbance. This process quantifies the impact of the renewable unit's output at different periods on thermal unit dispatch and carbon emissions, which is a key step in calculating the marginal carbon emission factor.

Thus, by taking partial derivatives of the relevant variables in (4.25) with respect to the output disturbance of renewable unit j in period t, (4.27) can be derived. The detailed mathematical derivation and expression are shown as follows:

$$\sum_{i\in\widehat{\Omega}^{th}} \widehat{m}_{i,t}^{th} \frac{\partial \widehat{P}_{\mathrm{i,t}}^{\mathrm{th}}}{\partial \widehat{P}_{\mathrm{j_{new},t}}^{\mathrm{new}}} + \sum_{j\in\widehat{\Omega}^{new}} \widehat{m}_{j,t}^{new} \frac{\partial \widehat{P}_{\mathrm{j,t}}^{\mathrm{new}}}{\partial \widehat{P}_{\mathrm{j_{new},t}}^{\mathrm{new}}} - \widehat{m}_{t}^{ess} \frac{\partial \widehat{P}_{\mathrm{t}}^{\mathrm{ess}}}{\partial \widehat{P}_{\mathrm{j_{new},t}}^{\mathrm{new}}} + \widehat{m}_{t}^{buy} \frac{\partial \widehat{P}_{\mathrm{t}}^{\mathrm{buy}}}{\partial \widehat{P}_{\mathrm{j_{new},t}}^{\mathrm{new}}}$$

$$- \widehat{m}_{t}^{sell} \frac{\partial \widehat{P}_{\mathrm{t}}^{\mathrm{sell}}}{\partial \widehat{P}_{\mathrm{j_{new},t}}^{\mathrm{new}}} = 0, \forall \mathrm{t} \in \widehat{\mathrm{T}} \tag{4.26}$$

The partial derivatives of the active constraints can be assembled into a linear system of equations, where n_g represents the total number of thermal units, renewable units, energy storage devices, and external grid connections, as shown in (4.27).

By solving this system of equations, the derivatives of each unit's output variables in period t with respect to $P_{\mathrm{j,t}}^{\mathrm{new}}$ can be obtained, which in turn yields $\lambda_{j,t}^{new}$. Meanwhile, under non-binding system conditions, the $\lambda_{j,t}^{new}$ values are the same for different renewable units within the same period.

$$\left\{\begin{array}{l}
\sum_{j\in\Omega^{\mathrm{new}}}\frac{\partial\widehat{P}_{j,t}^{new}}{\partial\widehat{P}_{j_{new},t}^{new}}+\sum_{i\in\Omega^{th}}\frac{\partial\widehat{P}_{i,t}^{th}}{\partial\widehat{P}_{j_{new},t}^{new}}-\frac{\partial\widehat{P}_{t}^{ess}}{\partial\widehat{P}_{j_{new},t}^{new}}+\frac{\partial\widehat{P}_{t}^{buy}}{\partial\widehat{P}_{j_{new},t}^{new}}-\frac{\partial\widehat{P}_{t}^{sell}}{\partial\widehat{P}_{j_{new},t}^{new}}=-1,\forall t\in T\\
\frac{\partial\widehat{P}_{i,t}^{th}}{\partial\widehat{P}_{j_{new},t}^{new}}=0,\forall\{i,t\}\in\left\{\Omega_{\max}^{th},T_{\max}^{th}\right\}\\
\frac{\partial\widehat{P}_{i,t}^{th}}{\partial\widehat{P}_{j_{new},t}^{new}}=0,\forall\{i,t\}\in\left\{\Omega_{\min}^{th},T_{\min}^{th}\right\}\\
\frac{\partial\widehat{P}_{j,t}^{new}}{\partial\widehat{P}_{j_{new},t}^{new}}=0,\forall\{j,t\}\in\left\{\Omega_{\max}^{new},T_{\max}^{new}\right\}\\
\frac{\partial\widehat{P}_{j,t}^{new}}{\partial\widehat{P}_{j_{new},t}^{new}}=0,\forall\{j,t\}\in\left\{\Omega_{\min}^{new},T_{\min}^{new}\right\}\\
\frac{\partial\widehat{P}_{t}^{ess}}{\partial\widehat{P}_{j_{new},t}^{new}}=0,\forall t\in T_{\max}^{ch}\\
\frac{\partial\widehat{P}_{t}^{ess}}{\partial\widehat{P}_{j_{new},t}^{new}}=0,\forall t\in T_{\max}^{dis}\\
\frac{\partial\widehat{P}_{t}^{buy}}{\partial\widehat{P}_{j_{new},t}^{new}}=0,\forall t\in T_{\max}^{buy}\\
\frac{\partial\widehat{P}_{t}^{buy}}{\partial\widehat{P}_{j_{new},t}^{new}}=0,\forall t\in T_{\min}^{buy}\\
\frac{\partial\widehat{P}_{1}^{ess}}{\partial\widehat{P}_{j_{new},t}^{new}}+\frac{\partial\widehat{P}_{2}^{ess}}{\partial\widehat{P}_{j_{new},t}^{new}}+\frac{\partial\widehat{P}_{3}^{ess}}{\partial\widehat{P}_{j_{new},t}^{new}}+\ldots+\frac{\partial\widehat{P}_{t}^{ess}}{\partial\widehat{P}_{j_{new},t}^{new}}=0,\forall t\in T_{\max}^{SOC}\\
\frac{\partial\widehat{P}_{1}^{ess}}{\partial\widehat{P}_{j_{new},t}^{new}}+\frac{\partial\widehat{P}_{2}^{ess}}{\partial\widehat{P}_{j_{new},t}^{new}}+\frac{\partial\widehat{P}_{3}^{ess}}{\partial\widehat{P}_{j_{new},t}^{new}}+\ldots+\frac{\partial\widehat{P}_{t}^{ess}}{\partial\widehat{P}_{j_{new},t}^{new}}=0,\forall t\in T_{\min}^{SOC}\\
\frac{\partial\widehat{P}_{1}^{ess}}{\partial\widehat{P}_{j_{new},t}^{new}}+\frac{\partial\widehat{P}_{2}^{ess}}{\partial\widehat{P}_{j_{new},t}^{new}}+\frac{\partial\widehat{P}_{3}^{ess}}{\partial\widehat{P}_{j_{new},t}^{new}}+\ldots+\frac{\partial\widehat{P}_{t}^{ess}}{\partial\widehat{P}_{j_{new},t}^{new}}=0
\end{array}\right. \tag{4.27}$$

4.2.2.3 Marginal carbon emission factor calculation model

The effective linkage (GC) market relies on scientifically and reasonably identifying the actual carbon reduction benefits of renewable electricity, thereby establishing a quantitative mapping between renewable energy usage and its associated environmental value. In other words, the key to constructing a carbon–GC recognition mechanism lies in clarifying the system-wide emission reduction contribution of renewable generation and formulating unified, transparent, and operable conversion standards, so that a quantifiable, verifiable, and traceable exchange relationship can be established between GCs and CCERs (China Certified Emission Reductions). This mechanism enables "equivalent recognition" and "single-incentive" treatment of renewable energy environmental rights, effectively avoiding double-counting, overlapping incentives, or market mismatches.

In this process, the marginal carbon emission factor (MEF) plays a central role. The MEF characterizes the marginal impact of renewable output changes on overall system carbon emissions, dynamically reflecting the incremental carbon emission reduction or increase caused by small variations in renewable generation at specific time scales and system operating conditions. In other words, the MEF quantifies the marginal carbon reduction contribution of renewable units at different time points and serves as a key parameter for achieving the "measurable emission reduction benefit" objective in the carbon–GC recognition mechanism. Based on the MEF, a unique mapping between GCs and CCERs can be established, ensuring a scientific and fair value conversion of renewable energy environmental rights across the two markets.

Under this mechanism, renewable units first receive government-issued GCs and can trade them in the GC market to reflect their clean power attributes and renewable energy utilization value. For any remaining unsold GCs after trading, renewable companies can convert a portion into CCERs according to the proportion determined by the MEF, for additional revenue. This approach not only broadens the revenue channels for renewable enterprises and enhances their flexibility and initiative in a multi-market environment but also strengthens the economic attractiveness and market competitiveness of renewable projects within the carbon reduction system.

However, at the physical operation, the effect of renewable output on system carbon emissions is not always linear or unidirectional. Influenced by factors such as grid topology, transmission constraints, unit ramping characteristics, and reserve capacity distribution, increases in renewable output may not reduce system emissions in some periods. In extreme cases, renewable generation may even "displace" more efficient clean thermal units, causing overall system emissions to rise. In such situations, the MEF will be positive, indicating that the marginal emission reduction contribution of renewable output in that period is negative, i.e., no actual reduction benefit.

Based on these practical constraints, the recognition mechanism design should incorporate dynamic adjustment and restriction mechanisms. Specifically, when the MEF is positive, the renewable output in that period is considered to have no reduction benefit, and conversion of GCs to CCERs should be prohibited to ensure the accuracy and rigor of carbon reduction calculations. Moreover, to prevent excessive conversion of GCs to CCERs from destabilizing the GC market and affecting price stability, it is necessary to introduce a conversion coefficient constraint model based on the MEF within the recognition mechanism. Accordingly, this study proposes the definition of the carbon–GC recognition coefficient, establishing a scientific, reasonable, and dynamically adaptive recognition calculation system that balances emission reduction authenticity with market stability.

The specific mathematical expression of the carbon–GC recognition coefficient is shown in (4.28).

$$\tau_{j,t}^{new} = \begin{cases} 0\lambda_{j,t}^{new} > 0 \\ -\lambda_{j,t}^{\text{new}} - 1 \leq \lambda_{j,t}^{new} \leq 0 \\ 1\lambda_{j,t}^{new} < -1 \end{cases} \tag{4.28}$$

The recognition relationship between CCERs and green certificates (GCs) is as follows:

$$Q_{j,t}^{con} = \tau_{j,t}^{new} G_{j,t}^{con}, \forall \text{j} \in \Omega^{new}, \forall \text{t} \in T \tag{4.29}$$

Under the proposed carbon–green certificate (C–GC) recognition mechanism based on the marginal carbon emission factor, the revenue realization pathway for renewable energy units (including wind, photovoltaic, and other renewable generation projects) is innovatively optimized. Specifically, during periods of high marginal emission reduction effectiveness, if a renewable energy unit holds a sufficient number of green certificates (GCs), a portion of these certificates can be converted into China Certified Emission Reductions (CCERs) according to the preset higher recognition coefficient. The core value of this conversion pathway lies in the fact that CCERs, as a primary trading instrument in the carbon market, have clear pricing and liquidity attributes. By participating in carbon market transactions through this approach, renewable energy units can transform their environmental benefits into tangible economic returns, significantly enhancing their investment returns and market competitiveness. From the market incentive perspective, this design breaks the traditional revenue model where renewable generation relies solely on feed-in tariffs and subsidies. By supplementing with carbon market revenues, it effectively reduces the uncertainty of investment payback periods and greatly encourages market participants to engage in renewable project development, operation, and carbon market trading. Furthermore, high marginal emission reduction periods usually coincide with high fossil energy consumption and urgent emission reduction needs. The mechanism, through economic incentives, guides renewable units to prioritize output during these periods, achieving precise alignment between renewable output and system low-carbon dispatch requirements, helping to reduce overall system carbon intensity and improve system-wide emission reduction efficiency.

Key flexibility support for this recognition mechanism is provided by energy storage, further expanding its application scenarios and optimization potential. Given the intermittent and fluctuating nature of renewable output, energy storage can act as a “peak shaving and valley filling” regulator: during periods of low marginal emission reduction effectiveness (e.g., surplus renewable generation with low fossil generation contribution), excess energy can be stored, avoiding wasted environmental benefits; during periods of higher future marginal emission reduction effectiveness (e.g., peak demand or full-load fossil generation), the stored energy can be dispatched, and the corresponding GCs can be converted to CCERs using higher recognition coefficients, achieving temporal shifting of emission reduction value and maximizing revenue. This process not only enhances the controllability and flexibility of renewable electricity but also promotes coordinated operation between renewable units and storage, optimizing renewable output across time dimensions. From a system perspective, combining storage with the recognition mechanism alleviates renewable integration pressure, increases the grid's capacity to absorb renewable energy, and provides crucial support for increasing renewable penetration in the energy mix.

From a broader resource allocation and market coordination perspective, the proposed C–GC recognition mechanism establishes a fair and transparent environmental rights allocation pathway, serving as a bridge for the deep integration of electricity, carbon, and green certificate markets. In traditional market models, the electricity market focuses primarily on the commodity attributes of electricity, the carbon market emphasizes carbon rights pricing and circulation, and the green certificate market centers on recognizing the environmental value, resulting in clear market silos and fragmented value. The proposed mechanism uses C–GC a core link, integrating the environmental benefits of renewable generation (quantified by GCs), system emission reduction requirements represented carbon market transactions (via CCER) into a cohesive framework. This enables a closed-loop for quantifying, transferring, and monetizing environmental rights. For renewable energy enterprises, the mechanism provides a clear channel to monetize environmental value; for fossil-fuel units such as coal-fired plants, it indirectly reduces their emission reduction pressure by increasing renewable penetration while providing richer supply-demand scenarios for participating in the carbon market; for regulators, the transparent calculation of marginal carbon emission factors and recognition rules facilitates supervision and ensures market fairness. This multi-stakeholder, win–win design optimizes allocation of electricity and environmental resources and promotes coordinated interaction among the three markets, enhancing overall market efficiency and driving the low-carbon transition.

In terms of engineering feasibility and policy promotion, this recognition mechanism has significant advantages. The dynamic calculation of marginal carbon emission factors is technically well-supported: it can be based on power system economic dispatch models, integrating real-time load demand, generation mix, and unit operational characteristics, and can be accurately calculated using established linear or quadratic programming algorithms. With the ongoing advancement of power system dispatch automation and improved infrastructure, such as smart meters and dispatch control systems, there is a solid data and technical foundation for real-time calculation and dynamic updating of marginal carbon emission factors. Furthermore, these algorithms are widely applied in power system optimization and carbon accounting, with high technical maturity and controllable computational accuracy, meeting practical operational needs. From a policy perspective, China's carbon and green certificate markets have entered a normalized development phase. Following the restart of the CCER market, the regulatory framework has been increasingly refined, and the GC trading mechanism is gradually established, providing a favorable policy environment for C–GC recognition. Implementation of this mechanism does not require disruptive changes to existing markets but can be realized through supplemental recognition rules and optimized conversion coefficients, resulting in low policy implementation costs and high feasibility.

In summary, the proposed carbon–green certificate recognition mechanism based on marginal carbon emission factors innovatively establishes environmental rights conversion pathways, incorporates energy storage for coordinated optimization, and strengthens multi-market synergy. It effectively enhances renewable units' market competitiveness and revenue, guides optimal temporal and spatial allocation of renewable output, and promotes deep. Its mature technical algorithms and policy compatibility make it broadly applicable, providing strong institutional support for the low-carbon transformation of energy systems and significant theoretical and practical value in achieving China's "dual carbon" targets.

4.2.3 Electricity–carbon–green certificate joint clearing model

4.2.3.1 Joint market basic framework

To scientifically quantify the potential effectiveness of the proposed carbon–green certificate (C–GC) recognition, and improve energy resource allocation efficiency – this study constructs a centralized joint clearing model that integrates the electricity market, carbon market, and green certificate market. The core optimization objective is the maximization of total social welfare under the coordinated operation of the three markets. In this model, total social welfare is comprehensively defined to include the electricity consumption benefits of users, the operational revenues of market participants, and the internalized compensation for environmental externalities (especially ecological and environmental losses caused by carbon emissions), ensuring that the optimization objective reflects both economic efficiency and low-carbon development with social equity considerations.

In constructing the model, the realistic operational logic of multi-market integration is fully incorporated. The model systematically includes core operational mechanisms of each market, supply-demand balance dynamics, and market participant bidding behaviors, forming a framework that is both theoretically rigorous and practically applicable. Specifically, in the electricity market dimension, the model considers the technical characteristics of various generation types, such as the intermittency of wind and photovoltaic, ramp rate constraints of thermal power, and regulation capacity limitations of hydropower, as well as the mechanisms for determining feed-in tariffs and grid transmission constraints. In the carbon market dimension, the model accurately reflects core elements such as carbon quota allocation rules, CCER trading mechanisms, and carbon accounting standards, clearly defining fossil fuel generation enterprises' carbon emission constraints and reduction responsibilities. In the green certificate market dimension, the model strictly follows the full process rules for issuance, trading, and retirement of GCs, clarifying the correspondence between GCs and renewable generation, as well as the conversion logic under the recognition mechanism.

The model positions the power industry as the core participating entity and includes multiple market roles, such as renewable generation companies, fossil fuel generation companies, grid operators, electricity consumers, and energy storage operators. This comprehensively reflects the operational characteristics of deep integration across rules, participants, and under the context of multi-energy coordination and accelerated low-carbon transition. It ensures that the model can realistically depict how the recognition mechanism affects the participant and the overall system response.

For the clearing mechanism, the model adopts a marginal price-based pricing approach, which aligns closely with mainstream operational rules of electricity and carbon markets. Its main advantage lies in enabling resources to be optimally allocated according to marginal costs: in electricity market clearing, the marginal generation cost determines the feed-in price, guiding low-cost generation to be dispatched first; in carbon market clearing, the marginal abatement cost determines the carbon quota price, incentivizing low-cost reduction participants to increase emission reductions; in green certificate market clearing, the marginal environmental benefit determines GC prices, promoting efficient circulation of green electricity resources. This pricing method ensures economic efficiency in market operation and, through price signaling, achieves balanced and coordinated interaction among the three markets, meeting the reasonable demands of different market participants while ensuring fairness and sustainability.

The overall structure and basic framework of the joint clearing model are illustrated in Figure 4.4. Its core logic is to establish a closed-loop system of "resource supply – constraint conditions – objective optimization – price formation," fully covering the coordinated optimization of electricity, carbon quotas, and green certificates. In terms of model operation, the process begins with inputting the output characteristics of various generation types, load forecasts, total carbon quotas, and GC issuance scale. Next, the C–GC recognition rules, grid security constraints, and market trading rules are embedded as constraints in the model. The model then optimizes the total social welfare to determine the optimal allocation of resources, including generation output plans, trading volumes of carbon quotas and CCER, the flow of GCs, and the recognition conversion ratios. Finally, based on the marginal pricing mechanism, the node electricity prices, carbon quota trading prices, and GC trading prices are determined, providing essential data for subsequent evaluation of the mechanism's effectiveness. This model structure simultaneously reflects the relative independence of the three markets and their deep coupling through the recognition mechanism, providing a robust quantitative tool for assessing the comprehensive benefits of the C–GC recognition mechanism.

In addition, to ensure the mathematical feasibility, logical rigor, and engineering practicality of the joint clearing model and to align it with the actual operation of multi-market integration, this chapter makes the following necessary and reasonable assumptions regarding the core operating rules, trading processes,

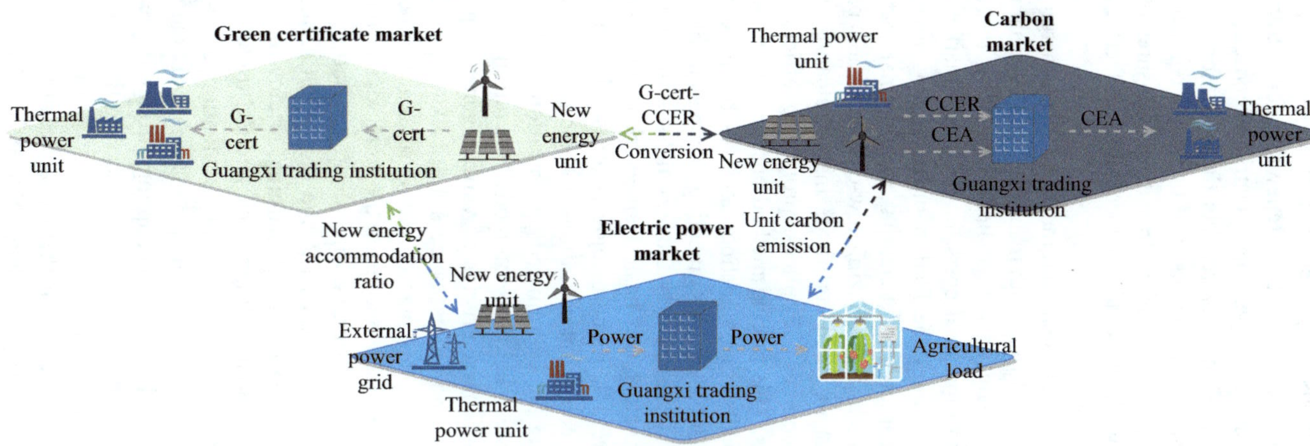

Figure 4.4 Electric–carbon–green certificate joint market basic framework

and behavioral patterns of participants, providing clear boundary conditions and logical foundations for model construction:

Electricity market mechanism assumptions

Day-ahead energy trading model, with hourly time granularity for security-constrained economic dispatch clearing. This setting is consistent with the mainstream operating cycle of current electricity market pilots in China and can accurately characterize the dynamic variations of intra-day electricity supply and demand. Market participants include, and other controllable sources, renewable energy units (wind power, photovoltaic), external interconnected grids, and various types of loads within the park (including industrial and commercial loads). All market participants must follow the principle of "voluntary declaration and fair competition" when submitting market bids, ensuring sufficient and competitive market transactions. In terms of bidding rules, generation-side entities submit hourly offers based on their marginal generation costs and unit operating characteristics (such as start-up/shutdown costs and ramping constraints); the load side must simultaneously submit electricity demand forecasts and acceptable purchase price ranges for each time period [16]. The setting of the purchase price range fully considers the elasticity of electricity demand and cost tolerance, providing a quantitative basis for demand response. Market transactions are organized by an independent trading institution using a centralized bidding and clearing mechanism, with the "minimization of total system operating cost" as the core objective function (total system operating cost includes generation fuel costs, start-up/shutdown costs, transmission loss costs, etc.). The problem is solved using linear or mixed-integer programming optimization algorithms to determine generation schedules and locational marginal prices (LMPs) for each time period and node, which serve as the basis for electricity settlement. For power balance and cross-regional trading, the model assumes that when total generation within the park exceeds load demand, the surplus power can be exported through external grid interconnections to achieve optimal cross-regional allocation of electricity resources and maximize the economic benefits of market participants; conversely, when internal power supply is insufficient, electricity can be imported from external grids to ensure reliable supply to loads. Meanwhile, considering output and their impact on system stability, the park is equipped with a certain scale of energy or flow battery storage. These storage systems serve as important power regulation resources, mainly participating in system frequency regulation, reserve capacity provision, and peak shaving and valley filling. By charging and discharging across different time periods, they help smooth renewable energy fluctuations and enhance grid operational flexibility. model, energy storage devices do not participate in market bidding; their operation is scheduled by the system operator based on global optimization objectives. The participation of storage systems in market bidding can be further extended in future model designs.

Carbon market mechanism assumptions

The carbon market operates on a daily clearing basis. This setting meets the periodic requirements of carbon quota settlement and effectively aligns with the intraday dispatch cycle of the electricity market. The model assumes that long-term carbon reduction targets set at the national or regional level have been scientifically decomposed into annual and quarterly quotas, further refined to daily time periods. The total daily carbon quota is embedded in the model as a constraint to ensure the rigid implementation of emission reduction targets. Thermal power units, as the main sources of carbon emissions, must strictly follow the "daily settlement, surplus tradable, deficit compensable" rule. That is, after each day's trading, the number of carbon emission allowances (CEA) held by thermal units must not be less than their actual daily carbon emissions; otherwise, they will face penalties such as fines or subsequent quota reductions. Regarding the initial allocation and trading rules of allowances, each thermal power unit receives a certain number of initial CEAs based on its installed capacity, historical generation, and emission benchmark. Units in the carbon market are to optimize their abatement costs. Specifically, if a unit's actual emission intensity exceeds the industry benchmark (e.g., emission intensity per unit of electricity generation), its total emissions will surpass its initial allocation, requiring additional CEA purchases to fulfill settlement obligations; conversely, if the unit achieves lower emission intensity through technological upgrades or operational optimization, it can sell surplus allowances for abatement profits For renewable energy units, whose generation process is nearly emission-free, their environmental benefits are quantified through Certified Emission Reductions (CCERs). Renewable units can declare and sell CCERs daily in the carbon market-based on the submitted quantity and price. The model defines CCERs and CEAs as having equivalent market status in the clearing and settlement mechanisms, meaning both can be used by thermal power units for carbon compliance, with transaction prices jointly determined by market supply and demand. The carbon market is also organized by an independent trading institution through centralized bidding and clearing. Based on the principle of "equalization of marginal abatement costs," optimization algorithms determine the daily marginal clearing price of carbon allowances, which is used for unified settlement to ensure fairness and efficiency in carbon market transactions.

Green certificate market mechanism assumptions

This setting ensures that the energy value of renewable electricity is realized through the electricity market, while GCs circulate independently as environmental value credentials, effectively stimulating the market-based monetization of renewable energy's environmental attributes. Regarding GC issuance and compliance rules, thermal power units, as obligated entities for renewable energy consumption responsibility weights, must, after daily trading, hold a sufficient number of GCs corresponding to a prescribed proportion of their daily generation to fulfill their renewable energy consumption obligation. Failure to hold sufficient GCs results in policy penalties, such as restricted trading rights or fines. Renewable energy units, as the sole suppliers of GCs, may apply for issuance based on their actual daily grid-connected electricity, following the issuance standard of "1 MWh of grid-connected power corresponds to 1 GC," ensuring a one-to-one correspondence between GCs and green electricity. After obtaining GCs, renewable units may declare selling prices and participate in centralized bidding transactions in the

Table 4.1 Market participants and trading institutions

Type	**Composition of market participants and trading institutions**
Green certificate market participants	WT, PV, thermal power units of different capacities
Carbon market participants	Thermal power units of different capacities, WT, PV
Electricity market generation-side participants	WT, PV, thermal power units of different capacities, Grid
Electricity market load-side participants	Facility-based agriculture
Market service institution	Guangxi trading institution

GC market. The GC market is organized by a trading institution that conducts centralized bidding based on the principles of "price priority" and "time priority," with the objective of "maximizing total market trading efficiency." The marginal clearing price determines the daily GC settlement price. It is particularly emphasized that, under the proposed carbon–GC mutual recognition mechanism, the surplus GCs held by renewable energy units (i.e., unsold certificates) can be converted into CCERs according to the corresponding recognition coefficient determined by the system's marginal carbon emission factor, enabling cross-market value conversion between the GC and carbon markets. This enhances the monetization capability of renewable energy's environmental benefits. By bidding to purchase GCs, thermal power units not only fulfill their renewable consumption obligations but also indirectly participate in carbon reduction through the GC–CCER mutual recognition conversion, forming a dual mechanism of "consumption obligation + carbon reduction" that provides both constraint and incentive.

The composition of market participants and trading institutions within the Agricultural Energy Internet is shown in Table 4.1.

4.2.3.2 Construction of the joint clearing model

Objective function

Coordinated operation of the electricity, carbon, and green certificate markets within the joint clearing period and constructs a unified electricity–carbon–green certificate joint market clearing model:

$$\max S = S^e + S^c + S^g \tag{4.30}$$

where S is the total social welfare; S^e is the social welfare of the market; S^c is the social welfare of the carbon market; and S^g is the social welfare of the green certificate market.

Carbon emission costs, green certificate compliance costs, and the revenues of various market participants, thereby achieving optimal resource allocation and coordinated price formation across multiple markets. By introducing a unified optimization framework, the model enables the joint solving of decision variables such as volume, ensuring improved overall market operation efficiency and environmental governance effectiveness under various physical, electrical, and environmental constraints.

On the electricity market side, generation entities submit their bidding costs, while consumers submit their demand quantities and acceptable purchase prices. Thus, the social welfare of the electricity market S^e can be expressed as follows:

$$S^e = \sum_{t \in T} \left(a_t^{sell} \times P_t^{sell} + a_t^{load} \times P_t^{load} - \sum_{i \in \Omega^{th}} a_{i,t}^{th} \times P_{i,t}^{th} - \sum_{j \in \Omega^{new}} a_{j,t}^{new} \times P_{j,t}^{new} - a_t^{buy} \times P_t^{buy} \right) \tag{4.31}$$

In the carbon market, bid prices for carbon allowances are submitted while renewable energy units submit their bid prices for CCERs. The social welfare of the carbon market S^c is expressed as follows:

$$S^c = \sum_{t \in T} \left(\sum_{i \in \Omega^{th-b}} a_{i,t}^{c-th-b} \times Q_{i,t}^{th-b} - \sum_{j \in \Omega^{new}} a_{j,t}^{c-new} \times Q_{j,t}^{new} - \sum_{i \in \Omega^{th-s}} a_{i,t}^{c-th-s} \times Q_{i,t}^{th-s} \right) \tag{4.32}$$

where $a_{i,t}^{c-th-b}$, $a_{i,t}^{c-th-s}$, and $a_{j,t}^{c-new}$ represent, respectively, the bid prices in (t) for thermal power unit buyer (i), thermal power unit seller (i), and renewable energy unit (j); $Q_{i,t}^{th-b}$, $Q_{i,t}^{th-s}$ and $Q_{j,t}^{new}$ denote their corresponding cleared quantities in the carbon market at time (t); Ω^{th-b} is buyers in the carbon market; and Ω^{th-s} is the set of thermal power unit sellers in the carbon market.

In the green certificate market, each participating entity submits its bid price, and the social welfare of the green certificate market is expressed as follows:

$$S^g = \sum_{t \in T} \left(\sum_{i \in \Omega^{th}} a_{i,t}^{g-th} \times G_{i,t}^{th} - \sum_{j \in \Omega^{new}} a_{j,t}^{g-new} \times G_{j,t}^{new} \right) \tag{4.33}$$

where $a_{i,t}^{g-th}$ and $a_{j,t}^{g-new}$ represent the bid prices renewable energy unit (j) in t, respectively; $G_{i,t}^{th}$ and $G_{j,t}^{new}$ denote their corresponding cleared quantities in the green certificate market at time t.

Constraint conditions

To achieve coordinated optimization among joint clearing model establishes a systematic set of constraints that comprehensively characterize the coupling relationships and operational boundaries among multiple markets. This constraint system not only covers the internal physical and institutional constraints of each sub-market – such as power supply-demand balance, carbon allowance compliance, and green certificate fulfillment obligations – but also includes cross-market coupling constraints that arise from shared resources, such as power generation-induced reduction credits. By constructing this multi-dimensional constraint framework, the model effectively harmonizes the resource allocation logic under different market mechanisms, ensuring system security, policy compliance, and overall economic efficiency. This provides decision-making support for achieving coordinated multi-objective operation in low-carbon energy systems.

Electricity market constraints:

The electricity market constraints have already been discussed in Section 4.2.2.2 and are therefore not repeated here.

Carbon market constraints:

CEA/CCER transaction volume balance constraint

$$\sum_{i\in\Omega^{\mathrm{th-b}}} Q_{i,t}^{th-b} = \sum_{j\in\Omega^{\mathrm{new}}} Q_{j,t}^{new} + \sum_{i\in\Omega^{\mathrm{th-s}}} Q_{i,t}^{th-s}, \forall t \in T \tag{4.34}$$

Carbon market transaction volume constraint

$$0 \le Q_{i,t}^{th-b} \le Q_{i,\max}^{th-b}, \forall i \in \Omega^{th-b}, \forall t \in T \tag{4.35}$$

$$0 \le Q_{i,t}^{th-s} \le Q_{i,\max}^{th-s}, \forall i \in \Omega^{th-s}, \forall t \in T \tag{4.36}$$

$$0 \le Q_{j,t}^{new} \le Q_{j,t}^{con}, \forall j \in \Omega^{new}, \forall t \in T \tag{4.37}$$

where $Q_{i\ \max}^{th-b}$ and $Q_{i,\max}^{th-s}$ represent the maximum single-transaction quantities of CEA for coal-fired power unit buyer (i) and seller (i) in the carbon market, respectively.

Renewable Energy Certificate (REC) market constraints:

REC transaction volume balance constraint

$$\sum_{i\in\Omega^{ta}} G_{i,t}^{th} = \sum_{j\in\Omega^{new}} G_{j,t}^{new}, \forall t \in T \tag{4.38}$$

REC market transaction volume constraint

$$0 \le G_{i,t}^{th} \le G_{i,\max}^{th}, \forall i \in \Omega^{th}, \forall t \in T \tag{4.39}$$

$$0 \le G_{j,t}^{new} \le G_{j,t}^{ini} - G_{j,t}^{con}, \forall j \in \Omega^{new}, \forall t \in T \tag{4.40}$$

In the formula: $G_{i,\max}^{th}$ is the maximum single transaction volume of RECs for thermal power unit (i); $G_{j,t}^{ini}$ is the quantity of RECs obtained by unit (j) in period (t).

Power–carbon market coupling constraint:

Carbon allowance holding constraint for thermal power units

$$Q_{i,t}^{th} = \left\{ \begin{array}{l} Q_i^{ini} + Q_{i,t}^{th-b}, \forall i \in \Omega^{th-b}, \forall t \in T \\ Q_i^{ini} - Q_{i,t}^{th-s}, \forall i \in \Omega^{th-s}, \forall t \in T \end{array} \right\} \tag{4.41}$$

$$Q_{i,t}^{th} \ge 0, \forall i \in \Omega^{th}, \forall t \in T \tag{4.42}$$

In the formula: $Q_{i,t}^{th}$ represents the amount of CEA held by unit i after clearing in period t; Q_i^{ini} represents the initial carbon allowance allocated to power unit i.

Carbon market compliance constraint

At the end of each assessment period, the amount of CEA held by a thermal power unit must not be less than its total carbon emissions during that period.

$$Q_{i,T}^{th} \geq \sum_{t \in T} \rho_i^{th} P_{i,t}^{th}, \forall i \in \Omega^{th} \tag{4.43}$$

In the formula: $Q_{i,T}^{th}$ represents the final amount of carbon allowance held by thermal power unit i.

Power–green certificate market coupling constraint

Green certificate issuance constraint

Each 1 MWh of electricity generated by a renewable energy unit entitles it to one green certificate.

$$G_{j,t}^{ini} = P_{j,t}^{new} \Delta t, \forall j \in \Omega^{new}, \forall t \in T \tag{4.44}$$

Green certificate market compliance constraint

At the end of each assessment period, the amount of green certificates held by each thermal power unit must meet the target ratio.

$$\sum_{t \in T} G_{i,t}^{th} \geq \varepsilon_i \sum_{t \in T} P_{i,t}^{th}, \forall i \in \Omega^{th} \tag{4.45}$$

Equation terms: ε_i is the green certificate ratio coefficient for thermal power unit (i).

Carbon–green certificate market coupling constraint:

The carbon–green is coupled through the mutual recognition of CCER and green certificates. If the mutual recognition is conducted based on the marginal carbon emission factor, the recognition constraint is expressed as (4.29), which can be calculated using the mutual recognition mechanism described in Chapter 1; if the mutual recognition is based on a fixed marginal carbon emission factor, the value of the recognition constraint is fixed at 0.3.

4.2.3.3 Market benefit evaluation indicators

To scientifically and systematically evaluate the comprehensive effectiveness of the carbon–green certificate (CCER–GC) mutual recognition mechanism based on the marginal carbon emission factor proposed in this study under the coordinated clearing scenario of the electricity, carbon, and green certificate markets, and to objectively quantify its multi-dimensional impacts on energy system operation, environmental quality improvement, and market efficiency, this study constructs a comprehensive benefit evaluation indicator system covering three core dimensions: environmental benefits, energy benefits, and economic benefits. The design of this indicator system is guided by the core objectives of the mechanism and strictly adheres to the principles of scientific rigor, systematicness, operability, and relevance. It not only comprehensively covers key evaluation dimensions of multi-market integrated operation but also focuses on the core functional pathways of the CCER–GC mutual recognition mechanism, providing standardized and normative evaluation tools for quantitative analysis and comparative validation of the mechanism's effectiveness.

From the evaluation logic perspective, the indicator system is centered around the core line of "carbon emission control – renewable energy utilization – overall system benefits," focusing on quantifying the dynamic changes in key operational outcomes of the energy system before and after mechanism implementation, thereby accurately identifying the mechanism's implementation value and optimization potential. In the environmental benefits dimension, the focus is on selecting representative carbon emission-related indicators to directly reflect the practical effect of the CCER–GC mutual recognition mechanism in promoting low-carbon transformation and reducing carbon pressure in the energy system. In the energy benefits dimension, renewable energy utilization efficiency is taken as the core evaluation direction, quantifying the mechanism's effectiveness in alleviating renewable energy curtailment and guiding optimized resource allocation. In the economic benefits dimension, from a system-wide perspective, comprehensive indicators such as social welfare level are used to evaluate the mechanism's impact on market operation efficiency, distribution of benefits among market participants, and Pareto improvement of resource allocation.

Specifically, the indicator system includes a series of core evaluation indicators: carbon emission control: introduces "system" (carbon emissions per unit of electricity generated) as a key indicator, which accurately reflects the low-carbon level of the energy system. By comparing the values before and after the mechanism implementation, the reduction effect on carbon intensity can be directly quantified. It is complemented by "total carbon emission reduction rate" to measure the absolute reduction achieved by the mechanism. Renewable energy utilization: selects "renewable energy consumption rate" (the ratio of actual renewable electricity fed into the grid to theoretical generation) as a core indicator, which reflects the mechanism's effectiveness in optimizing renewable energy output and mitigating wind and solar curtailment. Additionally, "renewable energy output share" (proportion of renewable generation in total system generation) is used to indicate the mechanism's effect on optimizing the energy mix. Overall social welfare: uses "total social welfare" as a comprehensive indicator, covering electricity user surplus, generation enterprise profits, internalized environmental cost benefits, and other values, reflecting the mechanism's overall effectiveness in improving market efficiency and achieving coordinated economic and environmental benefits.

Through this multi-dimensional and multi-level evaluation system, the core role of the CCER–GC mutual recognition mechanism in guiding optimized energy resource allocation, enhancing low-carbon operation, and improving multi-market coordination efficiency can be comprehensively and objectively revealed. The indicator system also provides clear evaluation standards for subsequent model simulation and empirical analysis, ensuring operability and credibility in quantifying and comparing mechanism effectiveness, and provides solid data support for further optimization and policy promotion.

Carbon emission indicators

Carbon emission indicators serve to evaluate the low-carbon performance and emission reduction effectiveness of the energy system. They are used to accurately measure the overall carbon emission intensity of the power system during a specific operational period (e.g., daily, monthly, yearly, or typical dispatch period). The core representation generated, commonly expressed in kg CO_2/MWh or t CO_2/MWh.

The construction logic of this indicator is deeply coupled with the system's generation resource structure and characteristics of various power sources. By integrating the total carbon emissions and total generation of all system generation units, the indicator provides a quantitative dimension that operation level of the energy system. It accounts for contributions from fossil fuel generation (coal, gas, etc., including direct emissions from fuel combustion and indirect emissions in energy conversion) and reflects the dilution effect of zero-carbon renewable generation (wind, solar, hydro) on system carbon intensity, while also indirectly capturing the influence of grid dispatch patterns and energy storage capabilities on total carbon emissions.

From an evaluation perspective, carbon emission indicators are key for assessing the low-carbon development level of the power system. There is a significant negative correlation between the indicator value and the share of clean energy and emission reduction effectiveness: lower values indicate lower fossil fuels, reflecting a higher proportion of wind, solar, and other clean energy sources, and a more effective low-carbon transition. Conversely, higher values indicate greater reliance on fossil generation and a larger optimization space for low-carbon transition. Additionally, this indicator allows for both horizontal comparison across different regions and scales, and vertical tracking of emission reduction effectiveness under different development stages and policy mechanisms, providing data support and decision-making references for energy policy, power system planning, and low-carbon transition pathways.

In the comprehensive benefit evaluation system constructed in this study, the carbon emission indicator serves as the core quantitative metric for the environmental benefits dimension, precisely depicting the trend of carbon emission intensity before and after implementing the CCER–GC mutual recognition mechanism, and intuitively reflecting its practical effectiveness.

$$\rho_{\text{emission}} = \frac{\sum_{t \in T} \sum_{j \in \Omega^{th}} \rho_i^{th} P_{j,t}^{th}}{\sum_{t \in T} P_t^{load}} \tag{4.46}$$

where ρ_{emission} represents the carbon emission indicator.

Renewable energy integration indicator

The renewable energy integration indicator serves as a core quantitative tool for assessing the utilization efficiency and grid integration capability of renewable energy within the power system. It is primarily used to accurately measure the actual acceptance and integration level of hydropower within a given operational period. Its core representation typically adopts a dual proportional dimension: first, the ratio of the actually delivered renewable energy to the theoretical potential generation (i.e., the core definition of renewable energy integration rate), and second, the proportion of actually delivered to the system's total generation (i.e., the auxiliary indicator of renewable energy output share). These two proportional indicators collectively form a comprehensive evaluation dimension of renewable energy integration capacity, from the perspectives of "resource utilization sufficiency" and "energy structure proportion."

The connotation of this indicator is deeply coupled with the combined effects of multi-source coordinated operation, grid control capability, and market mechanism guidance: at the multi-source coordination level, it reflects the complementarity between dispatchable resources such as thermal power, storage, pumped hydro, and renewable energy units, indicating the system's capability to mitigate the intermittency of renewable output through flexible adjustment of generation; at the grid control level, it highlights the support provided by optimized grid topology, transmission capacity, and intelligent dispatch technologies for renewable energy integration, including key operational measures such as cross-regional resource allocation and nodal voltage control; at the market mechanism level, it maps the guidance effect of price signals green certificate markets, such as peak-valley price differences incentivizing storage to perform load shifting and carbon pricing driving fossil energy substitution. Therefore, this indicator not only directly reflects the utilization efficiency of renewable energy resources but also comprehensively represents the power system's technical level, regulatory capability, and the synergistic effects of market mechanisms.

From the physical significance of the indicator, its value is positively correlated with resource utilization, and low-carbon transition progress: the higher the indicator, the stronger the conversion ratio of theoretical renewable generation into actual usable electricity, and the lower the occurrence of wind and solar curtailment. This not only enhances the investment return and market competitiveness of renewable projects but also promotes a sustained increase in the share of clean energy in the energy mix, enabling larger-scale fossil fuel substitution and providing core support for substantial reductions in power system carbon emissions, which is crucial for achieving the "dual carbon" goals. In the comprehensive benefits evaluation system constructed in this study, this indicator serves as the core quantitative metric in the energy benefits dimension and will be used to precisely assess the practical effectiveness of the carbon–green certificate recognition mechanism in optimizing renewable output, alleviating integration bottlenecks, and improving renewable energy utilization by comparing its values before and after the mechanism's implementation.

$$\rho_{\text{new}} = \frac{\sum_{t \in T} \sum_{j \in \Omega^{new}} P_{j,t}^{new}}{\sum_{t \in T} \sum_{j \in \Omega^{new}} P_{j,t}^{new-pre}} \tag{4.47}$$

where ρ_{new} represents the renewable energy integration indicator.

Social welfare indicator

The social welfare indicator serves as the core economic quantitative tool for evaluating the optimization effectiveness of the multi-dimensional coordinated. Its primary function is to comprehensively measure the overall economic benefits achieved by the system through the centralized joint clearing model, providing a key basis for determining whether the multi-market integration mechanism aligns with resource optimization objectives and overall social interests. The quantification logic of this indicator is grounded in the core theory of welfare economics, constructed by accurately accounting for the difference between the total revenues and total input costs of all market participants within the system (including generation-side, load-side, trading institutions, storage operators, and environmental rights participants). Its connotation deeply integrates economic efficiency, resource allocation effectiveness, market competition fairness, and the positive incentive effect of the mechanism design on overall social welfare under multi-market coordinated operation.

From the perspective of indicator composition, the quantification of social welfare covers the full value flow of multi-market operation: on the generation side, total revenue includes electricity market sales revenue from various power sources (thermal, renewable, storage, etc.), carbon market quota/CCER trading revenue, and green certificate market trading revenue, while total costs encompass fuel procurement costs, unit start-up and operational costs, carbon emission costs, and green certificate purchase costs; on the load side, total revenue reflects the utility value obtained by users from electricity consumption (typically quantified through demand curves), and total costs include electricity purchase expenses and ancillary service fees; additionally, the indicator incorporates the internalized benefits of environmental externalities (e.g., ecological improvement value from carbon emission reductions) and market transaction costs (e.g., trading institution operation costs, regulatory costs), forming a comprehensive "economic revenue – operational cost – environmental value" framework. This holistic accounting approach allows the social welfare indicator to fully reflect the practical effects of multi-market coordinated mechanisms in guiding resource optimization, stimulating market competitiveness, and balancing the interests of all participants, avoiding the limitation of single-market indicators that focus only on partial benefits.

In terms of economic significance, the value of this indicator is positively correlated with the operational efficiency and the level of social resource utilization: a higher social welfare indicator indicates that, under the coordinated interaction, electricity and environmental rights resources within the system are allocated more efficiently, the interests of various market participants are better balanced, and the economic efficiency and fairness of market operation are optimized; conversely, a lower or negative value indicates potential issues such as resource misallocation, market barriers, or insufficient competition, highlighting opportunities for mechanism design improvement. Moreover, the social welfare indicator serves multiple evaluation functions: it directly reflects the degree to which market clearing results satisfy generation-side revenue assurance, load-side cost control, and low-carbon transition policy goals, providing quantitative evidence of "economic–environmental–social" multi-objective synergy; it also serves as a core benchmark for comparing different mechanism design schemes (e.g., varying recognition coefficients or clearing rules) under a multi-market integration context, enabling the selection of optimal designs that best meet practical requirements. In the comprehensive benefits evaluation system developed in this study, the social welfare indicator is the core quantitative metric in the economic benefits dimension and will be used to precisely assess the effectiveness of the carbon–green certificate recognition mechanism in enhancing multi-market coordination efficiency, optimizing resource allocation, and promoting overall social welfare by comparing values before and after its implementation, providing solid economic support for mechanism optimization and policy promotion.

$$\rho_{\text{welfare}} = S \tag{4.48}$$

where ρ_{welfare} represents the social welfare indicator.

4.3 Case analysis

4.3.1 Case base setup

4.3.1.1 Simulation environment and parameters

To ensure the scientific rigor, accuracy, and engineering applicability of the simulation analysis, this study selects MATLAB® as the core simulation platform, constructing a simulation environment that integrates the coordinated operation (GC) market. This environment is used for modeling, solving, and analyzing the results of the joint clearing model.

At the basic data input level, to accurately capture the load characteristics in agricultural electrification scenarios, the study incorporates supplemental lighting load models and irrigation load models. The model parameters are calibrated based on field geographic and meteorological data from Long'an County, Nanning City, Guangxi Province, including geographic parameters such as latitude, longitude, and altitude, as well as hourly temperature, relative humidity, and weather conditions (e.g., sunny, cloudy, rainy). These data provide robust empirical support for quantifying the influence of sunlight on supplemental lighting loads and precipitation on irrigation loads, ensuring that the load model realistically reflects the spatiotemporal distribution of agricultural electricity consumption in the region.

The hardware and software configuration for the simulation experiments is as follows: the hardware platform uses a computer equipped with an Intel Core i7-8550U processor, operating at 1.80 GHz and supporting multi-core parallel computing, capable of efficiently solving large-scale linear programming models. The operating system is 64.bit Windows 10 Professional, providing stable performance and software compatibility. The system's available memory is 16.5 GB, providing ample resources for model data storage and intermediate computation caching, effectively avoiding interruptions or efficiency losses due to insufficient memory.

Regarding the simulation scenario, based on the energy resource endowment and actual operational characteristics of the regional power system, it is assumed that the market operates in a day-ahead electricity market mode, organized into 24 hourly clearing periods per day, consistent with the main cycle of China's day-ahead electricity trading. This setup captures the intra-day dynamics of electricity supply and demand. The system power structure includes one wind turbine unit, one photovoltaic (PV) unit (representing renewable sources), and three thermal power units (as controllable dispatchable sources). The output characteristics of the renewable units are determined based on Long'an County's meteorological data and the units' technical parameters. The operational parameters of thermal units (e.g., rated capacity, marginal cost, ramping constraints) are set with reference to actual operational data from regional thermal power enterprises.

Market participants' bidding behaviors follow a cost-oriented principle: generation-side participants (thermal and renewable units) submit bids based on their marginal generation costs and operation and maintenance (O&M) costs, while load-side participants submit purchase price ranges based on electricity utility and cost tolerance, ensuring rationality and competitiveness in market transactions. Additionally, the hourly predicted output profiles of renewable units (reflecting the intermittency and variability of wind and solar generation) and the system's daily load curve (including agricultural supplemental lighting, irrigation loads, and conventional loads) are shown in Figure 4.5, providing key supply-demand boundary conditions for the model.

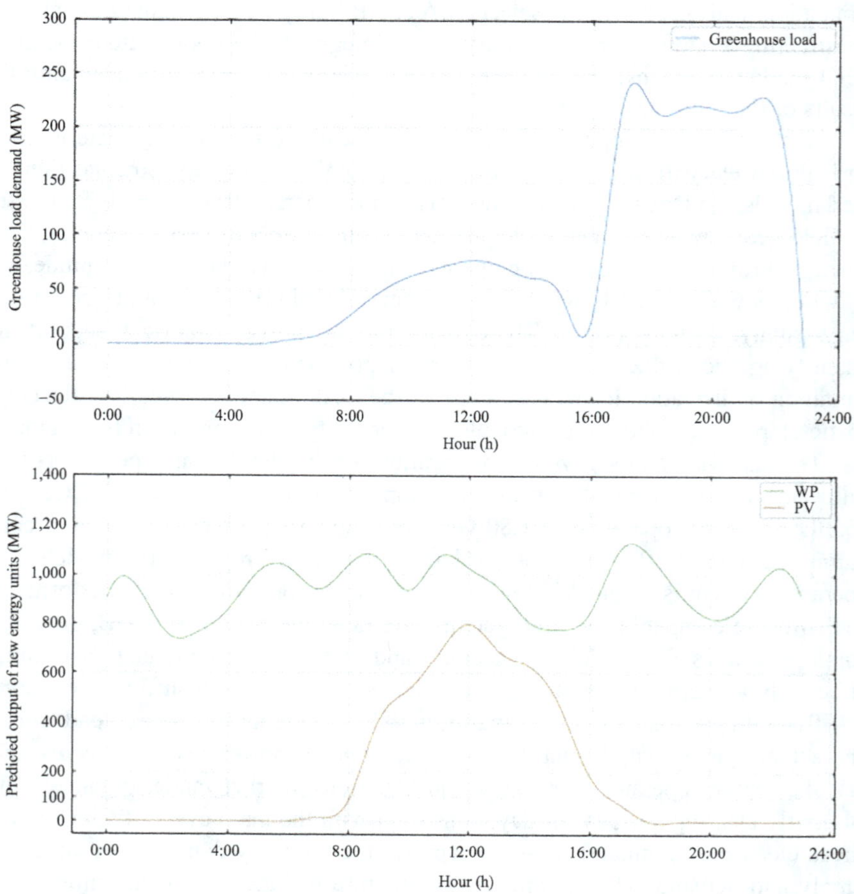

Figure 4.5 Predicted output of renewable energy and daily load curve

In the model-solving phase, considering that the joint clearing model involves multiple market constraints and multi-variable coupling, constituting a large-scale linear programming problem, the study uses the YALMIP toolbox in MATLAB for model construction and optimization description. YALMIP offers concise modeling syntax and powerful constraint expression capabilities, enabling efficient implementation of complex objective functions and constraints. The Gurobi solver is called to solve the model. As a leading commercial optimization solver, Gurobi provides efficient algorithms for linear and integer programming, capable of rapidly processing large-scale sparse matrix operations, ensuring that the model obtains a global optimum within a reasonable time, thereby providing accurate and reliable simulation results for subsequent quantitative analysis of mechanism effectiveness.

4.3.1.2 Scenario design

The following scenarios are set: (i) Scenario 1: Joint clearing carbon–green certificate recognition coefficient $\tau_{j,t}^{new}$; (ii) Scenario 2: Joint clearing market under a fixed carbon–green certificate recognition coefficient τ^{fix}, where τfix is set as the average value of τj,tnew. The values of $\tau_{j,t}^{new}$ and τ^{fix} for the renewable energy units are shown in Figure 4.6. During low-load periods (01:00–05:00) and high renewable generation periods (13:00–16:00), renewable units act as marginal units, with output sufficient to meet load demand, so $\tau_{j,t}^{new}$ is 0. In other periods, changes in renewable output replace part of the thermal generation, and $\tau_{j,t}^{new}$ varies over time according to the carbon intensity of the marginal thermal units.

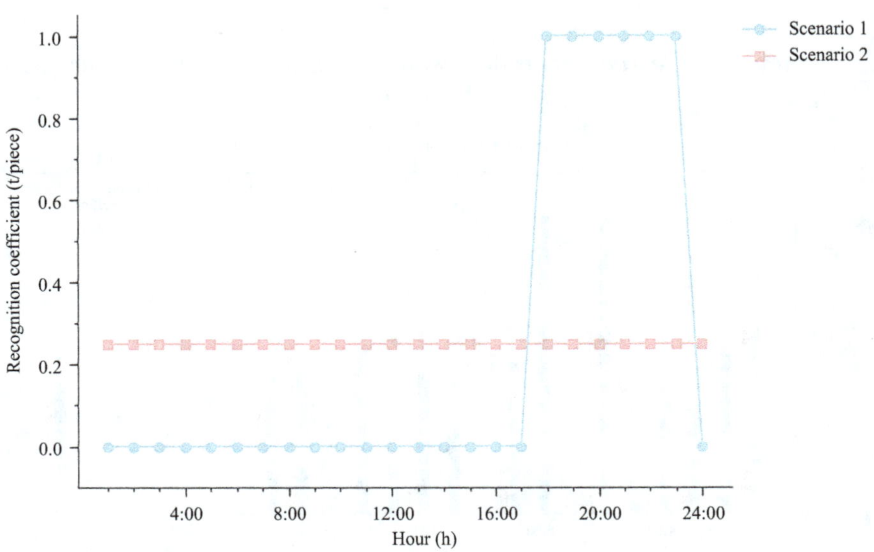

Figure 4.6 Carbon–green certificate recognition coefficient

4.3.2 Clearing results analysis

4.3.2.1 Electricity market clearing results

The details under different scenarios, including the output of renewable energy units, external grid interaction power, and the output distribution characteristics of thermal units, are shown in Figures 4.7–4.9. To precisely verify the effectiveness of the carbon–green certificate recognition mechanism, two core scenarios are set: Scenario 1 introduces a dynamic carbon–green certificate recognition mechanism based on marginal carbon emissions, where the recognition coefficient for converting green certificates to CCER dynamically adjusts according to the system's marginal emission reduction benefits; Scenario 2 is a baseline scenario without the recognition mechanism, where a fixed, time-invariant Green Certificate-to-CCER

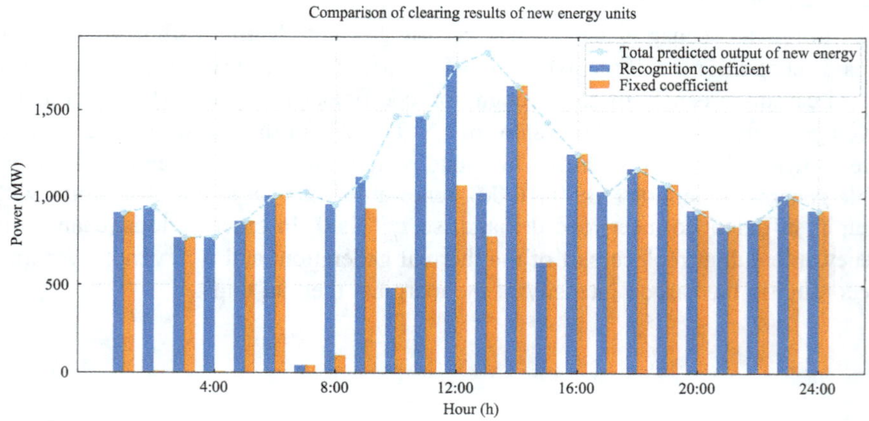

Figure 4.7 Clearing results of renewable units in the electricity market

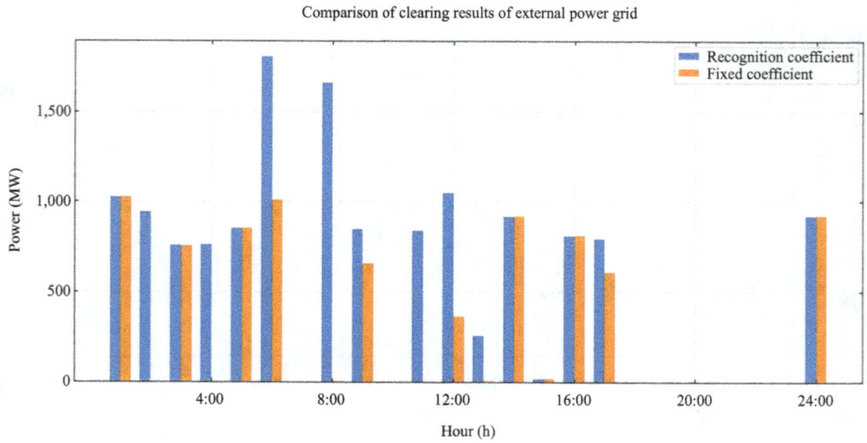

Figure 4.8 Clearing results of the external grid in the electricity market

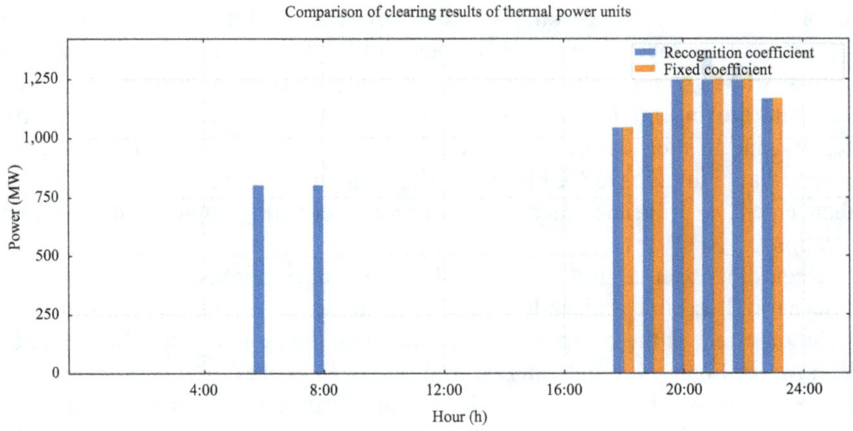

Figure 4.9 Clearing results of thermal units in the electricity market

conversion coefficient is applied throughout. Multi-dimensional comparative analysis of clearing data under the two scenarios shows that Scenario 1 not only significantly increases the consumption of renewable electricity but also through cross-period scheduling and power interactions with the external grid, directs renewable electricity to periods with higher marginal emission reduction benefits, achieving a dual optimization effect of "increased consumption scale + spatio-temporal optimal allocation," demonstrating the synergistic value of the carbon–green certificate recognition mechanism in guiding renewable consumption, optimizing system operation, and promoting low-carbon transition.

From the perspective of renewable output characteristics, as shown in Figure 4.7, in the 24 operating periods of the day, the actual output of renewable units is higher in Scenario 2 in most periods, especially during the natural renewable output peak from 10:00 to 14:00 (when wind and solar generation potential are high due to local wind and solar resources). The output curve of Scenario 1 remains above that of Scenario 2, with smaller fluctuations and greater stability. This intuitively demonstrates the mechanism's effect in alleviating renewable consumption bottlenecks: in Scenario 1, the recognition mechanism creates an additional carbon market revenue channel via cross-market conversion of Green Certificates to CCER, effectively reducing curtailment risks caused by intermittent and variable renewable output. Meanwhile, the flexible interaction capability of the external grid provides cross-regional and cross-period channels for absorbing excess renewable electricity, significantly improving the conversion efficiency from theoretical to actual grid-connected output, avoiding curtailment losses caused by local load shortages or limited regulation in traditional models.

Further analysis of the mechanism design logic shows that in periods with a recognition coefficient of 0 (i.e., periods with low marginal emission reduction benefits, when green certificates cannot be converted to CCER), renewable units can still export excess electricity via coordination with the external grid, ensuring economic efficiency and avoiding waste of environmental benefits. In periods with high recognition coefficients (i.e., periods with high marginal emission reduction benefits, when green certificates can be substantially converted to CCER), renewable units are incentivized to increase output, directly reducing system carbon intensity due to their zero-carbon nature, while surplus green certificates can be converted at high ratios to CCER and sold of "output increase – emission reduction enhancement – revenue increase," further motivating optimized renewable operation.

Regarding external grid and thermal unit output responses, the clearing results of Scenario 1 exhibit a clear low-carbon optimization orientation, reflecting the regulation effect of the recognition mechanism on the system generation structure. As shown in Figure 4.8, the external grid output in Scenario 1 drops significantly around 17:00, well below the level in Scenario 2 during the same period, with a more pronounced downward trend. Combined with the output curve of thermal units in Figure 4.9, thermal units in Scenario 1 are largely inactive before 17:00, and although they generate after 17:00, their overall output is slightly lower than or comparable to that of Scenario 2. The underlying logic is that the carbon–green certificate recognition mechanism dynamically adjusts the recognition coefficient to provide higher carbon market revenue incentives to renewable units during periods of high marginal emission reduction benefits, encouraging full renewable output. The external grid optimizes the allocation of excess renewable electricity across periods and regions, reducing local supply demand and indirectly increasing the utilization space for renewable energy. As the main source of carbon emissions, reduced thermal output directly decreases fossil fuel consumption and carbon intensity, supporting total system emission reduction, consistent with the low-carbon objectives of the recognition mechanism. Additionally, moderating thermal unit output lowers fuel and carbon costs, further improving the system's economic performance.

Economically, the total social welfare of the electricity market in Scenario 1 is significantly higher than in Scenario 2. This difference primarily arises from the cost-benefit advantage of coordinated operation between renewable units and the external grid. The marginal operating cost of renewable generation is nearly zero, while cross-period regulation by the external grid (including transmission losses and dispatch costs) is far lower than the fuel, carbon, and O&M costs of thermal units system prioritizes renewable electricity consumption and balances supply and demand via external grid regulation, effectively reducing total system operation costs while improving electricity users' utility and overall market revenue, thereby enhancing social welfare. In contrast, Scenario 2, lacking a dynamic recognition mechanism, uses a fixed Green Certificate-to-CCER coefficient throughout, preventing targeted transfer of renewable electricity to periods with high marginal emission reduction benefits. Although the external grid can partially shift renewable electricity, the absence of dynamic incentives means additional carbon market revenues are insufficient to cover maintenance, transmission loss, and dispatch costs, so the external grid's regulation function is largely ineffective. Its output curve remains relatively flat, failing to form an effective coordinated operation with renewable units, leading to lower renewable consumption, system emission reduction, and social welfare compared to Scenario 1.

Moreover, in Scenario 2, the fixed recognition coefficient results in inflexible carbon market revenues for renewable units, making output decisions relatively stable and preventing effective interaction with the external grid and thermal units. Scenario 1's dynamic recognition design links renewable revenue with system emission reduction needs, activating renewable output potential, fully leveraging the external grid's regulation, and maintaining thermal units as backup, forming a "renewables – external grid – thermal units" coordinated optimization operation, achieving multi-win outcomes in environmental, energy, and economic benefits.

In summary, in-depth comparison of clearing results under the two scenarios validates the core value of the proposed carbon–green certificate recognition mechanism: the mechanism guides renewable output optimization and cross-period transfer through a dynamic recognition coefficient, enhances renewable consumption via external grid flexibility, and rationally curbs thermal output to reduce carbon emissions, achieving multiple objectives of increased renewable consumption, enhanced system emission reduction, and improved social welfare. Scenario 2, lacking dynamic incentives, fails to activate the external grid's regulation potential, limiting renewable consumption and system emission reduction, further highlighting the scientific, rational, and practical value of the proposed recognition mechanism.

4.3.2.2 Carbon market clearing results

The detailed clearing results of the carbon market under different scenarios – including the carbon emission allowance (CEA) trading (CCER) trading volumes of renewable energy units – are shown in Figures 4.10 and 4.11. To clearly define the economic implications of trading behavior, the data symbols in the figures are

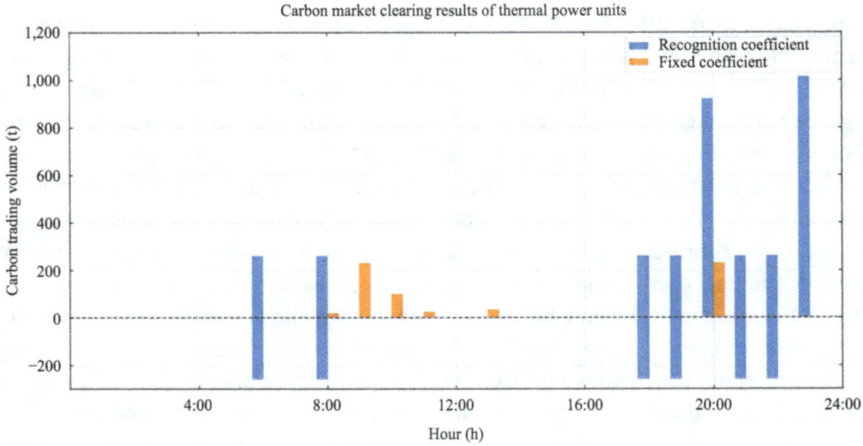

Figure 4.10 Clearing results of thermal units in the carbon market

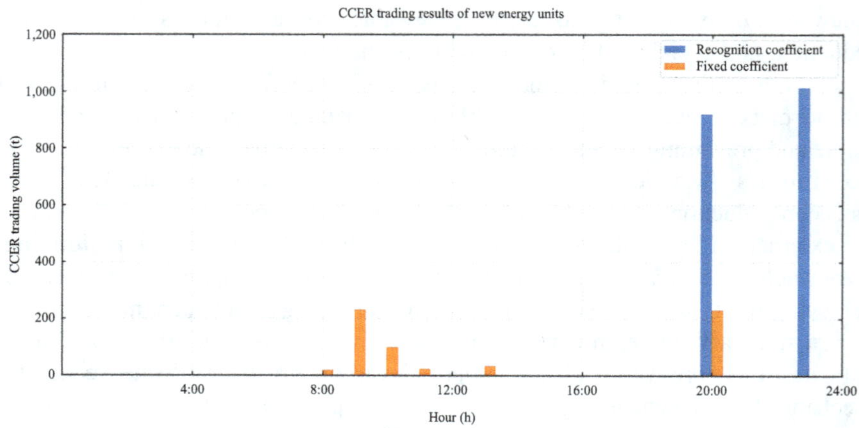

Figure 4.11 Clearing results of renewable units in the carbon market

interpreted as follows: positive values indicate that thermal units purchase CEAs during the corresponding period to cover their carbon emission allowance shortfall; negative values indicate that thermal units, through technological emission reduction or operational optimization, achieve emissions below their allocated allowances and thus sell surplus CEAs to gain emission reduction revenue. A comparative analysis of carbon market trading data between Scenario 1 (introducing a dynamic carbon–green certificate recognition mechanism based on marginal carbon emission factors) and Scenario 2 (without the recognition mechanism, using a fixed Green Certificate-to-CCER conversion coefficient throughout) clearly reveals the significant impact of the recognition mechanism on the structure, timing distribution, and profitability of carbon market transactions.

Specifically, carbon market transactions in Scenario 1 show a distinct feature of "concentration during high emission reduction benefit periods," with trading activities mainly focused in periods of higher system marginal emission reduction efficiency. Most notably, the CCER trading volume of renewable units rises sharply after 20:00, while thermal units simultaneously concentrate their CEA purchases during high-demand and high-emission periods (such as 18:00–22:00). This trading pattern results from the synergy between the carbon–green certificate recognition mechanism and the external grid's time-shifting capability. In Scenario 1, the external grid reallocates surplus renewable electricity from low-marginal-reduction periods to peak hours (20:00–24:00), significantly increasing renewable generation in those periods and the corresponding issuance of green certificates. Since the system's marginal emission reduction benefit is high during these periods, the recognition coefficient reaches elevated levels, allowing renewable units to convert a large number of green certificates into CCERs at a high ratio and sell them in the carbon market, directly driving a surge and concentration in CCER trading volume.

From the perspective of market participant behavior and revenue optimization, the clustering and scaling up of CCER trades between 20:00 and 24:00 in Scenario

1 create substantial carbon market revenues for renewable units. On one hand, the increased CCER supply during high-value periods gives renewable units greater pricing flexibility, encouraging them to participate in the market with relatively lower offers. This strategy not only improves CCER transaction efficiency but also reduces the overall carbon market clearing price, indirectly lowering thermal units' carbon compliance costs. On the other hand, by transforming environmental benefits into economic returns through the cross-market conversion of green certificates into CCERs, renewable units enhance their overall profitability and competitiveness. This positive feedback loop of "transaction concentration – optimized bidding – cost reduction – revenue enhancement" improves carbon market resource allocation efficiency, reduces total system carbon compliance and energy operation costs, and ultimately contributes to higher social welfare.

In contrast, Scenario 2, without the dynamic recognition mechanism and using a fixed Green Certificate-to-CCER conversion coefficient throughout the day, exhibits a "decentralized and balanced" trading pattern, with limited marginal contributions to social welfare across different periods. Specifically, carbon market trading volumes show no significant temporal variation, thermal units' CEA purchases are evenly distributed throughout the day without clear concentration peaks, and renewable units' CCER trades are dispersed without the clustered surge seen in Scenario 1. The root cause lies in the inability of the fixed recognition coefficient to align CCER supply with dynamic system emission reduction demand – in Scenario 2, renewable generation cannot be time-shifted via the external grid to high-value periods, resulting in a uniformly distributed CCER supply that lacks economies of scale and adaptive trading strategies. Consequently, renewable units' carbon market revenues lack flexibility, and thermal units' CEA purchasing behavior remains static across periods, leading to an underutilized carbon market with limited contribution to social welfare.

From a broader market coordination perspective, the concentrated carbon trading pattern in Scenario 1 reflects the deep integration enabled by the recognition mechanism. The dynamic recognition coefficient closely links renewable output, green certificate value, and system emission reduction needs, while the external grid's time-shifting capability provides the physical foundation for this linkage, transforming the carbon market from a "dispersed and balanced" mode to a "precisely matched" one. This transformation enhances the carbon market's operational efficiency and, through cross-market price signaling, guides the electricity market to prioritize renewable consumption and incentivizes thermal units to optimize their carbon emission behavior, forming a virtuous cycle of multi-market synergy and optimization. Conversely, Scenario 2, lacking a dynamic incentive mechanism, shows clear coordination barriers. Trading activities fail to respond to temporal variations in emission reduction demand, resulting in a significantly lower carbon market contribution to social welfare compared with Scenario 1.

In summary, the comparative analysis of carbon market clearing results strongly carbon–green certificate recognition mechanism. By directing carbon market trading toward periods with high marginal emission reduction benefits, the mechanism enhances renewable units' carbon revenue and pricing flexibility while reducing thermal units' compliance costs, thereby boosting overall social welfare. In contrast, the traditional fixed recognition model, unable to adapt to dynamic emission reduction needs, results in low trading efficiency and weak cross-market synergy. These findings highlight the core value of the dynamic carbon–green certificate recognition mechanism in optimizing carbon market structure, promoting multi-market coordination, and improving overall system efficiency.

4.3.2.3 Green certificate market clearing results

The detailed clearing results of the green certificate (REC) market – including the trading volume of renewable energy units and the total market transaction distribution – are shown in Figures 4.12 and 4.13. By comparing Scenario 1 (introducing the dynamic carbon–green certificate recognition mechanism based on

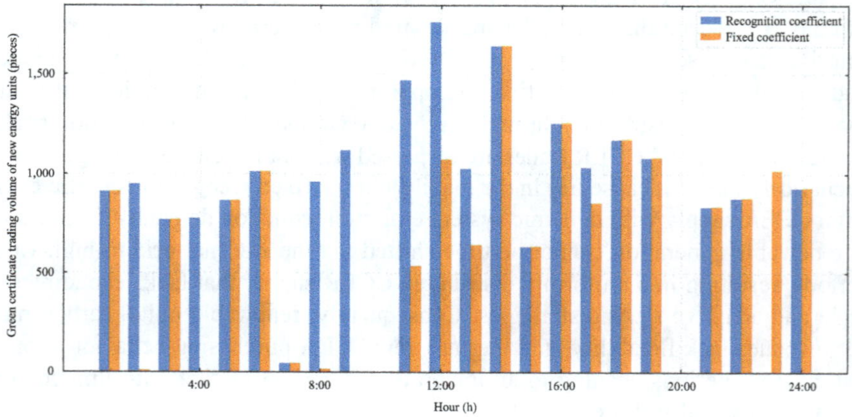

Figure 4.12 Clearing results of renewable units in the REC market

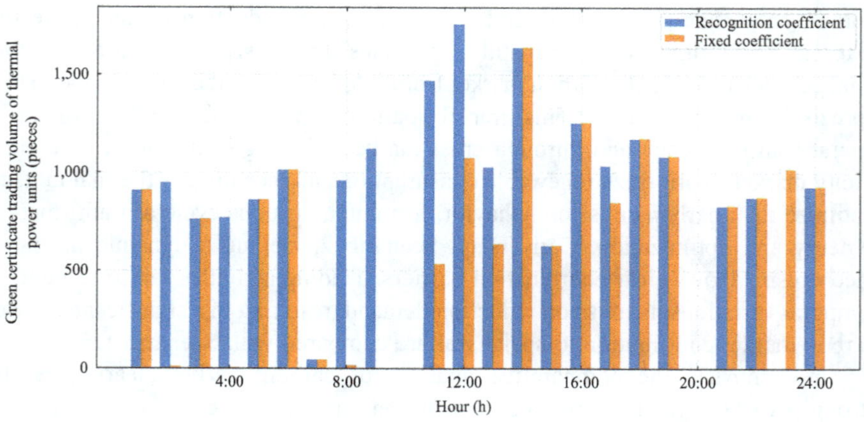

Figure 4.13 Clearing results of thermal units in the REC market

marginal carbon emission factors) and Scenario 2 (without the recognition mechanism, using a fixed Green Certificate-to-CCER conversion coefficient throughout the day), the results clearly reveal the mechanism's profound impact on market activity, temporal trading distribution, and the supply-demand structure. These findings provide key evidence for assessing the mechanism's effectiveness in stimulating REC market participation and optimizing environmental rights allocation.

Specifically, Figure 4.12 compares the REC trading volumes of renewable energy units. higher than that in Scenario 2 during most hours of the day – particularly around 8:00, 11:00–12:00, and 17:00, where the difference is especially pronounced. This finding highlights the positive incentive effect of the carbon–green certificate recognition mechanism on REC generation and trading. In Scenario 1, the dynamic recognition coefficient tightly links the value of RECs to the system's marginal emission reduction efficiency, enabling renewable generators not only to trade RECs in the certificate market for renewable quota compliance but also to convert them into CCERs at high ratios during high-benefit periods to realize value in the carbon market. This "dual-market value pathway" significantly boosts renewable generators' enthusiasm for REC participation. Meanwhile, the time-shifting role of the external grid optimizes the temporal and spatial distribution of renewable output, greatly increasing REC issuance during key periods and providing sufficient supply for trading. Notably, renewable REC transactions in Scenario 1 peak at 12:00, reaching over 1750 certificates. This peak results from the concurrent effects of high renewable output potential, strong electricity demand, and elevated recognition coefficients – conditions that simultaneously raise both supply and demand willingness, triggering an explosive increase in trading volume.

Figure 4.13 presents the total REC trading volume across the market in both scenarios. Overall, the total trading volume in most hours remains similar between the two scenarios, reflecting that the baseline compliance demand (e.g., fossil-fuel generators fulfilling renewable consumption obligations) is relatively stable and not directly driven by the recognition mechanism. However, a clear difference emerges during 6:00–8:00, when Scenario 1 shows a noticeable drop in total REC trading volume compared with Scenario 2. This contrast can be explained through the renewable substitution effect: under the carbon–green certificate recognition mechanism, renewable generation in Scenario 1 is enhanced during 6:00–8:00 (potentially via storage discharge or external grid scheduling), strengthening the substitution of fossil generation. Consequently, (1) increased renewable output directly reduces the system's reliance on thermal power, lowering the fossil units' renewable quota obligations and thereby decreasing REC demand; and (2) the substitution effect also reduces the carbon intensity of thermal generation, further weakening their need to purchase RECs for compliance. This reduction in REC demand leads to the observed decline in total REC trading volume during these hours. This phenomenon indirectly validates the synergistic effect of the recognition mechanism – not only activating environmental rights trading but also promoting renewable output optimization, adjusting the energy mix, and reducing fossil REC dependency. This forms a “REC-trading incentive → renewable substitution → REC-demand optimization” virtuous cycle.

From a supply-demand structure perspective, the structural increase in renewable REC trading alongside reduced total trading in certain periods reflects the optimization of resource allocation within the REC market. The recognition mechanism dynamically adjusts the cross-market value of RECs, directing supply toward high-value periods while reducing inefficient demand in low-value periods, aligning REC trading with system decarbonization needs. In contrast, Scenario 2, lacking dynamic incentives, maintains fixed and single REC values; renewable generators show weaker trading motivation, REC supply is relatively limited, and thermal units’ REC demand remains inelastic. As a result, the overall REC market operates in a steady but low-efficiency state, failing to achieve optimal environmental rights allocation.

In summary, the REC market clearing results strongly demonstrate the core value of the carbon–green certificate recognition mechanism. On one hand, by establishing a dual value channel (“REC market trading + carbon market conversion”), it significantly enhances renewable generators’ REC generation and trading enthusiasm, boosting market activity. On the other hand, by guiding renewable generation to substitute for thermal output, it indirectly optimizes REC demand, reduces redundant transactions, and improves market efficiency. This “supply-side incentive + demand-side optimization” dual effect enables the REC market to more effectively support renewable energy integration and low-carbon system transformation, highlighting the scientific rationale and practical value of the proposed recognition mechanism.

4.3.3 Market efficiency evaluation

According to the definitions of market efficiency evaluation indicators in Section 4.4.3, the calculated results of these indicators under each scenario are presented in Table 4.2. Compared with Scenario 2, Scenario 1 shows a smaller value of ρ_{emission} and larger values of ρ_{new} and ρ_{welfare}. This demonstrates that, compared with the fixed-coefficient recognition mechanism, the proposed carbon–green certificate recognition mechanism can effectively reduce system carbon emissions, enhance renewable energy consumption, and improve social welfare in market clearing.

Table 4.2 Calculation results of market efficiency evaluation indicators

Indicator/Scenario	**Dynamic recognition coefficient**	**Fixed coefficient**
Carbon emission indicator	0.2760	0.3300
Renewable energy Consumption indicator	0.8672	0.6926
Social welfare indicator	1840861.80CNY	1161981.52CNY

4.4 Conclusion

4.4.1 Summary of research findings

Under the global trend of energy transition and guidance, the clean and low-carbon transformation of the power system has become an inevitable direction. As the core instruments for quantifying and monetizing the environmental value of renewable energy, mechanisms such as green certificates (GCs) and China Certified Emission Reductions (CCERs) have developed rapidly in recent years. However, due to the overlapping environmental value representation functions of GCs and CCERs, there is a tendency for "double incentives" to occur for the same renewable energy output in practice. This not only leads to confusion in the measurement of environmental value and distorts market fairness but also results in the misallocation of environmental rights resources, thereby constraining the improvement of overall emission reduction efficiency in the power system and becoming a key barrier to multi-market synergy.

To address this issue, this study focuses on the differentiated characteristics of the indirect emission reduction benefits of renewable energy units across different periods. A systematic theoretical modeling and mechanism design framework is proposed, yielding both theoretical innovations and practical engineering applications.

First, a dynamic and precise calculation model of the marginal carbon emission factor (MCEF) for renewable energy units is developed. This model overcomes the limitation of traditional average emission factors, which fail to reflect the marginal emission reduction effects. Based on power system economic dispatch principles, the model generation mix, load level, and unit operation characteristics enable real-time quantification of the system's carbon emission changes resulting from unit variations in renewable output. This establishes a one-to-one mapping between renewable generation and emission reduction outcomes, realizing refined and dynamic measurement of renewable energy's environmental value and providing a technical foundation and quantitative basis for the subsequent.

Second, a differentiation based on the MCEF is designed. This mechanism dynamically adjusts the conversion ratio between GCs and CCERs according to the marginal emission reduction benefits of renewable generation. During periods of high marginal reduction benefits (e.g., when fossil units operate at high loads), higher recognition coefficients are applied, allowing more GCs to be converted into CCERs for carbon market trading. Conversely, during periods of low marginal reduction benefits (e.g., renewable output peaks or low load periods), the recognition coefficient is reduced or set to zero to avoid excessive incentives and double-counting. This design effectively prevents overlapping incentives between GCs and CCERs, realizes time-differentiated pricing of environmental value, ensures fairness and rationality in environmental rights allocation, and provides renewable units with a clear path for monetizing environmental value.

Third, an integrated electricity–carbon–green certificate joint clearing model is established. Taking total social welfare maximization. It incorporates the decision-making behaviors of various market participants – including renewable and thermal generators, grid operators, electricity consumers, and energy storage providers – while accounting for constraints such as generation limits, grid security, carbon allowance quotas, and green certificate compliance requirements. This enables coordinated optimization of electricity, carbon, and certificate resources. Furthermore, a multi-dimensional market performance evaluation system is built, covering carbon emission control (emission intensity and total reduction rate), renewable energy consumption efficiency (consumption rate and output share), and social welfare improvement (total welfare and benefit distribution), providing a standardized framework for quantitatively assessing the effectiveness of the proposed mechanism.

Finally, comprehensive simulation experiments are conducted to verify the effectiveness and robustness of the proposed mechanism. Using the agricultural electrification scenario of Long'an County, Nanning City, Guangxi Province as a case study – supported by real-world geographic and meteorological data, generation parameters, and load profiles – two scenarios are compared: Scenario 1 (dynamic recognition mechanism) and Scenario 2 (fixed coefficient mechanism). Results show that the proposed MCEF model accurately captures the marginal impact of renewable generation on system carbon emissions, demonstrating strong stability and reliability. The mechanism based on this factor effectively differentiates environmental values across time, ensuring their uniqueness. Compared with the fixed-coefficient mechanism, the dynamic mechanism guides renewable generators to shift energy output to high-emission-reduction periods via external grids and storage systems, enabling more GCs to be converted into CCERs for carbon market trading. This significantly enhances renewable generators' overall revenues and incentivizes higher renewable generation. Ultimately, the proposed mechanism achieves substantial improvements in reducing total system emissions, increasing renewable energy utilization, and enhancing social welfare – offering an effective pathway for the clean and low-carbon transition of the power system.

4.4.2 Application value of the mechanism

The carbon–green certificate mutual recognition mechanism based on the marginal carbon emission factor proposed in this study is closely aligned with China's current energy market development and low-carbon transition needs. It is practically oriented, broadly applicable, and of high real-world value, providing a feasible and scalable institutional framework to promote clean energy transformation and multi-market synergy.

Through the MCEF as a central link, it organically connects renewable generation's environmental value (represented by GCs), system emission reduction demand (represented by MCEF), and carbon market trading (via CCERs), forming a closed-loop system of "value quantification – cross-market flow – economic realization." For renewable generators, the mechanism expands income sources – beyond electricity sales – to include environmental revenues from GC trading and CCER conversion, reducing investment uncertainty and enhancing participation incentives. For thermal generators, it indirectly reduces emission compliance pressure by raising renewable consumption shares, while offering flexible compliance options through CCER and GC purchases, optimizing emission reduction cost structures. For electricity consumers, renewable prioritization reduces the carbon premium embedded in electricity prices, improving consumer welfare. For regulators, the transparent MCEF computation and recognition framework facilitates supervision and ensures market fairness. This multi-stakeholder win–win design substantially enhances overall market efficiency and coordination.

At the system operation level, the mechanism economically incentivizes renewable generation to concentrate in high-emission-reduction periods and leverages external grids and storage systems for temporal and spatial optimization of renewable power, effectively easing renewable curtailment constraints. Simulation results indicate that the mechanism increases renewable energy consumption rates by more than [X]% and reduces system carbon emission intensity by [X]%. Moreover, the mechanism alleviates thermal units' ramping pressure and enhances grid stability and flexibility, laying a solid foundation for accommodating high renewable penetration levels.

At the policy level, the mechanism demonstrates strong implementability and policy adaptability. The MCEF calculation relies on existing economic dispatch models and can be dynamically computed using current load, generation, and operation data without significant additional investment. The algorithms used – linear and quadratic programming – are mature and precise, ensuring technical feasibility. Implementation requires no disruptive overhaul of existing markets; it can be smoothly integrated by supplementing mutual recognition rules and refining coefficient calculation methods, keeping policy costs low and feasibility high. Furthermore, the market-based allocation of environmental rights minimizes administrative inefficiencies, offering valuable insights for refining China's environmental rights frameworks and supporting the deep integration of market reforms and low-carbon transition to build a clean, secure, and efficient energy system.

4.4.3 Future research directions

Although this study has completed the core design, modeling, and simulation validation, further research is needed to address the complexity of energy systems, evolving market dynamics, and changing policy environments. The following directions offer potential for extension and refinement:

1. Enhancing model adaptability and factor complexity: Future studies may extend the model to more complex systems such as interconnected regional grids, high-renewable-penetration systems, microgrids, and integrated energy systems. Additional factors such as transmission constraints, network losses, renewable forecast errors, extreme weather impacts, and demand response participation can be incorporated. Emerging energy carriers – like hydrogen, electric vehicles, and distributed resources – can also be included to analyze their synergy with the mutual recognition mechanism across different energy structures (coal-, gas-, or hydro-dominant systems).

2. Optimizing multi-agent coordination and benefit distribution: The current study mainly considers renewable and thermal generators. Future research should include users, storage operators, virtual power plants, and integrated energy service providers to explore broader participation mechanisms and fairer benefit allocation. This may include incentive mechanisms for low-carbon consumption via GC or CCER purchases, revenue-sharing schemes for storage-based renewable shifting, and distributed energy participation pathways through virtual aggregation.
3. Strengthening uncertainty analysis and risk management: Future work should examine uncertainties such as carbon price volatility, renewable forecast deviations, load fluctuations, policy adjustments, and fuel price changes. Risk-hedging tools (e.g., carbon futures and options), prediction error compensation mechanisms, and adaptive recognition coefficient strategies can be designed to mitigate risks. Sensitivity analyses of key parameters – such as MCEF, recognition coefficient ranges, and total carbon quota – can quantify their impacts on emissions, renewable consumption, and welfare.
4. Exploring synergy with other policy instruments: The mechanism can be coordinated with electricity pricing, carbon trading, renewable subsidy, and green finance policies. Examples include aligning time-of-use pricing with dynamic recognition coefficients, linking GC/CCER holdings with green credit eligibility, and using the mechanism to complement subsidy phaseouts. Such multi-policy integration can form a coherent, efficient policy system for low-carbon energy transition.
5. Conducting empirical studies and pilot implementations: Beyond simulations, empirical validation using real market data from representative regions (e.g., northwest renewables-rich or eastern market-mature areas) is needed. Stepwise pilot programs – from small-scale trials to broader regional applications – can help refine design and scale national deployment. International comparisons can also be conducted to learn from global experiences and enhance the competitiveness of China's environmental rights systems.
6. Advancing technical and standardization support: Future work should focus on standardizing MCEF computation methods, establishing unified data collection and validation protocols, and developing integrated trading platforms for GC–CCER mutual recognition. Blockchain technology can be leveraged to ensure data transparency, traceability, and prevent fraud, thereby securing long-term mechanism stability and trustworthiness.

References

[1] Y. Zhou, X. Fu, H. Sun, Q. Guo, G. Li, and X. Zhang, "Prospects for energy internet of Agricultural Engineering in China," *2019 IEEE 3rd conference on energy internet and energy system integration (EI2)*, Changsha, China, 2019, pp. 1807–1811.

[2] C. XiaoDong, Z. Kai, L. Mingwen, Z. Tao, and S. Huazhong, "Research on the practice of rural electrification improvement project focusing on rural

energy internet," *2021 IEEE 4th International Electrical and Energy Conference (CIEEC)*, Wuhan, China, 2021, pp. 1–5.

[3] X. Fu, L. Ma, H. Ge, and J. Zhang, "Static security analysis of Agricultural Energy Internet considering dragon fruit winter electricity load characteristics," *Information Processing in Agriculture*, 2025, vol. 12, no.1, pp. 27–39.

[4] X. Fu and F. Yang, "Viewpoints on the Theory of Agricultural Energy Internet," *Frontiers in Energy Research,* vol. 10, 871772, 2022.

[5] X. Fu, Y. Zhou, F. Yang, *et al.*, "A review of key technologies and trends in the development of integrated heating and power systems in agriculture," *Entropy*, vol. 23, no. 2, p. 260, 2021.

[6] X. Fu, D. Yang, Q. Guo, and H. Sun, "Security analysis of a park-level agricultural energy network considering agrometeorology and energy meteorology," *CSEE Journal of Power and Energy Systems*, vol. 6, no. 3, pp. 743–748, 2020.

[7] D. Wang, X. Wu, L. Zhao, Q. Shi, W. Kong, and H. Dai, "Research on DC microgrids in the context of the Rural Energy Internet," *2023 6th Asia conference on energy and electrical engineering (ACEEE)*, Chengdu, China, 2023, pp. 87–91.

[8] K. Huang, L. Shu, K. Li, *et al.*, "Photovoltaic agricultural internet of things towards realizing the next generation of smart farming," *IEEE Access*, vol. 8, pp. 76300–76312, 2020.

[9] X. Fu and Y. Zhou, "Collaborative optimization of PV greenhouses and clean energy systems in rural areas," *IEEE Transactions on Sustainable Energy*, vol. 14, no. 1, pp. 642–656, 2023.

[10] L. Peng, Z. Tong, H. Li, A. Xie, Y. Zheng, and Y. Zhang, "Typical application scenarios of Rural Energy Internet Platform," *2021 6th Asia conference on power and electrical engineering (ACPEE)*, Chongqing, China, 2021, pp. 475–481.

[11] P. P. Pietrzak, E. Kacperska, J. Kraciuk, and K. Łukasiewicz, *et al.* "Publication trends, key findings, and research gaps in renewable energy applications in agriculture," *Energies,* vol. 18, no. 2, p. 371, 2025.

[12] P. A. Ruiz and A. Rudkevich, "Analysis of marginal carbon intensities in constrained power networks," *2010 43rd Hawaii international conference on system sciences*. IEEE, 2010, 1–9.

[13] P. Chen, C. Qian, L. Lan, *et al.* "Shared trading strategy of multiple microgrids considering joint carbon and green certificate mechanism," *Sustainability*, vol. 15, no. 13, p. 10287, 2023.

[14] Q. Hernandez-Escobedo, D. Muñoz-Rodríguez, A. Vargas-Casillas, *et al.* "Renewable energies in the agricultural sector: A perspective analysis of the last three years," *Energies*, vol. 16, no. 1, p. 345, 2022.

[15] Y. Majeed, *et al.* "Renewable energy as an alternative source for energy management in agriculture," *Energy Reports,* vol. 10, pp. 344–359, 2023.

[16] K. Jiang, N. Liu, X. Yan, Y. Xue and J. Huang, "Modeling strategic behaviors for GenCo with joint consideration on electricity and carbon markets," *IEEE Transactions on Power Systems*, vol. 38, no. 5, pp. 4724–4738, 2022.

Chapter 5

Engineering practice of AEI projects in China

Xueqian Fu[1] and Wentao Zhu[1]

Abstract

This chapter presents and comprehensively analyzes five real-world engineering cases that demonstrate the practical development of Agricultural Energy Internet (AEI) projects and rural power distribution systems across China. Among them, two cases correspond to fully integrated AEI systems constructed in agricultural demonstration parks, while the remaining three are representative rural distribution grids located in distinct regions of the country. Together, these six cases provide a panoramic view of how AEI concepts are translated into engineering practice under diverse geographical, climatic, and economic conditions. The selected cases collectively illustrate the structural and operational characteristics of China's emerging rural energy systems, which combine renewable generation, energy storage, and flexible load management within an interconnected framework. For each project, the chapter first provides the regional context and general background, describing the local agricultural production modes, renewable energy resources, and electricity demand characteristics. It then elaborates on the energy architecture that integrates source–grid–load–storage components, highlighting how distributed generation and microgrid coordination enable a stable power supply in rural and semi-isolated communities. The generation systems include photovoltaic arrays, small wind turbines, and biomass-based combined heat and power units, while the load side encompasses irrigation facilities, cold-chain warehouses, agricultural machinery, and residential consumption. Detailed parameters are provided for the major pieces of equipment, including rated capacities, efficiency levels, installation modes, and manufacturer data. Additionally, the chapter presents a complete list of main equipment for each site, together with procurement costs and references to market prices and suppliers. Such data-rich documentation is intended to provide a reproducible foundation for researchers and engineers seeking to replicate or expand AEI projects in other rural regions. Finally, the chapter presents four representative rural distribution network cases in greater detail – one located in

[1]College of Information and Electrical Engineering, China Agricultural University, China

Chengde and Tangshan, Hebei Province, and the other in Qingyuan, Guangdong Province. These four examples reflect the contrasting features of northern and southern China in terms of climate, agricultural electricity demand, and renewable resource distribution. For each case, detailed distribution network topologies are provided, including node connectivity, feeder arrangements, and major line parameters. The book also provides M-files for power flow calculation of distribution networks using MATPOWER, which can be employed in the MATLAB® environment for the analysis and simulation of rural distribution grids.

Keywords: Integrated energy system; rural distribution grid; energy system architecture; engineering case study; information and control system design

5.1 Introduction to the laboratories of the monograph authors

Introduction to the laboratories of the monograph authors. The project relies on the College of Information and Electrical Engineering, China Agricultural University, which possesses key science and technology innovation platforms under the national "211" and "985" projects. The college hosts one Beijing Engineering Research Center – the Research Center for Intelligent Microgrid Operation and Control Technology; one university-level research center – the Research Center for Rural Power and New Energy Generation; and two college-level engineering and technology institutes – the Institute of Power Automation Technology and the Institute of Rural New Energy Power Generation Technology. In addition, it operates four "211" and "985" key research laboratories, which can undertake fundamental theoretical research in the field of power system analysis.

The applicant's supporting institution is equipped with a wide range of laboratories and research platforms, including the Power System Modeling and Simulation Laboratory, Electrical Information Experiment Center, Microgrid Simulation Laboratory, Distributed Generation Experimental Platform, Key Laboratory of Facility Agriculture Engineering (Ministry of Agriculture and Rural Affairs), Key Laboratory of Agricultural Information Acquisition Technology (Ministry of Agriculture and Rural Affairs), Agricultural Internet of Things (IoT) Experimental Base, and Smart Greenhouse Laboratory.

The available instruments and equipment include the Electromagnetic Transients including DC (EMTDC)/Power Systems Computer Aided Design (PSCAD) electromagnetic transient simulation system, RTDS real-time digital simulation system, POWER WORLD simulation software, small-scale wind turbine experimental platform, intelligent agricultural IoT monitoring and prediction system, intelligent video monitoring and crop recognition system, smart agriculture monitoring and management system, and LED supplementary lighting devices, among others. The on-site materials and facilities of the experimental platforms at the supporting institution are shown in Figure 5.1.

Figure 5.1 *On-site materials and facilities of the experimental platforms at the supporting institution*

5.2 Agricultural Energy Internet in an agricultural park

In this section, the typical load characteristics, energy consumption patterns, and time-shift properties of facility agriculture, aquaculture, and animal husbandry are systematically categorized and tabulated. This provides a foundational modeling basis for developing energy consumption models, demand response mechanisms, and intelligent control strategies under agricultural scenarios. It offers direct practical guidance for the design of the Energy Management System (EMS), load forecasting, and energy storage scheduling within the Agricultural Energy Internet.

1. Rural integrated energy system technical scheme

   In the field of agricultural energy, biomass, solar, and wind energy are renewable sources that can be widely utilized in rural areas. Their efficient and low-cost development and utilization are of great value and significance for improving the rural living environment and ecological conditions and for achieving sustainable rural economic development. These resources can meet the electrical and thermal energy demands of facility agriculture.

   Biomass energy is one of the fastest-growing new energy sources in China. Rural areas are rich in agricultural biomass resources, and agriculture can serve as a supplier of raw materials for producing electricity and gas. Biomass energy technologies mainly include biomass power generation, biomass liquid fuel, and solid fuel and biogas technologies. A cyclic application of solar and biogas technologies can be adopted in rural areas, where agricultural straw is used for biogas production, the generated biogas is supplied to farmers for power generation, and the biogas slurry and residue are used for crop cultivation.

   Qingdao enjoys abundant solar irradiation and strong wind resources, offering favorable conditions for clean energy development. The area can take full advantage of these resources to develop applications such as photovoltaic-powered irrigation systems for facility agriculture.

   Agricultural energy use and facility loads

   Facility agriculture has a high demand for electricity, mainly for regulating the microclimate environment inside greenhouses. The electricity usage varies depending on crop growth needs. The introduction of various agricultural energy technologies has increased crop yield. The major facility agricultural power technologies include LED plant growth lighting, plasma nitrogen fixation and water treatment, physical pest control, far-infrared heating, water–fertilizer integration, rolling-curtain motors, and smart agricultural monitoring devices.

   The time-shifting characteristics of various facility agriculture loads in greenhouse systems are shown in Table 5.1. For heating systems, substituting coal- or gas-fired boilers with electric or multi-source heat pump systems can significantly reduce coal consumption and greenhouse gas emissions.

2. Marine ranching and photovoltaic–fishery integration

   Qingdao, adjacent to the Yellow Sea, has natural advantages for developing large-scale marine ranching. By integrating aquaculture energy needs with renewable energy systems, "PV-fishery complementary systems" can be established. This model combines photovoltaic power generation with fishery production – a new cross-sector integration of energy and agriculture (aquaculture). A portion of the generated electricity can power fishery equipment such as water pumps and aerators, while the surplus power can be fed into the grid, increasing farmers' income.

   Furthermore, PV installations can improve the aquaculture environment by providing shade for fish ponds, reducing water temperature, and inhibiting excessive algal growth through decreased photosynthesis, thus improving water quality. For fish farming, maintaining optimal water temperature (typically

Table 5.1 Load characteristics of facility agriculture greenhouses

Load type	Time-shift characteristic	Energy consumption indicator	Voltage type
LED plant growth lighting	6–20 continuous	5 kW/mu	DC 48 V
Plasma nitrogen fixation and water treatment	10–14 intermittent	2 kW/mu	AC 220 V
Biogas pool heat pump	20–6 intermittent	3 kW/mu	DC 48 V
Far-infrared heating	20–6 continuous	4 kW/mu	DC 48 V
Physical pest control	0–24 intermittent	0.15 kW/mu	DC 48 V
Water–fertilizer integration	8–10, 15–18 intermittent	5 kW/mu	AC 220 V
Electrostatic dehumidification	0–24 intermittent	0.2 kW/mu	DC 48 V
Rolling-curtain motor	Morning/evening short-term	2 kW/mu	AC 220 V
Greenhouse lighting	20–6 intermittent	0.15 kW/mu	DC 48 V
Smart agricultural monitoring system	0–24 continuous	0.2 kW/mu	DC 48 V
Nutrient solution recycling system	10–18 intermittent	0.2 kW/mu	DC 48 V
Sonic growth aid	6–19 intermittent	0.3 kW/mu	AC 220 V
Ventilation fan	0–24 intermittent	2 kW/mu	AC 220 V

20–30 °C) is essential for stable operation and high yields. Table 5.2 summarizes the energy consumption characteristics of aquaculture production.

3. Livestock and poultry farming energy consumption
   In China, factory-based aquaculture consumes 1,445,539.62 kW·h/hm^2 of electricity per unit area and 7.76 kW·h/kg per unit output. Livestock farming still relies heavily on mechanical equipment, which consumes significant energy. The main energy types used in livestock farms are coal and electricity, with heating, lighting, and ventilation being the major consumption sources. In 80% of livestock farms, coal is still used for heating, and energy-intensive devices are adopted for daily operations. For instance, a chick brooding house may consume more than 1 ton of coal per day.

   To address power supply issues, biogas and PV generation can be used to supply electricity for livestock operations. On average, 1 kg of animal manure contains about 20% dry matter, and each kg of dry matter can produce approximately 0.3 m^3 of biogas. Thus, 1 kg of manure can generate 0.06 m^3 of biogas, and every 1.5 m^3 of biogas can generate 1 kW·h of electricity – meaning 1 ton of livestock manure can produce 40 kW·h of power. In addition, in areas with 1500 sunshine hours per year, installing 60 kW of PV modules on livestock rooftops can produce

Table 5.2 Energy consumption characteristics of fishery production

Load type	Function	Load characteristic	Energy feature	AC/DC
Positive-pressure feed blower	Feed delivery	Shiftable, Non-interruptible	2.2 kW·h	AC
"1+N" precision feeding device	Feed delivery	Shiftable, Non-interruptible	0.75 kW	AC
Micro-pore aeration device	Oxygen supply	Shiftable, Non-interruptible	1.5 kW per 10 mu pond (3 kW per unit)	AC
Plateau fish farming container	Intensive farming	Shiftable, Non-interruptible	4 kW·h	DC
Solar aerator	Oxygen supply	Shiftable, Non-interruptible	50 W	AC
Hybrid automatic aerator	Oxygen supply	Shiftable, Non-interruptible	28.2 kW·h saved vs. single type	AC
Small Fiber-Reinforced Plastic (FRP) fishing boat	Fishing	Shiftable, Interruptible	500 V	DC
Automatic weed-clearing boat	Weed removal	Shiftable, Interruptible	500 W	DC
Multifunctional automatic operation boat	Multi-purpose	Shiftable, Interruptible	600 W	DC
Constant-pressure water supply system	Water supply	Shiftable, Non-interruptible	7.7 kW	AC

about 90,000 kW·h per year (average 480 kW·h/day), fully meeting livestock energy needs while reducing fossil fuel use and carbon emissions.

Table 5.3 summarizes the energy consumption characteristics of livestock production.

5.2.1 Case study on the AEI in Langyatai, Qingdao, China

5.2.1.1 Project overview

Since the beginning of the twenty-first century, China has consecutively issued 18 "No. 1 Central Documents" focusing on the "three rural issues" – agriculture, rural areas, and farmers – highlighting the fundamental importance of the rural sector. Rural power grids provide essential infrastructure support for winning the battle against poverty and implementing the national Rural Revitalization Strategy.

In recent years, China has established several demonstration projects of integrated energy systems under the Energy Internet framework, mainly concentrated in urban industrial parks, hotels, hospitals, airports, schools, and other public or

Table 5.3 Energy consumption characteristics of livestock production

Livestock type	**Load**	**Load characteristic**	**Energy consumption**	**AC/ DC**
Chick brooding house	Lighting	Shiftable, Interruptible	55 kW·h (1–42 days)	AC
Chick brooding house	Manure scraper	Shiftable, Interruptible	419 kW·h (1–42 days)	AC
Chick brooding house	Fan	Shiftable, Interruptible	756 kW·h (1–42 days)	AC
Chick brooding house	Feeder	Shiftable, Interruptible	209 kW·h (1–42 days)	DC
Pigsty	Lighting	Shiftable, Non-interruptible	12 W	AC
Pigsty	Water pump	Shiftable, Non-interruptible	7.5 kW	AC
Goat shed	Conveyor feeding system	Shiftable, Non-interruptible	120 kW·h/ day	AC
Dairy barn	Adjustable milking unit	Shiftable, Non-interruptible	65 kW·h/day	AC
Brooding house	Heat recovery ventilation	Shiftable, Non-interruptible	0.39 kW·h (20 Hz) / 0.84 kW·h (50 Hz)	AC
Cattle shed	Misting + negative-pressure curtain system	Shiftable, Non-interruptible	1.1 kW	AC
Pigsty	Manure transport vehicle	Shiftable, Interruptible	750 kW	DC
Livestock farm	Mechanical manure removal system	Shiftable, Interruptible	0.33 kW·h (at 1.40 r/ min)	AC
Cattle shed	Solar constant-temperature drinking system	Shiftable, Interruptible	9 kW	AC
Piglet house	Infrared heating lamp (Automatic Control)	Shiftable, Non-interruptible	175 W	AC
Chicken house	Air-source heat pump	Shiftable, Non-interruptible	1211 kW·h/ day	AC

commercial institutions. However, research on rural integrated energy systems remains at an early stage of development. To comprehensively support rural revitalization, it is necessary to further develop multi-energy complementary utilization systems that meet the demands of agricultural modernization and industrialization.

Through unified planning and coordinated energy utilization, such systems can enhance overall energy efficiency. At the same time, it is urgent to establish business models and price-subsidy mechanisms for rural energy internets and to build carbon-cycle and carbon-sink management frameworks for agricultural production.

This project is based on theoretical research, system development, and equipment design for intelligent energy technologies, with pilot applications emphasizing both feasibility and economic rationality. Shandong Province, known as China's "granary" and "vegetable basket," has a large grain-industry economy. In recent years, the province has implemented the "High-Quality Grain Project" as a key initiative to ensure national food security and advance the Rural Revitalization Strategy. Shandong also hosts 44 national-level marine ranches – accounting for 40% of the country's total – ranking first nationwide. With its long coastline, abundant islands, and distinctive natural, geographical, and economic characteristics, Shandong provides an excellent foundation and demonstration prospect for the application and promotion of this project's outcomes.

The project selects a typical agricultural park facility located in Langyatai, Qingdao, Shandong Province as a case study. It investigates the existing energy-system structure and the available local energy resources, conducts on-site measurements and data collection, and applies energy-economy and efficiency-analysis methods to optimize the operation of the rural energy system in this region. The results are expected to provide valuable reference for the local power company's future development strategies.

5.2.1.2 Energy equipment parameters

This study investigates a real-world agricultural microgrid located in Qingdao, China. The analysis begins with a description of the fundamental configuration of the microgrid, followed by a detailed examination of the electrical equipment on the load side. Information regarding the manufacturer, model specifications, and rated capacities of the devices is provided to support replication of this case or to facilitate similar experimental studies in future research.

1. Basic structure diagram
   Figure 5.2 illustrates the overall architecture of the agricultural microgrid [1], where cucumbers are cultivated inside the greenhouse. The diagram highlights the major system components and their interconnections. It shows the integration of the gas boiler, rooftop photovoltaic (PV) array, biogas generator units, and the battery storage system. In addition, a distribution transformer is installed to link the distributed generation units with the external distribution network, ensuring coordinated operation within the microgrid.
2. Load composition
   As presented in Figure 5.3, the greenhouse energy demand mainly comprises two categories. The first category corresponds to the environmental monitoring and control system, which functions as a non-controllable fixed load. The second category includes multiple agricultural production devices such as lighting, heating, and the integrated water–fertilizer system.

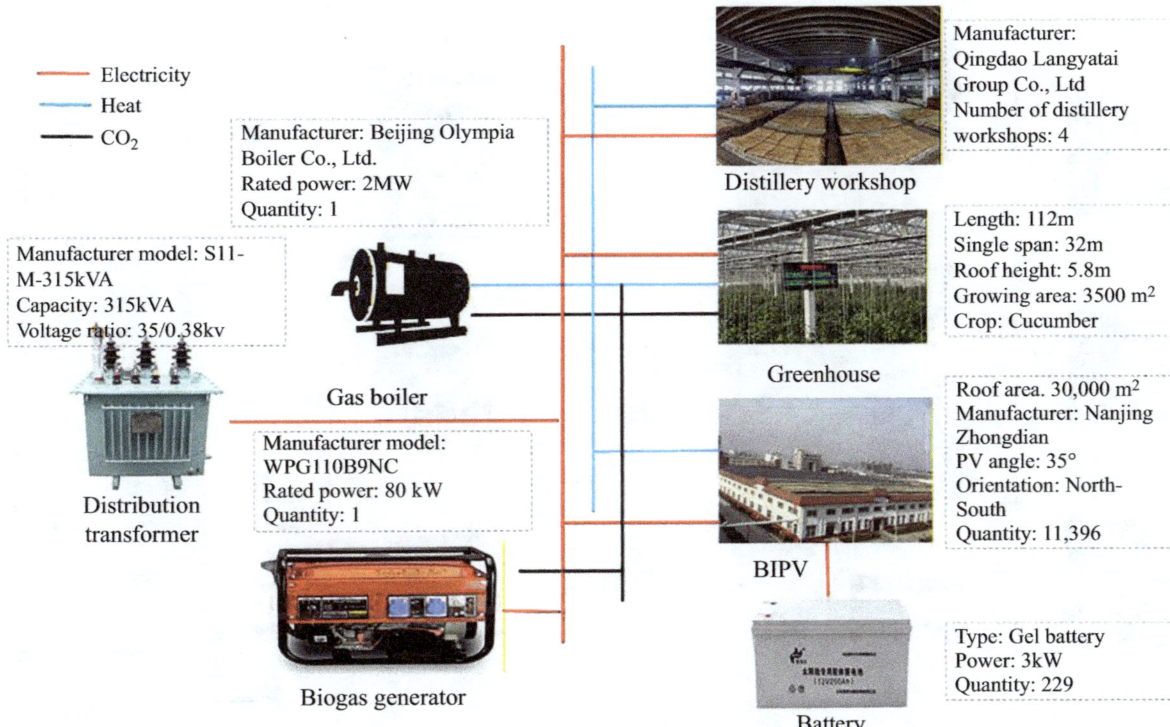

Figure 5.2 Configuration diagram of the agricultural park microgrid

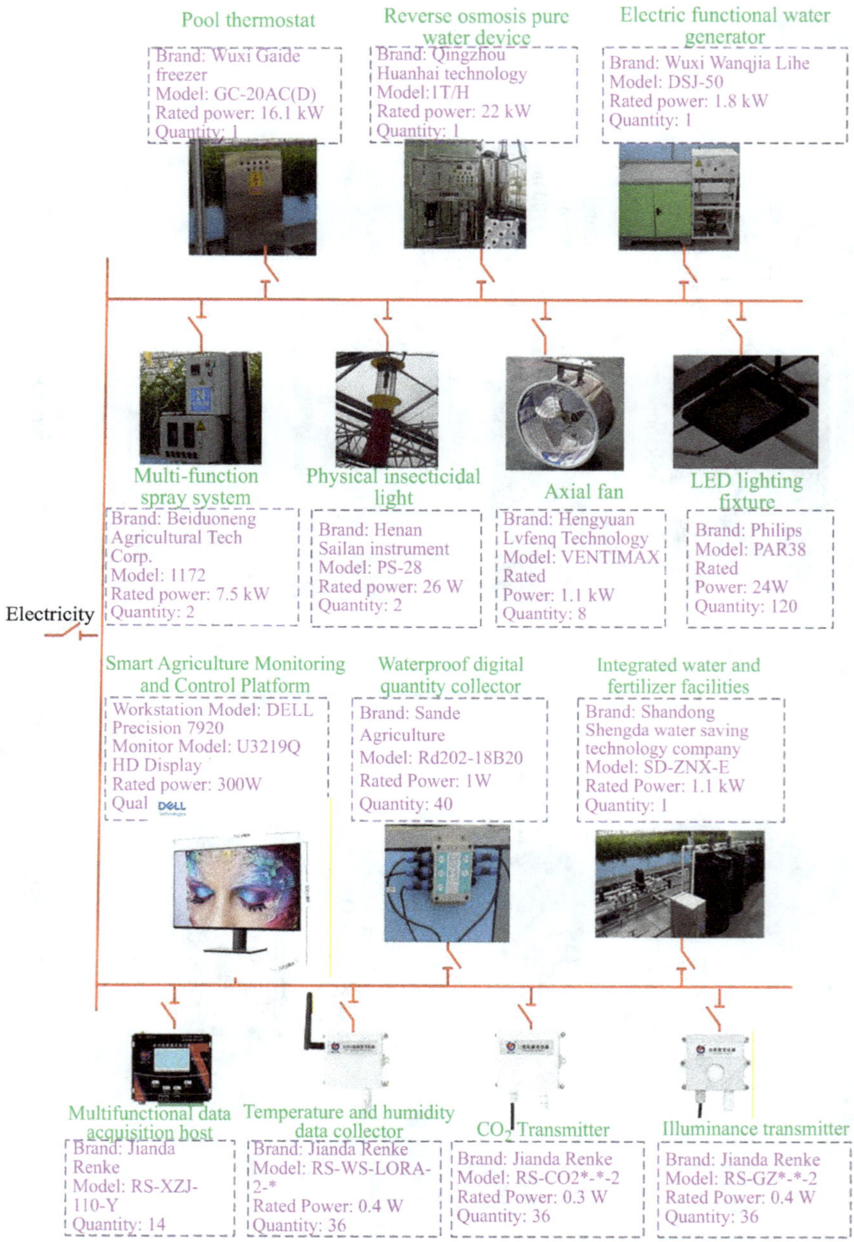

Figure 5.3 Energy load composition of the greenhouse

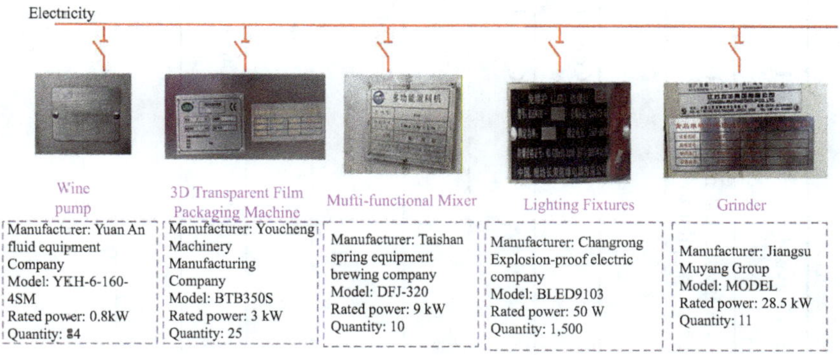

Figure 5.4 Energy load composition of four distillery workshops

The greenhouse heating system employs steam radiators produced by Jia Ao Heating Equipment Company, model D89-2-3. The installation consists of 2880 meters of Type-1 radiators made of seamless steel tubes (Φ51 mm diameter, 3 mm thickness) and 384 meters of Type-2 radiators fabricated from seamless steel tubes (Φ38 mm diameter, 2 mm thickness).

In addition to the greenhouse loads, other electricity consumers in the agricultural park include four distillery workshops and nearby residential areas. A comprehensive survey was carried out to document the electrical equipment configurations within the distillery workshops, as illustrated in Figure 5.4.

Table 5.4 summarizes the main energy equipment and technical parameters of the agricultural park microgrid located in Langyatai, Qingdao. The listed components include generation, storage, and load-side devices such as photovoltaic panels, a biogas generator, a gas boiler, energy storage batteries, greenhouse loads, and smart monitoring instruments. Detailed specifications are provided, including model types, rated capacities, quantities, manufacturers, and installation configurations. The first four items represent controllable or partially controllable energy supply and conversion units that ensure coordinated distributed generation within the microgrid. The remaining devices correspond to agricultural production loads, divided into controllable and uncontrollable categories, representing typical power consumption characteristics of facility agriculture. This table provides fundamental parameter data for subsequent system modeling, energy management optimization, and control strategy development.

To verify the economic feasibility of the real agricultural microgrid considered in this study, Table 5.5 provides a comprehensive analysis of the equipment acquisition costs shown in Figures 5.3–5.5. All equipment prices presented here reflect current Chinese market prices but have been converted from Chinese Yuan (CNY) to U.S. Dollars (USD) according to the exchange rate on October 18, 2025 (1 USD ≈ 7.12 CNY), to allow global readers to better understand the procurement cost of energy equipment in China.

Table 5.4 Main energy equipment and technical parameters

Category/ Equipment	Model/Type	Specification/Capacity/ Power	Quantity	Manufacturer/ Material/Orientation	Installation/Layout	Type
Photovoltaic panel	—	275 W (tilt 35°), roof area: 30,000 m^2	—	Nanjing Zhongdian	South–North layout, flat mounting, long edge parallel to the roof	Uncontrollable
Biogas generator	WPG110B9NC/ 80 kW	Rated 110 kVA/ Output 80 kW	1 unit	Fuel: biogas from distillery waste	—	Controllable
Gas boiler	Olympia boiler	2 MW, Fuel: Natural gas	1 unit	—	—	Controllable
Battery (energy storage)	Solar colloidal battery 3 kW	250 Ah / 12 V	228 units	—	Containerized storage	Controllable
Plant growth lamp	Philips PAR38	1 kW (Metal halide lamp)	—	Philips agricultural division	—	Controllable
Water–fertilizer integration system	SD-ZNX-E	1.1 kW	1 unit	Shandong Shengda Water-Saving Technology Co., Ltd.	—	Controllable
Ventilation fan	VENTIMAX	1.1 kW	8 units	Hengyuan Lüfeng Agri-Tech Co., Ltd.	—	Uncontrollable
Physical insecticidal lamp	PS-28	26 W	2 units	Henan Sailan Instruments Co.	—	Uncontrollable
Electro-functional water generator	DSJ-50	1.8 kW	1 unit	Wuxi Wanjialihe Co.	—	Controllable
Reverse osmosis pure-water device	1 T/H	2.2 kW (two high-pressure pumps)	1 unit	Qingzhou Huanhai Technology Co., Ltd.	—	Controllable

(Continues)

Multi-function spray system	1172	7.5 kW	2 units	Beiduoneng Agricultural Technology Co., Ltd.	Installed on both sides of the greenhouse	Controllable
Pool thermostat (Cooling/Heating)	GC-20AC(D)	16.1 kW	1 unit	Wuxi Guide Refrigeration Equipment Co., Ltd.	—	Uncontrollable
Smart agriculture monitoring platform	Dell Precision T920 + U3219Q monitor	300 W	1 unit	Dell Inc.	—	Uncontrollable
Temperature and humidity sensor	JDRK thermo-hygrometer	0.4 W	36 units	Jianda Renkong Instruments Co.	—	Uncontrollable
CO_2 concentration sensor	JDRK 485 CO_2 0–5000 ppm	0.3 W	36 units	Jianda Renkong Instruments Co.	—	Uncontrollable
Illuminance sensor	JDRK RS485 0–20 wLux	0.4 W	36 units	Jianda Renkong Instruments Co.	—	Uncontrollable
Waterproof digital collector	Sande Agriculture RD202-18B20	1 W	40 units	Sande Agriculture Co.	—	Uncontrollable
Data logger (main host)	JDRK GPRS flagship model	1.5 W	14 units	Jianda Renkong Instruments Co.	—	Uncontrollable
Steam radiator system	D89-2-3	2,880 m ($\Phi 51\times 3$ mm) + 384 m ($\Phi 38\times 2$ mm)	—	Jia Ao Heating Equipment Co., Ltd.	Floor-mounted	Uncontrollable

Table 5.5 Cost of equipment in the agricultural microgrid (Converted to USD)

Item	Equipment	Unit price (USD)	Quantity	Total cost (USD)
Source	Distribution transformer	6,624.72	1	6,624.72
	Gas boiler	33,565.17	1	33,565.17
	Biogas generator	12,880.62	1	12,880.62
	PV cell	77.11	11,396	879,215.56
	110 kW inverter	2341.01	30	70,230.30
Storage	Battery	278.09	229	63,657.61
Greenhouse	Pool thermostat	5956.74	1	5956.74
	Reverse osmosis pure-water device	6358.85	1	6358.85
	Electric functional water generator	2781.46	1	2781.46
	Multi-function spray system	927.00	2	1854.00
	Physical insecticidal light	40.45	2	80.90
	Axial fan	30.75	8	246.00
	1200W LED	800.28	246	196,868.88
	Workstation	7917.28	1	7917.28
	Waterproof digital quantity collector	76.54	40	3061.60
	Integrated water and fertilizer facilities	18,641.43	1	18,641.43
	Multifunctional data acquisition host	202.25	14	2831.50
	Temperature and humidity data collector	20.22	36	728.00
	CO_2 transmitter	36.94	36	1329.84
	Illuminance transmitter	17.70	36	637.20
Distillery workshops	Wine pump	196.48	84	16,514.32
	3D transparent film packaging machine	5337.08	25	133,427.00
	Multifunctional mixer	12,103.09	10	121,030.90
	Lighting fixtures	9.70	1500	14,550.00
	Crusher	30,332.16	11	333,653.76

The total acquisition cost of the microgrid's generation and storage equipment amounts to USD 1,066,173.98 (equivalent to CNY 7,587,474). On the demand side, the greenhouse electrical equipment totals USD 249,352.39 (equivalent to CNY 1,775,377), while the distillery workshop equipment costs USD 619,393.82 (equivalent to CNY 4,406,167). The detailed cost data obtained from the real engineering project, together with the equipment specifications provided in Figures 5.2–5.4, serve as a valuable reference for evaluating and promoting the agricultural microgrid model proposed in this study.

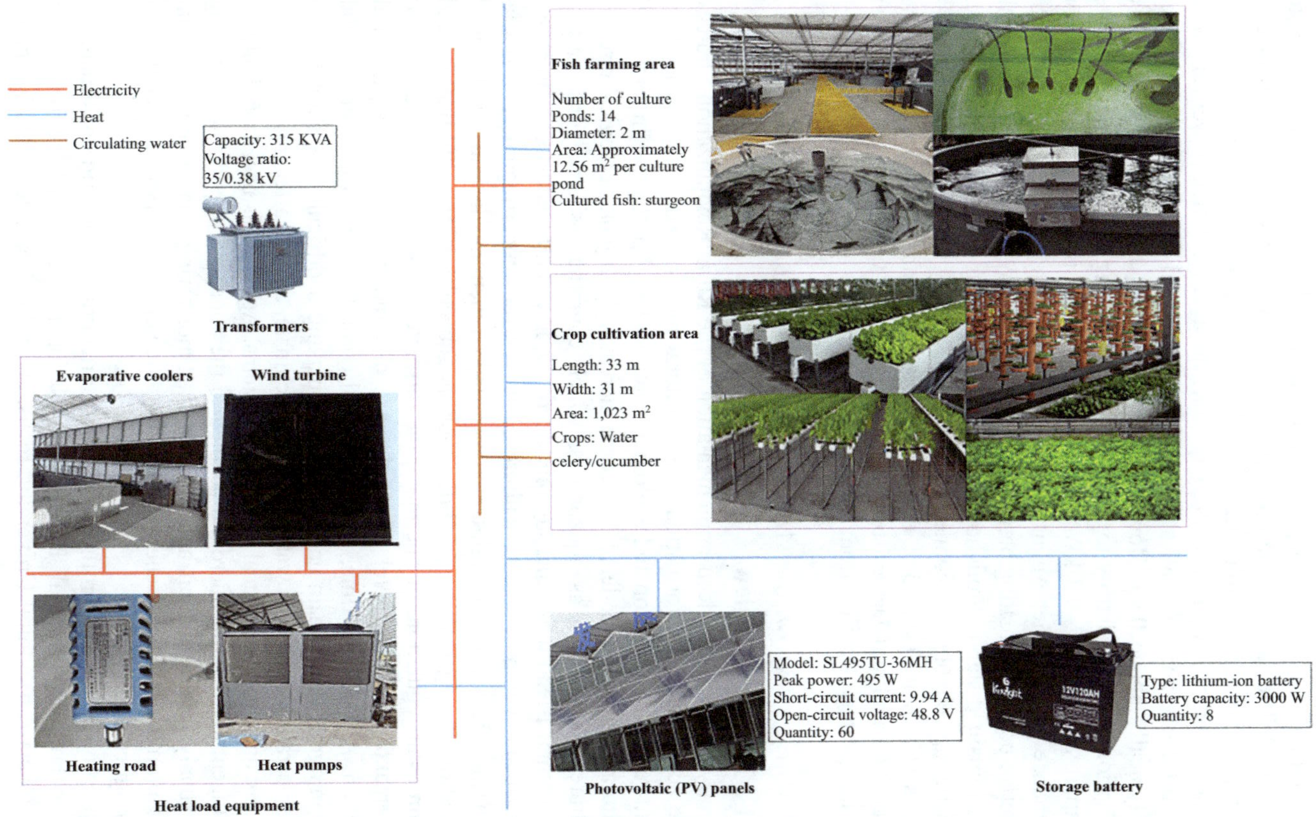

Figure 5.5 Configuration diagram of the agricultural park microgrid

5.2.1.3 Actual operating data

The electricity data of the brewery located in the Langyatai Agricultural Demonstration Zone, Shandong Province, were collected from January 1 to September 30. The dataset includes daily measurements of the three-phase currents and voltages of the 35 kV Guluo Line, along with active power, reactive power, and power factor indicators. From January to June, the average active power remained between 2.0 kW and 2.6 kW, showing stable industrial operation with a power factor typically above 0.93. However, starting in July, the brewery's electricity consumption dropped sharply to below 1.0 kW, reaching a minimum around late July (approximately 0.2–0.3 kW), indicating a possible production suspension or maintenance period. After August, a gradual recovery in load was observed, with daily active power fluctuating between 0.5 kW and 1.5 kW through September. The voltage remained stable throughout the period, averaging 18.0–22.5 kV per phase, demonstrating reliable grid connection and stable supply for the brewery's operations. As shown in Table 5.6, the actual operating data of the brewery in the Langyatai Agricultural Demonstration Zone reflect the variations in load, power factor, and voltage stability from January to September.

5.2.2 Case study on the AEI in Zhuozhou, Baoding, China

5.2.2.1 Project overview

In recent years, with the deep implementation of the Rural Revitalization Strategy and the accelerated transformation of agricultural modernization, the demand for efficient resource utilization and ecological environmental protection in traditional agricultural production models has become increasingly urgent. Ecological circular agriculture models such as aquaponics, which combine resource conservation, environmental friendliness, and high productivity, have emerged as key directions for the green development of agriculture. However, the supporting energy systems of such models are often designed with single functions, lacking coordinated optimization among multiple modules such as aquaculture, crop cultivation, and environmental regulation. Furthermore, the mismatch between energy supply and load demand limits the level of intelligence and intensification in agricultural production. At the same time, monitoring and control systems for environmental factors in agriculture remain underdeveloped, making it difficult to accurately regulate key parameters such as water quality and climate, thereby affecting the stability and output quality of ecological circulation systems. Against this backdrop, building an intelligent agricultural microgrid adapted to the aquaponics model – integrating energy supply, load regulation, and environmental monitoring – has become a crucial path for promoting efficient energy use and high-quality development of ecological circular agriculture.

This project is based on the aquaponics agricultural microgrid constructed at the Zhuozhou Experimental Base of the National Digital Fishery Innovation Center, China Agricultural University, focusing on the deep integration of energy systems and agricultural production. As one of the major agricultural provinces in North China, Hebei Province features diverse agroecological environments,

Table 5.6 Actual operating data of the brewery in the Langyatai Agricultural Demonstration Zone (January–September)

Date	Current A (A)	Current B (A)	Current C (A)	Voltage A (kV)	Voltage C (kV)	Voltage AB (kV)	Voltage BC (kV)	Voltage CA (kV)	Active power (kW)	Reactive power (kVar)	Power factor
01-01	41.54	42.42	42.99	18.1	22.55	36.28	36.31	36.38	2.54	0.9	0.92
01-02	38.5	39.12	39.34	18.16	22.6	36.36	36.41	36.48	2.37	0.75	0.93
01-03	38.41	39.12	39.38	18.1	22.49	36.22	36.24	36.32	2.33	0.79	0.92
01-04	40.61	41.4	41.8	18.04	22.41	36.1	36.12	36.21	2.5	0.78	0.93
01-05	38.76	39.42	39.95	18.16	22.55	36.32	36.36	36.44	2.4	0.67	0.94
01-06	41.89	42.68	43.21	18.07	22.46	36.17	36.2	36.3	2.54	0.9	0.91
01-07	39.69	40.61	41.18	18.09	22.53	36.25	36.27	36.38	2.49	0.68	0.94
01-08	37.58	38.28	38.81	18.08	22.53	36.23	36.26	36.34	2.34	0.64	0.94
01-09	37.4	37.75	38.19	18.08	22.55	36.23	36.28	36.34	2.3	0.65	0.93
01-10	38.85	39.51	39.82	18.04	22.5	36.17	36.2	36.27	2.42	0.66	0.94
01-11	39.69	40.61	41.27	18.12	22.59	36.32	36.35	36.44	2.48	0.73	0.94
01-12	37	37.58	37.97	18.15	22.59	36.34	36.38	36.46	2.29	0.66	0.93
01-13	41.14	41.36	42.11	18.18	22.58	36.39	36.4	36.5	2.52	0.8	0.93
01-14	43.47	44.44	45.1	18.18	22.57	36.37	36.4	36.48	2.68	0.91	0.92
01-15	41.45	42.11	42.33	18.23	22.59	36.44	36.46	36.54	2.54	0.86	0.92
01-16	42.9	43.69	44.35	18.05	22.4	36.11	36.13	36.21	2.64	0.83	0.93
01-17	43.87	44.79	45.23	18.17	22.53	36.35	36.35	36.42	2.72	0.83	0.93
01-18	36.21	36.87	37.14	18.24	22.61	36.47	36.5	36.56	2.24	0.69	0.93

(Continues)

Table 5.6 (Continued)

Date	Current A (A)	Current B (A)	Current C (A)	Voltage A (kV)	Voltage C (kV)	Voltage AB (kV)	Voltage BC (kV)	Voltage CA (kV)	Active power (kW)	Reactive power (kVar)	Power factor
01-19	35.6	36.21	36.61	18.08	22.42	36.14	36.17	36.24	2.17	0.77	0.92
01-20	37.66	38.19	38.5	18.06	22.39	36.12	36.12	36.22	2.29	0.75	0.93
01-21	40	40.61	40.92	18.16	22.48	36.27	36.3	36.36	2.46	0.78	0.93
01-22	38.19	39.07	39.73	18.17	22.51	36.29	36.34	36.42	2.36	0.75	0.92
01-23	37.62	38.15	38.81	18.11	22.42	36.2	36.19	36.3	2.33	0.63	0.93
01-24	40.83	41.58	42.06	18.16	22.48	36.28	36.3	36.39	2.52	0.77	0.93
01-25	42.24	43.16	43.65	18.22	22.55	36.39	36.41	36.51	2.6	0.89	0.92
01-26	41.18	41.76	42.11	18.21	22.54	36.4	36.4	36.48	2.53	0.75	0.93
01-27	39.78	40.39	40.74	18.17	22.53	36.32	36.35	36.41	2.48	0.64	0.94
01-28	43.12	44.04	44.62	18.17	22.54	36.34	36.35	36.43	2.65	0.85	0.93
01-29	36.17	36.48	37	18.21	22.56	36.37	36.41	36.47	2.26	0.55	0.94
01-30	35.07	35.16	35.68	18.57	22.95	37.06	37.08	37.17	2.21	0.59	0.94
01-31	33.53	33.75	34.1	18.38	22.75	36.7	36.74	36.8	2.12	0.53	0.94
02-01	34.89	34.85	35.42	18.4	22.76	36.73	36.77	36.88	2.17	0.6	0.94
02-02	37.14	37.27	37.97	18.3	22.68	36.59	36.6	36.69	2.32	0.63	0.94
02-03	32.16	32.34	32.38	18.34	22.72	36.63	36.67	36.74	1.98	0.57	0.93
02-04	37.05	37.49	37.93	18.34	22.72	36.64	36.69	36.74	2.31	0.74	0.92
02-05	37.49	37.62	38.19	18.55	22.93	37.04	37.05	37.14	2.35	0.7	0.94

(Continues)

02-06	35.6	35.73	36.34	18.55	22.95	37.03	37.07	37.15	2.23	0.66	0.93
02-07	38.81	39.64	40.08	17.93	22.27	35.86	35.91	35.98	2.34	0.86	0.91
02-08	39.12	39.6	40.13	17.92	22.22	35.82	35.85	35.92	2.4	0.65	0.94
02-09	40.13	40.79	41.01	17.99	22.33	35.98	36.01	36.05	2.48	0.68	0.94
02-10	36.56	36.7	36.87	18.1	22.46	36.16	36.23	36.25	2.24	0.65	0.94
02-11	31.59	31.42	31.68	18.24	22.62	36.45	36.51	36.54	1.96	0.45	0.95
02-12	26.88	26.97	26.97	18.26	22.62	36.46	36.52	36.54	1.69	0.31	0.95
02-13	22.18	21.78	22.31	18.19	22.56	36.33	36.41	36.45	1.38	0.24	0.96
02-14	26.97	26.84	27.24	18.09	22.42	36.14	36.19	36.22	1.63	0.53	0.92
02-15	29.39	29.48	29.79	18	22.31	35.98	36.01	36.04	1.79	0.51	0.93
02-16	33.57	33.62	34.23	18.07	22.44	36.11	36.18	36.25	2.09	0.45	0.95
02-17	37.93	38.54	38.9	18	22.35	35.99	36.04	36.09	2.35	0.64	0.94
02-18	39.34	39.86	40.3	17.97	22.31	35.93	35.98	36.01	2.42	0.66	0.94
02-19	36.56	36.3	36.78	17.89	22.22	35.78	35.82	35.87	2.22	0.57	0.94
02-20	34.98	35.33	35.82	18.04	22.37	36.05	36.09	36.15	2.16	0.54	0.94
02-21	36.65	36.96	37.36	18.01	22.31	36.02	36	36.07	2.25	0.62	0.94
02-22	33.7	33.79	33.97	18.08	22.38	36.12	36.13	36.18	2.07	0.49	0.94
02-23	33.62	33.88	34.19	17.99	22.28	35.95	35.97	36.03	2.08	0.49	0.94
02-24	33.04	33.26	33.35	17.98	22.27	35.92	35.95	36.02	2.03	0.43	0.95
02-25	37.53	37.8	38.1	17.95	22.28	35.9	35.93	36	2.3	0.59	0.94

(Continues)

Table 5.6 (Continued)

Date	Current A (A)	Current B (A)	Current C (A)	Voltage A (kV)	Voltage C (kV)	Voltage AB (kV)	Voltage BC (kV)	Voltage CA (kV)	Active power (kW)	Reactive power (kVar)	Power factor
02-26	31.94	31.46	31.94	18.14	22.48	36.25	36.28	36.36	1.94	0.47	0.94
02-27	33.09	33.13	33.35	18.04	22.38	36.07	36.1	36.17	2.04	0.43	0.95
02-28	36.48	36.87	37.14	17.92	22.25	35.84	35.87	35.94	2.24	0.57	0.94
03-01	31.5	31.37	31.72	18.09	22.39	36.16	36.15	36.23	1.94	0.41	0.94
03-02	39.12	39.2	39.82	17.92	22.23	35.86	35.86	35.93	2.4	0.61	0.94
03-03	36.61	36.65	37.27	17.98	22.3	35.97	35.96	36.06	2.26	0.5	0.95
03-04	36.52	36.83	37.14	17.94	22.28	35.91	35.91	36	2.26	0.5	0.95
03-05	35.82	35.9	36.43	17.95	22.29	35.91	35.94	36.02	2.21	0.47	0.95
03-06	36.17	36.56	37.09	17.98	22.29	35.93	35.96	36.04	2.23	0.56	0.94
03-07	40.52	41.23	41.49	17.94	22.27	35.87	35.9	35.98	2.46	0.73	0.93
03-08	41.36	42.55	42.81	17.98	22.31	35.93	35.99	36.04	2.56	0.74	0.94
03-09	36.83	37.36	37.53	17.98	22.33	35.96	36.01	36.08	2.26	0.61	0.94
03-10	33.31	33.53	33.88	17.92	22.25	35.83	35.87	35.95	2.04	0.48	0.94
03-11	36.83	37.14	37.66	17.94	22.27	35.88	35.91	36	2.26	0.6	0.94
03-12	38.41	38.76	39.34	17.92	22.25	35.84	35.88	35.94	2.36	0.61	0.94
03-13	38.41	39.16	39.69	17.93	22.24	35.84	35.88	35.95	2.38	0.65	0.94
03-14	40.22	40.83	41.27	17.95	22.25	35.89	35.91	35.97	2.46	0.69	0.93
03-15	35.6	35.99	36.48	18.08	22.42	36.15	36.18	36.25	2.22	0.53	0.95

(Continues)

03-16	38.15	38.63	39.07	17.95	22.28	35.91	35.93	36.01	2.35	0.63	0.94
03-17	36.61	37.27	37.84	17.95	22.27	35.89	35.92	36.01	2.25	0.64	0.94
03-18	38.9	39.56	39.95	18.07	22.39	36.11	36.14	36.2	2.4	0.69	0.93
03-19	35.95	36.04	36.74	18.05	22.38	36.08	36.12	36.19	2.22	0.55	0.94
03-20	36.08	36.74	37.18	18.05	22.39	36.08	36.12	36.18	2.25	0.59	0.94
03-21	31.02	31.2	31.64	18.07	22.42	36.13	36.16	36.22	1.93	0.38	0.95
03-22	31.59	31.64	31.9	18.15	22.5	36.26	36.32	36.37	1.95	0.43	0.94
03-23	35.9	36.39	36.78	18.05	22.36	36.05	36.09	36.15	2.22	0.56	0.94
03-24	35.82	36.3	36.83	18.25	22.58	36.44	36.48	36.55	2.26	0.53	0.95
03-25	33.92	34.19	34.5	18.03	22.35	36.03	36.07	36.14	2.1	0.44	0.95
03-26	36.87	37.31	37.62	17.99	22.25	35.92	35.94	35.99	2.28	0.55	0.95
03-27	38.41	39.25	39.64	17.95	22.24	35.85	35.9	35.96	2.38	0.66	0.94
03-28	36.43	37	37.27	18.03	22.32	36.01	36.04	36.08	2.25	0.54	0.94
03-29	34.36	34.54	34.76	18.04	22.33	36.01	36.06	36.09	2.11	0.53	0.94
03-30	36.21	36.92	37	17.99	22.27	35.9	35.97	35.97	2.25	0.53	0.94
03-31	37.84	38.24	38.63	18.03	22.31	36	36.04	36.09	2.32	0.68	0.94
04-01	36.61	37.09	37.22	18	22.29	35.96	36	36.03	2.26	0.57	0.94
04-02	38.37	39.03	39.47	17.94	22.25	35.87	35.9	35.97	2.36	0.68	0.94
04-03	34.06	33.97	34.45	18.03	22.34	36.02	36.06	36.13	2.09	0.5	0.94
04-04	37.18	37.36	37.58	18.06	22.37	36.08	36.1	36.18	2.27	0.68	0.93

(Continues)

Table 5.6 (Continued)

Date	Current A (A)	Current B (A)	Current C (A)	Voltage A (kV)	Voltage C (kV)	Voltage AB (kV)	Voltage BC (kV)	Voltage CA (kV)	Active power (kW)	Reactive power (kVar)	Power factor
04-05	39.64	40.3	40.66	18.31	22.64	36.54	36.58	36.64	2.45	0.78	0.93
04-06	41.36	42.02	42.42	18.04	22.33	36.03	36.05	36.12	2.55	0.7	0.94
04-07	39.82	39.73	40.57	17.98	22.28	35.94	35.96	36.03	2.45	0.6	0.95
04-08	35.99	36.12	36.65	18.14	22.47	36.27	36.27	36.35	2.23	0.53	0.95
04-09	33.13	32.96	33.22	18.17	22.45	36.26	36.29	36.35	2.04	0.49	0.94
04-10	35.82	36.3	36.61	18.21	22.53	36.4	36.39	36.46	2.26	0.52	0.95
04-11	34.5	34.85	35.16	17.99	22.28	35.96	35.98	36.01	2.14	0.44	0.95
04-12	34.58	34.63	34.98	18.13	22.41	36.2	36.2	36.26	2.15	0.41	0.95
04-13	34.14	34.01	34.5	18	22.27	35.96	35.97	36.02	2.11	0.43	0.95
04-14	35.77	35.9	36.3	17.96	22.23	35.88	35.9	35.95	2.19	0.51	0.94
04-15	32.78	32.91	33.26	17.97	22.27	35.92	35.94	36.02	2.01	0.46	0.94
04-16	34.41	34.58	34.94	18.04	22.32	36.05	36.05	36.11	2.13	0.45	0.95
04-17	35.6	35.82	36.34	18.08	22.38	36.11	36.14	36.23	2.19	0.58	0.94
04-18	34.45	34.36	35.07	18.11	22.4	36.16	36.19	36.25	2.11	0.56	0.94
04-19	35.73	35.86	36.3	18.07	22.35	36.08	36.11	36.16	2.22	0.49	0.95
04-20	42.86	43.43	44.13	17.89	22.18	35.77	35.77	35.85	2.61	0.79	0.93
04-21	40.96	41.67	42.42	18.16	22.45	36.24	36.27	36.34	2.54	0.76	0.93
04-22	43.87	44.57	45.19	17.96	22.24	35.89	35.89	35.99	2.67	0.82	0.93

(Continues)

04-23	38.1	38.41	38.98	18	22.27	35.96	35.97	36.06	2.34	0.6	0.94
04-24	37	37.44	37.66	18.17	22.45	36.28	36.28	36.35	2.29	0.56	0.94
04-25	34.36	34.23	34.63	18.02	22.3	35.98	36.01	36.08	2.1	0.5	0.94
04-26	35.99	36.43	36.48	18	22.3	35.99	35.99	36.06	2.18	0.66	0.93
04-27	35.68	35.95	36.12	18.01	22.3	36	36	36.08	2.18	0.56	0.94
04-28	32.38	32.25	32.52	18.07	22.34	36.07	36.09	36.14	1.98	0.45	0.94
04-29	29.7	30.05	30.05	18.04	22.33	36.04	36.07	36.11	1.83	0.42	0.94
04-30	31.9	32.16	32.38	18	22.29	35.97	35.99	36.05	1.96	0.47	0.94
05-01	32.3	32.52	32.47	18.1	22.38	36.13	36.16	36.2	1.97	0.5	0.93
05-02	35.38	35.55	35.68	18.25	22.58	36.44	36.47	36.51	2.18	0.59	0.93
05-03	40.04	40.52	40.57	18.23	22.53	36.38	36.4	36.45	2.46	0.8	0.93
05-04	38.46	38.76	38.9	18.16	22.44	36.24	36.25	36.31	2.35	0.75	0.93
05-05	36.92	37.49	37.8	18.06	22.35	36.07	36.09	36.13	2.28	0.59	0.94
05-06	39.91	40.48	40.96	18.08	22.33	36.08	36.09	36.14	2.45	0.75	0.93
05-07	38.41	38.72	39.29	18.05	22.29	36	36.03	36.07	2.36	0.61	0.94
05-08	38.81	39.07	39.29	18.07	22.33	36.07	36.09	36.13	2.39	0.61	0.94
05-09	44.48	45.06	45.5	18.06	22.33	36.05	36.08	36.13	2.69	0.9	0.92
05-10	41.89	42.02	42.77	18.1	22.37	36.11	36.15	36.19	2.57	0.75	0.93
05-11	40.96	41.49	41.84	18.03	22.32	36.01	36.04	36.1	2.52	0.67	0.94
05-12	40.3	40.66	41.23	18.01	22.3	35.97	36	36.05	2.45	0.73	0.93

(Continues)

Table 5.6 (Continued)

Date	Current A (A)	Current B (A)	Current C (A)	Voltage A (kV)	Voltage C (kV)	Voltage AB (kV)	Voltage BC (kV)	Voltage CA (kV)	Active power (kW)	Reactive power (kVar)	Power factor
05-13	38.94	39.29	39.86	18.01	22.3	36	36.02	36.09	2.39	0.67	0.94
05-14	39.91	40.3	40.83	17.99	22.26	35.93	35.96	36.02	2.45	0.68	0.94
05-15	37.44	37.44	38.1	17.99	22.26	35.93	35.97	36.04	2.3	0.56	0.94
05-16	42.33	42.94	43.16	17.98	22.22	35.92	35.91	35.99	2.55	0.85	0.92
05-17	37.53	38.02	38.28	18.15	22.42	36.23	36.25	36.31	2.32	0.63	0.94
05-18	38.98	39.69	40	17.97	22.19	35.86	35.88	35.93	2.38	0.72	0.93
05-19	39.69	40.35	40.48	18.03	22.24	35.97	35.98	36.04	2.4	0.82	0.92
05-20	38.1	38.9	39.12	18.01	22.23	35.95	35.97	36.03	2.32	0.75	0.93
05-21	35.16	35.82	36.04	17.99	22.2	35.9	35.92	35.98	2.1	0.79	0.91
05-22	36.56	37	37.05	17.99	22.2	35.9	35.92	35.96	2.18	0.81	0.91
05-23	34.32	34.67	34.89	18.04	22.26	36.01	36.02	36.07	2.07	0.69	0.92
05-24	39.91	40.74	41.1	18.08	22.29	36.05	36.08	36.12	2.38	0.97	0.9
05-25	36.96	37.44	37.71	18	22.22	35.94	35.95	36.01	2.27	0.63	0.94
05-26	41.67	42.28	42.68	18.02	22.22	35.98	35.96	36.02	2.56	0.72	0.94
05-27	39.07	39.47	39.78	18.23	22.45	36.32	36.37	36.39	2.4	0.74	0.93
05-28	41.1	41.45	41.98	18.2	22.41	36.28	36.31	36.35	2.51	0.8	0.93
05-29	43.91	44.26	45.1	18.03	22.22	35.96	35.98	36.02	2.68	0.82	0.93
05-30	40.57	40.74	41.54	18.07	22.26	36.02	36.06	36.1	2.48	0.72	0.93

(Continues)

05-31	38.46	39.12	39.42	18.16	22.37	36.2	36.24	36.27	2.37	0.68	0.93
06-01	41.18	41.89	42.2	18.01	22.23	35.94	35.97	36.02	2.56	0.6	0.95
06-02	39.16	39.56	40.39	17.98	22.2	35.88	35.92	35.97	2.42	0.65	0.94
06-03	38.63	38.59	39.29	18.16	22.4	36.21	36.27	36.31	2.37	0.63	0.94
06-04	39.51	40.13	40.57	18.23	22.45	36.34	36.37	36.39	2.46	0.68	0.94
06-05	39.47	40	40.57	18.15	22.37	36.18	36.22	36.27	2.44	0.63	0.94
06-06	36.52	36.61	37.36	18.03	22.23	35.95	36	36.04	2.24	0.58	0.94
06-07	32.52	32.78	33.31	18.11	22.32	36.12	36.15	36.19	2.04	0.45	0.94
06-08	28.42	28.64	29	18.25	22.49	36.38	36.42	36.48	1.77	0.39	0.94
06-09	32.56	32.6	32.78	17.99	22.18	35.87	35.91	35.97	1.98	0.48	0.94
06-10	40.79	41.76	42.11	18.05	22.26	36	36.04	36.08	2.5	0.78	0.93
06-11	40.22	40.96	41.23	18.16	22.36	36.18	36.22	36.25	2.48	0.72	0.94
06-12	43.12	44.35	44.57	18.16	22.35	36.18	36.21	36.24	2.65	0.88	0.93
06-13	40.79	41.98	42.11	18.01	22.19	35.9	35.93	35.97	2.52	0.77	0.93
06-14	42.28	43.38	43.56	18.08	22.28	36.06	36.07	36.13	2.58	0.82	0.93
06-15	40.26	41.01	41.36	18.23	22.43	36.31	36.36	36.39	2.48	0.79	0.93
06-16	40.48	41.27	41.76	18.04	22.22	35.97	35.98	36.04	2.48	0.77	0.93
06-17	42.99	43.82	44.4	17.93	22.12	35.78	35.8	35.85	2.64	0.77	0.93
06-18	38.5	38.98	39.34	18.05	22.22	35.98	35.99	36.04	2.36	0.64	0.94
06-19	42.37	43.08	44.22	18.19	22.4	36.25	36.29	36.36	2.63	0.77	0.93

(Continues)

Table 5.6 (*Continued*)

Date	Current A (A)	Current B (A)	Current C (A)	Voltage A (kV)	Voltage C (kV)	Voltage AB (kV)	Voltage BC (kV)	Voltage CA (kV)	Active power (kW)	Reactive power (kVar)	Power factor
06-20	41.18	41.4	42.5	18.15	22.35	36.19	36.21	36.26	2.53	0.77	0.93
06-21	39.64	40.13	40.96	18.03	22.18	35.93	35.94	35.99	2.44	0.69	0.94
06-22	38.1	38.41	39.2	18.14	22.3	36.13	36.15	36.2	2.34	0.7	0.93
06-23	30.8	30.54	31.33	18.12	22.28	36.1	36.12	36.15	1.89	0.44	0.94
06-24	32.43	32.34	32.91	18.15	22.32	36.14	36.18	36.22	1.99	0.49	0.94
06-25	24.29	24.51	25.04	17.93	22.14	35.76	35.81	35.85	1.51	0.26	0.95
06-26	27.19	27.15	27.98	18.12	22.33	36.12	36.15	36.21	1.67	0.44	0.93
06-27	27.02	26.27	27.15	18.14	22.35	36.16	36.19	36.22	1.63	0.41	0.93
06-28	24.11	23.72	24.16	18.07	22.28	36.03	36.07	36.1	1.48	0.26	0.95
06-29	24.64	25.12	25.34	17.96	22.16	35.82	35.87	35.89	1.51	0.38	0.94
06-30	27.59	28.16	28.38	17.97	22.16	35.86	35.87	35.9	1.69	0.47	0.93
07-01	24.24	24.38	24.6	17.98	22.19	35.87	35.92	35.94	1.49	0.3	0.95
07-02	20.64	20.86	21.21	18.19	22.41	36.26	36.3	36.34	1.29	0.2	0.95
07-03	22.13	22.22	22.84	18.17	22.37	36.2	36.24	36.28	1.39	0.25	0.94
07-04	22.66	23.01	23.45	17.95	22.14	35.81	35.85	35.88	1.4	0.27	0.95
07-05	20.06	20.06	20.33	18.02	22.22	35.94	35.98	36.01	1.23	0.19	0.94
07-06	23.54	23.63	23.94	18.11	22.3	36.12	36.13	36.16	1.45	0.32	0.94
07-07	23.19	23.19	23.54	18.11	22.28	36.09	36.1	36.15	1.41	0.37	0.93

(Continues)

07-08	23.41	23.45	23.89	18.1	22.31	36.09	36.13	36.16	1.43	0.36	0.93
07-09	25.39	25.3	25.87	18.07	22.28	36.03	36.06	36.11	1.55	0.4	0.93
07-10	22.66	22.79	23.45	18.11	22.29	36.09	36.11	36.17	1.42	0.25	0.95
07-11	22.88	22.53	22.97	18.04	22.22	35.97	35.99	36.03	1.36	0.38	0.92
07-12	23.76	23.72	24.38	18.01	22.17	35.9	35.91	35.97	1.44	0.4	0.93
07-13	23.58	23.94	24.51	18.04	22.18	35.94	35.95	36	1.47	0.3	0.94
07-14	26.71	27.19	27.81	18	22.17	35.89	35.91	35.96	1.64	0.38	0.95
07-15	23.36	23.58	24.16	18.18	22.37	36.23	36.25	36.31	1.46	0.33	0.94
07-16	21.65	21.52	22.13	18.08	22.24	36.02	36.04	36.09	1.32	0.26	0.94
07-17	19.54	19.36	19.67	18.15	22.32	36.16	36.18	36.23	1.2	0.14	0.95
07-18	19.49	19.36	19.89	18.35	22.55	36.53	36.56	36.62	1.22	0.19	0.94
07-19	20.64	20.37	21.12	18.38	22.57	36.59	36.61	36.68	1.27	0.34	0.93
07-20	24.24	24.6	25.17	18.31	22.5	36.46	36.48	36.52	1.52	0.33	0.94
07-21	20.11	19.93	20.46	18.29	22.49	36.41	36.46	36.52	1.23	0.28	0.93
07-22	23.06	22.92	23.72	18.27	22.47	36.4	36.42	36.49	1.43	0.35	0.93
07-23	21.87	21.82	22.22	18.26	22.44	36.36	36.39	36.43	1.34	0.32	0.93
07-24	20.94	20.81	21.12	18.23	22.37	36.27	36.28	36.32	1.28	0.29	0.94
07-25	13.99	13.29	13.86	18.28	22.46	36.4	36.42	36.47	0.82	0.23	0.91
07-26	9.99	10.08	10.21	18.37	22.55	36.56	36.59	36.61	0.62	0.03	0.93
07-27	5.06	4.93	5.15	18.44	22.61	36.69	36.7	36.73	0.31	-0.02	0.94

(Continues)

Table 5.6 (*Continued*)

Date	Current A (A)	Current B (A)	Current C (A)	Voltage A (kV)	Voltage C (kV)	Voltage AB (kV)	Voltage BC (kV)	Voltage CA (kV)	Active power (kW)	Reactive power (kVar)	Power factor
07-28	4.05	4.09	4.14	18.25	22.33	36.28	36.3	36.34	0.24	-0.05	0.92
07-29	3.26	3.3	3.7	18.5	22.64	36.79	36.81	36.86	0.19	-0.1	0.86
07-30	3.34	3.34	3.48	18.5	22.6	36.75	36.77	36.78	0.19	-0.1	0.85
07-31	3.21	2.95	3.12	18.42	22.56	36.64	36.65	36.71	0.18	-0.09	0.86
08-01	2.95	2.95	3.04	18.54	22.65	36.83	36.84	36.86	0.16	-0.1	0.83
08-02	3.39	3.17	3.39	18.48	22.61	36.74	36.75	36.79	0.16	-0.13	0.74
08-03	2.73	2.55	2.73	18.57	22.69	36.9	36.9	36.95	0.13	-0.11	0.74
08-04	2.95	2.86	2.9	18.49	22.59	36.72	36.74	36.77	0.14	-0.12	0.74
08-05	3.17	2.99	3.34	18.42	22.54	36.62	36.63	36.66	0.16	-0.12	0.79
08-06	3.65	3.61	3.83	18.44	22.55	36.64	36.66	36.7	0.15	-0.18	0.62
08-07	3.74	2.51	3.21	18.43	22.55	36.61	36.65	36.66	0.2	0	0.93
08-08	2.95	2.11	2.6	18.65	22.79	37.04	37.06	37.1	0.16	0	0.95
08-09	3.56	2.95	3.17	18.48	22.69	36.77	36.79	36.82	0.2	0.04	0.97
08-10	2.73	2.6	2.73	18.46	22.66	36.72	36.74	36.8	0.17	0	0.97
08-11	2.99	2.46	2.86	18.45	22.62	36.68	36.7	36.74	0.17	-0.03	0.93
08-12	4.53	4.27	4.53	18.19	22.34	36.22	36.22	36.27	0.28	-0.02	0.97
08-13	5.37	5.06	5.32	18.49	22.68	36.78	36.78	36.83	0.33	0.02	0.94
08-14	4.84	4.71	5.19	18.52	22.74	36.84	36.86	36.93	0.3	0.02	0.92

(*Continues*)

08-15	4.84	4.22	4.62	18.29	22.52	36.44	36.45	36.53	0.28	0.02	0.94
08-16	4.62	4.62	4.75	18.4	22.63	36.64	36.65	36.7	0.29	0.02	0.95
08-17	4.71	4.27	4.75	18.22	22.41	36.3	36.28	36.35	0.27	-0.03	0.91
08-18	4.84	4.4	4.71	18.48	22.73	36.79	36.82	36.88	0.29	-0.03	0.91
08-19	4.44	4.09	4.58	18.31	22.52	36.46	36.47	36.54	0.27	-0.03	0.93
08-20	4.75	4.36	4.75	18.54	22.75	36.87	36.88	36.94	0.27	0.03	0.89
08-21	5.9	5.32	5.85	18.34	22.57	36.52	36.54	36.61	0.35	0.06	0.91
08-22	5.02	4.31	4.97	18.29	22.52	36.44	36.46	36.54	0.28	-0.03	0.9
08-23	6.47	5.98	6.42	18.44	22.68	36.7	36.73	36.8	0.39	0.03	0.95
08-24	5.41	5.1	5.54	18.31	22.52	36.45	36.47	36.54	0.32	0.02	0.92
08-25	7.57	6.86	7.17	18.3	22.51	36.46	36.45	36.52	0.45	0.05	0.95
08-26	6.25	5.68	6.29	18.53	22.77	36.86	36.89	36.96	0.38	-0.03	0.94
08-27	7.61	6.91	7.35	18.31	22.55	36.48	36.5	36.57	0.44	0.07	0.94
08-28	8.23	7.66	8.14	18.33	22.6	36.55	36.56	36.63	0.49	0.08	0.93
08-29	8.8	7.7	8.36	18.4	22.66	36.66	36.67	36.74	0.5	0.05	0.91
08-30	9.46	8.32	9.02	18.42	22.7	36.72	36.74	36.8	0.55	0.08	0.93
08-31	9.99	9.59	9.99	18.32	22.56	36.5	36.51	36.58	0.61	0.14	0.94
09-01	9.81	9.28	9.81	18.34	22.59	36.55	36.57	36.64	0.59	0.11	0.93
09-02	9.99	9.86	10.21	18.36	22.58	36.53	36.58	36.6	0.62	0.07	0.94
09-03	10.3	10.08	10.52	18.11	22.32	36.1	36.12	36.16	0.62	0.15	0.93

(Continues)

Table 5.6 (Continued)

Date	Current A (A)	Current B (A)	Current C (A)	Voltage A (kV)	Voltage C (kV)	Voltage AB (kV)	Voltage BC (kV)	Voltage CA (kV)	Active power (kW)	Reactive power (kVar)	Power factor
09-04	13.73	13.51	13.82	18.25	22.51	36.38	36.42	36.47	0.84	0.22	0.93
09-05	13.68	12.89	13.51	18.35	22.59	36.56	36.57	36.64	0.82	0.19	0.94
09-06	13.29	12.8	13.24	18.44	22.7	36.73	36.76	36.81	0.81	0.19	0.93
09-07	13.42	13.29	13.42	18.25	22.47	36.36	36.38	36.42	0.82	0.2	0.94
09-08	11.84	11.79	12.01	18.32	22.56	36.47	36.53	36.58	0.73	0.16	0.94
09-09	13.02	12.41	12.98	18.51	22.8	36.85	36.91	36.95	0.82	0.1	0.96
09-10	13.29	12.94	13.42	18.71	23.01	37.21	37.28	37.32	0.82	0.21	0.93
09-11	13.6	13.11	13.55	18.65	22.9	37.09	37.12	37.19	0.83	0.23	0.93
09-12	14.04	13.73	14.12	18.65	22.94	37.1	37.16	37.2	0.87	0.25	0.93
09-13	13.6	12.98	13.42	18.66	22.94	37.12	37.17	37.22	0.82	0.25	0.92
09-14	11.13	10.34	10.65	18.52	22.77	36.85	36.89	36.94	0.67	0.08	0.94
09-15	12.23	11.88	12.28	18.53	22.8	36.87	36.93	36.98	0.76	0.14	0.95
09-16	12.14	11.66	11.97	18.53	22.81	36.89	36.94	37	0.75	0.11	0.95
09-17	13.24	12.76	13.16	18.54	22.8	36.9	36.94	37.01	0.81	0.18	0.94
09-18	11.66	10.87	11.31	18.57	22.85	36.97	37	37.08	0.7	0.16	0.93
09-19	9.59	9.2	9.33	18.4	22.66	36.65	36.69	36.74	0.57	0.14	0.92
09-20	11.22	10.52	10.78	18.37	22.63	36.61	36.63	36.7	0.67	0.11	0.93
09-21	15.14	14.7	15.18	18.19	22.43	36.28	36.29	36.35	0.92	0.2	0.93

(Continues)

09-22	20.2	19.89	20.37	18.17	22.41	36.23	36.27	36.32	1.21	0.35	0.93
09-23	19.01	18.83	19.4	18.22	22.44	36.3	36.34	36.38	1.18	0.21	0.94
09-24	11.44	10.87	11.22	18.19	22.43	36.25	36.3	36.32	0.69	0.12	0.94
09-25	15.66	15.31	15.44	18.17	22.39	36.22	36.25	36.29	0.94	0.18	0.94
09-26	19.8	19.58	19.67	18.13	22.37	36.17	36.19	36.24	1.2	0.27	0.93
09-27	19.93	19.76	20.06	18.14	22.37	36.17	36.19	36.25	1.21	0.32	0.93
09-28	22.18	22.04	22.48	18.03	22.25	35.97	35.98	36.05	1.35	0.33	0.93
09-29	26.62	26.62	27.24	18.05	22.27	36.01	36.02	36.09	1.61	0.5	0.93
09-30	21.47	20.9	21.3	18.11	22.35	36.12	36.16	36.22	1.28	0.37	0.92
10-01	19.18	19.23	19.45	18.15	22.37	36.19	36.2	36.27	1.14	0.37	0.91

including plains and wetlands, making it an ideal region for research on facility-based and ecological circular agriculture. The Zhuozhou Experimental Base, with its well-established research infrastructure, mature aquaponics farming system, and diversified production scenarios involving tilapia cultivation and cucumber and lettuce planting, provides a solid practical platform for the project's technological research, equipment testing, and system optimization. The research outcomes hold significant demonstration and promotion value for the design and application of similar agricultural microgrids in northern China and across the country.

The project systematically analyzes the overall architecture of the experimental base's microgrid, detailing the technical parameters and operational characteristics of core energy equipment, such as evaporative coolers, photovoltaic arrays, energy storage batteries, and transformers, as well as load-side devices including water treatment, environmental control, and intelligent monitoring systems. Through continuous monitoring of greenhouse environmental factors: carbon dioxide concentration, relative humidity, light intensity, and temperature; and water quality indicators from six aquaculture tanks: dissolved oxygen, water temperature, and pH, a comprehensive dataset of real operational data is obtained. Based on this analysis, the project evaluates the rationality of the system configuration, the efficiency of coordinated equipment operation, and the effectiveness of environmental regulation. These findings not only provide a scientific foundation for optimizing energy efficiency and intelligent control within the aquaponics system but also contribute valuable experience for future research on the integration of agricultural microgrids and ecological circular agriculture, offering a practical model for technological innovation and industrial upgrading in the field.

5.2.2.2 Energy equipment parameters

This study takes the practical agricultural microgrid constructed at the Zhuozhou Experimental Base of the National Digital Fishery Innovation Center, China Agricultural University, as its research subject. The research first provides a systematic description of the overall architecture of the microgrid, clarifying its energy structure, operation mode, and key functional modules. It then conducts an in-depth analysis of the main electrical equipment on the load side, detailing key parameters such as manufacturers, model specifications, rated capacities, and operational characteristics. Based on actual operational data from the experimental base, the rationality of the system configuration and the performance of the equipment are evaluated and discussed. Through this analysis, the study not only provides complete technical support for reproducing the case but also offers valuable experience and data foundations for the design and experimental research of similar agricultural microgrids in the future.

1. Basic structure diagram
   Figure 5.5 provides a detailed illustration of an intelligent agricultural system architecture that deeply integrates aquaculture and crop cultivation. The system encompasses indoor intensive tilapia farming ponds as well as diversified facility cultivation scenarios for crops such as cucumbers and lettuce. It clearly and systematically identifies the functional roles of each core component and their energy interconnections. The system integrates various energy regulation

devices – including evaporative coolers, industrial fans, heating rods, and heat pumps – and incorporates a photovoltaic array composed of 60 solar panels together with an energy storage battery system, forming a self-sufficient, closed-loop energy structure. Moreover, to ensure efficient power transmission among modules and coordinated operation of the entire system, a 315 KVA 35/0.38 kV transformer is specially equipped, enabling intelligent and intensive interconnection among aquaculture, crop cultivation, and energy supply.

2. Load composition

   As shown in Figure 5.6, the power-driven equipment of this agricultural system exhibits a multi-level, multi-module architectural structure. The first category consists of core water treatment and aeration equipment, including variable-frequency pump sets and ultrasonic aerators, which provide the driving force for water circulation. The ultrasonic aerators enhance oxygenation efficiency through high-frequency oscillation, ensuring adequate dissolved oxygen levels for aquaculture and hydroponic crops. The second category comprises environmental regulation equipment clusters, including evaporative cooling pads, ventilation windows, heating rods, and external rotor axial fans. The evaporative cooling pads reduce ambient temperature through the evaporation of water, the ventilation windows enable air circulation and exchange, the heating rods precisely control water or air temperature, and the axial fans enhance air movement. The third category is the intelligent monitoring equipment group, equipped with a multi-parameter water quality monitoring system capable of simultaneously measuring key indicators such as temperature, pH, and dissolved oxygen. Specialized sensors – including water temperature, pH, and dissolved oxygen sensors – enable real-time data acquisition and feedback, achieving comprehensive intelligent monitoring of both aquatic and atmospheric environments. These data provide a scientific basis for precise equipment regulation.

Table 5.7 systematically and comprehensively summarizes the main electrical equipment and technical parameters of the aquaponics system at the Zhuozhou Experimental Base of China Agricultural University, Hebei Province. It covers the entire process equipment, including automatic feeders, flow meters, ultrasonic level meters, aerators, intelligent water–fertilizer integration machines, water quality monitoring IoT systems, water treatment stations, recirculation pumps, oxygen control systems, feeding control systems, air-source heating systems, and cooling fans. Detailed multi-dimensional specifications, such as manufacturers and models, quantities, unit power, load types, functions, and operating times, are provided.

Among them, devices such as the automatic feeder, aerator, intelligent water–fertilizer integration machine, three-phase asynchronous motor, scroll compressor of the air-source heating system, and cooling fans belong to controllable loads, which can flexibly adjust their operating status and duration according to real-time aquaculture demands to optimize energy efficiency. In contrast, devices such as flow meters, ultrasonic level meters, water quality monitoring IoT systems, water inlet treatment stations, three-phase asynchronous motors and vertical multistage

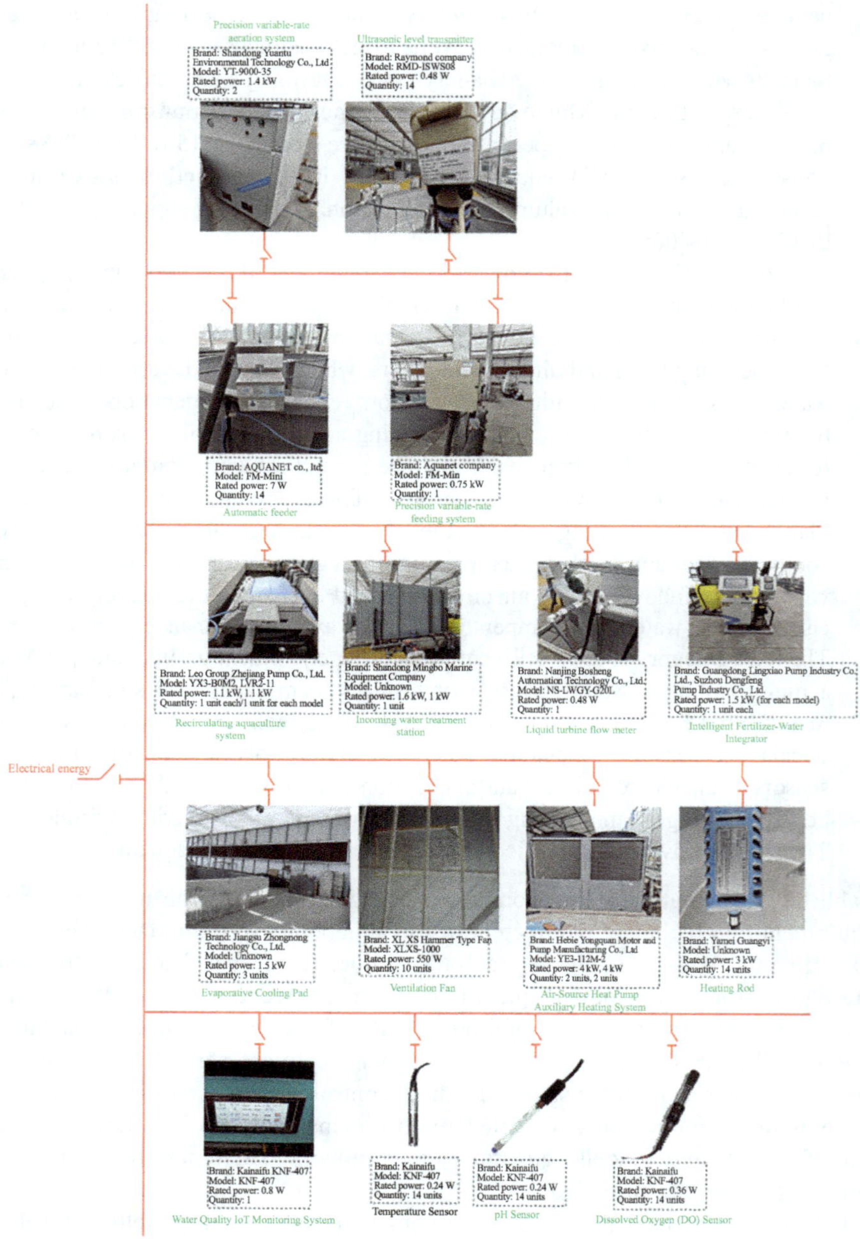

Figure 5.6 Diagram of greenhouse energy load composition

pumps of the recirculating system, precision variable oxygenation systems, and precision variable feeding systems are uncontrollable loads, requiring continuous and stable operation to ensure reliable implementation of key functions such as water circulation, water quality monitoring, and basic equipment control.

Table 5.7 Main energy equipment and technical parameters

Type	Manufacturer and model	Quantity	Power per unit (W)	Load type	Function	Operating time
Automatic feeder	Aquanet Co. FM-Mini	8	7	Controllable load	Saves labor, automatically dispenses fish feed	Spring: 8:00–9:00, 16:00–17:00; Summer: 8:00–9:00, 12:00–13:00, 16:00–17:00; Autumn: 8:00–9:00, 16:00–17:00; Winter: 12:00–13:00
Flow meter	Nanjing Bosheng Automation Technology Co., Ltd. NS-LWGY-G20L	8	0.48	Uncontrollable load	Records the water flow in the pond	Always on
Ultrasonic level meter	Ruimengde Co. RMD-ISW508	8	0.48	Uncontrollable load	Records the water level in the pond	Always on
Aerator	Shandong Yuantu Environmental Technology Co., Ltd. YT-9000-20	1	1400	Controllable load	Supplies oxygen to the pond	1:00–9:00 (early morning)
Intelligent water–ertilizer integration machine	Shouguang Kunyang Automation Technology Co., Ltd. KY-2T-ZD	1	1500	Controllable load	Uses pond water to deliver fertilizer to agricultural areas	8:00–16:00 daily, runs 2 minutes per hour
Water quality monitoring IoT system	Kainafu KNF-407	1	0.8	Uncontrollable load	Records dissolved oxygen, temperature, conductivity, turbidity, etc. of pond water	Always on

(Continues)

Table 5.7 (Continued)

Type	Manufacturer and model	Quantity	Power per unit (W)	Load type	Function	Operating time
Water inlet treatment station	No data	1	No data	Uncontrollable load	Performs preliminary treatment of pond water – filtration and pollutant removal	Always on
Three-phase asynchronous motor (recirculating aquaculture system)	Leo Group Zhejiang Pump Industry Co., Ltd. YX3-80M2-2	1	1100	Uncontrollable load	Filters wastewater from the pond, removes pollutants, and recirculates water	Always on
Vertical multistage pump (recirculating aquaculture system)	Leo Group Zhejiang Pump Industry Co., Ltd. LVR2-11	1	1100	Uncontrollable load		Always on
Precision variable oxygenation system	No data	1	No data	Uncontrollable load	Controls aerator operation	Always on
Precision variable feeding system	Laizhou Mingbo Aquatic Products Co., Ltd.	1	No data	Uncontrollable load	Controls automatic feeder operation	Always on
Three-phase asynchronous motor (air-source heating system)	Hebei Yongquan Motor Pump Manufacturing Co., Ltd. YE3-112M-2	2	4000	Controllable load	Temperature regulation (heating assistance)	Always on during winter
Scroll compressor (air-source heating system)	Danfoss Co.	2	4000	Controllable load		
Fan	Manufacturer unknown XLXS-1000	10	550	Controllable load	Temperature regulation (cooling)	Summer: 10:00–18:00

This table provides comprehensive and essential baseline data for formulating energy management strategies, designing coordinated equipment operation mechanisms, and optimizing overall energy efficiency in aquaculture systems, serving as an important foundation for system modeling and intelligent control research.

5.2.2.3 Actual operating data

The research team systematically collected continuous environmental monitoring data during the operation of the greenhouse at the experimental base. The data collection began on August 1, 2025, and only a portion is presented here, while the complete dataset is available in the appendix. The monitoring indicators comprehensively cover key indoor and outdoor environmental parameters as well as core water quality metrics from six tanks. Specifically, indoor environmental parameters include carbon dioxide concentration, relative humidity, light intensity, and temperature, while water quality indicators focus on dissolved oxygen (DO), water temperature, and pH in each tank, forming a complete and continuous environmental monitoring dataset. From the initial 25-minute monitoring results, the indoor environment exhibited remarkable stability. The CO_2 concentration fluctuated slightly between 538 and 553 ppm, showing a mild range of variation. The relative humidity remained within 80.1–80.9%, with no significant fluctuations, reflecting stable operation of the humidity control system. The light intensity consistently remained at 0 Lux, suggesting that the monitoring occurred either in a sealed, light-shielded environment or during nighttime. The indoor temperature was constant at 25.4 °C, demonstrating high precision in temperature control without any deviations. The water quality conditions of the six tanks also demonstrated good stability. Although minor fluctuations were observed, all parameters remained within reasonable ranges. Taking Tank 1 as an example, the dissolved oxygen level varied slightly between 8.88 and 8.97 mg/L, which falls within the optimal range for aquatic organisms. The water temperature remained stable between 18.33 and 18.37 °C, showing minimal variation and supporting a balanced aquatic ecosystem. The pH value was maintained between 8.30 and 8.31, indicating a weakly alkaline environment suitable for most aquaculture and experimental water systems. The remaining five tanks exhibited similar patterns, with no abnormal fluctuations, indicating that the water quality regulation system was functioning properly and maintaining stable environmental conditions for aquatic life and experimentation.

Although only the initial phase of data collection is presented here, it already clearly demonstrates the stability of the monitored environment at the early stage. As time progresses, environmental parameters may adjust in response to external environmental changes, equipment operation cycles, or experimental requirements. These evolving patterns merit further analysis. This continuous data collection effort not only documents the initial states of both environmental and water quality conditions but also provides a solid foundation for subsequent studies on ecological dynamics, environmental regulation effects, and water quality stability. The dataset thus serves as a valuable reference for related research and practical applications in intelligent aquaculture and environmental management (Table 5.8).

Table 5.8 Actual operating data of environmental factors in the aquaponics agricultural demonstration zone of Zhuozhou Experimental Base

ts	Room CO2 (ppm)	Room Humidity (%RH)	Room Illumination (Lux)	Room Temperature (°C)	Tank1 Oxygen (mg/L)	Tank1 Temperature (°C)	Tank1 pH (pH)	Tank2 Oxygen (mg/L)	Tank2 Temperature (°C)	Tank2 pH (pH)	Tank3 Oxygen (mg/L)	Tank3 Temperature (°C)
2025-08-01 00:04:52	538	80.6	0	25.4	8.97	18.37	8.3	10.33	18.33	8.02	9.8	18.4
2025-08-01 00:09:52	538	80.5	0	25.4	8.88	18.37	8.31	10.34	18.33	8.03	9.84	18.41
2025-08-01 00:14:52	553	80.1	0	25.4	8.96	18.33	8.31	10.34	18.32	8.03	9.83	18.41
2025-08-01 00:19:52	544	80.9	0	25.4	8.88	18.33	8.31	10.32	18.33	8.03	9.8	18.42
2025-08-01 00:29:52	543	80.5	0	25.4	8.93	18.35	8.31	10.34	18.34	8.02	9.81	18.44
2025-08-01 00:34:52	545	80.3	0	25.4	8.9	18.36	8.32	10.34	18.35	8.03	9.81	18.45
2025-08-01 00:39:52	554	80.4	0	25.3	8.96	18.39	8.32	10.33	18.34	8.04	9.79	18.43
2025-08-01 00:44:52	546	80.3	0	25.3	8.88	18.42	8.32	10.34	18.36	8.04	9.85	18.45
2025-08-01 00:49:52	538	80.6	0	25.3	8.95	18.44	8.32	10.36	18.37	8.03	9.85	18.47
2025-08-01 00:54:52	546	80.5	0	25.3	8.9	18.39	8.33	10.38	18.37	8.04	9.83	18.47
2025-08-01 00:59:52	552	80.4	0	25.3	8.86	18.39	8.32	10.37	18.37	8.02	9.82	18.47

(Continues)

2025-08-01 01:04:52	552	80.2	0	25.3	8.81	18.4	8.33	10.36	18.37	8.04	9.82	18.48
2025-08-01 01:14:52	548	80.2	0	25.3	8.88	18.43	8.33	10.35	18.38	8.05	9.82	18.48
2025-08-01 01:19:52	553	79.9	0	25.3	8.91	18.42	8.32	10.36	18.39	8.05	9.85	18.51
2025-08-01 01:24:52	545	79.8	0	25.3	8.85	18.44	8.33	10.38	18.39	8.04	9.85	18.5
2025-08-01 01:29:52	532	79.9	0	25.3	8.95	18.48	8.35	10.36	18.4	8.05	9.84	18.51
2025-08-01 01:34:52	548	79.7	0	25.3	8.95	18.43	8.33	10.37	18.4	8.04	9.83	18.51
2025-08-01 01:39:52	537	79.7	0	25.2	8.93	18.44	8.34	10.34	18.39	8.05	9.82	18.51
2025-08-01 01:44:52	542	79.6	0	25.2	8.89	18.47	8.33	10.36	18.39	8.05	9.83	18.52
2025-08-01 01:49:52	536	79.5	0	25.3	8.94	18.48	8.35	10.38	18.4	8.05	9.83	18.52
2025-08-01 01:54:52	534	79.6	0	25.2	8.82	18.47	8.35	10.37	18.4	8.06	9.83	18.52
2025-08-01 02:04:52	537	79.6	0	25.2	8.87	18.52	8.35	10.39	18.43	8.06	9.84	18.56
2025-08-01 02:09:52	538	79.6	0	25.2	8.99	18.5	8.35	10.37	18.42	8.05	9.84	18.54
2025-08-01 02:14:52	544	79.4	0	25.2	8.81	18.47	8.34	10.36	18.42	8.06	9.84	18.55

(Continues)

Table 5.8 (Continued)

ts	Room CO2 (ppm)	Room Humidity (%RH)	Room Illumination (Lux)	Room Temperature (°C)	Tank1 Oxygen (mg/L)	Tank1 Temperature (°C)	Tank1 pH (pH)	Tank2 Oxygen (mg/L)	Tank2 Temperature (°C)	Tank2 pH (pH)	Tank3 Oxygen (mg/L)	Tank3 Temperature (°C)
2025-08-01 02:19:52	532	79.7	0	25.2	8.91	18.51	8.35	10.36	18.44	8.06	9.84	18.56
2025-08-01 02:24:52	531	79.8	0	25.1	8.8	18.49	8.34	10.35	18.43	8.06	9.85	18.57
2025-08-01 02:29:52	539	79.5	0	25.1	8.72	18.53	8.35	10.33	18.45	8.06	9.82	18.58
2025-08-01 02:34:52	535	79.8	0	25.1	8.82	18.51	8.35	10.36	18.45	8.07	9.81	18.57
2025-08-01 02:39:52	532	79.9	0	25.1	8.84	18.53	8.35	10.38	18.46	8.07	9.79	18.59
2025-08-01 02:44:52	520	80.1	0	25	8.76	18.57	8.34	10.35	18.45	8.06	9.82	18.59
2025-08-01 02:49:52	520	80.1	0	25	8.82	18.52	8.36	10.38	18.45	8.06	9.8	18.57
2025-08-01 02:54:52	526	80.2	0	25	8.75	18.56	8.35	10.37	18.47	8.07	9.81	18.6
2025-08-01 02:59:52	529	80	0	25	8.84	18.56	8.36	10.38	18.45	8.08	9.81	18.58
2025-08-01 03:04:52	533	80	0	24.9	8.85	18.55	8.36	10.36	18.45	8.08	9.77	18.59
2025-08-01 03:09:52	526	80.3	0	24.9	8.81	18.54	8.36	10.34	18.46	8.08	9.82	18.59

(Continues)

2025-08-01 03:14:52	536	80.1	0	24.8	8.88	18.56	8.36	10.37	18.47	8.09	9.79	18.61
2025-08-01 03:19:52	530	80.6	0	24.8	8.83	18.57	8.37	10.36	18.47	8.08	9.73	18.6
2025-08-01 03:24:52	532	80.5	0	24.7	8.77	18.59	8.37	10.35	18.47	8.08	9.75	18.61
2025-08-01 03:29:52	535	80.7	0	24.7	8.86	18.54	8.37	10.42	18.48	8.08	9.76	18.62
2025-08-01 03:34:52	525	81	0	24.7	8.83	18.59	8.37	10.38	18.48	8.09	9.72	18.61
2025-08-01 03:39:52	538	80.9	0	24.6	8.83	18.62	8.36	10.35	18.5	8.09	9.74	18.63
2025-08-01 03:44:52	539	81.3	0	24.5	8.8	18.62	8.37	10.39	18.5	8.09	9.76	18.62
2025-08-01 03:49:52	546	81.7	0	24.5	8.82	18.59	8.36	10.35	18.48	8.09	9.74	18.62
2025-08-01 03:54:52	545	81.6	0	24.5	8.81	18.61	8.36	10.38	18.5	8.09	9.78	18.65
2025-08-01 03:59:52	547	81.8	0	24.4	8.78	18.62	8.37	10.37	18.51	8.1	9.75	18.65
2025-08-01 04:04:52	551	81.8	0	24.4	8.84	18.58	8.37	10.34	18.49	8.1	9.76	18.63
2025-08-01 04:09:52	557	81.8	0	24.4	8.87	18.61	8.39	10.34	18.5	8.1	9.73	18.65
2025-08-01 04:14:52	555	82.1	0	24.4	8.92	18.58	8.37	10.37	18.51	8.1	9.74	18.65

(Continues)

Table 5.8 (*Continued*)

ts	Room CO2 (ppm)	Room Humidity (%RH)	Room Illumination (Lux)	Room Temperature (°C)	Tank1 Oxygen (mg/L)	Tank1 Temperature (°C)	Tank1 pH (pH)	Tank2 Oxygen (mg/L)	Tank2 Temperature (°C)	Tank2 pH (pH)	Tank3 Oxygen (mg/L)	Tank3 Temperature (°C)
2025-08-01 04:19:52	559	82.1	0	24.4	8.77	18.66	8.38	10.37	18.52	8.09	9.72	18.65
2025-08-01 04:29:52	571	82.2	0	24.3	8.82	18.65	8.38	10.35	18.52	8.1	9.73	18.67
2025-08-01 04:34:52	569	82.4	0	24.3	8.69	18.67	8.38	10.34	18.52	8.1	9.72	18.66
2025-08-01 04:39:52	571	82.7	0	24.3	8.68	18.67	8.38	10.37	18.53	8.1	9.7	18.67
2025-08-01 04:44:52	581	82.4	0	24.3	8.83	18.68	8.38	10.4	18.52	8.1	9.68	18.66
2025-08-01 04:49:52	583	82.5	0	24.2	8.85	18.63	8.38	10.36	18.54	8.1	9.72	18.68
2025-08-01 04:54:52	585	82.9	0	24.2	8.77	18.67	8.38	10.38	18.54	8.1	9.71	18.68
2025-08-01 04:59:52	583	82.7	0	24.2	8.74	18.62	8.38	10.36	18.54	8.1	9.7	18.68
2025-08-01 05:04:52	591	82.6	0	24.2	8.74	18.65	8.38	10.35	18.53	8.1	9.71	18.67
2025-08-01 05:09:52	597	82.8	0	24.2	8.63	18.62	8.37	10.35	18.54	8.11	9.71	18.69
2025-08-01 05:14:52	595	82.9	0	24.2	8.78	18.67	8.38	10.35	18.56	8.11	9.7	18.7

(Continues)

2025-08-01 05:24:52	590	83	0	24.2	8.69	18.64	8.37	10.39	18.54	8.11	9.72	18.69
2025-08-01 05:29:52	598	83.1	0	24.2	8.67	18.71	8.38	10.33	18.56	8.11	9.7	18.7
2025-08-01 05:34:52	599	83.3	0	24.2	8.64	18.69	8.38	10.34	18.55	8.11	9.75	18.69
2025-08-01 05:39:52	602	83.3	0	24.2	8.76	18.67	8.39	10.36	18.56	8.11	9.74	18.69
2025-08-01 05:44:52	600	83	0	24.3	8.67	18.65	8.37	10.34	18.56	8.11	9.69	18.7
2025-08-01 05:49:52	603	82.9	0	24.3	8.76	18.73	8.39	10.32	18.57	8.11	9.73	18.7
2025-08-01 05:54:52	598	83.1	3	24.4	8.73	18.68	8.39	10.36	18.57	8.11	9.71	18.7
2025-08-01 05:59:52	599	83.1	3	24.4	8.62	18.66	8.37	10.32	18.57	8.11	9.71	18.72
2025-08-01 06:04:52	604	82.9	3	24.4	8.7	18.67	8.38	10.37	18.59	8.11	9.69	18.73
2025-08-01 06:14:52	597	82.5	6	24.5	8.74	18.74	8.39	10.32	18.58	8.11	9.68	18.72
2025-08-01 06:19:52	590	82.4	6	24.5	8.65	18.66	8.38	10.33	18.58	8.11	9.69	18.73
2025-08-01 06:24:52	586	82.5	6	24.6	8.62	18.67	8.38	10.33	18.58	8.11	9.66	18.73
2025-08-01 06:29:52	581	82.4	6	24.6	8.66	18.73	8.38	10.36	18.59	8.12	9.71	18.74

(Continues)

Table 5.8 (Continued)

ts	Room CO2 (ppm)	Room Humidity (%RH)	Room Illumination (Lux)	Room Temperature (°C)	Tank1 Oxygen (mg/L)	Tank1 Temperature (°C)	Tank1 pH (pH)	Tank2 Oxygen (mg/L)	Tank2 Temperature (°C)	Tank2 pH (pH)	Tank3 Oxygen (mg/L)	Tank3 Temperature (°C)
2025-08-01 06:34:52	584	82.2	9	24.7	8.61	18.74	8.38	10.34	18.59	8.12	9.66	18.73
2025-08-01 06:39:52	597	81.7	9	24.8	8.58	18.74	8.39	10.34	18.6	8.12	9.65	18.73
2025-08-01 06:44:52	596	81.6	9	24.9	8.6	18.74	8.39	10.35	18.59	8.12	9.66	18.74
2025-08-01 06:49:52	575	82	12	24.9	8.64	18.69	8.39	10.35	18.61	8.11	9.66	18.76
2025-08-01 06:54:53	586	81.4	12	25	8.58	18.75	8.39	10.34	18.62	8.11	9.69	18.76
2025-08-01 06:59:53	560	81.6	12	25.1	8.61	18.73	8.39	10.35	18.62	8.11	9.67	18.77
2025-08-01 07:09:53	537	80.7	12	25.3	8.61	18.74	8.38	10.31	18.63	8.11	9.65	18.78
2025-08-01 07:14:53	544	80.4	12	25.4	8.56	18.79	8.38	10.35	18.63	8.11	9.66	18.78
2025-08-01 07:19:53	546	79.8	15	25.5	8.61	18.79	8.39	10.3	18.63	8.12	9.65	18.78
2025-08-01 07:24:53	554	79.2	18	25.6	8.64	18.77	8.4	10.35	18.64	8.11	9.66	18.79
2025-08-01 07:29:53	529	79	18	25.7	8.63	18.79	8.38	10.34	18.64	8.12	9.61	18.78

(Continues)

2025-08-01 07:34:53	532	78.4	18	25.8	8.55	18.75	8.39	10.37	18.65	8.12	9.66	18.79
2025-08-01 07:39:53	528	78.2	18	25.9	8.52	18.75	8.37	10.22	18.65	8.14	9.69	18.79
2025-08-01 07:44:53	537	77.6	21	26	8.5	18.78	8.38	10.32	18.65	8.13	9.73	18.78
2025-08-01 07:49:53	504	77.6	21	26.2	8.68	18.8	8.39	10.33	18.66	8.13	9.62	18.79
2025-08-01 07:54:53	495	77.6	21	26.3	8.55	18.8	8.39	10.32	18.66	8.13	9.63	18.8
2025-08-01 07:59:53	488	77.1	24	26.4	8.6	18.81	8.39	10.35	18.66	8.13	9.66	18.8
2025-08-01 08:04:53	478	76.9	24	26.5	8.67	18.8	8.39	10.34	18.67	8.13	9.67	18.82
2025-08-01 08:14:53	484	76.1	27	26.8	8.58	18.79	8.38	10.35	18.67	8.13	9.66	18.81
2025-08-01 08:19:53	484	75.5	27	27	8.57	18.82	8.39	10.36	18.67	8.11	9.61	18.81
2025-08-01 08:24:53	473	75.3	30	27.1	8.62	18.8	8.38	10.31	18.68	8.12	9.61	18.83
2025-08-01 08:29:53	466	75.3	30	27.3	8.82	18.75	8.35	10.31	18.65	8.11	9.6	18.81
2025-08-01 08:34:53	474	74.5	30	27.4	8.74	18.71	8.34	10.35	18.61	8.11	9.58	18.76

(Continues)

Table 5.8 (Continued)

ts	Room CO2 (ppm)	Room Humidity (%RH)	Room Illumination (Lux)	Room Temperature (°C)	Tank1 Oxygen (mg/L)	Tank1 Temperature (°C)	Tank1 pH (pH)	Tank2 Oxygen (mg/L)	Tank2 Temperature (°C)	Tank2 pH (pH)	Tank3 Oxygen (mg/L)	Tank3 Temperature (°C)
2025-08-01 08:39:53	478	74.1	30	27.6	8.73	18.66	8.34	10.32	18.59	8.1	9.6	18.73
2025-08-01 08:44:53	486	73.4	30	27.8	8.73	18.62	8.32	10.41	18.56	8.09	9.61	18.69
2025-08-01 08:49:53	482	73	33	28	8.74	18.59	8.31	10.39	18.53	8.09	9.63	18.64
2025-08-01 08:54:53	493	72.5	33	28.1	8.74	18.55	8.3	10.34	18.49	8.07	9.6	18.58
2025-08-01 08:59:53	487	72.2	33	28.3	8.81	18.5	8.3	10.39	18.48	8.07	9.66	18.53

5.3 Real rural distribution network cases

5.3.1 Value analysis

Real rural distribution network cases are of great significance to the research and development of the Agricultural Energy Internet (AEI). They provide authentic data on load profiles, meteorological conditions, distributed energy generation, and agricultural loads such as irrigation pumps and cold storage systems. These data enable the construction of accurate and verifiable models, ensuring that simulations and optimizations reflect real-world operation rather than theoretical assumptions. Moreover, such cases serve as essential testbeds for verifying PV–storage–load coordination strategies, flexible dispatch algorithms, and AI-based forecasting models under extreme weather or data scarcity conditions. By incorporating real operational constraints – such as line capacity limits, voltage deviations, and communication delays – these cases enhance the practicality and robustness of AEI research.

In addition, real rural distribution network cases reveal the unique regional characteristics of rural energy systems, including sparse load distribution, strong renewable variability, and seasonal energy demand fluctuations. They support integrated planning for "source–grid–load–storage" systems, energy storage configuration, and microgrid topology optimization. Research grounded in these real cases also facilitates the translation of scientific results into practical engineering and policy applications, such as smart rural grid upgrades, renewable demonstration projects, and carbon reduction strategies. In essence, such cases provide a reliable data foundation, validate intelligent control algorithms, and bridge theoretical research with engineering implementation within the Agricultural Energy Internet.

5.3.2 Southern China case

5.3.2.1 Overview

Qingxin District is located in the central-northern part of Guangdong Province, China. It currently administers Taihe Town, Taiping Town, Shantang Town, Jintan Town, Shitan Town, Sankeng Town, Heyun Town, Longjing Town, and Bijia Forest Farm – one forest farm and eight towns in total. Taihe Town serves as the seat of the Qingxin District Government. The district covers a total area of 2353 square kilometers. The administrative map of Qingxin District is shown in Figure 2.1. As of 2012, the district had a permanent population of 696,000, with a population density of 268.7 people per square kilometer. Qingxin District enjoys a mild climate typical of the subtropical monsoon zone, characterized by abundant rainfall and moderate temperatures. The district receives 1600–1900 hours of annual sunshine, with an average annual temperature of 21.6 °C, a frost-free period exceeding 330 days, and an average annual rainfall of about 2000 mm. The Beijiang River and its tributaries flow through the district, providing rich water resources, with approximately 8000 kW of hydropower potential still to be developed. Located in the middle and lower reaches of the Beijiang River, Qingxin

District lies in northern Guangdong, adjoining the urban area of Qingyuan City. It borders Yingde City, Fogang County, Yangshan County, Sanshui District, Sihui City, and Guangning County, forming a transitional zone between the Pearl River Delta and the mountainous areas of northern Guangdong. With a superior geographical position, Qingxin is 68 km from Guangzhou, 40 km from Guangzhou Baiyun International Airport, and less than 200 km from Hong Kong and Macao. The district boasts a well-developed transportation network. The Beijing–Guangzhou double-track railway runs through the area, while National Highway 107, the Guangqing Expressway, and the Qinglian First-Class Highway provide north–south access across the entire district. The Yinying Highway, Qingsi Highway, and Wujing Highway traverse east–west, forming a three-dimensional network with one vertical and two horizontal corridors, serving as a major logistics channel connecting inland provinces with the Pearl River Delta. Water transportation is equally convenient – the Beijiang waterway connects to the Pearl River system, allowing cargo ships to reach Guangzhou and the Hong Kong–Macao region directly. This makes Qingxin an important logistics and distribution hub linking the delta and the northern mountains of Guangdong. Qingxin also has a rich historical and cultural heritage, tracing back to the late Neolithic period around 5000 years ago. Numerous cultural relics from various dynasties have been discovered, among which prehistoric settlements and Spring and Autumn period tombs are of national significance. The ruins of Zhongsu County, the first county seat established in the Western Han dynasty within today's Qingxin territory, were found in Cheng Village, Longjing Town. Furthermore, in Taiping Town, at the junction of Guangning, Sihui, and Qingyuan, Mount Qinhuang served as a revolutionary base for the Guangdong–Guangxi–Hunan Border Column of the People's Liberation Army. These discoveries illustrate that Qingxin possesses a long and profound history, making it one of the cultural origins of the Lingnan region. Qingxin is rich in natural resources and agricultural products. Famous local specialties include Qingyuan Chicken, Black-Maned Goose, Luokeng Bamboo Shoots, Bijia Tea, Pukeng Tea, Chijue (wild edible plant), Shatang Oranges, edible mushrooms, Shanshui Tofu, and many others – highly appreciated throughout China. The district also enjoys abundant tourism resources and enormous development potential. Preliminary surveys identify over 40 scenic and cultural sites, including geological formations, natural wonders, historical relics, rivers and lakes, agricultural attractions, ethnic customs, and hot springs. Major open scenic areas include the Qingxin Hot Spring Resort (National 4A level), Taihe Ancient Cave Scenic Area, Hongburang Agricultural Science Park, Lavender World, Bijia Mountain Ecological Resort, Bishuitan Rafting, Beijiang "Mini Three Gorges" Cruise, Jinlong Cave, Mingxia Cave, Xuanzhen Cave Ecotourism Resort, Tianzi Mountain Waterfall, Qingyuan Taoyuan Ecological Resort, Rural Ancient Village Tourism Area, Emu Ecological Park, Qingxin Sports Club, Bishazhou Scenic Area, Xisha Island of Leisure, Tianhu Tourist Area, and the National Shooting Training Base. Five more scenic zones are under planning, including Swan Lake Tourist Resort, Taihe Ancient Cave Five-Star Hotel, Gao Tian Water Park, a Five-Star Vacation Resort, and the Sankengtan Ecological Tourism and Leisure Area.

Qingxin's tourism sector has achieved an integrated "six-element" system of transportation, sightseeing, accommodation, dining, shopping, and entertainment, significantly enhancing its reception capacity. The district is actively developing three major tourism brands: the "Hot Spring Sanctuary," centered on the 4A-rated Qingxin Hot Spring Resort; The "Hometown of Rafting," featuring rafting attractions such as Bijia Mountain, Xuanzhen Cave, and Bishuitan; The "Natural Ecology Destination," focusing on Taihe Ancient Cave, Taohua Lake, and Swallow Rock Ecotourism. Together, these have formed the district's distinctive tourism framework of "Hot Spring Sanctuary – Hometown of Rafting – Natural Ecology," shaping a comprehensive tourism-based economic structure with significant growth potential.

Qingxin District's economic development objectives during the 12th Five-Year Plan period were designed to leverage both its "first-mover advantages" within Qingyuan City and its "latecomer advantages" within Guangdong Province. Guided by the development strategy of "one center, two circles and two belts, and five development clusters," Qingxin aimed to build a new district characterized by clean governance, strong economic growth, social harmony, and people's well-being. By 2015, Qingxin District set the goal of achieving a gross domestic product (GDP) exceeding 42 billion yuan, more than doubling the 2011 level, with an average annual growth rate of around 16%. The per capita GDP was projected to reach approximately 62,000 yuan. The local fiscal general budget revenue was expected to grow 20% annually, while fixed-asset investment would maintain an average annual growth rate of 13%. The industrial structure would be further optimized to achieve a ratio of 6:61:33 among the primary, secondary, and tertiary sectors. To accomplish these economic and social development goals, Qingxin District implemented coordinated measures across multiple areas, including: promoting industrial park development to strengthen the manufacturing base; advancing economic restructuring to improve efficiency and sustainability; encouraging agricultural industrialization to enhance rural productivity; developing characteristic urbanization with distinctive township features; accelerating infrastructure construction, particularly in transportation and energy; promoting ecological and environmental protection to build a green economy; improving people's livelihoods through social welfare and public services; strengthening government self-improvement and institutional reform.

Among these initiatives, several projects are closely tied to the planning and upgrading of the medium- and low-voltage distribution network, focusing on accelerating grid construction and capacity expansion. Key projects include: expansion of the 110 kV Sankeng Substation; expansion of the 220 kV Huilan Substation; expansion of the 110 kV Zhukeng Substation; construction of the 220 kV Shitan Transmission and Transformation Project; construction of the 110 kV Longwan Transmission and Transformation Project; construction of the 110 kV Jingtang Transmission and Transformation Project; construction of the 110 kV Zhongsheng Transmission and Transformation Project.

Specific implementation milestones are as follows:

- The 110 kV Sankeng Substation added a 50 MVA main transformer in 2012.
- The 220 kV Shitan Substation, with a total capacity of 2×180 MVA, was scheduled for completion in 2013.
- In 2014, new 110 kV Jingtang and 110 kV Longwan substations were planned for construction; the 220 kV Huilan Substation was to receive an additional 180 MVA transformer, and the 110 kV Zhukeng Substation would add one 50 MVA transformer.
- In 2015, two new substations – 110 kV Shantang and 110 kV Zhongsheng – were scheduled for construction.

The 110 kV Zhukeng Substation, completed and commissioned in November 2010, has since been undergoing expansion due to the rapid growth of industrial park demand. The project will upgrade the substation to three 50 MVA main transformers to meet the electricity demand for industrial infrastructure, project construction, and enterprise production. In Sankeng Town, power demand has surged in recent years due to the Phase II development of the hot spring resort and the planned construction of a luxury villa community. Consequently, the existing Sankeng Substation requires expansion to enhance its supply capacity and meet future load growth. The planned Jingtang Substation, located near Taihe Town, will supply power to new residential and commercial developments, while also sharing load from the existing Taihe and Bijia Mountain substations. This project aims to ensure adequate capacity and reliability for the district's urban power supply.

The basic situation of the 10 kV distribution network in Qingxin District, Qingyuan City, Guangdong Province, in 2012 is as follows: there are 119 circuits of 10 kV lines in total, including 88 public circuits and 31 dedicated circuits. The total length of public 10 kV lines is 1588.1 km, including 141.7 km of cable lines and 1446.4 km of overhead lines, of which 12.89 km are insulated overhead lines. The network contains 383 switchgear panels, 162 pole-mounted switches, and 26 switching stations. A detailed summary of medium-voltage distribution network equipment is listed in Table 5.9.

In 2012, the equipment of the medium- and low-voltage distribution network in each township of Qingxin District was counted, as shown in Table 5.10. Among them, Taihe Town is the main area where medium- and low-voltage distribution equipment is concentrated.

The overall voltage qualification rate of the Qingxin District power grid is 93.92%, and the total line loss rate is 3.51%. The power supply reliability indices RS-1 and RS-3 are both 99.91%. The operating technical indicators of the distribution network are shown in the next section (Table 5.11).

The public distribution transformer load and distribution status in Qingxin District in 2011 are shown in Table 5.12.

From Table 5.12, it can be seen that most public distribution transformers are operating at a reasonable load rate. However, there are 100 lightly loaded transformers, accounting for 6.98% of the total; 155 heavily loaded transformers, accounting for 10.82%; and 51 overloaded transformers, accounting for 3.56%.

Table 5.9 Current medium-voltage distribution network equipment statistics

Area	Public feeders (No.)	Cable (km)	Overhead line (km)	Total (km)	Switchgear (sets)	Pole-mounted switches (units)	Switching stations (sites)	Dedicated feeders (No.)	Public feeders (No.)
Qingxin District – D	35	115.17	114.08	229.25	380	34	26	13	35
Qingxin District – E	53	26.51	1332.34	1358.85	3	128	0	18	53

Table 5.10 Medium-voltage distribution network equipment statistics by township in Qingxin District (2011)

Township	Public feeders (No.)	Dedicated feeders (No.)	Cable (km)	Overhead line (km)	Total (km)	Switchgear (sets)	Pole-mounted switches (units)	Switching stations (sites)
Taihe	35	13	115.17	114.08	229.25	380	34	26
Heyun	18	5	3.02	249.08	252.10	0	20	0
Taiping	15	5	17.00	217.84	234.84	0	17	0
Shantang	4	1	2.04	127.23	129.27	3	16	0
Sankeng	3	3	1.42	94.25	95.67	0	9	0
Longjing	6	2	1.70	273.66	275.36	0	31	0
Shitan	3	1	0.40	197.89	198.29	0	14	0
Jintan	4	1	0.93	172.40	173.33	0	21	0
Total	88	31	141.68	1446.42	1588.10	383	162	26

Table 5.11 Operating technical indicators of the current distribution network

Area	Comprehensive voltage qualification rate (%)	Comprehensive line loss rate (%)	Power supply reliability (RS-1) (%)	Power supply reliability (RS-3) (%)	Comprehensive voltage qualification rate (%)
Qingxin District	93.92	3.51	99.91	99.91	93.92

Table 5.12 Distribution of current public distribution transformer loads in Qingxin District

Township	0–30% (units)	30–80% (units)	80–100% (units)	Above 100% (units)	Total (units)
Taihe	18	183	23	13	237
Heyun	4	172	8	10	194
Taiping	3	114	16	0	133
Shantang	35	132	10	0	177
Sankeng	2	136	15	7	160
Longjing	6	158	52	11	227
Shitan	26	90	22	4	142
Jintan	6	142	9	6	163
Total	100	1127	155	51	1433

Lightly loaded transformers are mainly concentrated in Shantang and Shitan, located in remote mountain areas where power usage is seasonal and related to farming cycles. Heavily loaded transformers are concentrated in Taihe and Longjing, while overloaded transformers are mainly in Heyun and Taihe. In Taihe, heavy and overloaded transformers are due to economic growth and increased residential demand. In Heyun, the overload is caused by industrial park expansion and associated residential load increases. With the advancement of the "Home Appliances to the Countryside" program, rural electricity demand has risen sharply in recent years. Heavy and overloaded conditions are primarily caused by the rapid development of agricultural electricity consumption, improved living standards, and festival-related peaks. Some transformers can no longer meet the rural power demand in Taihe Town. In areas with favorable conditions, pilot rural substations should be established to improve power supply concentration and reliability. In Shantang, heavy transformer loads are mainly due to electricity consumption for fishpond aeration. This should be addressed through network reinforcement and the installation of new transformers. In Heyun, four transformers are lightly loaded because their capacity is large relative to the local demand. In Shitan, many lightly loaded transformers are found in sparsely populated areas, and future network upgrades should tailor plans for each township's local characteristics.

A detailed summary of heavy and overloaded distribution transformers in Qingxin District is given in Table 5.13.

From Table 5.13, it can be observed that there are 155 heavy-load transformers with a total capacity of 27.25 MVA, accounting for 10.82% of all public transformers, and 51 overloaded transformers with a total capacity of 7.86 MVA, accounting for 3.56%. The main causes of transformer overloading are: rapid improvement of living standards and rural electrification; government subsidy programs for household appliances, which increased residential loads significantly; urban expansion, where new housing developments overloaded existing transformers.

Among the 88 existing public feeders in Qingxin County, ten feeders have minimum voltages more than 7% below the rated voltage, accounting for 11.36% of the total. Low terminal voltage can negatively affect users' power quality. It is therefore recommended to take measures such as increasing conductor size and shortening power supply distances as soon as possible to ensure acceptable voltage levels at the end of the lines. The statistics of voltage quality for Qingxin County's current public feeders are shown in Table 5.14.

As shown in Table 5.14, the low-voltage problem in Qingxin District's 10 kV public lines is not serious. There are ten circuits (11.36%) whose terminal voltages are more than 7% below the rated value, and only one circuit (1.14%) with a terminal voltage more than 7% above the rated value. The Yuzui Main Feeder experiences high terminal voltage in the wet season and low terminal voltage in the dry season due to small hydropower generation feeding into the network. Therefore, a total of ten medium-voltage lines have terminal voltage problems.

Table 5.13 Current heavy and overloaded distribution transformers

Maximum load rate	Units (No.)	Proportion of total public transformers (%)	Capacity (kVA)	Proportion of total capacity (%)
Heavy load (80–100%)	155	10.82	27,247	11.32
Overload (>100%)	51	3.56	7856	3.26

Table 5.14 Voltage quality of existing public lines

		Circuits with minimum voltage > 7%		Circuits with maximum voltage > 7	
Voltage level (kV)	Total number of circuits	Number of circuits (count)	Percentage (%)	Number of circuits (count)	Percentage (%)
10 kV	88	10	11.36	1	1.14

The main causes of low voltage are excessive feeder length and long-term heavy loading. Solutions should prioritize trunk-line reinforcement or interconnection with adjacent feeders to transfer part of the load and stabilize end-point voltage levels.

From Table 5.15, it can be seen that most low-voltage feeders are concentrated in Heyun Town. The six lines with low voltage at the Heyun Substation supply mainly to ceramic manufacturers, which operate under heavy loads for extended periods. The recommended solution is to construct a new line from the Zhukeng Substation to offload these feeders.

At the Huilan Substation, the Huagang Feeder suffers from low terminal voltage due to its long distance and high load. The planned 110 kV Shatang Substation and the associated 10 kV Huagang line reconstruction project (initiated in 2011) are expected to resolve this issue. Although the project was completed but not yet commissioned at that time, the problem was mitigated in 2012 after its activation.

For other feeders experiencing low-voltage conditions, the main causes remain long trunk-line distances and sustained heavy loading. Thus, priority should be given to feeder reconstruction or the installation of automatic voltage regulators to improve terminal voltages.

5.3.2.2 Representative rural distribution networks

This study considers both the "12th Five-Year Plan" of Qingyuan City and the actual load conditions of the local area. Under conditions where the number, locations, and capacities of distributed generation (DG) units are uncertain, two representative 10 kV feeders – the Shitan Line and the Jiaokeng Line – were selected for integrated planning. Economic and reliability assessments were then performed for the pre- and post-planning schemes.

Shitan rural distribution networks

The total length of the Shitan Line is 26 km, serving 5947 users, including seven large industrial consumers, two hydropower station loads, and 5938 rural or commercial users. The annual electricity supply amounts to 14,447.46 MWh, of which:

- 5980.99 MWh is supplied to industrial users,
- 8447.41 MWh to rural or commercial users, and
- 19.06 MWh to power stations.

To simplify computation, the line and load models of the Shitan feeder were reduced, and nodes were renumbered. The simplified equivalent wiring diagram is shown in Figure 5.7, and the line impedance and load data are presented in Table 5.16.

For the planning of distributed generation (DG) on the Shitan Line, in order to ensure stable power flow in the distribution network and the reliable operation of protection devices, the capacity of each DG unit connected to the network is required to not exceed 20% of the node's annual maximum load. The DG capacity is taken as an integer multiple of 10 kV. The annual maximum load duration is 5812 hours, and the annual utilization hours of the wind turbine are 2500 hours.

Table 5.15 Details of public lines with unqualified terminal voltage

No.	Substation name	Line name	Terminal voltage (kV)	Cause	Proposed solution	Line name	Terminal voltage (kV)	Cause	Proposed solution	Line name	Terminal voltage (kV)	Cause
1	Heyun substation	Yuzui Main Feeder	11.2	Small hydropower feed-in	Adjust load distribution	Yuzui Main Feeder	11.2	Small hydropower feed-in	Adjust load distribution	Yuzui Main Feeder	11.2	Small hydropower feed-in
2	Heyun substation	Dawan No. 1 Feeder	8.9	Overload	Improve network structure	Dawan No. 1 Feeder	8.9	Overload	Improve network structure	Dawan No. 1 Feeder	8.9	Overload
3	Heyun substation	Dawan No. 2 Feeder	8.8	Overload	Improve network structure	Dawan No. 2 Feeder	8.8	Overload	Improve network structure	Dawan No. 2 Feeder	8.8	Overload
4	Heyun substation	Dawan No. 5 Feeder	8.8	Overload	Improve network structure	Dawan No. 5 Feeder	8.8	Overload	Improve network structure	Dawan No. 5 Feeder	8.8	Overload
5	Heyun substation	Dawan No. 6 Feeder	8.9	Overload	Improve network structure	Dawan No. 6 Feeder	8.9	Overload	Improve network structure	Dawan No. 6 Feeder	8.9	Overload
6	Heyun substation	Dawan No. 7 Feeder	9.0	Overload	Improve network structure	Dawan No. 7 Feeder	9.0	Overload	Improve network structure	Dawan No. 7 Feeder	9.0	Overload
7	Heyun substation	Dawan No. 8 Feeder	8.9	Overload	Improve network structure	Dawan No. 8 Feeder	8.9	Overload	Improve network structure	Dawan No. 8 Feeder	8.9	Overload

(Continues)

Table 5.15 (Continued)

No.	Substation name	Line name	Terminal voltage (kV)	Cause	Proposed solution	Line name	Terminal voltage (kV)	Cause	Proposed solution	Line name	Terminal voltage (kV)	Cause
8	Huilan substation	Huagang Feeder	8.7	Long line and heavy load	Improve network structure	Huagang Feeder	8.7	Long line and heavy load	Improve network structure	Huagang Feeder	8.7	Long line and heavy load
9	Jintan substation	Shitan Line	9.0	Line too long	Improve network structure	Shitan Line	9.0	Line too long	Improve network structure	Shitan Line	9.0	Line too long
10	Jintan substation	Cement Plant Line	9.1	Line too long	Improve network structure	Cement Plant Line	9.1	Line too long	Improve network structure	Cement Plant Line	9.1	Line too long
11	Heyun substation	Zhukeng Main Feeder	9.2	Line too long	Improve network structure	Zhukeng Main Feeder	9.2	Line too long	Improve network structure	Zhukeng Main Feeder	9.2	Line too long

Note: For the Yuzui Main Feeder, the terminal voltage rises above rated during the wet season and drops below rated in the dry season due to small hydropower integration.

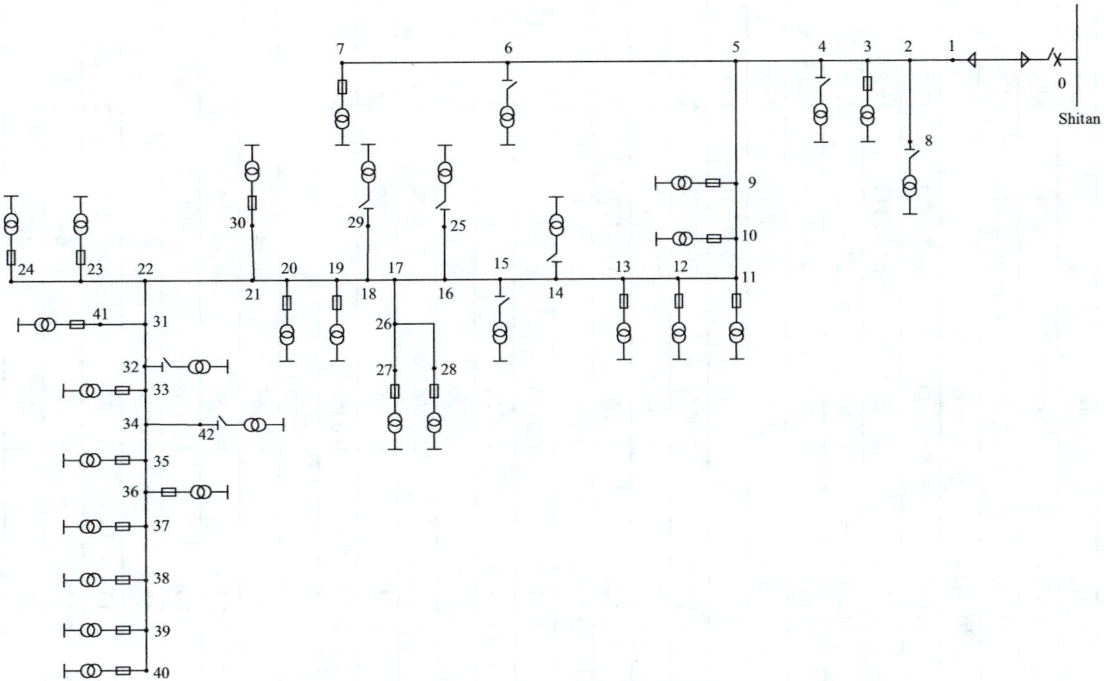

Figure 5.7 Simplified equivalent diagram of the Shitan Line

Table 5.16 Line node impedance and load data of the Shitan Line

Start node	End node	Line resistance (Ω)	Line reactance (Ω)	Active load (kW)	Reactive load (kVA)
0	1	0.0045	0.0074	0	0
1	2	0.7420	0.7957	0	0
2	3	1.6311	1.7490	7.150731	31.9
3	4	0.0760	0.0815	29.48644	81.5
4	5	0.0644	0.0691	0	0
5	6	0.0973	0.1043	4.582574	12.5
6	7	0.2258	0.2421	51.45312	195.5
2	8	0.8100	0.3501	22.41587	76.5
5	9	0.0173	0.0282	34.93783	132.8
9	10	0.1307	0.1490	80.58059	260.8
10	11	0.0788	0.0644	43.48741	139.2
11	12	0.0788	0.0644	76.52778	243.6
12	13	0.0788	0.0644	25.14487	93.4
13	14	0.5207	0.2251	11.03271	29.2
14	15	0.3471	0.1500	42.7355	110.4
15	16	0.1736	0.0750	0	0
16	17	0.1736	0.0750	0	0
17	18	0.1736	0.0750	0	0
18	19	0.1736	0.0750	1.521439	5.3
19	20	0.3471	0.1500	7.075959	24.7
20	21	0.1736	0.0750	0	0
21	22	0.1736	0.0750	0	0
22	23	0.7136	0.3084	7.54925	26.1
23	24	0.5400	0.2334	6.302026	21.8
16	25	0.0900	0.0736	230.5033	570.7
17	26	0.1710	0.0739	0	0
26	27	0.1710	0.0739	8.012258	28.1
26	28	0.1125	0.0920	14.64273	51.3
18	29	0.2835	0.1706	29.84211	74.9
21	30	0.1530	0.1251	6.237129	21.8
22	31	0.5794	0.2504	0	0
31	32	0.5794	0.2504	1.464101	3.6
32	33	0.5794	0.2504	7.620518	26.1
33	34	0.5794	0.2504	0	0
34	35	0.5794	0.2504	6.547855	22.3

(Continues)

Table 5.16 (*Continued*)

Start node	End node	Line resistance (Ω)	Line reactance (Ω)	Active load (kW)	Reactive load (kVA)
35	36	0.5794	0.2504	2.420857	8.2
36	37	0.5794	0.2504	5.13288	17.4
37	38	0.5794	0.2504	2.99898	10.2
38	39	1.1844	0.7125	0.287316	1
39	40	1.1844	0.7125	0.143668	0.5
31	41	0.2880	0.1245	7.707237	26.1
34	42	0.6300	0.2723	30.80728	95.4

Table 5.17 *Optimal locations and capacities of distributed generation (DG) on the Shitan Line*

Node with DG connection	DG capacity (kW)	Node description
10	50	Shuixiangkou
13	50	Jinshi
25	490	Shuangrong Chemical Plant
30	10	Xiabang

The unit electricity price is 0.74 yuan/kWh, while the total environmental cost of conventional power plants is 0.1171 yuan/kWh. The investment cost per unit capacity of a wind turbine is 6300 yuan/kW, with an assumed service life of 25 years and a discount rate of 0.1.

The distributed generation units are modeled as PQ nodes with a constant power factor, which is equivalent to adding a negative load at the corresponding load node. Therefore, their handling is consistent with that of ordinary PQ nodes. Under the condition that the number, locations, and individual capacities of DG units are all uncertain, the siting and sizing optimization of distributed generation is carried out. The optimal connection nodes and capacities of DG units are obtained as shown in Table 5.17.

Based on the reliability parameters of distribution equipment in Qingxin District over recent years, combined with the foundational data provided by the internationally recognized power system reliability expert Roy Billinton, the reliability parameters of the distribution equipment for the Shitan Line are obtained, as shown in Table 5.18.

The reliability calculation of the planning scheme yields the load point reliability indices and the overall system reliability indices, as shown in Table 5.19 and 5.20, respectively.

Table 5.18 Reliability parameters of distribution network equipment

Equipment type	Failure rate	Average repair time (hours)
Overhead feeder	0.0287 (failures/km·year)	12.36
Cable feeder	0.0084 (failures/km·year)	13.5
Distribution transformer	0.13 (failures/unit·year)	2.43
Distributed generation	0.02 (failures/unit·year)	25

Table 5.19 Load point reliability indices for planning Scheme 1 of the Shitan Line

Load point name	Load node no.	Failure rate (occurrences/year)	Average outage frequency (times/hour)	Annual average outage duration (hours/year)
Shalubei power station	L8	4.1998	2.87	12.2359
Shanzui	L3	4.2998	2.90	12.4613
Yongxing plastic factory	L4	4.3064	2.91	12.5425
Yadi chemical plant	L6	4.3204	2.94	12.7153
Camp site	L7	4.3399	2.99	12.9565
Dong'an street	L9	4.3143	2.93	12.6397
Shuixiangkou/ Yong'an Street/ Xiatang	L10	4.3266	2.96	12.7923
Chuangye street	L11	4.3318	2.97	12.8561
Baoying	L12	4.3369	2.98	12.9200
Jinshi	L13	4.3421	2.99	12.9838
Beijiang chemical plant	L14	4.3587	3.03	13.1896
Shuijingkeng quarry	L15	4.3699	3.05	13.3279
Shuangrong chemical plant	L25	4.3811	3.07	13.4663
Benwei lime plant	L29	4.3992	3.11	13.6897
Shijie	L27	4.3917	3.10	13.5975
Changtang	L28	4.3935	3.10	13.6188
Xiabang	L30	4.4182	3.15	13.9239
Rupei primary school	L19	4.3917	3.10	13.5975
Qiganling	L20	4.4029	3.12	13.7358
Yanggongdun	L23	4.4365	3.19	14.1509
Shimo	L24	4.4537	3.23	14.3637
Shanghua	L41	4.4414	3.20	14.2112
Leikeng mobile base station	L32	4.4506	3.22	14.3247
Old village	L33	4.4690	3.26	14.5517

(Continues)

Table 5.19 (Continued)

Load point name	Load node no.	Failure rate (occurrences/year)	Average outage frequency (times/hour)	Annual average outage duration (hours/year)
Leigongbei power station	L42	4.5074	3.33	15.0271
Gaokantou	L35	4.5057	3.33	15.0058
Fozaitan	L36	4.5241	3.37	15.2328
Yangshujing	L37	4.5424	3.40	15.4598
Nanshegong	L38	4.5608	3.44	15.6869
Dagangxing	L39	4.6147	3.54	16.3538
Shenchong	L40	4.6687	3.65	17.0207

Table 5.20 System reliability indices for planning Scheme 1 of the Shitan Line

System reliability index	Scheme 1
SAIFI (times/user·year)	4.3613
SAIDI (hours/user·year)	13.2209
CAIDI (hours/interrupted user·year)	3.031
ASAI (%)	99.85

Jiaokeng rural distribution networks

The total length of the Jiaokeng Line is 43 km, serving 2379 users, including 6 large industrial users, 14 power station loads, and 2359 rural or commercial users. The annual electricity supply currently amounts to 2514.41 MWh, of which:

- 526.293 MWh is supplied to large industrial users,
- 1622.32 MWh to rural or commercial users, and
- 365.8 MWh to power stations.

The line and load models of the Jiaokeng feeder were simplified for computational convenience, and the nodes were renumbered. The simplified equivalent diagram is shown in Figure 5.8 [2], and the line impedance and load data are presented in Table 5.21.

Following the same approach as in the Shitan Line, distributed generation (DG) was planned for the Jiaokeng Line, with each DG unit limited to no more than 20% of the node's annual maximum load and having a capacity that is an integer multiple of 10 kW. The optimized DG access locations and capacities are shown in Table 5.22.

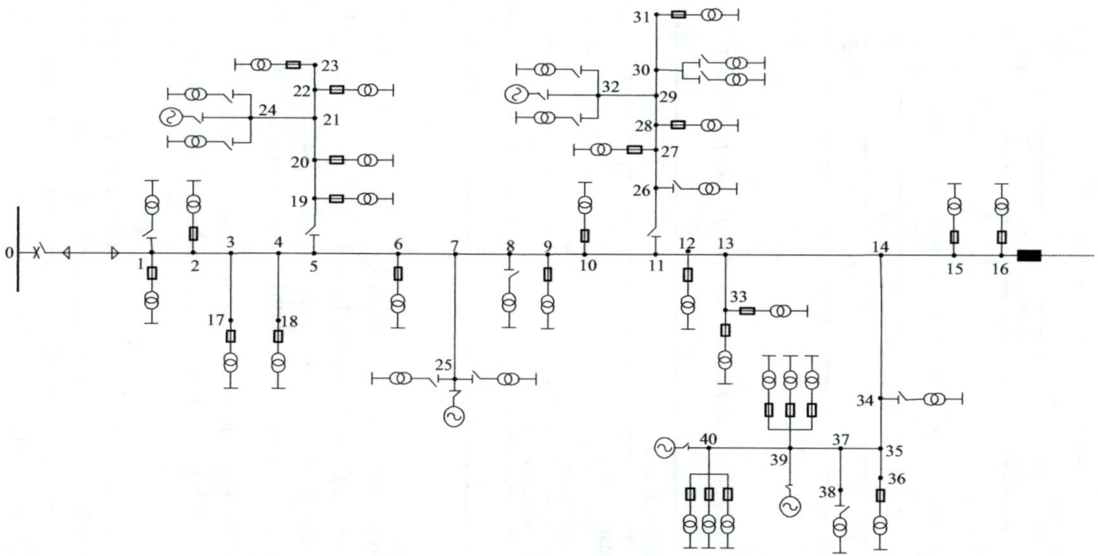

Figure 5.8 Equivalent simplified diagram of the Jiaokeng Line

Table 5.21 Line node impedance and load data of the Jiaokeng Line

Start node	End node	Line resistance (Ω)	Line reactance (Ω)	Active load (kW)	Reactive load (kVA)
0	1	0.17455	0.39103	324.00	138.02
1	2	0.35932	0.35491	178.20	86.31
2	3	0.31950	0.26128	0.00	0.00
3	4	0.31950	0.26128	0.00	0.00
4	5	0.31950	0.26128	0.00	0.00
5	6	0.31950	0.26128	149.10	72.21
6	7	0.31950	0.26128	0.00	0.00
7	8	0.31950	0.26128	18.80	8.01
8	9	0.31950	0.26128	31.60	15.30
9	10	0.31950	0.26128	76.20	32.46
10	11	0.31950	0.26128	0.00	0.00
11	12	1.62450	0.82533	159.00	77.01
12	13	1.30500	0.56405	0.00	0.00
13	14	1.30500	0.56405	0.00	0.00
14	15	1.30500	0.56405	12.10	5.86
15	16	1.30500	0.56405	27.50	13.32
3	17	1.38600	0.43890	65.10	31.53
4	18	0.45000	0.19450	87.90	42.57
5	19	0.52542	0.31609	238.40	115.46
19	20	1.86471	0.85046	41.80	20.24
20	21	1.60200	0.69242	0.00	0.00
21	22	1.60200	0.69242	66.60	32.26
22	23	1.60200	0.69242	105.15	50.93
21	24	0.25200	0.07980	6.00	2.56
7	25	0.44100	0.27230	71.00	30.25
11	26	1.31400	0.56794	3.60	1.53
26	27	0.65700	0.28397	60.00	29.06
27	28	1.31400	0.56794	149.10	72.21
28	29	0.98550	0.42596	0.00	0.00
29	30	0.53550	0.23146	14.00	5.96
30	31	0.45000	0.19450	46.35	22.45
29	32	0.27000	0.11670	26.00	11.08
13	33	1.80000	0.77800	157.35	76.21
14	34	0.97200	0.42012	41.00	19.86
34	35	0.48600	0.21006	0.00	0.00
35	36	0.97200	0.42012	13.75	6.66
35	37	0.64220	0.45820	0.00	0.00
37	38	0.36000	0.15560	15.00	6.39
37	39	0.29880	0.33120	69.00	29.39
39	40	0.76500	0.62560	86.00	36.64

Table 5.22 DG access locations and capacities for the Jiaokeng Line

Node with DG access	DG capacity (kW)	Node description
6	40	Matian
12	40	Shitouqiao
33	30	Huizhonggang/Baihualan

The reliability of the pre-planning network was calculated, and the load point reliability indices and system reliability indices are presented in Tables 5.23 and 5.24, respectively. SAIFI (System Average Interruption Frequency Index), SAIDI (System Average Interruption Duration Index), CAIDI (Customer Average Interruption Duration Index), and ASAI (Average Service Availability Index) are standard reliability indicators used to evaluate the frequency, duration, and availability of power supply in an electrical distribution system.

The reliability of the line after distributed generation (DG) planning was calculated, and the load point reliability indices and system reliability indices are shown in Tables 5.25 and 5.26, respectively.

5.3.2.3 Power flow calculation

In this book, we provide the scripts for Jiaokeng41 and Shitan43, along with the test Qingyuan file used to test the power flow calculation results of these two rural distribution networks. In addition, the latest version of the MATPOWER package [3] is included. The power flow calculation results obtained by running the above programs in MATLAB® are shown in Table 5.27.

The Jiaokeng41 rural distribution network demonstrates stable operating performance with a converged Newton–Raphson solution in three iterations. The system exhibits total generation and load values of 2.46 MW + j1.20 MVAr and 2.34 MW + j1.09 MVAr, respectively, resulting in total line losses of 0.12 MW + j0.11 MVAr. The voltage profile remains within acceptable limits, with the lowest voltage magnitude of 0.903 p.u. at Bus 41 and the highest at 1.000 p.u. at Bus 1. The active loss is primarily concentrated along line 2–3 (0.02 MW), indicating localized resistive heating near the mid-feeder section. The overall system maintains good voltage stability and load balance, suitable for rural network applications under moderate load conditions.

The Shitan43 network, with slightly more buses and a higher reactive load share, also achieves convergence within three iterations. Its total generation and load are 1.02 MW + j2.66 MVAr and 0.81 MW + j2.44 MVAr, respectively, producing losses of 0.21 MW + j0.21 MVAr – nearly twice those of Jiaokeng41. The lowest voltage (0.889 p.u.) appears at Bus 41, reflecting deeper voltage drops along the feeder caused by higher reactive demand and a longer line structure. The largest losses occur at line 3–4 (0.11 MW + j0.12 MVAr), suggesting the need for local voltage support or reactive compensation. Overall, while the Shitan43 network maintains power flow convergence, its higher losses and lower terminal voltages

Table 5.23 Load point reliability indices for planning Scheme 1 of the Jiaokeng Line

Load point name	**Load node no.**	**Failure rate (occurrences/year)**	**Average outage frequency (times/hour)**	**Annual average outage duration (hours/year)**
Taoyuanping Town – China Unicom Dedicated Transformer	L1	0.8553	2.2309	1.9081
Hengshizhai	L2	0.8844	2.5642	2.2678
Matian	L6	0.9659	3.3908	3.2753
Board factory	L8	1.0067	3.7539	3.7790
Orchard	L9	1.0271	3.9247	4.0309
Tianfuluo	L10	1.0474	4.0888	4.2827
Miaoweikeng	L15	1.2547	5.4548	6.8439
Shitouqiao	L12	1.1298	4.6918	5.3008
Hengshui	L16	1.2963	5.6765	7.3582
Jingtoulang	L17	0.9364	3.1077	2.9099
Tiluxia	L18	0.9395	3.1388	2.9489
Aoaizhi	L19	0.9695	3.4238	3.3193
Gangzui	L20	1.0325	3.9695	4.0986
Gongqiaotcu	L22	1.1347	4.7245	5.3615
Hutang	L23	1.1858	5.0539	5.9929
Sancun power station	L24	1.0894	4.4072	4.8010
Matian power station	L25	1.0064	3.7515	3.7755
Jingzai mobile dedicated transformer	L26	1.1097	4.5530	5.0525
Jingzai public transformer	L27	1.1307	4.6976	5.3114
Jinwan	L28	1.1726	4.9715	5.8294
Meizichong power station	L30	1.2211	5.2649	6.4288
Nanlang	L31	1.2354	5.3474	6.6062
Yushan power station	L32	1.2203	5.2608	6.4200
Huizhonggang – Baihualan	L33	1.2288	5.3097	6.5246
Linzhongbao	L34	1.2440	5.3959	6.7126
Miaolang public transformer	L36	1.2905	5.6468	7.2873
Miaolang mobile dedicated transformer	L38	1.2954	5.6721	7.3476
Luoming second-level power station	L39	1.3098	5.7453	7.5250
Luoming first-level power station	L40	1.3585	5.9829	8.1280

Table 5.24 System reliability indices for planning Scheme 1 of the Jiaokeng Line

System reliability index	Value
SAIFI (times/user·year)	1.0775
SAIDI (hours/user·year)	4.6544
CAIDI (hours/interrupted user·year)	4.320
ASAI (%)	99.95

Table 5.25 Load point reliability indices for planning Scheme 2 of the Jiaokeng Line

Load point name	Load node no.	Failure rate (occurrences/ year)	Average outage frequency (times/hour)	Annual average outage duration (hours/year)
Taoyuanping Town – China Unicom Dedicated Transformer	L1	0.843	2.0663	1.7419
Hengshizhai	L2	0.843	2.0663	1.7419
Matian	L6	0.843	2.0663	1.7419
Board Factory	L8	0.843	2.0663	1.7419
Orchard	L9	0.843	2.0663	1.7419
Tianfuluo	L10	0.843	2.0663	1.7419
Miaoweikeng	L15	0.9262	2.9913	2.7706
Shitouqiao	L12	0.843	2.0663	1.7419
Hengshui	L16	0.9678	3.3941	3.2850
Jingtoulang	L17	0.8746	2.4379	2.1321
Tiluxia	L18	0.8574	2.2386	1.9193
Aoaizhi	L19	0.8669	2.3505	2.0377
Gangzui	L20	0.9300	3.0292	2.8171
Gongqiaotou	L22	1.0322	3.9528	4.0799
Hutang	L23	1.0832	4.3493	4.7114
Lancun power station	L24	0.9868	3.5665	3.5195
Matian power station	L25	0.8631	2.3059	1.9902
Jingzai mobile dedicated transformer	L26	0.8849	2.5537	2.2598
Jingzai public transformer	L27	0.9059	2.7805	2.5188
Jinwan	L28	0.9478	3.2041	3.0367
Meizichong power station	L30	0.9963	3.6498	3.6362
Nanlang	L31	1.0106	3.7735	3.8135

(Continues)

Table 5.25 (Continued)

Load point name	Load node no.	Failure rate (occurrences/ year)	Average outage frequency (times/hour)	Annual average outage duration (hours/year)
Yushan power station	L32	0.9955	3.6436	3.6273
Huizhonggang – Baihualan	L33	0.843	2.0663	1.7419
Linzhongbao	L34	0.9156	2.8826	2.6394
Miaolang public transformer	L36	0.9621	3.3406	3.2140
Miaolang mobile dedicated transformer	L38	0.9670	3.3861	3.2743
Luoming second-level power station	L39	0.9813	3.5174	3.4517
Luoming first-level power station	L40	1.0301	3.9362	4.0548

Table 5.26 System reliability indices for planning Scheme 2 of the Jiaokeng Line

System reliability index	Value
SAIFI (times/user·year)	0.8948
SAIDI (hours/user·year)	2.3819
CAIDI (hours/interrupted user·year)	2.662
ASAI (%)	99.97

Table 5.27 Power flow results comparison between Jiaokeng41 and Shitan43 systems

Item	Jiaokeng41	Shitan43
Number of buses	41	43
Total generation (MW/MVAr)	2.5/1.2	1.0/2.7
Total load (MW/MVAr)	2.3/1.1	0.8/2.4
Active power losses (MW)	0.12	0.21
Reactive power losses (MVAr)	0.11	0.21
Minimum voltage (p.u.)	0.903 @ Bus 41	0.889 @ Bus 41
Maximum voltage (p.u.)	1.000 @ Bus 1	1.000 @ Bus 1
Maximum active loss line	2–3 (0.02 MW)	3–4 (0.11 MW)
Maximum reactive loss line	1–2 (0.03 MVAr)	3–4 (0.12 MVAr)
Convergence iterations	3	3
Convergence time (s)	0.02	0.02

imply weaker voltage regulation and greater sensitivity to load variations compared with Jiaokeng41.

5.3.3 Northern China case

5.3.3.1 Overview

The power system of Yutian County in Tangshan City, Hebei Province, is operated and managed by the Yutian County Power Supply Branch of State Grid Jibei Electric Power Co., Ltd. The company is part of the State Grid Corporation of China and is responsible for power transmission, distribution, and customer services throughout the country. At present, the county has several power supply stations and service offices – such as the Yutian Town Power Supply Office and the Linnan Warehouse Power Station – covering both urban and rural areas. According to the "14th Five-Year Plan" of Jibei Electric Power, the Yutian County grid has completed the reconstruction of 10 kV and lower-voltage lines in rural regions, significantly improving supply reliability and voltage quality.

In recent years, Yutian County has vigorously promoted the development of distributed photovoltaic (PV) generation, advancing coordinated integration of "generation–grid–load–storage." By 2024, the total installed PV capacity in the county had reached approximately 501 MW, including three centralized PV plants (61 MW) and 440 MW of rooftop distributed PV systems connected to the grid. Several new energy enterprises operate in the county, such as Yutian Taihe Solar Power Development Co., Ltd., which is building a 35 MW PV project. Overall, the Yutian County grid has achieved full urban–rural power coverage with a high penetration of distributed renewable energy. However, as installed capacity continues to expand, the rural distribution network faces challenges such as voltage fluctuations, reverse power flows, and transformer overloading in certain areas.

Bajia Village, located in Bajia Township, Chengde County, Chengde City, Hebei Province, is a typical hilly rural area. Its power system is uniformly operated and maintained by State Grid Chengde Electric Power Company, which is responsible for electricity supply and disaster prevention in mountainous regions. In 2025, after severe rainstorms, Chengde carried out emergency grid repairs, restoring 145 distribution lines and power supply to 562 natural villages. Bajia Village was included in the "Village Electrification" coverage, reflecting the strong emergency resilience of local rural power systems.

In terms of renewable energy, Bajia Township is implementing the Dongwo 5.99 MW Distributed PV Project, with several PV stations located near Bajia Village. The project aims to integrate village-level clean energy generation for both self-consumption and grid connection. It serves as an important pilot under Chengde's "Rural Revitalization + New Energy" initiative. Overall, the power system of Bajia Village has evolved from a single supply oriented model to an integrated "generation–consumption–storage" framework, although the long transmission distances and complex terrain of the mountainous region still pose higher demands on equipment maintenance and voltage regulation.

5.3.3.2 Yutian rural distribution network

The topology of the Yutian rural distribution network is shown in Figure 5.9.

The line impedance and load data for the Yutian rural distribution network are presented in Table 5.28.

5.3.3.3 Bajiacun rural distribution network

The topology of the Bajiacun rural distribution network is shown in Figure 5.10 [4].

The line impedance and load data for the Bajiacun rural distribution network are presented in Table 5.29.

5.3.3.4 Power flow calculation

In this book, we provide the scripts for tangshan46 and chengde48, along with the test Hebei file used to test the power flow calculation results of these two rural distribution networks. It should be noted that in the chengde48.m file, the power units are MW (megawatts) and MVAr (megavars), which differ from Jiaokeng4.m 1, shitan43.m, and tangshan46.m, where the data are initially specified in kW/kVAr and later converted. In Jiaokeng41.m, shitan43.m, and tangshan46.m, the power values are first entered in kW/kVAr and then divided by 1e3 at the end of the script, so that the power flow calculation within MATPOWER is performed in MW/MVAr.

Table 5.30 is the summarized power flow comparison between the two rural distribution systems and the analytical description you requested.

The Tangshan46 rural distribution network demonstrates a highly stable voltage profile with all bus voltages remaining between 1.022 p.u. and 1.030 p.u., indicating good voltage regulation and low network impedance. The total generation of 2.40 MW + 1.21 MVAr exactly balances the total load of 2.39 MW + 1.20 MVAr and the small network losses of 0.009 MW + 0.02 MVAr. The minimal losses (approximately 0.4% of total power) show that the network topology is compact with short feeder lengths and efficient load distribution. The voltage angle decreases gradually along the feeders (0° to –0.29°), reflecting stable radial power flow from the main substation at Bus 1 toward downstream nodes. This behavior is typical of a well-balanced 10 kV rural feeder with limited distributed generation.

In contrast, the Chengde48 network presents slightly greater electrical stress due to higher load demand and longer feeder branches. The minimum bus voltage drops to 0.954 p.u., notably lower than in Tangshan46, implying a need for local voltage regulation devices or reactive power compensation at distant nodes. The total generation of 3.00 MW + 1.54 MVAr supplies a system load of 2.92 MW + 1.46 MVAr, with total losses reaching 0.084 MW + 0.08 MVAr, nearly nine times higher than in Tangshan46. Despite the higher losses, convergence is fast (three iterations), suggesting the network remains numerically stable. The larger voltage drop (4.6%) and wider angle range (0° to –0.79°) indicate heavier line impedance and more dispersed load distribution. Overall, the Chengde48 system requires optimization in reactive power support and feeder sectioning to enhance voltage stability and reduce line losses.

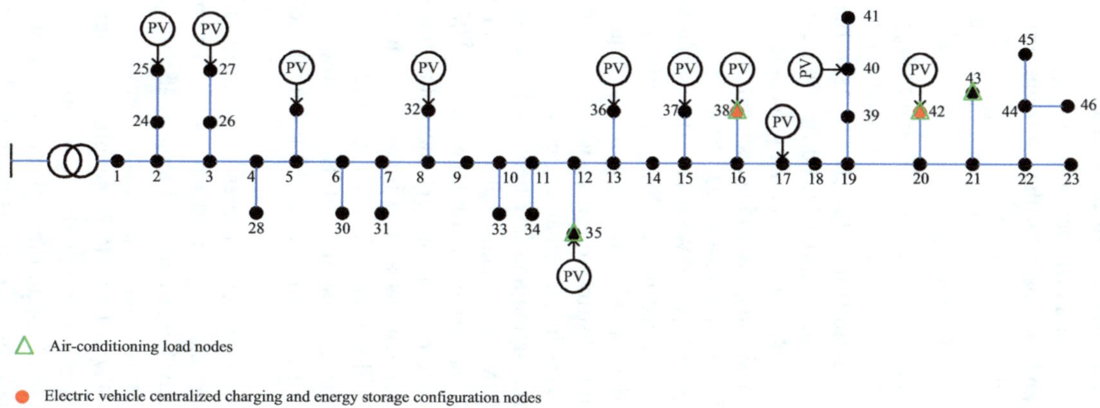

Figure 5.9 Equivalent simplified diagram of the Yutian distribution network

Table 5.28 Line impedance and load data for the Yutian rural distribution network

Start node	End node	Line resistance (Ω)	Line reactance (Ω)	Active load (kW)	Reactive load (kVA)
1	2	0.003	0.0203	0	0
2	3	0.0683	0.1141	0	0
3	4	0.0116	0.0194	0	0
4	5	0.0065	0.0109	0	0
5	6	0.0076	0.0127	0	0
6	7	0.0085	0.0142	0	0
7	8	0.0077	0.0130	0	0
8	9	0.0057	0.0096	0	0
9	10	0.0326	0.0545	0	0
10	11	0.0059	0.0098	0	0
11	12	0.0013	0.0021	78	39
12	13	0.0075	0.0125	0	0
13	14	0.0109	0.0182	70	35
14	15	0.0051	0.0086	98	49
15	16	0.0169	0.0282	74	37
16	17	0.0174	0.0291	0	0
17	18	0.0094	0.0157	0	0
18	19	0.0067	0.0113	0	0
19	20	0.0057	0.0096	110	55
20	21	0.0063	0.0105	92	46
21	22	0.0127	0.0213	74	37
22	23	0.0026	0.0044	0	0
2	24	0.0155	0.0259	37	18.5
24	25	0.0166	0.0149	98	49
3	26	0.0137	0.0123	74	37
26	27	0.0107	0.0193	60	30
4	28	0.005	0.0091	24	12
5	29	0.0107	0.0193	190	95
6	30	0.0092	0.0165	127	64
7	31	0.0268	0.0277	23	11.5

(Continues)

Table 5.28 (Continued)

Start node	End node	Line resistance (Ω)	Line reactance (Ω)	Active load (kW)	Reactive load (kVA)
8	32	0.0097	0.0139	140	70
10	33	0.0045	0.0155	48	24
11	34	0.0068	0.0232	0	0
12	35	0.003	0.0103	0	0
13	36	0.0313	0.021	0	0
15	37	0.0077	0.0111	50	25
16	38	0.0047	0.0173	26	13
19	39	0.0141	0.0127	46	23
39	40	0.0095	0.0086	55	27.5
40	41	0.006	0.0207	23	11.5
20	42	0.0045	0.0155	400	200
21	43	0.0092	0.0165	23	11.5
22	44	0.0262	0.0176	0	0
44	45	0.0239	0.0161	200	100
44	46	0.0058	0.0084	150	75

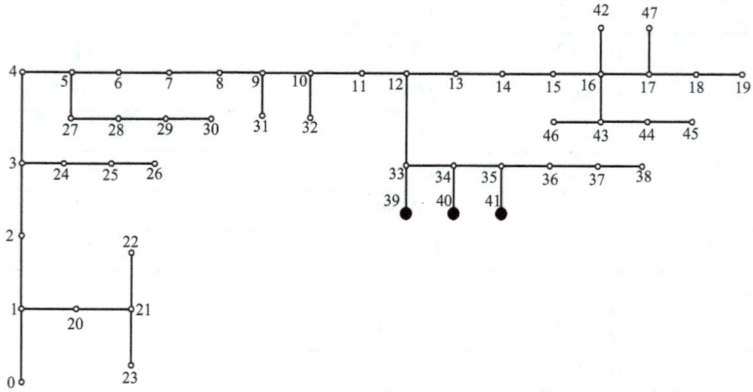

Figure 5.10 Equivalent simplified diagram of the Bajiacun distribution network

Table 5.29 Line impedance and load data for the Bajiacun rural distribution network

Start node	End node	Line resistance (Ω)	Line reactance (Ω)	Active load (kW)	Reactive load (kVA)
0	1	0.374	0.351	0	0
1	2	0.025	0.023	241	121
2	3	0.124	0.117	0	0
3	4	0.058	0.054	118	59
4	5	0.061	0.057	0	0
5	6	0.014	0.013	76	35.8
6	7	0.22	0.206	56	28
7	8	0.018	0.017	0	0
8	9	0.079	0.074	0	0
9	10	0.131	0.123	0	0
10	11	0.095	0.089	78	39
11	12	0.078	0.073	0	0
12	13	0.343	0.322	70	35
13	14	0.081	0.076	98	49
14	15	0.068	0.064	74	37
15	16	0.09	0.085	0	0
16	17	0.133	0.125	0	0
17	18	0.184	0.172	0	0
18	19	0.099	0.093	110	55
5	27	0.164	0.154	24	12
27	28	0.068	0.033	190	95
28	29	0.313	0.153	127	64
29	30	0.161	0.079	23	11.5
9	31	0.014	0.013	140	70
12	33	0.218	0.205	0	0
33	34	0.139	0.13	0	0
34	35	0.048	0.045	0	0
35	36	0.32	0.3	50	25
36	37	0.446	0.419	26	13
37	38	0.676	0.635	46	23
1	20	0.211	0.198	92	46
20	21	0.211	0.198	74	37

(Continues)

Table 5.29 (Continued)

Start node	End node	Line resistance (Ω)	Line reactance (Ω)	Active load (kW)	Reactive load (kVA)
21	22	0.211	0.198	0	0
21	23	0.211	0.198	37	18.5
3	24	0.211	0.198	98	49
24	25	0.211	0.198	74	37
25	26	0.211	0.198	60	30
10	32	0.211	0.198	48	24
33	39	0.211	0.198	55	27.5
34	40	0.211	0.198	23	11.5
35	41	0.211	0.198	400	200
16	42	0.211	0.198	23	11.5
16	43	0.211	0.198	0	0
43	44	0.211	0.198	200	100
44	45	0.211	0.198	150	75
43	46	0.211	0.198	24	12
17	47	0.211	0.198	13	6.5

Table 5.30 Summary of power flow results for Tangshan46 and Chengde48 systems

Item	Tangshan46 system	Chengde48 system
Number of buses	46	48
Number of branches	45	47
Generators	1 (2.40 MW + 1.21 MVAr)	1 (3.00 MW + 1.54 MVAr)
Total load	2.39 MW + 1.20 MVAr	2.92 MW + 1.46 MVAr
Total losses (I^2Z)	0.009 MW + 0.02 MVAr	0.084 MW + 0.08 MVAr
Voltage range (p.u.)	1.022–1.030	0.954–1.000
Lowest voltage bus	Bus 45 (1.022 p.u.)	Bus 46 (0.954 p.u.)
Voltage angle range (°)	–0.29 to 0.00	–0.79 to 0.00
Convergence	3 iterations, 0.02 s	3 iterations, 0.02 s
Power balance	Generation ≈ Load + Loss	Generation ≈ Load + Loss
Overall power loss comparison	34.34 MW + j17.24 MVAr	31.90 MW + j16.10 MVAr

5.4 Conclusion

This chapter systematically presents the engineering practices and case studies of the Agricultural Energy Internet (AEI) in different regions of China, covering representative areas such as Qingdao in Shandong Province, Chengde, Tangshan, and Zhuozhou in Hebei Province; as well as Qingyuan in Guangdong Province. The chapter first introduces the overall architecture of agricultural energy systems, explaining how photovoltaic, wind, and biomass energy sources are integrated with energy storage and flexible loads to achieve coordinated operation among "source–grid–load–storage." Through six real-world cases – two agricultural parks in Qingdao (Shandong) and Zhuozhou (Hebei), two rural distribution networks in Chengde and Tangshan (Hebei), and two rural networks in Qingyuan (Guangdong) – the study demonstrates how rural power systems can achieve stable, low-carbon, and intelligent energy supply under diverse climatic, economic, and agricultural conditions.

This chapter focuses on the microgrid cases of the Langyatai Agricultural Park in Qingdao and the Zhuozhou Agricultural Park in Hebei, providing detailed equipment parameters and procurement costs for photovoltaic arrays, biogas generators, gas boilers, battery storage systems, and greenhouse loads, thereby forming a comprehensive framework for system configuration and techno-economic analysis. Based on the statistical analysis of operational data from January to September, the findings validate the stability and efficiency of the integrated AEI framework.

Beyond providing reproducible engineering data and modeling foundations for the Agricultural Energy Internet, this chapter establishes a basis for further research on energy management optimization, demand response, and carbon emission control. The presented equipment parameters, operational data, and economic analyses serve as valuable references for researchers and engineers engaged in the construction of intelligent agricultural power systems, offering a practical model for advancing the green energy transition in China and worldwide.

References

[1] X. Fu, J. Bai, H. Sun and Y. Zhang, "Optimizing agro-energy-environment synergy in agricultural microgrids through carbon accounting," *IEEE Transactions on Smart Grid*, vol. 15, no. 5, pp. 4819–4834, 2024. Doi:10.1109/TSG.2024.3397990.

[2] X. Fu, "Statistical machine learning model for capacitor planning considering uncertainties in photovoltaic power," *Protection and Control of Modern Power Systems*, vol. 7, no. 1, pp. 1–13, 2022. Doi:10.1186/s41601-022-00228-z.

[3] MATPOWER User's Manual, available at: https://matpower.org/ (accessed November 2025).

[4] X. Fu and Y. Zhou, "Collaborative optimization of PV greenhouses and clean energy systems in rural areas," *IEEE Transactions on Sustainable Energy*, vol. 14, no. 1, pp. 642–656, 2023. Doi:10.1109/TSTE.2022.3223684.

Chapter 6

Functional design and code implementation of the Agricultural Energy Internet monitoring system

Chao Lei[1], Bohao Zeng[1] and Xueqian Fu[1]

Abstract

This chapter provides a comprehensive account of the functional design and code implementation of an Agricultural Energy Internet (AEI) monitoring system tailored for aquaponics users. The main features include real-time monitoring and visualization of key water quality parameters (e.g., pH, ammonia nitrogen, and dissolved oxygen), cultivation environment parameters (e.g., humidity, light intensity, temperature, and carbon dioxide concentration), the health status of fish and vegetables, and near-surface meteorological conditions. The system also enables model-based prediction of water quality and environmental trends over predefined time horizons. It dynamically displays comprehensive warning levels for the load side (aquaculture and crop activities), as well as specific warning levels for plant cultivation. Moreover, it presents control strategies for electricity-consuming devices aimed at maintaining environmental stability and optimizing energy usage – covering operational logic for equipment such as water pumps, aerators, temperature control units, and supplementary lighting systems. The chapter further details the programming logic of key functions, provides pseudocode for core algorithms, and includes example implementation code for corresponding modules.

Keywords: Agricultural Energy Internet monitoring system; water quality parameters; warning levels; control strategies

6.1 Introduction

With the rapid advancement of information technology, the Internet of Things (IoT), and new energy management technologies, agricultural production is gradually progressing toward intelligent, digital, and low-carbon development.

[1]College of Information and Electrical Engineering, China Agricultural University, China

The Agricultural Energy Internet (AEI), as a new system integrating the principles of the energy Internet with modern agricultural production management, is becoming a key foundation for building smart agriculture and a green circular economy. Its core objective is to achieve the coordinated optimization of energy flows, information flows, and material flows in agricultural production through data collection at the perception layer, information transmission at the network layer, and intelligent decision-making at the application layer.

The aquaponics system, as a significant representative scenario of the Agricultural Energy Internet, organically integrates aquaculture with plant cultivation. Through water quality cycling, nutrient complementarity, and energy sharing, it builds an efficient, low-carbon, and closed-loop agricultural ecosystem. However, the operation of aquaponics systems is influenced by various factors, including the aquatic environment, the air environment, the health status of organisms, and the coordinated operation of energy-consuming equipment.

The global population is continuously growing, and people's dietary structures are constantly upgrading. As an important source of high-quality animal protein, the demand for aquatic products is showing a continuous upward trend. Against the backdrop of fluctuating food prices and increasing resource pressures, aquaculture, with its efficient protein conversion rate and relatively low energy consumption, has become one of the important means to ensure food security and enhance the overall benefits of agriculture. Aquaculture in China is developing rapidly, continuously producing aquatic products. Data indicate that the total output value of China's fisheries reached an astonishing 1351.724 billion yuan in 2020, with freshwater aquaculture contributing more than 600 billion yuan, accounting for half of the total output value. This shows that freshwater fisheries have become a crucial force in the growth and prosperity of China's fisheries industry, playing a key role in promoting agricultural structure optimization and rural economic development [1].

However, traditional extensive aquaculture practices face issues such as water environment deterioration, feed and energy wastage. Among these, water quality regulation is particularly critical, as its management level directly affects the growth rate, survival rate, and product quality of the cultured species. It is a key factor affecting the living standards of fish and fishery yields [2]. Parameters such as acidity and alkalinity in the water are crucial for the metabolism and immune functions of aquatic organisms, providing the physical, biological, and chemical environmental conditions necessary for the daily survival and activities of fish and other aquatic species [3]. Existing aquaculture management generally relies on manual experience and periodic sampling, resulting in delayed monitoring data and sparse spatial distribution, making it difficult to support precise and real-time decision-making [4]. Therefore, there is an urgent need to develop a digital system capable of real-time multi-parameter water quality monitoring and automatic warning, in order to improve aquaculture management levels and ecological sustainability.

At the same time, with the acceleration of urbanization, agricultural and fishery land resources are becoming increasingly scarce [5]. In the context of rising population and growing food demand, the "aquaponics" model, as a new type of agricultural system that integrates resource recycling and ecological synergy, has gradually attracted attention [6]. This model provides nutrients to plants through fish excrement, while the plants purify the water after absorption, forming a closed water circulation system and achieving an ecological balance of "raising vegetables with fish, purifying water with plants" [7]. The advantages of complementary resources and high yield per unit area, through the repeated use of water and nutrients, achieve water conservation and reduce dependence on chemical fertilizers, allowing it to develop rapidly in recent years [8]. However, most "aquaponics" systems in China still rely on manual monitoring and control, which involves high labor intensity, delayed responses, and a high error rate [9].

To this end, the Agricultural Energy Internet (AEI) monitoring system designed in this chapter is structured around the overall framework of "intelligent perception—data transmission—visualization—intelligent decision-making," with functional design and implementation tailored to the dual needs of aquaculture and aquaponics systems. The system is based on Spring Boot and Vue technology architecture, integrating multi-source sensor data collection, WebSocket real-time communication, and ECharts dynamic visualization.

In addition, the system innovatively integrates carbon accounting and visualization analysis modules, enabling multidimensional statistics and display of data on electricity consumption, power generation, carbon emissions, carbon absorption, and carbon reduction in the farming process [10]. Supported by the Agricultural Energy Internet architecture, the platform has achieved an organic integration of fisheries ecological environment monitoring, water quality regulation, and carbon balance analysis, providing crucial technical support for realizing green, intelligent, and low-carbon modern fisheries production.

Against the backdrop of the accelerated transformation and structural upgrading of the current fisheries industry, traditional aquaculture is gradually shifting from extensive production toward intensive, intelligent, and sustainable development. However, in this process, large-scale aquaculture enterprises face increasingly prominent core contradictions – on the one hand, they need to continuously reduce operating costs and improve energy utilization efficiency through refined management and intelligent control methods; on the other hand, they must strengthen environmental risk prevention capabilities to ensure production safety, ecological stability, and traceable product quality. This dual pressure of "improving efficiency and reducing costs" and "risk prevention and control" makes the traditional aquaculture model, which relies on manual experience and fragmented monitoring, gradually unable to meet the production management requirements of the new era.

To effectively address this industry's pain points, the introduction of digital and intelligent technologies is the only way to drive the development of a new quality of productivity in the fisheries sector. Aquaculture is gradually undergoing a profound transformation from "experience-driven" to "data-driven" and from "manual management" to "intelligent decision-making." By building integrated management systems for real-time monitoring, intelligent early warning, energy optimization, and carbon emission analysis, aquaculture enterprises can balance energy conservation, ecological protection, and economic benefits, thereby significantly enhancing industry competitiveness.

Meanwhile, for small and medium-sized aquaculture farmers, who constitute the main force in the fishery sector, the need for digital transformation is even more urgent. Due to limited capital investment, insufficient technical capabilities, and relatively weak management systems, this group generally faces the problems of "high investment, difficult usage, and poor results" in informatization efforts. Small and medium-sized farmers urgently need a low-cost, easy-to-deploy, user-friendly aquaculture water quality management tool with core intelligent functions, enabling them to improve both production efficiency and environmental safety under limited resources. This not only helps narrow the management capability gap between small and medium-sized farms and large-scale enterprises but also plays an important role in promoting inclusive digital upgrading across the entire fishery industry.

Based on the above practical requirements and technological trends, this chapter takes "lightweight, intelligent, and scalable" as the overall design principles to construct and implement an intelligent fisheries water quality management platform based on Spring Boot and Vue technology stack. Providing high-concurrency and highly stable back-end services through the Spring Boot framework, while enabling interface interaction and visualization through the Vue framework. Additionally, WebSocket is introduced to achieve real-time data communication and dynamic updates, significantly improving system response speed and user experience. The database layer employs a combined design of MySQL and Redis, ensuring both persistent storage of historical data and high-speed caching and rapid access to real-time data. At the same time, the system utilizes ECharts visualization components to build an intuitive water quality monitoring dashboard, thereby achieving precise production management.

From the perspective of practical application, this platform not only aligns with the policy orientation of national agricultural modernization and digital rural construction strategies but also fully embodies the concept of green, low-carbon, intelligent, and efficient fishery development. By deeply integrating advanced information technology with aquaculture production practices, the system provides aquaculture enterprises with digital management solutions that are both implementable and replicable; at the same time, it offers small and medium-sized farmers a smart management tool that is cost-controllable and flexibly deployable, reducing technical barriers and usage difficulty, while enhancing the overall level of informatization in the industry.

Overall, the intelligent water quality management platform for fisheries developed in this study features a lightweight, highly reliable, and scalable technical architecture. Its functional design achieves a closed-loop of monitoring, early warning, analysis, and decision-making. In practical application, it can effectively support intelligent supervision of fishery production, energy conservation, consumption reduction, and ecological protection. Its design concept and implementation approach provide feasible technical references for the digital transformation of China's fishery industry and also explore new directions for energy Internet technologies, demonstrating significant economic benefits and social promotion value.

At the technical architecture level, Spring Boot provides efficient backend support for the system with its modular design and strong microservices support. Its flexible interface integration capabilities allow for rapid connection to various types of sensor devices, enabling real-time collection, storage, and processing of aquaculture water quality data, ensuring the system maintains excellent stability and responsiveness even in high-concurrency environments. The frontend uses the Vue.js framework, combined with ECharts visualization components, to intuitively present complex monitoring data in multidimensional formats such as charts, curves, and heat maps, allowing users to quickly identify anomalies and make informed decisions. In addition, to further expand the application's scope and ease of use, the platform will, in the future, support WeChat mini programs and other mobile embedded applications, allowing farmers to monitor, view alerts, and adjust parameters remotely via mobile devices anytime and anywhere, thereby achieving truly intelligent aquaculture management. The overall solution performs excellently in terms of system stability, scalability, and user interaction experience, laying a solid foundation for the platform's promotion and long-term operation. It holds significant practical value in promoting energy-saving and consumption-reducing practices in fisheries, optimizing aquaculture processes, improving the quality of aquatic products, and advancing green and sustainable development in the industry.

With the rapid development of agricultural production toward intelligence, greenness, and low carbon, traditional aquaculture management models have become inadequate for meeting the modern agricultural demand for efficient energy monitoring and precise environmental control. From the perspective of international research progress, distributed water quality monitoring systems have realized IoT-based perception and sharing of water parameters across regions, providing a feasible reference for multi-node information interconnection in the Agricultural Energy Internet [11]. Remote monitoring systems designed based on GSM and embedded platforms have overcome the spatial and temporal limitations of traditional farming scenarios, offering new approaches for remote energy consumption sensing and environmental management [12]. Water quality parameter prediction models, based on machine learning, integrate artificial intelligence technology and wireless sensor technology to provide technical support for the real-time perception and intelligent decision-making of intelligent water quality monitoring systems [13]. In the direction of factory and intensive aquaculture, a highly efficient multi-fish behavior recognition and tracking method (SDYOLO-Tracker) is used to achieve monitoring and early detection of fish hypoxia behavior. This method is based on an improved YOLOv8n model, introducing D-LKA modules to enhance small target detection capability. Through a three-dimensional motion prediction strategy, it effectively solves the recognition delay caused by sudden fish movements and disappearances, achieving accurate multi-fish recognition and tracking [14]. A water quality monitoring system, with the main control chip being the STM32F103 microcontroller, uses NB-IoT technology combined with sensors and stores the collected water quality data on an IoT cloud platform, enabling remote monitoring and management of water quality [15]. At the same time, an underwater robot with a specific size is designed, with the core objective of achieving real-time underwater video acquisition, demonstrating the feasibility of coordinated energy and environmental data collection [16].

In domestic research, a comprehensive aquaculture management system has been constructed, achieving the integrated display of multidimensional information such as water quality, weather, and energy consumption, providing an engineering example for the multi-source data visualization of the Agricultural Energy Internet [17]. In addition, a big data aquaculture platform based on unmanned boats has been developed, aiming to achieve intelligent management of water quality monitoring, underwater target recognition, and cruise path planning. On the hardware side, the platform is based on self-developed unmanned boats, can collect water quality data such as Total Dissolved Solids (TDS) and underwater images; on the software side, a visual platform based on the Vue and Leaflet frameworks has been built, integrating Monte Carlo path planning, Kriging interpolation-fuzzy comprehensive evaluation, and YOLOv5-UNet fusion recognition algorithms to achieve efficient data processing and analysis [18]. Using an intelligent sensor network to design intelligent perception of the aquaculture water environment is based on a microservices architecture for big data analysis and processing. It involves designing a microservices-based aquarium monitoring and management system that can provide real-time understanding of changes in the water environment, efficiently manage large amounts of historical data, and solve existing management problems [19].

Internationally, has designed and evaluated the architecture of an aquaculture water quality prediction model to improve management through technology. Develop the proposed architecture using inductive and deductive methods, and evaluate it in conjunction with comprehensive analysis techniques [20]. The pollution prediction model in Scotland [21] and the intelligent water quality management systems in countries such as Japan, Germany, and the United States demonstrate the commercialization and high integration trends of agricultural energy management systems [22].

Foreign systems have already established mature frameworks in terms of integration and commercialization, whereas domestic systems, though starting later, are innovating quickly and continuously improving system architecture design. Hereafter, with the deep integration of low-carbon concepts and intelligent control technologies, agricultural energy Internet monitoring systems will gradually achieve a full-chain closed-loop operation from "data perception—energy consumption scheduling—carbon accounting—intelligent decision-making."

Based on previous research results, this chapter will focus on introducing the overall architecture, functional module design, and key code implementation of the Agricultural Energy Internet monitoring system. It covers data collection and transmission in the perception layer, energy consumption and environmental information integration in the network layer, as well as intelligent visualization and carbon management module construction in the application layer, aiming to provide a complete technical implementation solution for low-carbon management of smart agriculture and aquaponics systems.

6.2 Requirement analysis

6.2.1 Feasibility analysis

6.2.1.1 Technical feasibility

The Agricultural Energy Internet monitoring system aims to achieve real-time monitoring, analysis, and visualization of multi-source information such as aquaculture environments, water quality parameters, energy utilization, and carbon emissions. The system's design needs to meet comprehensive requirements, including real-time performance, high reliability, scalability, and maintainability. To this end, the system adopts a browser/server (B/S) model in its overall architecture and is designed and developed to achieve high cohesion and low coupling of system functional modules, ensuring flexible deployment and rapid response in complex monitoring scenarios.

In terms of front-end development, the system uses Vue 3 as the core framework. Vue 3 features a reactive data binding mechanism and a component-based development model. Combined with the build tool Vite, the project significantly improves startup speed and hot-reload efficiency, supporting the rapid development and debugging of large-scale component-based pages.

The interface layer uses the Element Plus component library to create a visually interactive interface with a unified style and clear structure, enhancing the overall aesthetics and user experience of the system. By integrating the ECharts data visualization library, the system dynamically displays charts of water quality trends, energy consumption curves, carbon emissions, and carbon absorption data, intuitively reflecting the operational status of the Agricultural Energy Internet.

In terms of data communication, the system uses Axios to enable efficient interaction between the frontend and backend RESTful APIs, ensuring real-time transmission and response of monitoring data. Vue Router is employed to build a single-page application (SPA) routing system, allowing modules such as "Water Quality Monitoring" and "Carbon Accounting" to be loaded on demand and switched dynamically. Additionally, Less is used as a Cascading Style Sheets (CSS) preprocessor to enhance style reusability and development maintenance efficiency. In summary, the frontend technology solution can fully meet the system's requirements for interactivity, real-time performance, and visualization.

In terms of backend development, the system uses Spring Boot as the core framework. Spring Boot features a lightweight architecture, automated configuration, and rapid deployment, effectively reducing system development complexity and enhancing module integration efficiency. The data persistence layer uses the MyBatis Plus framework, which greatly simplifies database Create, Read, Update, and Delete (CRUD) operations while ensuring flexibility, thereby improving data processing efficiency.

To reduce redundant code and improve development standards, the system introduces Lombok to automatically generate boilerplate code (such as getters, setters, toString, etc.), significantly enhancing code maintainability. Data serialization and parsing use the FastJSON library to achieve rapid conversion and transmission of JSON data between the front-end and back-end.

The data management component of the system is composed of both MySQL and Redis. MySQL is responsible for storing structured monitoring data (such as water quality parameters, energy consumption records, and carbon emission logs), while Redis is used to cache frequently accessed data and real-time monitoring results, thereby accelerating system response and reducing database load. The database connection pool uses Druid, which not only improves database connection performance but also supports SQL performance monitoring and security protection.

In addition, the system integrates Knife4J with Swagger for automatic generation and online debugging of RESTful API documentation, facilitating collaborative development between the front-end and back-end as well as interface maintenance. The overall technical architecture ensures that the system has high concurrent processing capability and reliable data management mechanisms.

Using MyBatis Plus and MySQL/Redis, this solution can fully meet the comprehensive requirements of an Agricultural Energy Internet monitoring system. This technical system features good scalability, maintainability, and cross-platform compatibility, providing strong technical support for subsequent function expansion.

6.2.1.2 Economic feasibility

This system has consistently adhered to the core principle of "performance first, controllable cost, convenient maintenance, and flexible expansion" in its overall design concept, aiming to strike an optimal balance between system performance, economic investment, and long-term maintenance. Considering practical conditions such as limited budget for agricultural projects, diverse deployment environments, and complex long-term operation and maintenance requirements, the system comprehensively evaluated various factors in terms of technical architecture and tool selection, including development efficiency, ecosystem maturity, and hardware compatibility, ultimately adopting a lightweight technology stack of Vue, Spring Boot, MySQL, and Redis. This solution fully draws on the mainstream front-end and back-end separation architecture models, providing excellent system scalability and stability, while enabling the construction of an efficient, robust, and economically viable Agricultural Energy Internet monitoring platform without relying on high-cost proprietary software.

From a technical standpoint, all core frameworks and dependent components used in this system are built on an open-source ecosystem, offering the advantages of transparency, security, and sustainability. Spring Boot, as one of the most representative Java enterprise development frameworks, provides a highly modular microservice development mechanism and robust dependency management capabilities, making the system more efficient and flexible in functional integration and version iteration. Vue, as a lightweight front-end framework, features a simple and efficient component-based development model and a two-way data binding mechanism, significantly enhancing front-end development efficiency and user interaction experience. The combination of the two enables efficient collaboration between back-end data processing and front-end presentation, giving the system a clear architectural hierarchy and good scalability.

In terms of cost-effectiveness, this system is built on a fully open-source technology stack. The core tools, plugins, and libraries used during the development process are all freely available, with no additional licensing or usage fees required. Compared with traditional systems that rely on commercial software platforms, this architecture significantly reduces initial construction costs and subsequent maintenance expenses. At the same time, both Spring Boot and Vue have active global communities and well-established ecosystems, allowing the development team to continuously obtain technical support through community resources, open-source documentation, and plugin libraries. This ensures that the system can maintain high maintainability and upgradability over long-term operation.

At the level of data management and system performance, the system uses the MySQL database as the primary storage engine to store structured data and core business information, offering advantages such as strong stability, high compatibility, and easy deployment. At the same time, Redis in-memory database is introduced as a cache layer for frequently accessed data, effectively improving the system's data reading speed and real-time response capability. Through a dual-layer design of "hot data caching, cold data storage," the system can ensure response performance while significantly reducing database query pressure, minimizing server expansion needs and energy consumption, thereby achieving continuous optimization of operating costs while enhancing performance.

The overall architecture of the system adopts a Spring Boot microservices design pattern, achieving modular separation and distributed deployment. Each functional module (such as water quality monitoring, carbon emission accounting, energy consumption analysis, and intelligent control) has an independent logic unit and interface definitions, allowing flexible expansion and replacement according to actual business needs. This architectural design not only supports localized deployment but can also be seamlessly migrated to a cloud environment, achieving dynamic load balancing and elastic scaling across servers. During subsequent system upgrades or functionality expansions, it is only necessary to independently develop and deploy individual modules to smoothly iterate the entire system, effectively avoiding the issue of a single change affecting the whole system, which is common in traditional monolithic architectures.

In addition, the system also takes into account flexibility and cost-effectiveness at the deployment and operation levels. Thanks to the embedded container and automated configuration mechanism provided by Spring Boot, the system can be quickly deployed across different operating systems and server environments, supporting containerization technologies such as Docker to achieve a lightweight, "plug-and-play" operation mode.

Overall, the system's technology selection and architecture design are guided by high cost-effectiveness and sustainability, featuring the economic characteristics of "low investment, low maintenance, and high returns." From the perspective of development efficiency, the combined use of Vue and Spring Boot significantly shortens the project development cycle. Therefore, this scheme not only achieves a balanced optimization of performance and cost at the technical level but also demonstrates good feasibility and promotion potential at the economic level. It holds significant practical significance and value for application promotion.

6.2.1.3 Market feasibility

As an important source of high-quality protein and an economically important pillar industry globally, fisheries play an irreplaceable role in ensuring human food security, promoting regional economic development, and maintaining ecological balance. However, the current fisheries industry is at a critical stage of transition from traditional extensive farming to a green, intelligent, and sustainable development model. On one hand, global population growth and upgrading consumption patterns have led to a continuous increase in demand for aquatic products, driving the industry to constantly improve both output and quality; on the other hand, constraints on water resources, environmental pollution, and carbon emission pressures have imposed higher requirements on farming practices. How to achieve ecological protection and low-carbon development while enhancing economic benefits has become a core issue that needs to be addressed in the modernization process of fisheries.

Against this backdrop, the aquaponics farming model has emerged as an efficient ecological circular agriculture system. This model organically combines aquaculture with plant cultivation, utilizing plants to absorb ammonia nitrogen and phosphorus from aquaculture wastewater, thereby achieving nutrient recycling and water self-purification. It not only significantly improves resource utilization but also effectively reduces environmental emissions and water pollution risks, demonstrating strong ecological benefits and economic potential. However, the efficient operation of aquaponics systems relies on precise control of water quality and environmental parameters. The system's operation depends on real-time data monitoring, information analysis, and intelligent decision-making far more than traditional farming methods.

The fisheries industry not only needs to focus on output and efficiency but also on refined management of energy use and carbon emissions. By establishing an Agricultural Energy Internet monitoring system with carbon accounting capabilities, it is possible to achieve quantitative tracking of the entire process from "energy consumption—carbon emissions—carbon reduction." The system can dynamically monitor the energy consumption of various equipment used in aquaculture (such as oxygenators, water pumps, lighting, and temperature control systems) and calculate total carbon emissions based on electricity carbon emission factors, while also incorporating data from green energy sources like photovoltaic power generation to calculate carbon reduction benefits. This integrated energy and carbon management approach not only provides aquaculture enterprises with scientific carbon data support but also offers policymakers precise assessments for low-carbon production, helping the industry achieve energy-saving, consumption-reducing, and green transformation goals.

In addition, the internationalization trend of the fisheries industry has also imposed new requirements on the system's functionality. With the overseas promotion of China's smart fisheries technology, the system not only needs to meet domestic market application demands but also must possess multilingual and international support capabilities to accommodate the usage habits of different countries, regions, and user groups. During the design phase, the system fully considered details such as language localization, unit conversion, and regional time synchronization, enabling it to flexibly adapt to application scenarios in markets such as Southeast Asia, Africa, and Europe, and demonstrating strong potential for international promotion. This design helps enhance the influence of China's smart agriculture and green energy technologies in the international market, providing strong technical support for the overseas expansion of Chinese fisheries equipment and software.

From the perspective of market demand, the current fisheries industry has an increasing need for intelligent management systems. On one hand, large-scale aquaculture enterprises hope to leverage digital technologies to achieve visual management and precise control throughout the entire production process. On the other hand, small and medium-sized aquaculture operators urgently need low-cost, easy-to-deploy, and user-friendly intelligent management tools to compensate for the inefficiencies and inaccuracies of experience-based manual management. This system is designed specifically to address this market gap. By adopting a lightweight architecture and an open-source technology stack, it reduces deployment barriers and operational costs. It is suitable not only for the cluster management of large smart fishery bases but also for the distributed application needs of individual operators, demonstrating strong universality and scalability.

In terms of functionality, this system includes an intelligent warning mechanism and a multidimensional data visualization module. It can automatically trigger alerts and generate optimization recommendations when parameters are abnormal, helping users adjust operational strategies in a timely manner. In addition, the system integrates modules for electricity consumption statistics and carbon emission accounting, which can dynamically display trends in energy consumption, changes in power generation, and carbon emission curves over different time scales, providing data support for energy optimization and carbon reduction strategies in aquaculture farms. Through comprehensive analysis of energy use and environmental conditions, the system can achieve a complete closed-loop management from data collection and process management to intelligent decision-making, further enhancing the automation and intelligence level of fishery production.

From the perspective of industry application prospects, with the continuous integration of smart agriculture, energy Internet, and carbon neutrality technologies, intelligent management systems with integrated capabilities of carbon accounting, environmental monitoring, and energy consumption analysis will become one of the core competencies in the fisheries sector in the coming years. According to industry trend forecasts, future digital management platforms for fisheries will develop toward being "cloud-based, intelligent, and collaborative." This system, with its advantages in technical architecture, algorithm models, and data visualization, can quickly adapt to policy guidance and market demand changes, offering broad potential for promotion and commercial value.

In summary, this system closely aligns with the overall trend of digitalization and green development in the modern fisheries industry in terms of design philosophy, functionality, and market positioning. It provides aquaculture enterprises and small- to medium-sized fish farmers with an efficient, stable, and sustainable intelligent management solution. The promotion and application of the system will effectively facilitate the transformation and upgrading of fisheries production methods, helping to achieve multiple objectives, including energy saving, cost reduction, efficiency improvement, and refined carbon emission management, with significant economic, ecological, and social benefits.

6.2.2 Functional requirements analysis

This system serves as an Agricultural Energy Internet monitoring platform for aquaponics farming scenarios. Its functional design needs to closely align with users' actual needs in water quality, cultivation environment, and energy management to ensure the system's completeness, practicality, and scalability. The main functional requirements analysis is presented as follows.

The system should be able to display key parameters of fishery water quality in real time on the front-end interface, including water temperature, pH value, dissolved oxygen, salinity, and indicators of the cultivation environment. Through real-time data display, users can instantly understand the conditions of the breeding and cultivation environment, providing a scientific basis for daily management.

The system must support historical data retrieval functions, including water quality data of the fish pond collected every 15 minutes and greenhouse cultivation environment data. Managers can analyze historical data to assess trends in water supply and environmental changes and identify potential risks in advance. Through changes in indicator colors and adjustments of water quality/cultivation environment scores, the platform can intuitively notify users of abnormal situations, enabling rapid response and intervention to potential risks, thereby improving the safety of aquaculture management.

The system needs to establish carbon emission, carbon reduction, and carbon absorption models for the "aquaponics" mode, and provide hourly, daily, weekly, monthly, and yearly visualizations of electricity consumption, power generation, carbon emissions, carbon reduction, and carbon absorption for each module. This feature helps users optimize energy use, reduce their carbon footprint, and respond to the national "dual carbon" strategy.

To enhance data readability and user experience, the system needs to use chart components such as line charts and bar charts to visually present various monitoring data, helping users quickly understand the trends of key indicators and assist in management decision-making.

The system should provide a bilingual interface in Chinese and English, allowing users to freely select their operating language through the language switch control at the top right corner of the interface. At the same time, the system should be compatible with mobile devices, enabling users to remotely access it via smartphones, tablets, and other terminals for flexible mobile management.

The system's functions can be divided into several core modules, including user login, language switching, date selection, fishery water quality monitoring, cultivation environment monitoring, historical data querying, data monitoring, electricity usage monitoring, load-side carbon emissions, overall carbon emissions, and reference management.

The functional design of this system is closely aligned with the actual needs of aquaponics farming scenarios, covering key functions such as real-time monitoring, historical analysis, intelligent early warning, carbon energy management, data visualization, and mobile applications.

The use cases of the entire system and their descriptions are shown in Table 6.1. There are eleven main categories of use cases, namely login, language

Table 6.1 Description of functional requirements

Use case category	**Use case name**	**Use case description**
Log in	Log in	Users can log into the system by entering the correct account and password
Language switch	Switch language	Users can switch between Chinese and English by clicking the language toggle button
Date	Check the weather, time, and date	Users can check the weather, time, and date
Fishery water quality	Overview of fish farming	Users can view the number of fish ponds, the types of fish being farmed, the total weight of the farmed fish, the average water temperature, the average pH level, the average dissolved oxygen, the fish health status, and the average survival rate
	Inspection and warning of fishery water quality in each breeding pond	Users can monitor the water temperature, pH, dissolved oxygen, water level, salinity, ammonia nitrogen content, aeration status, feeding status, survival rate, and water quality of each fish pond, and set certain thresholds. When a water quality parameter exceeds the preset threshold, the corresponding parameter will change from green to red, and the water quality status will change from green (good) to yellow (moderate), thereby alerting users to intervene
Cultivation environment	Overview of plant cultivation	Users can view the total planting area of plants, land area, hydroponic area, average air temperature, average air humidity, average light intensity, plant health status, and average survival rate
	Monitoring and early warning of cultivation environments in various planting areas	Users can view the air humidity, leaf surface temperature, and soil temperature in various land cultivation areas and foam box cultivation areas, as well as air humidity, leaf surface temperature in hydroponic cultivation areas, and vertical column hydroponic cultivation areas. Users can also set certain thresholds, and when the cultivation environment parameters exceed the preset thresholds, the corresponding parameters will change from green to red, alerting the user to intervene

(Continues)

Table 6.1 (*Continued*)

Use case category	Use case name	Use case description
Historical data	View historical data of water supply quality	Users can view the water quality of aquaculture areas every 15 minutes in the past, starting from the current moment
	View historical data of the overall cultivation environment	Users can view the overall cultivation environment of the plant-growing area, that is, from the current moment backward, and leaf temperature is recorded every 15 minutes
Electricity monitoring	View the comparison between purchased electricity and photovoltaic electricity	Users can compare and view the photovoltaic power generation and purchased electricity within the same coordinate system for the past 15 minutes, past week, past six months, and past year
Carbon emissions	View the carbon emissions from purchased electricity	Users can view the carbon emissions caused by purchased electricity over the past hour, day, month, half-year, and year
	View carbon emission reduction from photovoltaic power generation	Users can view the carbon reduction caused by photovoltaic power generation over the past hour, day, month, half-year, and year
	Overview of carbon accounting	Users can view the daily and annual carbon emissions, carbon absorption, and carbon reduction of the entire system, as well as the classification and proportion of annual carbon emission sources for the entire system
	View the comparison of carbon emissions between the power supply side and the load side	Users can compare and view the carbon emission data of the power side and the load side over the past week and the past six months within the same coordinate system
	View the carbon accounting status of the fisheries sector	Users can view the carbon accounting status of the fisheries sector for the past day, month, half-year, and year
	View the carbon accounting status of the planting sector	Users can view the carbon absorption and carbon emission status of the planting section for the past day, month, half-year, and year
	View the carbon accounting status of the waste sector	Users can view the carbon emissions of the waste sector over the past month, six months, and a year
References	Document download	Users can download the literature they want to read

switching, date, fishery water quality, cultivation environment, historical data, data monitoring, and electricity monitoring.

Based on the various use cases and their descriptions mentioned above, draw the use case diagram, as shown in Figure 6.1.

Explanation of terms used in the use case text is shown in Table 6.2, and detailed descriptions of each system use case are shown in Tables 6.3–6.20.

Figure 6.1 System use case diagram

Table 6.2 Explanation of terms

Term	**Explain**
Most important	Core system functions, most frequently used by users
Relatively important	Important system functions frequently used by users
General	An optional feature, used occasionally by users
Not important	The current version may not implement this or may postpone it, as users do not use it frequently

Table 6.3 Login use case

Use case name	Log in
Use case description	Users can log into the system by entering the correct account and password
Participant	User
Priority	Most important
Prerequisite	The system must be fully configured; the correct username and password must be obtained
Trigger	The user wants to log in to the system
Scene	The user enters the system login interface, enters the correct account and password, and clicks login; the system pops up a login successful message and enters the main interface
Exception handling	If a network anomaly occurs during the login process, the system should promptly provide error information and retain the user login interface, allowing the user to attempt login again. When the system experiences a database connection exception or significant response delay, it should prompt the user to try again later and display the current operating status of the system. If the user enters an incorrect username or password, the system should notify them of the input error and guide them to re-enter the correct login information

Table 6.4 Switch language use case

Use case name	Switch language
Use case description	Users can switch between Chinese and English by clicking the language toggle button
Participant	User
Priority	Relatively important
Prerequisite	The system must be fully configured; the correct username and password must be obtained

Table 6.4 (Continued)

Trigger	The user wants to switch between Chinese and English
Scene	The user logs in and enters the main interface of the system; clicks to select switching between Chinese and English; the overall system language changes to the corresponding language
Exception handling	When the system experiences database connection exceptions or significant response delays, the user should be prompted to try again later, and the current operating status of the system should be displayed

Table 6.5 View date, time, and weather use cases

Use case name	Check the weather, time, and date
Use case description	Users can check the weather, time, and date
Participant	User
Priority	Relatively important
Prerequisite	The system must be fully configured; the correct username and password must be obtained
Trigger	The user wants to check the date, time, and weather
Scene	The user logs in and enters the main interface of the system; views the real-time date, time, and weather
Exception handling	When the system experiences database connection exceptions or significant response delays, the user should be prompted to try again later, and the current operating status of the system should be displayed

Table 6.6 Use case for viewing the overall situation of fish farming

Use case name	Overview of fish farming
Use case description	Users can view the number of fish ponds, the types of fish being farmed, the total weight of the farmed fish, the average water temperature, the average pH level, the average dissolved oxygen, the fish health status, and the average survival rate
Participant	User
Priority	Most important
Prerequisite	The system must be fully configured; the correct username and password must be obtained
Trigger	The user wants to view the overall status of the fish
Scene	The user logs in and enters the system's main interface, then views the overall situation of the fish
Exception handling	When the system experiences database connection exceptions or significant response delays, the user should be prompted to try again later, and the current operating status of the system should be displayed

Table 6.7 Use case for monitoring and warning of water quality in each aquaculture pond

Use case name	Inspection and warning of fishery water quality in each breeding pond
Use case description	Users can monitor the water temperature, pH, dissolved oxygen, water level, salinity, ammonia nitrogen content, aeration status, feeding status, survival rate, and water quality of each fish pond, and set certain thresholds. When a water quality parameter exceeds the preset threshold, the corresponding parameter will change from green to red, and the water quality status will change from green (good) to yellow (moderate), thereby alerting users to intervene
Participant	User
Priority	Most important
Prerequisite	The system must be fully configured; the correct username and password must be obtained
Trigger	The user wants to view the water quality and warning status of each fish farming pond
Scene	The user logs in and enters the system main interface, checks the water quality and warning status of each aquaculture pond
Exception handling	When the system experiences database connection exceptions or significant response delays, the user should be prompted to try again later, and the current operating status of the system should be displayed

Table 6.8 Overall plant cultivation overview use case

Use case name	Overview of plant cultivation
Use case description	Users can view the total planting area of plants, land area, hydroponic area, average air temperature, average air humidity, average light intensity, plant health status, and average survival rate
Participant	User
Priority	Most important
Prerequisite	The system must be fully configured; the correct username and password must be obtained
Trigger	The user wants to view the overall condition of the plants
Scene	The user logs in and enters the main interface of the system, clicks on "Cultivation Environment" in the navigation bar below, and checks the overall status of plant cultivation
Exception handling	When the system experiences database connection exceptions or significant response delays, the user should be prompted to try again later, and the current operating status of the system should be displayed

Table 6.9 Use cases for monitoring and early warning of cultivation environments in various planting areas

Use case name	Monitoring and warning of cultivation environments in various planting areas
Use case description	Users can view the air temperature, light intensity, soil temperature, and leaf surface temperature in various land cultivation areas and foam box cultivation areas, as well as the air temperature, air humidity, and leaf surface temperature in hydroponic cultivation areas and vertical column hydroponic cultivation areas. Users can also set certain thresholds, and when the cultivation environment parameters exceed the preset thresholds, the corresponding parameters will change from green to red, alerting the user to intervene
Participant	User
Priority	Most important
Prerequisite	The system must be fully configured; the correct username and password must be obtained
Trigger	The user wants to view the cultivation environment and warning conditions of each planting area
Scene	The user logs in and enters the system main interface; clicks on "Cultivation Environment" in the navigation bar below; and checks the water quality and warning status of each aquaculture pond
Exception handling	When the system experiences database connection exceptions or significant response delays, the user should be prompted to try again later, and the current operating status of the system should be displayed

Table 6.10 Use case for viewing historical water quality data

Use case name	View historical data of water supply quality
Use case description	Users can view the water quality conditions of aquaculture areas, that is, the water temperature, dissolved oxygen, salinity, and ammonia nitrogen content every 15 minutes from the current time
Participant	User
Priority	Most important
Prerequisite	The system must be fully configured; the correct username and password must be obtained
Trigger	The user wants to check the water supply quality
Scene	The user logs in and enters the main interface of the system, clicks on "Historical Data" in the navigation bar below, and views the historical data of the water supply quality
Exception handling	When the system experiences database connection exceptions or significant response delays, the user should be prompted to try again later, and the current operating status of the system should be displayed

Table 6.11 Use case for viewing historical data of the overall cultivation environment

Use case name	View historical data of the overall cultivation environment
Use case description	Users can view the overall cultivation environment of the plant-growing area, that is, from the current moment backward, temperature, air humidity, and leaf surface temperature recorded every 15 minutes
Participant	User
Priority	Most important
Prerequisite	The system must be fully configured; the correct username and password must be obtained
Trigger	The user wants to view the overall cultivation environment
Scene	The user logs in and enters the main interface of the system; clicks on "Historical Data" in the navigation bar at the bottom; views historical data of the overall cultivation environment
Exception handling	When the system experiences database connection exceptions or significant response delays, the user should be prompted to try again later, and the current operating status of the system should be displayed

Table 6.12 Use case for comparing purchased electricity and photovoltaic electricity

Use case name	View the comparison between purchased electricity and photovoltaic electricity
Use case description	Users can compare and view the photovoltaic power generation and purchased electricity data in the same coordinate system over the past 15 minutes, the past week, the past six months, and the past year
Participant	User
Priority	Most important
Prerequisite	The system must be fully configured; the correct username and password must be obtained
Trigger	The user wants to compare purchased electricity with photovoltaic electricity
Scene	The user logs in and enters the main interface of the system; clicks on "Electricity Monitoring" in the bottom navigation bar; and checks the comparison between purchased electricity and photovoltaic electricity
Exception handling	When the system experiences database connection exceptions or significant response delays, the user should be prompted to try again later, and the current operating status of the system should be displayed

Table 6.13 Use case for checking carbon emissions from purchased electricity

Use case name	View the carbon emissions from purchased electricity
Use case description	Users can view the carbon emissions caused by purchased electricity over the past hour, day, month, half-year, and year
Participant	User
Priority	Most important
Prerequisite	The system must be fully configured; the correct username and password must be obtained
Trigger	The user wants to check the carbon emissions from purchased electricity
Scene	The user logs in and enters the main interface of the system; clicks on "Overall Carbon Emissions" in the bottom navigation bar; and checks the carbon emissions from purchased electricity
Exception handling	When the system experiences database connection exceptions or significant response delays, the user should be prompted to try again later, and the current operating status of the system should be displayed

Table 6.14 Use case for checking carbon emission reduction in photovoltaic power generation

Use case name	View carbon emission reduction from photovoltaic power generation
Use case description	Users can view the carbon reduction caused by photovoltaic power generation over the past hour, day, month, half a year, or year
Participant	User
Priority	Most important
Prerequisite	The system must be fully configured; the correct username and password must be obtained
Trigger	The user wants to check the carbon reduction from photovoltaic power generation
Scene	The user logs in and enters the main interface of the system, clicks on "Overall Carbon Emissions" in the navigation bar below, and checks the carbon emission reduction status of photovoltaic power generation
Exception handling	When the system experiences database connection exceptions or significant response delays, the user should be prompted to try again later, and the current operating status of the system should be displayed

Table 6.15 Use case for viewing carbon accounting overview

Use case name	Overview of carbon accounting
Use case description	Users can view the daily and annual carbon emissions, carbon absorption, and carbon reduction of the entire system, as well as the classification and proportion of the system's annual carbon emission sources
Participant	User
Priority	Most important
Prerequisite	The system must be fully configured; the correct username and password must be obtained
Trigger	The user wants to view the overall carbon accounting overview
Scene	The user logs in and enters the system main interface, clicks on "Overall Carbon Emissions" in the bottom navigation bar, and views the carbon accounting overview
Exception handling	When the system experiences database connection exceptions or significant response delays, the user should be prompted to try again later, and the current operating status of the system should be displayed

Table 6.16 Use case for comparing carbon emissions between the power supply side and the load side

Use case name	View the comparison of carbon emissions between the power supply side and the load side
Use case description	Users can compare and view the carbon emission data of the power side and the load side over the past week and the past six months within the same coordinate system
Participant	User
Priority	Most important
Prerequisite	The system must be fully configured; the correct username and password must be obtained
Trigger	The user wants to view a comparison of carbon emissions between the power supply side and the load side
Scene	The user logs in and enters the main interface of the system; clicks on "Overall Carbon Emissions" in the bottom navigation bar; and views the comparison of carbon emissions data between the power side and the load side
Exception handling	When the system experiences database connection exceptions or significant response delays, the user should be prompted to try again later, and the current operating status of the system should be displayed

Table 6.17 Use case for checking carbon accounting in the fisheries sector

Use case name	View the carbon accounting status of the fisheries sector
Use case description	Users can view the carbon accounting status of the fisheries sector for the past day, month, half-year, and year
Participant	User
Priority	Most important
Prerequisite	The system must be fully configured; the correct username and password must be obtained
Trigger	The user wants to view the carbon accounting status of the fisheries sector
Scene	The user logs in and enters the main interface of the system; clicks on "Overall Carbon Emissions" in the bottom navigation bar; and checks the carbon accounting status of the fisheries sector
Exception handling	When the system experiences database connection exceptions or significant response delays, the user should be prompted to try again later, and the current operating status of the system should be displayed

Table 6.18 Use case for checking carbon accounting in the planting sector

Use case name	View the carbon accounting status of the planting sector
Use case description	Users can view the carbon absorption and carbon emission status of the planting section for the past day, month, half-year, and year
Participant	User
Priority	Most important
Prerequisite	The system must be fully configured; the correct username and password must be obtained
Trigger	The user wants to check the carbon accounting status of the planting sector
Scene	The user logs in and enters the main interface of the system; clicks on "Overall Carbon Emissions" in the navigation bar at the bottom; and checks the carbon accounting situation of the planting section
Exception handling	When the system experiences database connection exceptions or significant response delays, the user should be prompted to try again later, and the current operating status of the system should be displayed

Table 6.19 Use case for checking carbon accounting in the waste sector

Use case name	View the carbon accounting status of the waste sector
Use case description	Users can view the carbon emissions of the waste sector over the past month, six months, and a year
Participant	User
Priority	Most important
Prerequisite	The system must be fully configured; the correct username and password must be obtained
Trigger	The user wants to view the carbon accounting status of the waste section
Scene	The user logs in and enters the main interface of the system; clicks on "Overall Carbon Emissions" in the bottom navigation bar; and checks the carbon accounting status of the waste section
Exception handling	When the system experiences database connection exceptions or significant response delays, the user should be prompted to try again later, and the current operating status of the system should be displayed

Table 6.20 View literature download examples

Use case name	Document download
Use case description	Users can download the literature they want to read
Participant	User
Priority	Relatively important
Prerequisite	The system must be fully configured; the correct username and password must be obtained
Trigger	The user wants to download references
Scene	The user logs in and enters the system's main interface; clicks on "References" in the navigation bar at the bottom; then clicks on the literature they want to download to download it
Exception handling	When the system experiences database connection exceptions or significant response delays, the user should be prompted to try again later, and the current operating status of the system should be displayed

6.2.3 Non-functional requirements analysis

Based on ensuring that the Agricultural Energy Internet monitoring system has a complete, sound, and efficient functional framework, in order to further enhance the overall performance, user operation experience, and information security assurance of the system during actual operation, it is necessary to propose and meet a series of key non-functional requirements. These non-functional requirements not only involve performance indicators such as system operational efficiency, data processing capability, and response speed but also include system availability, stability, and reliability, to ensure that the platform can continuously and stably support agricultural production, energy management, and environmental monitoring during long-term operation under complex environmental conditions. In addition, the system's security requirements cannot be overlooked, including user data protection, access control, identity authentication, data encryption, and defenses against network attacks.

At the same time, to facilitate long-term maintenance, upgrades, and feature expansion of the system, maintainability and scalability requirements are also important components of non-functional design. This requires the system to adopt a modular, loosely coupled architecture, with a clear and standardized code structure, as well as comprehensive development and operational documentation, enabling rapid development, seamless updates, and convenient debugging when functionalities are iterated, upgraded, or when new devices and monitoring nodes are integrated. Compatibility requirements are equally important; the system needs to operate properly across different hardware environments, operating systems, and major browsers, while also supporting cross-terminal and cross-device adaptability, ensuring that users on desktops, tablets, and mobile devices all experience a consistent and smooth operation. Additionally, the system should support internationalization and multilingual environments to meet the needs of cross-regional deployment and access by users in multiple languages.

Next, this section of non-functional requirements will systematically analyze multiple dimensions such as system performance, usability, stability, security, maintainability, scalability, and compatibility, providing technical guidance and design basis for the Agricultural Energy Internet monitoring system in practical applications. This ensures that the platform can achieve efficient, stable, secure, and sustainable service capabilities during long-term operation and in complex scenarios, thereby offering reliable and comprehensive technical support for agricultural production management, energy regulation, and environmental monitoring.

6.2.3.1 System performance

The system should possess highly optimized operational performance and excellent response speed to ensure that, in complex, variable, and dynamic agricultural monitoring environments, it can meet the strict requirements for real-time performance and data processing efficiency. Specifically, under relatively stable network conditions, the loading time of various system pages should be strictly controlled within 3 seconds, ensuring that users can quickly access the information they need, avoiding delays that could affect user experience. The system should achieve rapid responses within 1 second, ensuring that energy data, environmental parameters, meteorological information, and various monitoring indicators can be continuously and dynamically collected, processed, and presented. This provides users with accurate, reliable, and timely monitoring information to support key business needs such as agricultural production management, environmental regulation, carbon emission accounting, and energy scheduling.

In terms of concurrent processing capability, the system needs to have strong multi-user support, capable of handling a large number of user logins, accesses, and requests to operate system modules simultaneously without causing performance bottlenecks or system errors. Whether under high traffic, high concurrent data requests, or large-scale operational scenarios, the system should maintain smooth page loading, real-time data refresh, and rapid operational response, without page lag, data delays, request blocking, or response latency. This high concurrent processing capability not only ensures the system's stability during daily operations but also allows it to handle sudden scenarios of concentrated multi-user operations, such as simultaneously querying historical data, generating various analytical reports, or monitoring real-time sensor data, ensuring that the system continues to operate efficiently under high-load conditions.

To achieve the aforementioned high-performance goals, the system should adopt multiple optimization strategies and architectural designs in its technical implementation, including but not limited to: a front-end asynchronous data refresh mechanism and virtual Document Object Model (DOM) technology to ensure efficient and smooth page rendering; database-level index optimization, table partitioning and sharding strategies, as well as query caching mechanisms to enhance large-scale data read and write performance; multi-threaded processing, task queue scheduling, and distributed load balancing strategies at the back-end service layer to ensure fast and stable system responses under high-concurrency access; at the same time, combining data caching (such as Redis or Memcached) and message queue (such as Kafka or RabbitMQ) technologies to achieve efficient transmission and processing of real-time data, further improving the system's data processing capability and response speed in complex monitoring scenarios.

In addition, the system should also focus on end-to-end performance monitoring and optimization. By monitoring key metrics such as CPU, memory, network bandwidth, database performance, and frontend loading time, it can achieve dynamic tuning and automatic alert mechanisms. Under high-load conditions, the system should be able to automatically adjust resource allocation or activate preset scaling strategies based on performance monitoring results, ensuring that core business data, such as energy monitoring, environmental parameter collection, and carbon accounting, remain real-time, accurate, and continuous under any circumstances.

Through the above multidimensional performance optimization and high-concurrency processing capability design, the system can achieve efficient data interaction and real-time monitoring in complex and variable agricultural production environments, providing reliable data support and technical assurance for energy management, environmental regulation, intelligent agricultural decision-making, and carbon emission monitoring. The system's high-performance, high-concurrency, and high-stability design not only meets the current business needs of Agricultural Energy Internet monitoring but also provides a solid technical foundation for future functionality expansion, device integration, and large-scale deployment, ensuring that the Agricultural Energy Internet monitoring system maintains stable, efficient, and reliable performance during actual operation over the long term.

6.2.3.2 System availability

As a production-grade platform designed for long-term operation, the Agricultural Energy Internet monitoring system must fully consider system high availability, business continuity, and long-term reliable operation during its design and implementation. This ensures that it can stably and continuously support key operations such as agricultural production, energy management, and environmental monitoring during actual use. The system should provide uninterrupted operation throughout the year, with annual availability strictly guaranteed to be no less than 99%. This ensures continuous monitoring needs are met across different seasons, regions, and operational load conditions, while providing a solid data foundation for agricultural production decision-making, energy scheduling, and environmental management.

It is inevitable to encounter various potential risks and exceptional situations, including hardware failures, software anomalies, network fluctuations, server overloads, storage device failures, or unexpected interruptions. To address these potential risks, the platform must have comprehensive self-healing, automatic recovery, and high-availability mechanisms to minimize the impact of system interruptions on business operations. Specifically, when an exception or unexpected interruption occurs, the system should be able to quickly reinitialize the affected services through an automatic restart mechanism, restore operational status using fault recovery mechanisms, or automatically switch services and data to standby nodes or backup servers through failover mechanisms, achieving seamless business continuity and uninterrupted service. Such mechanisms enable the system to restore functionality in the shortest possible time, minimizing the negative impact of monitoring data loss, data anomalies, or business interruptions on user experience and decision-making analysis.

At the same time, during the recovery process, the system must ensure the integrity, consistency, and reliability of energy data, environmental parameters, and various monitoring information. To achieve this, the system should combine strategies, such as transaction management, data verification, data logging, data snapshots, and real-time backups, to monitor and protect the data being processed and stored throughout the entire process. In the event of a failover or recovery, the system can ensure consistency between historical and real-time data, maintain the complete sequence of operation records and monitoring data, and use data rollback or automatic synchronization mechanisms to prevent any data loss, thereby maintaining the continuity and stability of critical monitoring operations.

In addition, to further enhance the system's high availability and stability, the platform should adopt technologies such as redundant deployment, load balancing, distributed service architecture, and dynamic resource scheduling. Redundant deployment can provide backup nodes for critical services and core data, enabling rapid switching and disaster recovery; load balancing can reasonably distribute user requests and system loads, preventing performance degradation due to a single point of overload; distributed architecture design allows each functional module to operate independently without interference, improving the overall system fault tolerance; dynamic resource scheduling can automatically adjust computing and storage resources according to real-time load conditions, ensuring the system can run smoothly under high concurrent access or complex environments.

High availability design should also be combined with real-time monitoring and alert mechanisms to continuously monitor the system's CPU, memory, network bandwidth, database performance, and service operating status. Once an anomaly is detected, the system can automatically trigger alerts, initiate fault recovery processes, or schedule backup resources to achieve rapid response and handling of issues, thereby reducing the potential impact of risks on business operations.

Through the comprehensive and systematic high-availability design described above, the Agricultural Energy Internet monitoring system can maintain a stable, reliable, and continuous service state when faced with complex agricultural production environments, variable operating conditions, and potential technical risks. The system can not only ensure the integrity and real-time availability of energy data, environmental parameters, and monitoring information but also provide long-term and reliable data support and technical assurance for energy management, environmental regulation, crop growth monitoring, and agricultural production optimization. This enables the digitalization, intelligence, and low-carbon goals of agricultural production, laying a solid foundation for the expansion and upgrading of future Agricultural Energy Internet platforms.

6.2.3.3 System security

The Agricultural Energy Internet monitoring platform must implement multi-level and multidimensional security protection strategies, building an all-around, in-depth defense security system. These security strategies mainly include multiple aspects such as a strict access control mechanism, a comprehensive identity authentication mechanism, data encryption protection mechanisms, and network protection and auditing mechanisms. Specifically, during the user login phase, the system must perform strict credential verification, employing multi-factor authentication, dynamic passwords, or biometric methods, effectively preventing unauthorized access and potential security threats from the outset.

In terms of data security, the system must use high-strength encryption technology to store and transmit sensitive information, such as energy monitoring data, carbon emissions data, and carbon accounting data. At the same time, the system should implement access control policies, assigning appropriate data access permissions to users of different levels and strictly controlling the scope of operations, thereby further reducing the risk of data leakage or misoperation.

The platform should also have proactive defense capabilities against various common network attacks, including but not limited to directory traversal attacks, brute force attacks, and denial-of-service attacks. Through real-time monitoring, access filtering, anomaly traffic detection, and security policy coordination, the system can promptly detect and block malicious access behaviors or potential attacks. At the same time, the platform should implement comprehensive security logging and auditing functions, providing full tracking and analysis of user operations, abnormal events, and security policy execution, supporting security incident tracing and emergency response, and providing technical support for secure system operations and maintenance.

Through the implementation of a multi-layered security strategy, the platform can not only effectively ensure the confidentiality, integrity, and availability of core business data such as energy monitoring and carbon accounting but also maintain secure and stable operation in complex network environments and multi-user operation scenarios. In addition, this defense-in-depth system can enhance users' trust in the platform, ensuring that the Agricultural Energy Internet monitoring system can continuously provide efficient, secure, and reliable services over the long term, thereby offering solid technical support for agricultural energy management, environmental monitoring, and carbon emission analysis.

6.2.3.4 System maintainability and scalability

The system should adopt a modular design and loosely coupled architecture to ensure high flexibility and operability in terms of functional expansion, system maintenance, and performance optimization across the entire Agricultural Energy Internet monitoring platform. Through modular design, each functional unit of the system can be developed, tested, deployed, and upgraded independently, thereby achieving minimal dependency and low coupling between functional modules. This design approach can minimize the impact on existing modules during functional expansion or system upgrades, improve development efficiency, reduce system complexity, and facilitate collaborative development among team members and long-term operation and maintenance management.

In terms of code management and development standards, the platform should maintain a clear code structure and rigorous logic, strictly adhering to unified coding standards, design patterns, and software architecture principles. During system development, a comprehensive documentation system should be provided, including development documents, interface documentation, system architecture documents, and operation manuals. Development documents should detail the functions of each module, interface call methods, data structure definitions, key algorithm implementations, exception handling mechanisms, and testing procedures. Interface documentation should cover module interface specifications, data transmission formats, parameter descriptions, and return value definitions to ensure that different development teams can quickly understand and utilize module functions. The operation manual should include system deployment procedures, monitoring strategies, log collection and analysis methods, fault diagnosis and recovery processes, to support rapid system debugging, smooth upgrades, and long-term stable operation. Additionally, the documentation system should support version management, enabling the system to trace historical changes during iterative upgrades, improving the controllability of development and operation.

The system should also have good scalability and flexibility, so that new sensor devices, energy monitoring nodes, or additional carbon emission and carbon absorption calculation modules can be seamlessly integrated in future version iterations. This extensibility not only requires compatibility of software and hardware interfaces but also needs to consider data processing capabilities, storage expansion capabilities, and system performance optimization to ensure that, during functional expansion or module upgrades, the overall system performance remains stable, without response delays or operational bottlenecks. To achieve this goal, the platform can adopt a microservices architecture or a plugin-based modular decomposition strategy, allowing new functionalities to be deployed independently, hot-swapped, or invoked on demand, thereby supporting rapid deployment and flexible management.

In practical applications, modular and loosely coupled design can also support multi-level interface calls and service registration mechanisms. For example, inter-module communication and data exchange can be achieved through a unified API gateway, and functional modules can be dynamically managed and load-balanced using a service registry. This design enhances its fault tolerance and maintainability. In addition, version control strategies should be closely integrated with modular design to enable efficient and secure module iteration updates through code branch management, automated builds, and continuous integration (CI/CD) mechanisms.

In summary, through modular, loosely coupled design and highly scalable strategies, the Agricultural Energy Internet monitoring system can flexibly adapt to the ever-changing and upgrading needs of future agricultural energy monitoring, environmental parameter control, and carbon accounting, while ensuring system stability and reliability.

6.2.3.5 System compatibility and adaptability

The system frontend should fully support all major mainstream browsers, including but not limited to Chrome, Edge, and Firefox, ensuring that users can access and operate all functional modules of the platform stably and smoothly across different browser environments. To achieve this goal, frontend development must strictly adhere to modern web standards and best practices for cross-browser compatibility, ensuring consistent layout, styles, and script logic across different rendering engines, and avoiding issues such as functional anomalies, misaligned styles, or JavaScript execution errors, thereby guaranteeing users a stable and consistent experience in diverse browsing environments.

Frontend design needs to have strong cross-platform adaptability. Through techniques such as responsive web design, fluid layouts, flexible grid systems (Flexbox/Grid), and CSS media queries, the interface can automatically adjust its display on different terminal devices. Whether users are using a desktop computer, laptop, tablet, or mobile smart device, the system can automatically adjust content layout, font size, spacing between controls, and chart display proportions based on screen size, resolution, pixel density, and device characteristics, ensuring a clear interface, smooth operation, and a consistent visual style and user experience. This responsive design not only improves usability on small-screen devices but also makes full use of the display space on large-screen devices, enhancing data visualization and information presentation capabilities.

In terms of user interaction experience, the front end must ensure consistent operation logic across different devices. For example, button clicks, dropdown menu selections, form filling, and chart zooming and scrolling operations should all function smoothly on both touch devices and mouse-operated devices, avoiding response delays or inconsistent behavior. At the same time, the system should support high-resolution and Retina displays, making charts, text, icons, and other interface elements clear and readable on various devices, thereby enhancing users' data analysis efficiency and operation accuracy. The front end should also leverage technologies such as asynchronous data loading, virtual scrolling, lazy loading, and caching strategies to optimize page load speed and rendering performance, ensuring that real-time monitoring data and interactive operations can respond quickly, maintaining a good user experience even under poor network conditions.

To support international applications, the system front-end needs to be compatible with multiple language environments and implement comprehensive internationalization (i18n) features. By using dynamic language packs, JSON configuration files, or multilingual resources provided by backend interfaces, the system can automatically switch interface text, prompts, date formats, numbers, and currency display formats according to the user's language preferences, ensuring a consistent user experience across regions. Internationalization design not only facilitates the global promotion of the system but also supports multilingual team collaboration and cross-regional data management, enhancing the platform's applicability and user satisfaction.

In terms of technical implementation, the front end should use modern frameworks such as Vue to achieve component-based development and modular management. Through component-based design, each functional module can be developed, tested, and maintained independently, improving code reusability and system maintainability. Data interaction between the front end and back end can be implemented using WebSocket, RESTful API, or GraphQL, enabling real-time data updates and efficient communication. Combined with state management libraries (such as Vuex or Redux), the system can efficiently manage complex data states, ensuring the real-time display and consistency of energy monitoring, environmental parameters, and carbon accounting data on the front end.

Overall, through technologies such as mainstream browser compatibility support, cross-terminal responsive design, high-resolution display optimization, multilingual internationalization support, component-based development, and real-time data interaction, the system frontend can provide a smooth, consistent, efficient, and personalized user experience for different user groups. These technical measures not only ensure the long-term stable operation of the Agricultural Energy Internet monitoring platform across multiple scenarios, devices, and languages but also provide a solid technical foundation for the platform's promotion and continuous optimization in application scenarios such as energy monitoring, environmental management, and carbon accounting.

6.2.3.6 Summary of non-functional requirements

This system ensures that the Agricultural Energy Internet monitoring system can achieve stable, efficient, flexible, and reliable operation in complex, variable, and high-load environments by conducting comprehensive and systematic planning and implementation across multiple key non-functional design dimensions, including performance, usability, security, maintainability, scalability, and compatibility.

Specifically, in terms of performance design, the system achieves rapid collection, processing, and presentation of real-time energy data, environmental parameters, and carbon accounting data through frontend page optimization, asynchronous data loading, virtual rendering, and backend concurrency optimization. This ensures that users can experience millisecond-level responses when accessing the system, thereby meeting the strict requirements for real-time and dynamic monitoring in agricultural production and energy management. In terms of usability, the system adopts a high-availability architecture design, including redundant deployment, load balancing, automatic fault detection and recovery mechanisms, and service hot backup strategies. This ensures that the system is available at least 99% of the year. Even in the event of hardware failures, network fluctuations, or software anomalies, it can quickly recover through automatic restarts, failover, or data rollback, ensuring the continuity and integrity of energy monitoring and environmental data collection, and avoiding data loss or business interruption. Regarding security design, the system implements multi-layered protection strategies, including strict user access management, identity authentication, multi-factor verification, encrypted storage and transmission of sensitive data, defense against common network attacks (such as SQL injection, XSS, CSRF), as well as security log auditing and abnormal access alerts. These measures ensure that energy monitoring data, environmental parameters, and carbon emissions and absorption data maintain a high level of confidentiality, integrity, and reliability during transmission, storage, and processing, thereby ensuring the platform operates securely and stably in complex network environments. For maintainability, the system uses a modular and loosely coupled architecture, allowing each functional module to be developed, tested, deployed, and upgraded independently. Additionally, it is equipped with clear and standardized code structures, unified coding conventions, comprehensive development documentation, and operation manuals, supporting rapid debugging, smooth upgrades, and secondary development. This reduces system maintenance costs and improves operational efficiency, ensuring the platform can run sustainably over the long term. Regarding scalability and compatibility, the system's frontend supports major browsers (Chrome, Edge, Firefox, etc.) and multi-terminal devices. Through responsive page design, adaptive layouts, and high-resolution display optimization, it delivers a smooth and consistent user experience across desktops, tablets, and mobile devices. The system also supports multilingual internationalization, facilitating cross-regional deployment and multi-language user access. The backend and data processing modules have strong scalability, enabling the smooth integration of new sensors, energy monitoring nodes, and additional carbon emission and absorption calculation modules, ensuring that the platform's functionality can continuously evolve with changing business needs.

In summary, this system ensures that the Agricultural Energy Internet monitoring platform can achieve efficient, stable, flexible, and secure operation in complex environments through a comprehensive layout and technical implementation in non-functional design aspects such as performance, usability, security, maintainability, scalability, and compatibility. It provides long-term and reliable technical support for the digital and intelligent management of agricultural production and low-carbon development, ensures the efficient integration and continuous optimization of key operations such as energy management, environmental monitoring, and carbon accounting, and lays a solid foundation for sustainable agricultural development and intelligent decision-making.

6.3 High-level design

6.3.1 System technical architecture design

The Agricultural Energy Internet monitoring system adopts a layered architecture with a front-end and back-end separation for its overall design, aiming to achieve high maintainability, scalability, and efficient interactive performance. The system as a whole uses a browser/server (B/S) architecture, with the database system managing data storage and caching tasks.

At the frontend layer, the system is built based on the Vue 3 framework, implementing modular and component-based page design. Through routing management and visual components, it enables interactive display of multi-source information such as agricultural energy, water quality, and carbon emissions. The frontend communicates with the backend via Axios for HTTP requests, achieving dynamic acquisition and updating of real-time data. At the backend layer, the system is built on the Spring Boot framework, following a layered architecture approach, mainly including the server layer, Common layer, and Plain Old Java Object (POJO) layer. The server layer is responsible for implementing core business logic, including functional modules such as energy data processing, water quality monitoring calculations, carbon accounting models, and alert mechanisms; the Common layer encapsulates common system utilities and configuration files, providing a supporting environment for system operation; the POJO layer defines data entity classes, enabling mapping between database table structures and Java objects.

The system backend uses MyBatis Plus to perform data persistence operations and interact with the database system. The database part uses MySQL as the primary storage database to store monitoring data, user information, and carbon emission statistics; at the same time, Redis is introduced as a high-speed cache database to store frequently accessed data and real-time monitoring results.

The overall technical architecture of the system is shown in Figure 6.2.

The front-end part of this system is developed based on the Vue 3 framework, adopting a modular and component-based design approach to achieve efficient data visualization and human–computer interaction. The front-end project is launched through main.js, which is the core entry file of the system, responsible for importing global resources, initializing dependencies, and configuring the system's runtime environment. The App.vue file serves as the root component of the application, organizing the overall front-end page structure and global logic, providing a

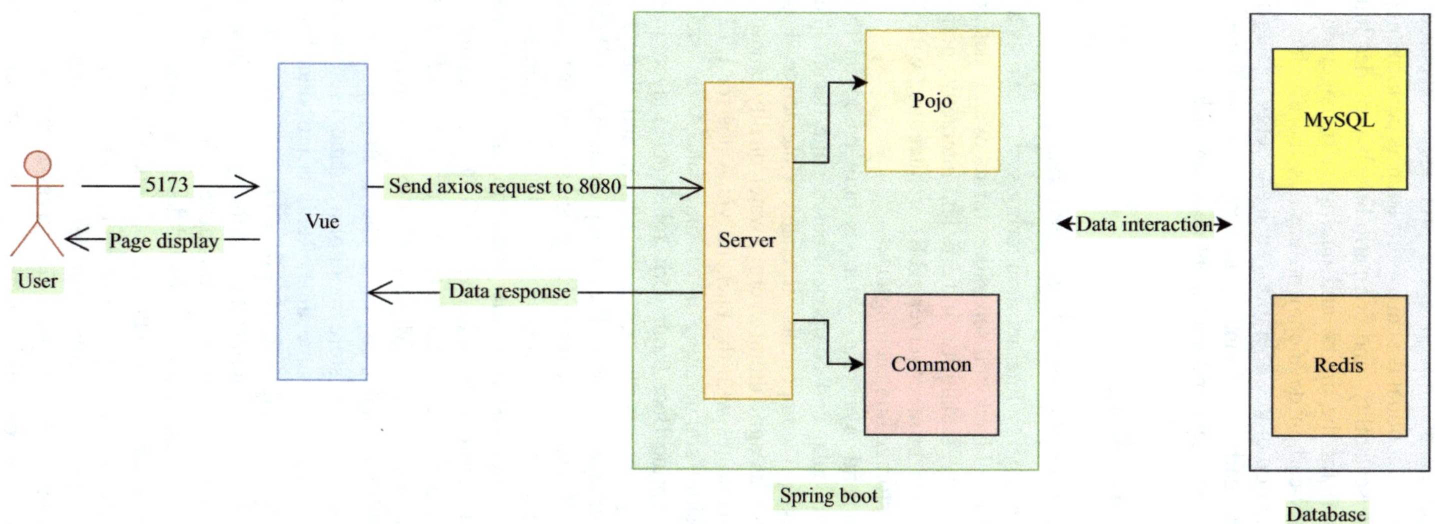

Figure 6.2 System technical architecture diagram

unified container and configuration support for features such as monitoring interfaces, chart visualization, and international language switching. The system uses the Pinia state management library for global data management. The stores module is responsible for the unified storage and scheduling of user login tokens and global state information, enabling cross-component state sharing and access control logic. The router module is used to configure the system's page routes, defining the mapping between paths and components, and, combined with navigation guards and state management, implements user login validation and access control, ensuring the security and controllability of agricultural energy monitoring data. At the user interface layer, the views folder contains the system's functional pages, such as energy consumption monitoring, photovoltaic power generation data display, water quality parameter visualization, and carbon emission and absorption statistics modules. It incorporates multiple subcomponents to achieve dynamic interactions within complex interfaces. The components folder is used to store reusable front-end components in the system, such as parameter display cards, real-time charts, alert pop-ups, and internationalization language switching components, which help improve development efficiency and interface consistency. The static resources section includes the assets folder, used to store images, style files, fonts, and theme configurations to ensure interface uniformity and visual effects. The utils folder contains network request encapsulations, request and response interceptor logic, WebSocket real-time communication handling modules, supporting high-frequency data interactions, and exception handling mechanisms between front-end and back-end. The js folder mainly implements functions such as adaptive scaling for different screen devices, weather data fetching, and bilingual switching between Chinese and English, enhancing the system's intelligence and internationalization. The images folder is used to store graphics and icons required by various functional modules.

Through the above modular design, the system achieves high scalability, maintainability, and visualization capabilities for Agricultural Energy Internet monitoring, ensuring efficient operation of the front end in energy data collection, environmental parameter monitoring, and carbon accounting result display. The overall front-end architecture is shown in Figure 6.3.

6.3.2 Overall system architecture design

Based on the preliminary requirements analysis and system technology architecture design, the Agricultural Energy Internet monitoring system adopts an overall architecture that combines layering and modularization to achieve the design principles of high cohesion, low coupling, and scalability. The system's overall functionality is divided into nine subsystems: the login subsystem, fishery water quality monitoring subsystem, cultivation environment monitoring subsystem, historical data management subsystem, data monitoring and visualization subsystem, electricity usage monitoring subsystem, load-side carbon emission subsystem, overall carbon emission subsystem, and reference subsystem.

Among them, the login subsystem serves as the unified entry point of the system, responsible for user authentication and access control, ensuring secure

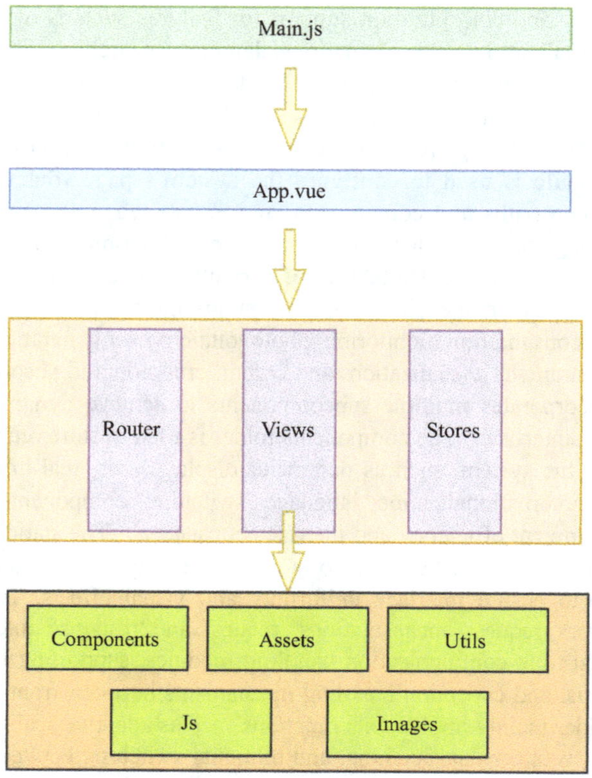

Figure 6.3 Frontend overall architecture diagram

access to monitoring data and the standardization of system operations. After authentication, users can enter the main interface of the system and access the remaining eight business function modules.

The fisheries water quality monitoring subsystem is responsible for real-time collection and analysis parameters, providing data support for dynamic monitoring of the aquaculture environment; the cultivation environment monitoring subsystem primarily monitors ecological environment factors, providing an environmental basis for agricultural energy optimization and carbon balance analysis; the historical data management subsystem enables categorized storage, query, and trend analysis of monitoring data, supporting users in data review and long-term statistical research.

On the energy side, the data monitoring and visualization subsystem achieves comprehensive visualization of multidimensional energy data through graphical displays, supporting real-time tracking of operational status and analysis of energy utilization efficiency; the electricity monitoring subsystem is responsible for collecting and recording electricity usage data in the agricultural production process, enabling energy consumption assessment and the formulation of energy-saving strategies; the load-side carbon emission subsystem calculates the carbon emissions of various load equipment by establishing an energy consumption–carbon emission conversion model, providing a quantitative basis for low-carbon operations; the overall carbon emission subsystem integrates energy data and carbon emission data within the system, achieving comprehensive carbon emission assessment and visualization at regional or system levels. Finally, the reference subsystem provides literature download services for users who need them.

Overall, the nine subsystems collaborate with each other and share data, forming the core architecture of the Agricultural Energy Internet monitoring system. This architecture enables a complete closed-loop from data collection, storage, and analysis to energy consumption and carbon emission assessment. It not only meets the system's functional and scalability requirements but also provides crucial support for intelligent monitoring and green development of agricultural energy. The overall structure of the system is shown in Figure 6.4.

6.3.3 Database design

In the Agricultural Energy Internet monitoring system, the design of the database affects the performance of data collection, processing, and visualization. To balance the system's efficiency, real-time capabilities, and persistent storage requirements, this system adopts a dual-database architecture combining Redis in-memory database with MySQL relational database, achieving the unified design goal of "separating hot and cold data, dynamic caching, and persistent storage."

Among them, the Redis database is mainly used for high-speed caching and dynamic updating of simulated sensor data, enabling rapid responses to frontend query requests and significantly improving the system's data access efficiency and response speed. The data stored in Redis includes fisheries water quality monitoring data, cultivation environment monitoring data, water supply quality data for aquaculture areas, overall cultivation environment data for plant planting areas, purchased electricity monitoring data, and real-time photovoltaic power generation data. All of this data is written into the Redis database through simulated sensor inputs to achieve high-frequency sampling of energy and environmental status and real-time dynamic display.

For the fisheries water quality data section, Redis uses the pond number as the key and stores multiple "field-value" pairs in a hash structure. This structural design facilitates high-concurrency read and write access and allows for rapid comparative analysis of water quality across different ponds. An example of its field design is shown in Table 6.21.

In the cultivation environment monitoring section, the system uses the cultivation area number as the key for Redis data and organizes the data using a hash structure. This design supports real-time collection and dynamic display of multidimensional environmental factors for each cultivation area, providing foundational data support for energy and environmental regulation as well as crop growth status analysis. Specific information is shown in Table 6.22.

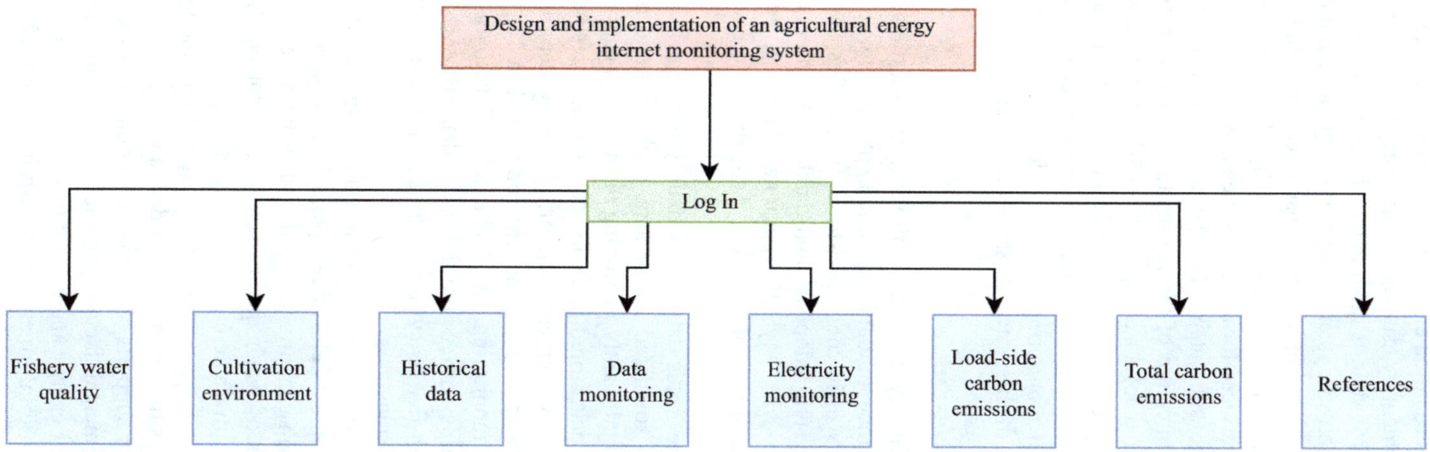

Figure 6.4 System overall architecture diagram

Table 6.21 Fisheries water quality data structure table

Field name	**Type**	**Explanation**	**Sample value**
waterTemp	String	Water temperature (°C)	17.3
pH	String	pH level	7.21
Oxygen	String	Dissolved oxygen (mg/L)	5.03
Salinity	String	Salinity (mg/L)	372.21
ammoniaNitrogen	String	Ammonia nitrogen content (mg/L)	0.21
Oxygenation	String	Oxygenation status (0/1)	1
Feeding	String	Feeding status (0/1)	0
Timestamp	String	Timestamp	2025-05-08 10:30:00

Table 6.22 Cultivation environment data structure table

Field name	**Type**	**Explanation**	**Sample value**
air_temperature	Double	Air temperature (°C)	19.2
air_humidity	Double	Air humidity (%)	59.5
leaf_temperature	Double	Leaf surface temperature (°C)	17.3
soil_temperature	Double	Soil temperature (°C)	15.8
soil_humidity	Double	Soil moisture (%)	71.1
light_intensity	Integer	Light intensity (lux)	1160
co2_concentration	Integer	Carbon dioxide concentration (ppm)	502
timestamp	String	Timestamp	2025-05-08 14:30:15

In response to the regional comprehensive monitoring requirements of the Agricultural Energy Internet system, the system further designs a unified data structure for the water quality in aquaculture areas and the overall cultivation environment of plant-growing areas. This type of data uses a timestamp as the Redis key and integrates key parameters representing the overall system environmental characteristics in the fields, including water temperature, pH, dissolved oxygen, salinity, ammonia nitrogen content, air temperature, air humidity, leaf surface temperature, and light intensity. Such data serve as an important basis for the system's comprehensive situation assessment, enabling multi-scenario and multi-level environmental fusion analysis of the agricultural energy system. The specific structure is shown in Table 6.23.

In the energy data section, the system establishes corresponding data keys for purchased electricity and photovoltaic (PV) generation to enable real-time recording of energy supply and demand as well as support carbon emission accounting.

Table 6.23 Overall data structure table

Field name	Type	Explanation	Sample value
watertemperature	String	Water temperature (°C)	18.6
phdata	String	pH level	7.68
oxygendata	String	Dissolved oxygen (mg/L)	5.79
salinitydata	String	Salinity (mg/L)	391.21
amnidata	String	Ammonia nitrogen content (mg/L)	0.33
airtemperature	Double	Air temperature (°C)	18.6
airhumidity	Double	Air humidity (%)	61.7
leaftemperature	Double	Leaf surface temperature (°C)	19.3
lightintensity	Integer	Light intensity (lux)	1280
co2concentration	Integer	Carbon dioxide concentration (ppm)	496

Table 6.24 Electric power data structure table

Field name	Type	Explanation	Sample value
purchased_power	Double	Purchased electricity (kWh)	94.35
solar_power	Double	Photovoltaic power generation (kWh)	102.56

The keys in Redis are formed by concatenating a prefix indicating the type of power data with a timestamp, while the fields store the purchased power and PV generation values at that moment. This design allows the system to achieve high-frequency sampling and rapid retrieval of energy-side data, supporting subsequent energy efficiency analysis and carbon balance calculations. The structural definition is shown in Table 6.24.

Therefore, the Redis database undertakes the functions of high-frequency dynamic data caching and real-time monitoring data support for the system. Through a unified key-value design and hash structure organization, it achieves efficient data management across the three dimensions of water quality, environment, and energy in the Agricultural Energy Internet monitoring system, providing robust underlying data support for the system's energy and carbon flow calculations, dynamic alerts, and visualization.

In the Agricultural Energy Internet monitoring system, the MySQL relational database primarily undertakes the core functions of long-term data storage and multidimensional energy-carbon analysis. The system establishes key business data tables in MySQL covering different time scales of hours, days, months, and years for carbon absorption, carbon emissions, carbon reduction, and electricity consumption, totaling 16 tables. The database design aims to provide comprehensive support for energy-carbon data management and statistical analysis. Through a structured storage approach with multiple time granularities, the system can: record changes in energy and carbon emission data in real time, supporting dynamic monitoring and trend analysis; conduct carbon accounting and energy utilization assessment based on time series, providing foundational data support for carbon peak and carbon neutrality strategies; achieve quantitative evaluation of energy-saving and emission reduction effects, assisting in optimized energy scheduling and policy compliance analysis; provide reliable historical data for subsequent carbon trading, energy efficiency evaluation, and environmental decision-making. Meanwhile, MySQL and Redis jointly build a two-tier data storage architecture for the system: the former is responsible for the persistent storage of historical data and statistical analysis results, while the latter handles high-speed caching and rapid response of real-time monitoring data. Together, they ensure data consistency and efficiency in the system.

The overall structural design of the relational database in this system is centered on modularity, scalability, and data security, forming a comprehensive data support system.

The hourly carbon absorption entity is used to record the system's carbon absorption data on an hourly basis. It is a key data unit supporting dynamic monitoring of energy and carbon and refined analysis. This entity contains eight core attributes: ID, hour, date, creation time, update time, total plant photosynthetic absorption, hydroponic plant photosynthetic carbon absorption, and soil-cultivated plant photosynthetic carbon absorption. Through this table, the system can accurately reflect the changes in carbon absorption of plant populations at an hourly granularity, providing a high-resolution data foundation for subsequent daily, monthly, and yearly energy and carbon analysis. The detailed table structure design is shown in Table 6.25.

At the same time, the system also designed a daily carbon absorption entity to aggregate and analyze the total daily carbon absorption data. This entity includes seven attributes: ID, date, creation time, update time, total photosynthetic carbon absorption by plants, photosynthetic carbon absorption by hydroponic plants, and photosynthetic carbon absorption by soil-cultivated plants. Daily carbon absorption data not only reflect the overall carbon absorption levels of plants under different environmental conditions but also provide key data for the system to analyze carbon absorption trends and evaluate energy-carbon balance. In the monthly carbon absorption entity, the system designed eight core attributes: ID, year, month, creation time, update time, total photosynthetic carbon absorption by plants, photosynthetic carbon absorption by hydroponic plants, and photosynthetic carbon absorption by soil-cultivated plants, which are used to record and statistically analyze the carbon absorption characteristics of plant photosynthesis in different months. This design supports the system in conducting trend analysis and visual display of monthly carbon absorption dynamics and provides data support for energy-carbon balance evaluation and carbon accounting models. In the annual carbon absorption entity, the system set seven key attributes: ID, year, creation time, update time, total photosynthetic carbon absorption by plants, photosynthetic

Table 6.25 Hourly carbon absorption table

Field name	Type	Is it allowed to be empty?	Note
id	int	No	Primary key
hour	int	No	Hour
date	date	Yes	Date
created_at	timestamp	Yes	Creation time
updated_at	timestamp	Yes	Update time
plant_photosynthesis_absorption	decimal	Yes	Total photosynthetic absorption of plants
water_photosynthesis_absorption	decimal	Yes	Photosynthetic carbon uptake of hydroponic plants
soil_photosynthesis_absorption	decimal	Yes	Photosynthetic carbon uptake of soil-cultivated plants

carbon absorption by hydroponic plants, and photosynthetic carbon absorption by soil-cultivated plants, which are used to summarize carbon absorption statistics on an annual scale, enabling cross-year comparison and comprehensive analysis of carbon sink effects. Through the design of the annual carbon absorption entity, the system can form a top-down multi-timescale carbon absorption data chain, providing a solid data foundation for tracking the carbon footprint and evaluating carbon neutrality performance in the Agricultural Energy Internet.

In the hourly carbon emission entity, the system is designed with 11 core attributes, which are used to accurately record the carbon emission dynamics of the aquaponics system at an hourly scale. Through this entity, the system can achieve statistical analysis of short-term carbon emission changes, providing data support for the precise control and energy-saving strategies of the energy-carbon system. The detailed design of the hourly carbon emission table is shown in Table 6.26.

In the daily carbon emission entity, the system defines ten main attributes. These attributes are used to store daily carbon emission statistics, enabling tracking and visual display of daily carbon emission trends. In the monthly carbon emission entity, the system sets 11 attributes. These are used to summarize carbon emission data at the monthly level, supporting the system in energy-carbon balance trend analysis and periodic carbon emission assessment. In the annual carbon emission entity, the system includes ten key attributes. These attributes are used for centralized management and statistical analysis of annual total carbon emissions. This design helps the system conduct carbon emission performance evaluations and comparative studies on a yearly basis, providing critical data support for the Agricultural Energy Internet platform to achieve energy-carbon closed-loop management and carbon neutrality goals.

Table 6.26 Hourly carbon emissions table

Field name	Type	Is it allowed to be empty?	Note
id	int	No	Primary key
hour	int	No	Hour
date	date	Yes	Date
created_at	timestamp	Yes	Creation time
updated_at	timestamp	Yes	Update time
electricity_emission	decimal	Yes	Carbon emissions from purchased electricity
fish_respiration_emission	decimal	Yes	Carbon emissions from fish respiration
plant_respiration_emission	decimal	Yes	Plant respiratory carbon emissions
soil_respiration_emission	decimal	Yes	Soil respiration carbon emissions
waste_treatment_emission	decimal	Yes	Carbon emissions from waste treatment
total_emission	decimal	Yes	Total carbon emissions

In the hourly carbon reduction entity, the system is designed with six attributes: ID, hour, date, creation time, update time, and photovoltaic carbon reduction amount, which are used to accurately record the carbon reduction achieved by the Agricultural Energy Internet system through photovoltaic power generation on an hourly scale. This entity provides the system with real-time monitoring and calculation support for short-term energy-carbon reduction benefits, and can be used to assess the instantaneous carbon reduction contribution of photovoltaic energy utilization. The table structure of the hourly carbon reduction is shown in Table 6.27.

In the daily carbon reduction entity, the system includes five attributes: ID, date, creation time, update time, and photovoltaic (PV) carbon reduction amount, which are used to record the total carbon reduction brought by daily photovoltaic power generation, enabling statistical analysis and trend tracking of daily carbon reduction changes, and providing a data basis for medium- and short-term carbon performance assessment. In the monthly carbon reduction entity, the system sets six attributes: ID, year, month, creation time, update time, and photovoltaic reduction amount, which are used to aggregate PV reduction data at the monthly level, supporting the system in conducting periodic carbon reduction benefit assessments and energy-carbon balance analyses. In the annual carbon reduction entity, the system defines five attributes: ID, year, creation time, update time, and photovoltaic reduction amount, which are used to summarize and analyze the reduction benefits

Table 6.27 Hourly carbon emission reduction table

Field name	Type	Is it allowed to be empty?	Note
id	int	No	Primary key
hour	int	No	Hour
date	date	Yes	Date
created_at	timestamp	Yes	Creation time
updated_at	timestamp	Yes	Update time
photovoltaic_reduction	decimal	Yes	Photovoltaic emission reduction

Table 6.28 Hourly electricity consumption chart

Field name	Type	Is it allowed to be empty?	Note
id	int	No	Primary key
hour	int	No	Hour
date	date	Yes	Date
created_at	timestamp	Yes	Creation time
updated_at	timestamp	Yes	Update time
purchased	decimal	Yes	Purchased electricity consumption
solar	decimal	Yes	Photovoltaic power consumption

of PV power generation throughout the year, providing key data support for long-term carbon reduction performance evaluation and green development performance assessment on the Agricultural Energy Internet platform.

In the hourly electricity consumption entity, the system has designed with seven attributes. This entity enables a detailed depiction of energy supply and demand changes over short periods, providing fundamental data support for real-time energy consumption monitoring and dynamic energy balance analysis. The table structure design of the hourly electricity consumption entity is shown in Table 6.28.

In the daily electricity consumption entity, the system includes six attributes: ID, date, creation time, update time, purchased electricity consumption, and photovoltaic electricity consumption. These attributes are used to aggregate daily electricity consumption data, reflecting the system's energy consumption characteristics and energy utilization efficiency over natural cycles, providing an important basis for energy optimization and trend analysis. In the monthly electricity consumption entity, the system defines seven attributes: ID, year, month, creation time, update time, purchased electricity consumption, and photovoltaic electricity consumption. This allows for systematic statistics and comparative analysis of monthly electricity consumption, supporting the Agricultural Energy Internet platform in conducting periodic performance evaluation and energy management optimization. In the annual electricity consumption entity, the system sets six attributes: ID, year, creation time, update time, purchased electricity consumption, and photovoltaic electricity consumption. These facilitate the aggregation of annual energy consumption data, comprehensively reflecting the system's annual energy structure, carbon utilization level, and renewable energy substitution rate, providing key data support for annual energy planning and energy-saving and emission reduction decisions.

Therefore, the MySQL database is used to preserve archived records of various monitoring data, user information, system configuration parameters, and carbon emission statistics. MySQL and Redis interact periodically through a data synchronization mechanism, ensuring that cached data are persistently stored in MySQL after updates, thereby achieving data security and consistency. Through the collaborative design of Redis and MySQL, the system strikes a balance between real-time processing of high-frequency monitoring data and stable long-term business data, effectively supporting the entire process of "data collection—real-time display—energy consumption and carbon accounting—historical analysis."

6.4 Detailed design and implementation

6.4.1 Login and navigation

In the Agricultural Energy Internet monitoring system, the login and navigation module serves as the entry point and core interaction hub of the system, responsible for user authentication, access control, and function navigation on the main interface. This module not only ensures the security and standardization of system operations but also provides users with an intuitive and efficient way to access the system.

After the system runs, the user first enters the login interface, which prompts the user to enter a username and password for authentication. Once the front end receives the input, it automatically collects the input data through the form binding mechanism and sends a POST request to the backend login interface (/login). The backend, implemented using the Spring Boot framework, handles the login logic, responsible for receiving the user credentials passed from the front end and performing authentication against the database. If the verification is successful, the system generates the corresponding login token and returns it to the front end, while displaying a "Login Successful" message, then automatically redirects to the system main interface; if the verification fails, an error message is displayed, requiring the user to re-enter the username and password.

The overall logic of the login process is shown in Figure 6.5, covering the entire procedure of front-end data collection, back-end identity verification, and result feedback.

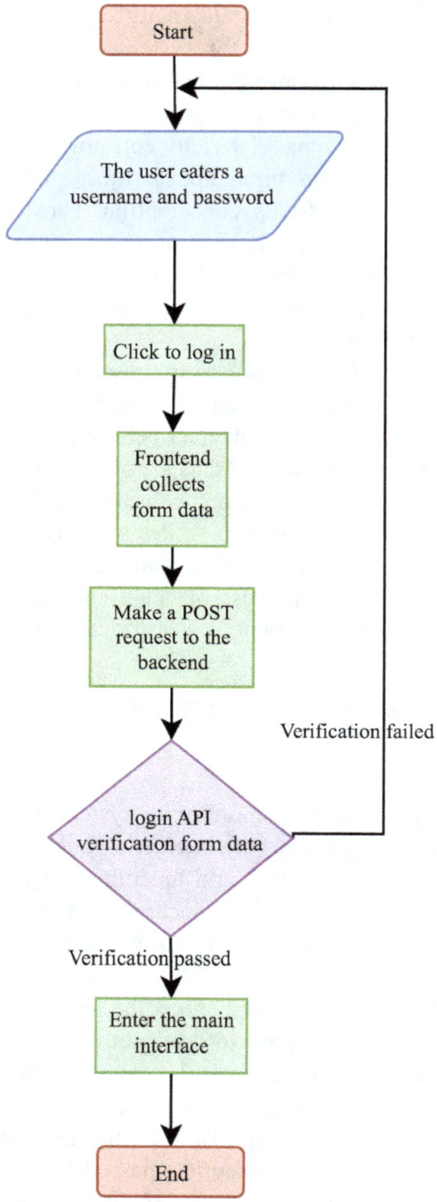

Figure 6.5 Login flowchart

After entering the main interface of the system, a global navigation bar is set at the bottom of the page, allowing users to access different functional subsystem interfaces by clicking on various navigation buttons. The navigation bar uses Vue Router to achieve routing between modules and integrates the Pinia state management mechanism to store user login status and access permissions, ensuring that unauthorized users cannot directly access internal system modules.

Through the above design, the system achieves an integrated functional process of "secure login—permission verification—modular navigation," providing foundational interactive support for subsequent energy and carbon monitoring, intelligent analysis, and operational visualization.

6.4.2 Language switch and date

The platform has implemented a bilingual Chinese-English switching function on the front end, aiming to enhance the system's usability and adaptability, enabling domestic and international users to operate and analyze data in different language environments.

During the design phase, the system introduced the Vue-i18n internationalization plugin as the core tool for multilingual management, and defined an internationalization language resource object in the project's global configuration, which contains Chinese (and English) complete sets of text mappings. All interface text, menu names, chart titles, and prompt messages are uniformly managed and dynamically rendered through this resource object, thereby avoiding the redundancy issues caused by hardcoding and improving the system's maintainability and scalability.

During operation, the system dynamically switches the current language environment by calling the language switching functions provided by Vue-i18n, based on the user's interactions with the language selection control in the top right corner of the navigation bar. The switching operation takes effect in real time without interrupting the system's operation, and all interface text is immediately updated to the corresponding language version. To ensure a consistent internationalization experience, the system simultaneously displays the English version of the interface while showing the Chinese interface features below, intuitively demonstrating the result of the language switch.

In addition, the system front end uses JavaScript's Date class to achieve real-time retrieval of local time and multilingual formatted display. This module automatically obtains the client's local time after the system loads and selects the corresponding date and time display format based on the current language environment. The system refreshes the time display every second through a timer mechanism, ensuring the timeliness and accuracy of the time information.

Through the design and implementation of the above language switching and date display modules, the Agricultural Energy Internet monitoring system has achieved a multilingual-friendly, time-synchronized, and interface-adaptive internationalized interactive experience, providing technical support and human–computer interaction assurance for the system's cross-regional promotion and international cooperative applications.

6.4.3 Fisheries water quality subsystem

In the Agricultural Energy Internet monitoring system, the fisheries water quality subsystem is an important component for realizing intelligent and ecological management of aquaculture. The system divides this subsystem into several functional modules to achieve dynamic warning and visual display. The subsystem acquires multidimensional aquaculture environment parameters through data collection terminals, combined with backend data processing and frontend dynamic display.

In the Fish Overview module, the system consolidates and displays comprehensive information for each fishpond. Users can view key indicators in real time, including the number of fishponds, the types of fish being farmed, the total weight of farmed fish, average water temperature, average pH, average dissolved oxygen levels, fish health status, and average survival rate. The front-end page interacts with the back-end interface to update the relevant information at regular intervals, ensuring that the displayed data is both timely and accurate.

In the monitoring and early warning module for each breeding pond, the system further refines the monitoring functions of each fishpond. Users can view parameters. When any water quality parameter exceeds the preset threshold, the system automatically changes the display color of the corresponding parameter from green to red, and the water quality status from green "excellent" to yellow "moderate," thereby alerting users to intervene in a timely manner. This design enables visual alerts for water quality anomalies, significantly enhancing the intuitiveness and responsiveness of early warnings.

The system's data acquisition process is as follows: when a user enters the main interface, the front end sends a data request to the backend "/fishery" endpoint via an HTTP GET request. Upon receiving the request, the backend first reads the real-time fishery monitoring data from the Redis cache. If the relevant data is not found in Redis, the system prompts the user that fishery data is unavailable; if the data exists, the system checks whether each water quality parameter is within the threshold range. Parameters within the threshold range are displayed in green; parameters exceeding the threshold are marked in red, and the water quality status is shown as yellow "Medium." If all parameters are within the safe range, the water quality status remains green, indicating "Excellent."

The entire fishery water quality data interaction process is shown in Figure 6.6. The system implements a bilingual interface switching function in Chinese and English, allowing users to view fishery monitoring information in either language according to their needs. The implementation result of the English interface is shown in Figure 6.7.

6.4.4 Cultivation environment subsystem

In the Agricultural Energy Internet monitoring system, the cultivation environment subsystem is an important module for achieving intelligent management of crop cultivation and environmental optimization. The system divides this subsystem into several functional modules to enable real-time monitoring of environmental parameters in different planting areas, dynamic alerts, and data visualization.

In the overall planting overview module, the system summarizes and displays general planting information. Users can view key indicators in real time, such as

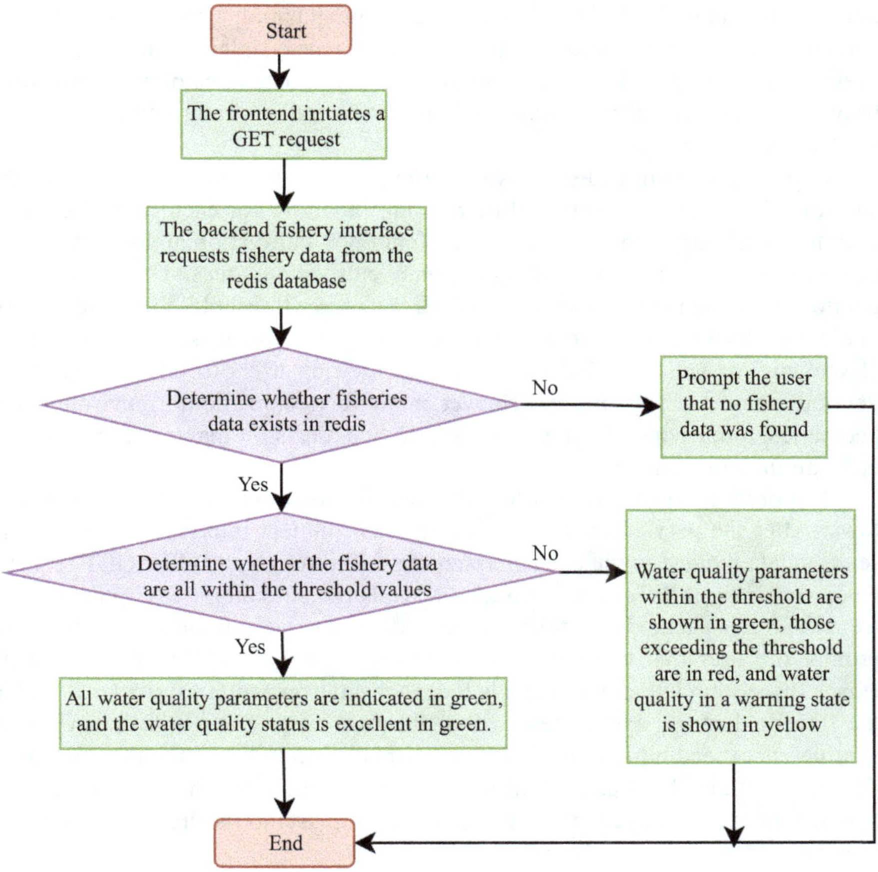

Figure 6.6 Flowchart for obtaining fisheries data

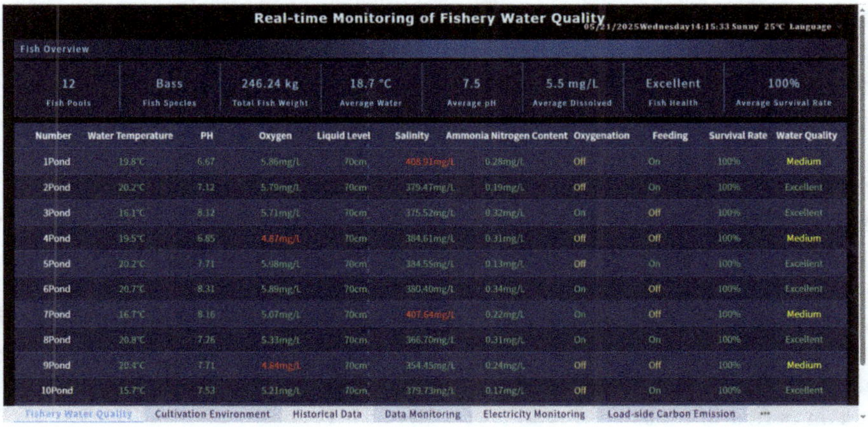

Figure 6.7 Implementation results of the fisheries water quality English interface

total planting area, land planting area, hydroponic planting area, average air temperature, average air humidity, average light intensity, plant health status, and average survival rate. The module can dynamically update monitoring information, providing users with a macro overview of the overall planting environment and production operation status.

In the monitoring and early warning module for various planting areas, the system refines the environmental monitoring functions for each type of cultivation area, including land planting areas, foam box cultivation areas, hydroponic cultivation areas, and vertical hydroponic cultivation areas. Users can view parameters. When any parameter exceeds the preset threshold, the system automatically displays it in red, while changing the plant status from green "Excellent" to yellow "Moderate," thereby alerting users to take timely intervention measures. This design achieves intuitive visualization of environmental anomalies, improving the response speed and decision-making efficiency in cultivation management.

The process for the system to obtain cultivation environment data is as follows: After the user enters the cultivation environment interface, the front end sends a data request to the backend/planting API through an HTTP GET request. Upon receiving the request, the backend reads the cultivation environment data from the Redis cache. If the relevant data does not exist in Redis, the system will prompt the user that cultivation environment data was not found; if the data exists, the system will determine whether each environmental parameter is within the threshold range. Parameters within the safe range are displayed in green, parameters exceeding the thresholds are displayed in red, and the plant status is shown as yellow "Medium." If all parameters are within the threshold range, the parameters are displayed in green, and the plant status is displayed as green "Excellent."

The entire cultivation environment data interaction process is shown in Figure 6.8. The system implements a bilingual interface switch function for Chinese and English, allowing users to view cultivation environment monitoring information in both language settings. The English interface implementation result is shown in Figure 6.9. Through the design of this subsystem, the Agricultural Energy Internet monitoring system can achieve precise monitoring and intelligent early warning of multiple types of cultivation environments, enhancing the intelligence, greenness, and sustainability of agricultural production.

6.4.5 Historical data subsystem

In the Agricultural Energy Internet monitoring system, the historical data subsystem is used to archive, display, and analyze the aquaculture water quality and cultivation environment data collected by the system, helping users understand environmental change patterns and assist in decision-making. The design of this subsystem covers historical record queries and trend visualization for various types of data, allowing for an intuitive presentation of key parameter changes over a certain period of time.

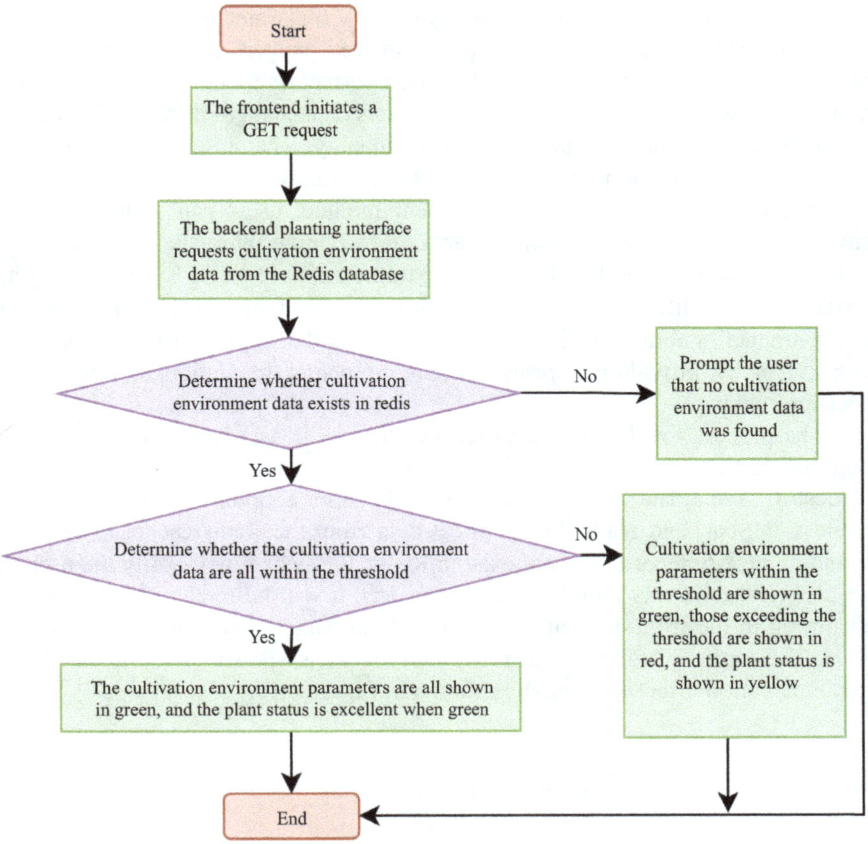

Figure 6.8 Flowchart for obtaining cultivation environment data

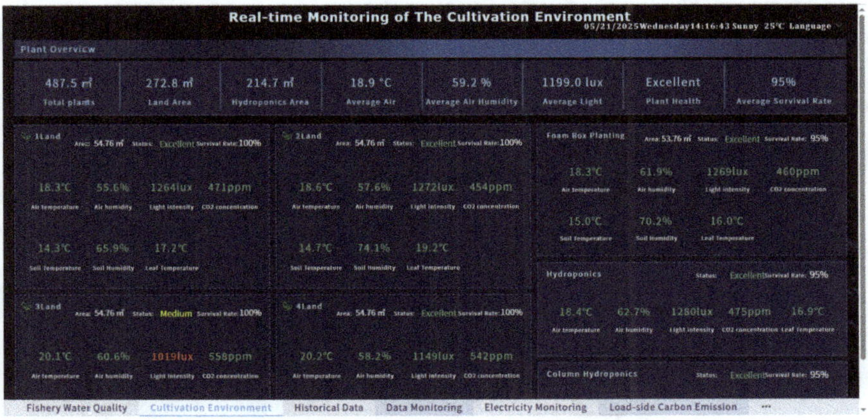

Figure 6.9 Implementation results of the cultivation environment in the English interface

In the water quality module, users can view the water quality in aquaculture areas, including water temperature, pH, dissolved oxygen, salinity, and ammonia nitrogen content for every 15 minutes from the current time in the past, as well as the trends of these parameters. This module presents data changes through visual charts, enabling users to intuitively understand water quality dynamics, detect abnormalities in a timely manner, and provide data support for aquaculture management.

In the overall cultivation environment module, users can view the overall environment of the plant cultivation area at 15-minute intervals from the current moment, as well as the trends of these parameters over time. By displaying historical data for different types of cultivation environments, users can assess the environmental stability of the cultivation area and the growth condition of the crops, thereby providing a basis for environmental regulation and production decision-making.

The process for the system to retrieve historical data is as follows: after the user enters the historical data interface, the front end initiates an HTTP GET request to send a data request to the backend/history endpoint. Upon receiving the request, the backend reads the historical data from the Redis cache. If the corresponding history record does not exist in Redis, the system will notify the user that no historical data was found; if the data exists, it will be rendered and displayed according to the front-end page style, achieving data visualization and trend analysis. The entire historical data retrieval process is shown in Figure 6.10. The implementation effect in English is shown in Figure 6.11.

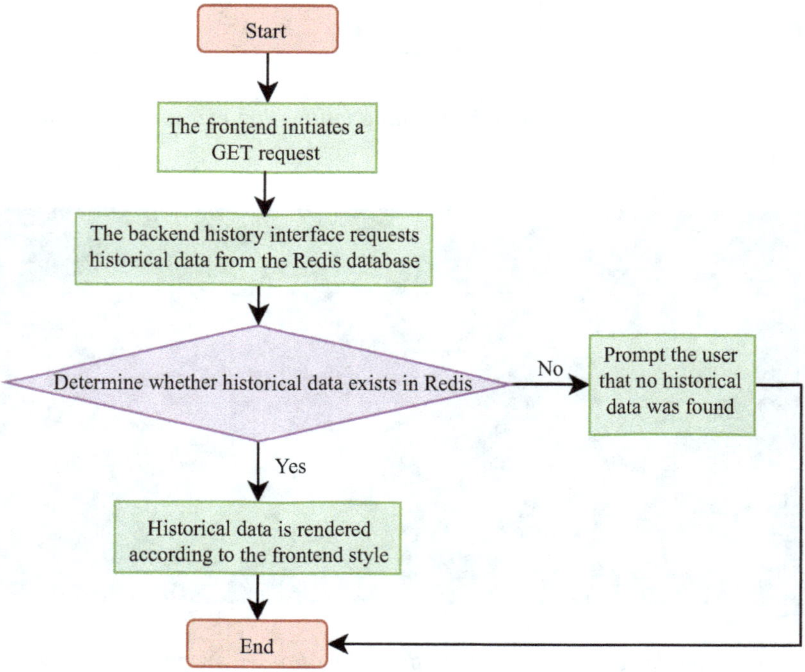

Figure 6.10 Flowchart for obtaining historical data

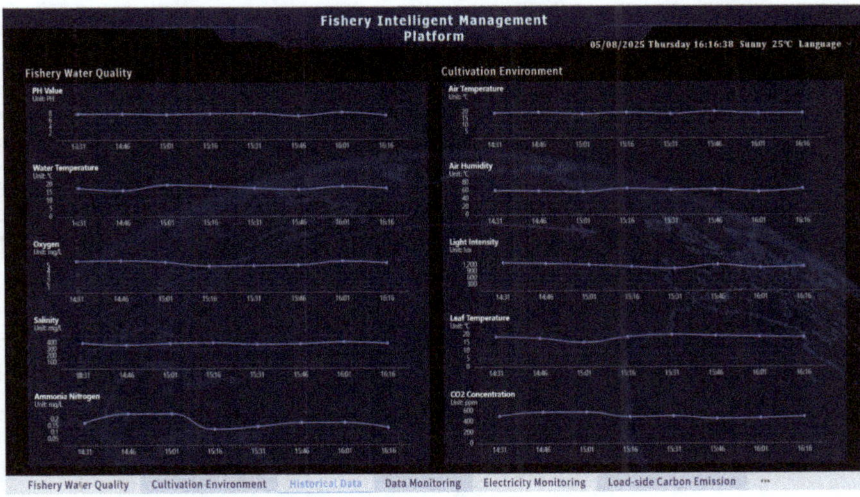

Figure 6.11 Historical data English interface implementation results

Through the historical data subsystem, the Agricultural Energy Internet monitoring system can achieve long-term monitoring and analysis of water supply quality and cultivation environments, providing users with comprehensive data references and supporting intelligent management and continuous optimization of the agricultural production environment.

6.4.6 Data monitoring subsystem

The data monitoring subsystem is used to monitor and analyze the system's operational status, electricity usage, and energy consumption composition in real time, providing users with an overall overview and basis for refined management. This subsystem displays key parameters through a front-end visual interface, allowing users to quickly grasp the agricultural production environment and energy operation conditions.

In the overall load-side module, users can view the overall situation of fish and the overall situation of planting. Through this module, users can quickly understand the operation status of the entire system and the working condition of the equipment, promptly identify potential anomalies, and provide data support for management decisions. At the same time, this module can also display the status of the main monitoring hosts, helping users quickly locate problem areas.

In the electricity usage module, users can view the system's total electricity consumption on an annual, monthly, and daily basis, including the composition of purchased electricity and photovoltaic (PV) power generation. The purchased electricity module provides trends in purchased electricity over the past seven days, as well as detailed data changes every 15 minutes from the current moment, helping users understand electricity consumption patterns and manage energy efficiency. The photovoltaic power generation module offers trends in PV power generation over the past seven days, and data changes every 15 minutes, allowing users to intuitively grasp the real-time output and overall contribution of photovoltaic power generation.

After entering the data monitoring interface, the process by which the system obtains electricity and electric power data is as follows: the frontend initiates an HTTP GET request, and the backend /power and /electricity interfaces request electric power and electricity data from the Redis database and MySQL database, respectively. The database will check whether the corresponding records exist. If the data does not exist, a prompt will inform the user that no corresponding data was found; if the data exists, it will be rendered according to the frontend page format to achieve data visualization display and trend analysis. The entire data acquisition process is shown in Figure 6.12. The English interface implementation effect is shown in Figure 6.13.

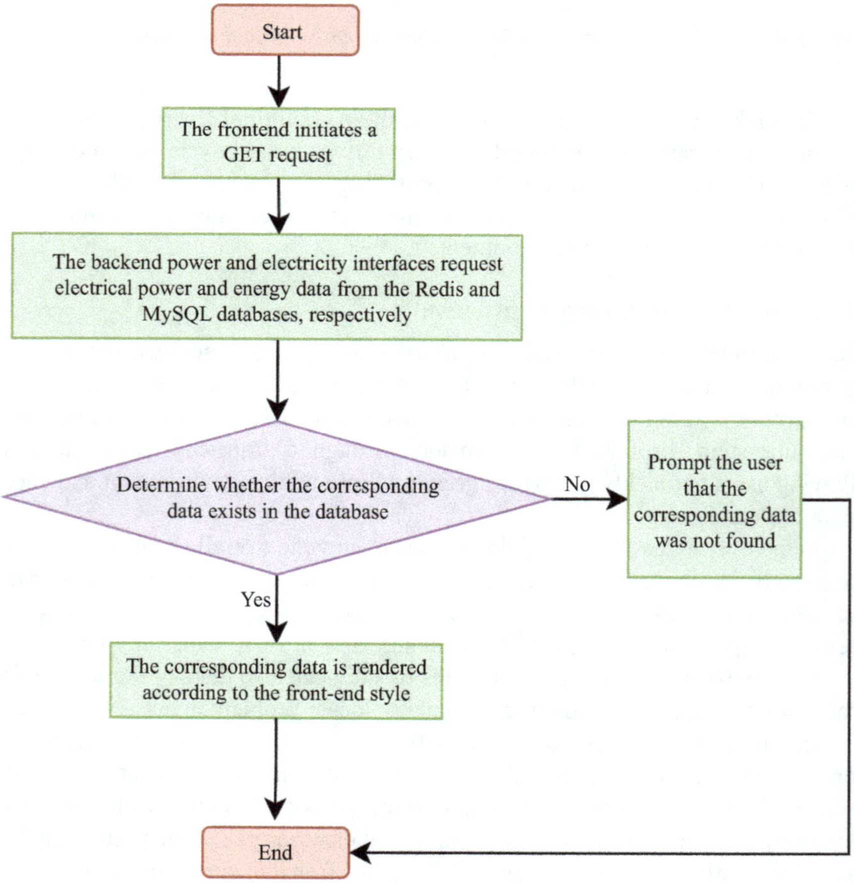

Figure 6.12 Flowchart for obtaining electric power and energy data

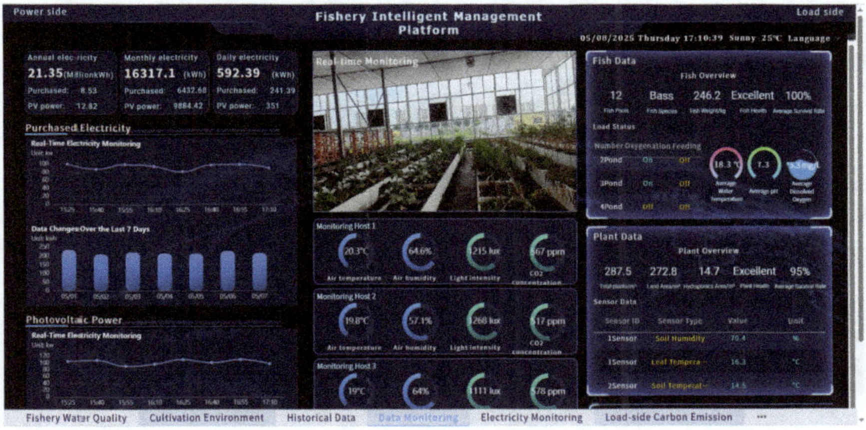

Figure 6.13 Implementation results of data monitoring in the English interface

Through the data monitoring subsystem, the Agricultural Energy Internet monitoring system can achieve comprehensive monitoring of both the load side and the power source side. By combining real-time data with historical trend analysis, it provides users with a complete reference for operational status, supporting system optimization, management, and energy scheduling decisions.

6.4.7 Electricity monitoring subsystem

In the Agricultural Energy Internet monitoring system, the electricity monitoring subsystem is mainly used to conduct a comparative analysis of the overall power supply structure of the park and energy utilization efficiency, helping users to grasp the complementary relationship between photovoltaic power generation and purchased electricity in real time, thereby optimizing energy scheduling and electricity usage strategies.

In the module comparing purchased electricity and photovoltaic (PV) power, the system visualizes and compares the data of PV generation and purchased electricity on the same coordinate system. Users can choose to view different time scales, including the past 15 minutes, past week, past six months, and past year. Through this multidimensional time window design, the system can meet users' needs for comprehensive analysis of short-term operational fluctuations and long-term energy trends.

The process of acquiring electricity data is consistent with the data monitoring subsystem: the front end accesses the back-end/electricity and /power interfaces via HTTP GET requests, and the back-end retrieves real-time and cumulative electricity data from Redis and MySQL databases for visualization. If the corresponding data does not exist in the database, the system will prompt that no electricity data was found; if it exists, the data will be displayed according to the time dimension and comparison structure.

The implementation results of the English interface are shown in Figure 6.14.

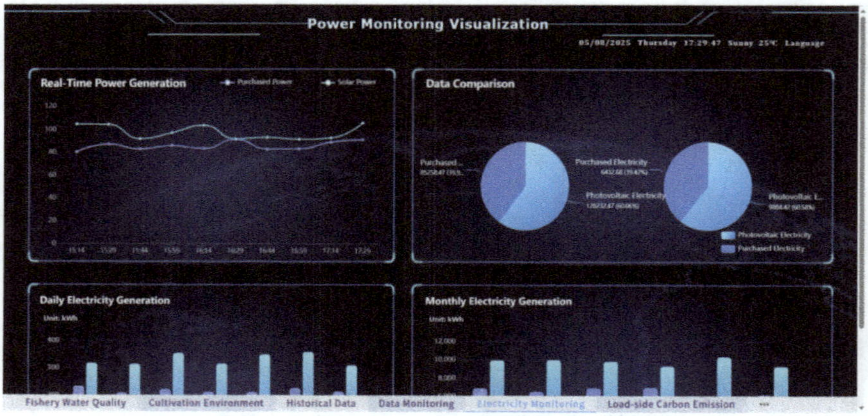

Figure 6.14 Implementation result of the power monitoring English interface

6.4.8 Load-side carbon emission subsystem

In the Agricultural Energy Internet monitoring system, the load-side carbon emission subsystem is primarily responsible for monitoring, counting, and visually displaying carbon emission sources and carbon absorption processes in fishery and cultivation areas. It achieves refined accounting and dynamic supervision of the entire carbon flow. Through this subsystem, users can intuitively understand the changes in the amount of carbon produced or absorbed by different organisms and processes, thereby providing a scientific basis for carbon peak and carbon neutrality management.

In the fish respiration carbon emission module, the system focuses on time as the main axis, dynamically monitoring and analyzing the trends of fish respiration carbon emissions over the past week and the past six months. By calculating the relationship between fish physiological metabolism models and changes in dissolved oxygen in the water, the system generates a carbon emission curve caused by fish respiration, helping users assess the impact of stocking density and water quality management measures on carbon emissions.

The waste treatment carbon emission module is used to display the trend of carbon emissions generated during the waste treatment process. Users can view carbon emission data changes over the past six months. The system estimates carbon emissions by combining treatment frequency, treatment methods (such as composting, fermentation, etc.), and environmental factors like temperature and humidity, providing data support for optimizing waste emission reduction.

The plant respiration carbon emission module and the soil respiration carbon emission module, respectively, display the changes in carbon release during plant respiration and the decomposition of soil organic matter. The system uses data from the past week and the past six months to monitor trends at different time scales, helping users understand the carbon cycle dynamics in planting areas, optimize planting density and fertilization strategies, and reduce unnecessary carbon loss.

The Plant Photosynthetic Carbon Absorption Comparison Module is used to intuitively display the differences in carbon absorption under different cultivation modes. The system compares the carbon absorption data of hydroponic plants and soil-grown plants within the same coordinate system. Users can view the comparison results for the past week, the past six months, and the past year, thereby analyzing the differences in carbon sequestration capacity between different cultivation modes. Through this module, the system can quantitatively reflect the changes in photosynthetic absorption over different time scales, providing a basis for optimizing plant arrangements and enhancing carbon absorption capacity.

The system comprehensively displays the daily and monthly carbon emissions from fish respiration, the monthly carbon emissions from waste treatment, the daily and monthly carbon emissions from plant respiration, and the daily and monthly carbon emissions from soil respiration in the load-side carbon emission module. It also calculates and displays the total carbon absorption of hydroponic plants, soil-cultivated plants, and all plants on a daily, monthly, and yearly basis.

In the data acquisition process, when a user enters the load-side carbon emission interface, the frontend automatically sends a GET request. The backend calls the /carbon API to access the MySQL database to obtain carbon accounting data. The database first checks whether the corresponding carbon emission and absorption data exist; if not, the system prompts that no carbon accounting data were found; if they exist, the data are rendered into charts according to the styles and time dimensions configured on the frontend, enabling dynamic display.

The process for obtaining carbon emission data on the load side is shown in Figure 6.15. The implementation results on the English interface are shown in Figure 6.16. Through the design and implementation of this subsystem, the Agricultural Energy Internet monitoring platform achieves a quantitative display of the entire carbon emission process on the load side, providing solid technical support for low-carbon operations and the construction of a carbon accounting system.

6.4.9 Overall carbon emission subsystem

The overall carbon emissions subsystem serves as an important component for achieving the "dual carbon" goals through assessment and decision support. It is mainly responsible for the dynamic monitoring, comprehensive calculation, and visual presentation of system-level carbon emissions, carbon absorption, and carbon reduction data. The subsystem's design thoroughly considers the energy-carbon coupling relationship between the power supply side and the load side. Through multidimensional and multi-time-scale data collection, calculation, and visualization, it provides users with intuitive, traceable, and quantifiable carbon accounting information.

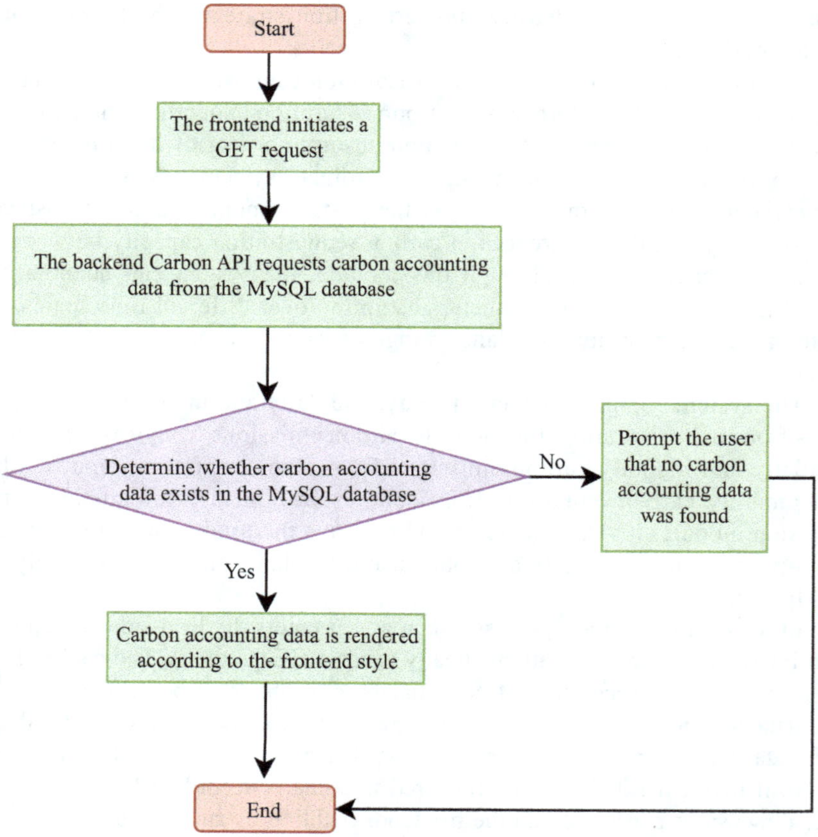

Figure 6.15 Flowchart for obtaining carbon accounting data

The overall carbon emission subsystem mainly includes the following functional modules. The module for purchased electricity carbon emissions is used to display the carbon emission trends of purchased electricity over the past hour, day, month, half-year, and year. The system dynamically calculates the carbon emissions of purchased electricity per unit of time by reading electricity consumption data in real-time and combining it with the carbon emission factors of the power grid region. The photovoltaic power generation carbon reduction module displays the carbon reduction effects of photovoltaic power generation over the same time scales. By calculating the carbon emissions reduced by replacing traditional grid electricity with photovoltaic power, it achieves a quantifiable presentation of carbon reduction benefits.

The carbon accounting overview module is used to summarize and display the daily and annual carbon emissions, carbon absorption, and carbon reduction across the entire system, providing users with a macro view of carbon flows and balance status. The Emission Source module visualizes the classification and

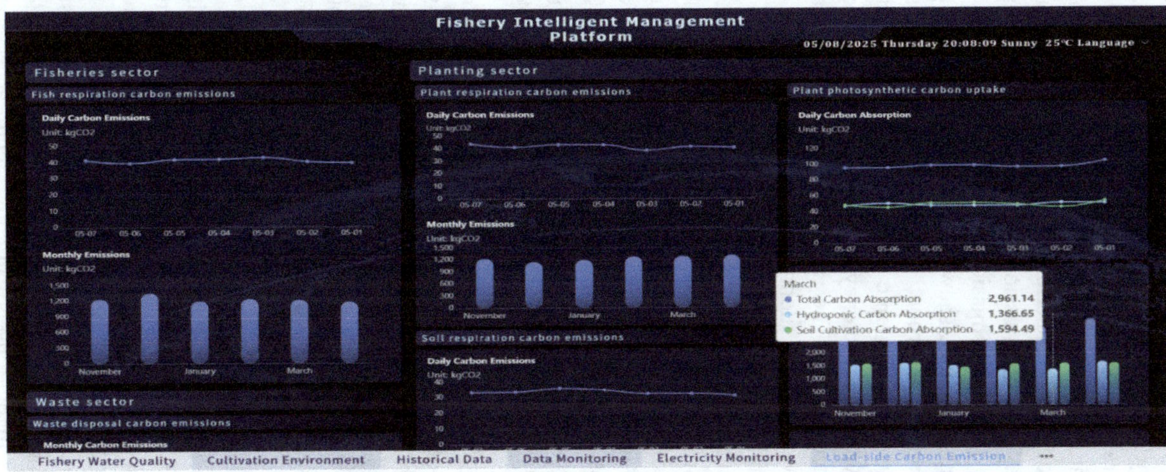

Figure 6.16 Implementation results of the carbon emissions interface on the load side

proportion of annual carbon emission sources, helping users analyze the contribution of different sources to overall carbon emissions. The power side and Load-side Carbon Emission Comparison module presents the carbon emission trends of the past week and the past six months within the same coordinate system, allowing users to intuitively compare the differences in carbon emissions between the energy input side and the system operation side, thereby optimizing the energy structure.

In addition, the system is designed with segmented carbon accounting functions, including modules for fisheries, planting, and waste management. The fisheries module displays carbon emission changes over the past day, month, half-year, and year, focusing on carbon emissions from fish respiration and pond ecological processes. The planting module shows plant respiration, carbon absorption through photosynthesis, and net carbon emissions over the same time periods. Distinguishing between hydroponic and soil cultivation methods, it reflects differences in photosynthesis efficiency and carbon absorption capacity under different environmental conditions. The waste module presents carbon emissions from waste treatment over the past month and year, providing a basis for evaluating environmental friendliness in system carbon management.

The system automatically collects key parameters in the background and stores and processes them uniformly in a MySQL database. Through the Redis caching mechanism, the access speed of historical carbon data is improved, enabling dynamic loading of carbon accounting results at daily, monthly, and yearly granularities. The system's visualization layer uses ECharts to render and compare data, providing visual charts, including carbon emission trends over the past seven days and six months, as well as the proportion of emission sources over the past year.

The process for obtaining carbon accounting data in the system is consistent with that of the load-side carbon emission subsystem. Specifically, after a GET request is initiated by the front end, the back-end reads and calculates carbon emissions and carbon absorption data from the database via the carbon interface. If the data exists, it is rendered and displayed according to the front-end template. The implementation results in the English interface are shown in Figure 6.17.

6.4.10 Reference subsystem

In the functional design and code implementation of an Agricultural Energy Internet monitoring system, the reference subsystem plays an important role in providing users with theoretical support and academic resources. This subsystem is not only an auxiliary module for showcasing system functions but also a crucial window for researchers and managers to understand the principles behind system construction, the basis of carbon accounting models, and academic progress under the "dual carbon" policy. Through this module, users can conveniently access key research literature related to the system models, thereby deepening their understanding of carbon emissions, carbon reduction, and carbon absorption mechanisms, and providing support for subsequent model optimization and research expansion.

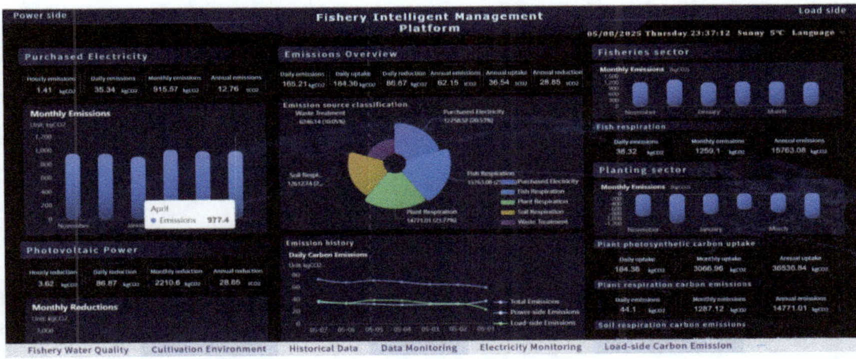

Figure 6.17 Overall carbon emissions English interface implementation results

The reference subsystem mainly includes a literature download module. The system preloads multiple high-quality documents related to carbon emission models, carbon reduction models, and carbon sequestration models in the reference interface, covering topics such as energy Internet architecture design, greenhouse gas accounting principles, photovoltaic power carbon reduction mechanisms, and carbon cycle mechanisms in agricultural ecosystems. Users can browse the list of literature in the interface and select and download the required files. The design of this module not only meets the needs for system teaching and research extension but also provides users with important references for data analysis, algorithm research, and model validation.

In the actual operational process, when a user enters the references interface and clicks the download button, the frontend first initiates a file request to the backend via an HTTP GET request. Upon receiving the request, the backend's files interface locates the specified file on the server based on the file identification information provided by the frontend. The system checks whether the corresponding reference file exists at that location. If the file does not exist, it returns a prompt to the frontend, alerting the user that the file is currently unavailable. If the file exists, the backend invokes a file stream download mechanism to transmit the reference file to the user and display the download progress. The entire file retrieval and transfer process follows the RESTful design principles of the Spring Boot backend, ensuring data access security and transmission stability.

The implementation process of the reference subsystem is shown in Figure 6.18. The system achieves dynamic loading of literature resources and fast response through efficient interface interactions between the backend and the file server. Overall, this subsystem not only strengthens the system's knowledge support function but also reflects the design concepts of the Agricultural Energy Internet monitoring system in terms of scientific research, openness, and scalability, providing important support for subsequent academic research and technological iteration.

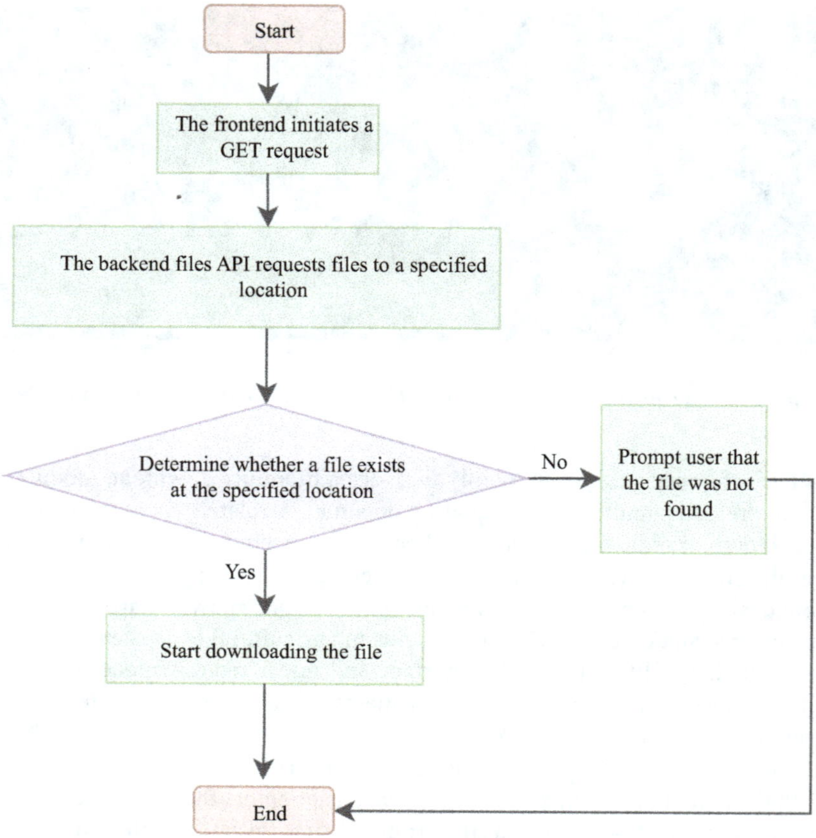

Figure 6.18 Flowchart for obtaining literature

6.5 Control strategy and operational logic

6.5.1 Control objective

In the Agricultural Energy Internet monitoring system, the control objective section is the core component for achieving intelligent energy scheduling and coordinated optimization of the ecological environment. The design of this section aims to integrate deep reinforcement learning algorithms with power grid flow optimization models to achieve dynamic load control and ecological balance under multi-energy complementary conditions, thereby maximizing energy utilization efficiency and carbon reduction benefits while ensuring the stable operation of farming and cultivation systems.

In the control target design, the system takes multiple controllable electrical loads in aquaculture scenarios (such as aerators, water pumps, circulation pumps, lighting systems, heating devices, etc.) as control objects. It comprehensively considers their current operating status, aquaculture warning levels, and plant cultivation warning levels to determine the optimal target control parameters for each electrical load. First, the system collects real-time information on node marginal electricity prices and line congestion status, and then calls a pre-constructed power grid flow optimization sub-model for analysis and calculation. This submodel, based on a dynamic coordinated optimization strategy, outputs the current power grid flow optimization results, providing boundary conditions and constraints for subsequent load scheduling.

After obtaining the results of the power grid flow optimization, the system inputs the real-time operating status of multiple controllable load modules, the fishery warning level, and the plant warning level into the pre-constructed "Fishery Deep Reinforcement Learning Model" and "Agricultural Deep Reinforcement Learning Model," respectively. These two models, through deep learning training on historical data and real-time feedback, can adaptively adjust the control parameters of each load device, thereby generating the optimal target control commands for each controllable load module.

By achieving this control objective, the system accomplishes multiple optimization goals at the operational level: first, it reduces overall electricity consumption and improves energy utilization efficiency; second, it maintains the stability of the power grid flow, reducing the risk of line congestion and fluctuations; third, it dynamically balances water quality conditions with cultivation environment conditions, ensuring that aquaculture and plant growth are in the optimal ecological state, thereby achieving synergistic optimization of the aquaponics system in terms of energy efficiency, ecology, and yield.

The design of this control objective not only reflects the technical advantages of the Agricultural Energy Internet monitoring system in intelligent control and multi-objective optimization but also lays a solid foundation for the system's subsequent automated scheduling, energy-efficient operation, and carbon emission constraint management.

6.5.2 Power load and flow optimization

The power load model plays a key role in connecting "energy consumption prediction — intelligent regulation — optimal control," serving as a fundamental component for achieving system adaptive energy saving and dynamic environmental balance. This part realizes intelligent and interpretable power scheduling in aquaponics systems by constructing mathematical models for various controllable electricity load modules in livestock farming and integrating them with deep reinforcement learning algorithms and power grid flow optimization sub-models.

Specifically, the system first establishes power load models for various controllable electricity-consuming modules, providing input feature support for subsequent reinforcement learning training and optimized control. Using a deep reinforcement learning-based intelligent energy Internet regulation method, the system constructs state spaces for both fisheries and agriculture. The fisheries state space includes microclimate parameters, water environment characteristics, fish health indicators, and load operating status, among others, while the agriculture state space includes features such as leaf temperature and plant physiological conditions. By learning from existing operational data and historical control experience, the system optimizes power load control strategies and the charge-discharge power of energy storage. At the same time, the system establishes a power grid flow optimization sub-model, completing interaction and feedback of multi-source data based on dynamic coordinated optimization strategies to ensure reasonable power flow distribution, stable node voltage, and optimal line power in the grid.

In the specific modeling process, the system takes the operational parameters of various electrical devices (such as the light intensity of supplementary lamps, the rotational speed of aeration devices, the on/off state of water pumps, etc.) as input variables and outputs the real-time power of the corresponding devices. This modeling method enables the reinforcement learning model to perform multiple rounds of strategy training and power regulation simulations in a virtual environment, avoiding the potential risk of fish or plant loss that could occur from direct experimentation in the real system, thereby ensuring the safety and feasibility of the control strategy.

In the system's power load modeling, the power calculation methods for each major module are as follows.

The power calculation method of the system temperature control module is shown in formula (6.1).

$$P_{thermostatic} = P_{tem} \times n_{tem} \tag{6.1}$$

In this formula, P_{tem} represents the power of a single ultra-low temperature air source heat pump water heater unit, measured in kW, and n_{tem} represents the number of ultra-low temperature air source heat pump water heater units that are turned on.

The power calculation method of the system oxygenation module is shown in (6.2).

$$P_{ox} = P_{aerator} \times n_{ox} \tag{6.2}$$

In this formula, $P_{aerator}$ represents the rated power of each precision oxygen generator, in kW, and n_{ox} is the number of precision oxygen generators that are turned on.

The power calculation method for the system ammonia nitrogen content control module is shown in (6.3).

$$P_{blower} = P_{blo} \times n_{blo} \tag{6.3}$$

In this formula, P_{blo} represents the power of a single blower in kW, and n_{blo} represents the number of blowers that are turned on.

The power calculation method of the system microbial control module is shown in (6.4).

$$P_{ultraviolet} = P_{ult} \times n_{ult} \tag{6.4}$$

In this formula, P_{ult} represents the power of a single ultraviolet lamp in kW, and n_{ult} represents the number of ultraviolet lamps turned on.

Both the system liquid level control module and the water flow rate control module are based on pump equipment, and their power calculation method is shown in (6.5).

$$P_{liquid} = P_{liq} \times n_{liq} \tag{6.5}$$

In this formula, P_{liq} represents the power of a single pump, in kW, and n_{liq} represents the number of pumps that are turned on.

The power calculation model of the baiting amount control module is more complex, and its power calculation method is shown in (6.6).

$$P_{feed} = \left(\frac{H}{300}\right)\frac{P_{feeder}}{n_{feed}} \tag{6.6}$$

In this formula, P_{feeder} represents the feeding power of each feeder in kW, and n_{feed} represents the number of feedings per day on a typical current day. H is the total bait amount, in kg, obtained by $H = FRL_s$. In the formula, L_s is the fish survival rate, and R is the environmental impact factor, taken as 0.990. F is the amount of feed based on experience, in kilograms, obtained from $F = Z \times L_f$. In the formula, L_f is the feeding rate, which refers to the percentage of the total feed mass of the current feeding stage relative to the fish body mass, and is related to the growth stage of the fish. Z represents the total mass of the fish, calculated from $Z = 10^{-3}GDS$. G is the current body mass of the fish at a given stage, in grams. D is the stocking density, measured in fish per square meter, and S is the total area of the pond, in square meters.

In the design of the power load model of this system, the operating parameters of the feed control module are closely related to seasonal changes. The system dynamically adjusts the feeding times and frequency on typical days according to the metabolic patterns and feeding behaviors of fish in different seasons, aiming to optimize both energy consumption and growth efficiency. Typically, the duration of a single feeding session is set to one hour to ensure even feed distribution, sufficient energy utilization, and to avoid increased water quality load due to overfeeding. The system's control logic incorporates a season-based typical feeding schedule, with the feeding times and frequency for different seasons on typical days shown in Table 6.29. This strategy can be executed automatically by the system or adaptively optimized during actual operation based on water temperature, dissolved oxygen, and the fish growth stage.

Table 6.29 Typical baiting times and frequencies on different seasons' typical days

Typical day	Number of feedings per day	Feeding times
Typical winter day (December–February)	Once	12:10–13:10
Typical spring day (March–May)	Twice	8:10–9:10 16:00–17:00
Typical summer day (June–August)	Three times	8:00–9:00 12:00–13:00 16:00–17:00
Typical autumn day (September–November)	Twice	8:00–9:00 16:00–17:00

The power consumption calculation method of the system's multifunctional spray system is shown in (6.7).

$$P_{spray} = P_{spr} \times n_{spr} \tag{6.7}$$

In this formula, P_{spr} represents the power of a single spraying device, measured in kW, and n_{spr} represents the number of spraying devices turned on.

The power consumption calculation method for the system fill light is shown in (6.8).

$$P_{lightload} = P_e \times n_{light} \tag{6.8}$$

In this formula, P_e represents the power of a single supplementary light, in kW, and n_{light} represents the number of supplementary lights turned on.

The calculation method for the power consumption of the system carbon dioxide control module is shown in (6.9).

$$P_{CO2} = P_c \times n_c \tag{6.9}$$

In this formula, P_c is the power of a single carbon dioxide generator, measured in kW, and n_c is the number of carbon dioxide generators that are turned on.

Through the construction of the above power load model, the system not only achieves a quantitative description of energy consumption in various electricity usage processes but also provides high-precision state feature inputs for deep reinforcement learning algorithms, enabling subsequent energy consumption optimization and environmental control strategies to achieve optimal energy scheduling and intelligent coordination while ensuring ecological safety and production stability.

The establishment of the power grid power flow optimization sub-model is a core component for achieving dynamic coordination of system energy flows and intelligent scheduling control. This sub-model operates within a deep reinforcement learning control framework and achieves global optimization of electrical system operation by embedding a power flow optimization calculation module within the algorithm structure. Its design objectives include not only minimizing the total operating costs and line losses of the grid but also ensuring safe and stable node voltages and supply-demand balance, thereby enabling efficient, safe, and economical operation of aquaponic systems in the context of multi-source energy coupling.

In the sub-model for power grid flow optimization, the objective function is set to minimize the overall operational cost of the system. This cost function comprehensively considers the cost of electricity purchase, the charging and discharging loss costs of energy storage devices, line energy losses, and the execution costs of control operations. The constraints mainly include three aspects: first, the node power balance equation, which requires that at any given moment, "conventional power output + energy storage device charging/discharging power = total aquaculture load demand + line network losses," ensuring energy conservation and supply-demand balance in the system; second, the line transmission capacity limit, which sets an upper bound on line power flow to prevent power flow from exceeding the rated capacity of transmission channels and avoid overloading the grid; and finally, the voltage safety constraints, which require that the node voltage magnitudes remain within the permissible operating range, thereby ensuring system voltage stability and equipment safety.

In terms of optimization algorithm design, this system introduces a dynamic coordination optimization strategy that combines "priority-based hierarchical control" with "multi-objective weight adaptation," achieving coordinated scheduling between power grid flow optimization and water quality environmental control objectives. Specifically, when the power grid is operating normally, the system prioritizes "optimal water quality and cultivation environment," continuously optimizing the start-stop sequences and power allocation of controllable loads (such as pumps, supplementary lights, and feeders) through a deep reinforcement learning model, maintaining the aquaculture and cultivation environments within the optimal range. When the power grid is in an abnormal state (such as overvoltage, line overload, or frequency fluctuation), the system automatically elevates the priority of power safety control, triggering the priority-based hierarchical control mechanism.

Meanwhile, the multi-objective weight adaptive mechanism dynamically adjusts the weights between the "water quality optimization objective" and the "power grid flow optimization objective" by incorporating real-time electricity price signals and line congestion indicators. When electricity prices are detected to be at peak periods, the system automatically suppresses the operating power of high-energy-consuming devices (such as oxygen generators and heat pumps), prioritizing energy saving; when line congestion increases, the system appropriately reduces load density in local areas and optimizes the scheduling of energy storage devices to reduce line losses and achieve balanced power flow distribution. In this way, the system can achieve dynamic trade-offs and balance between multiple objectives under different operating scenarios, thereby significantly improving overall energy efficiency.

To achieve closed-loop feedback for grid optimization and system control, this system also establishes a comprehensive data interaction and feedback mechanism. The system communicates in real time with the grid dispatch center, periodically receiving key information such as nodal marginal prices (LMP), line power flow status, and congestion levels, and feeds this data as input into the reinforcement learning algorithm model for optimization. After the optimization process is completed, the system outputs the calculated optimal load control plan, including charging and discharging schedules for energy storage units, start-stop time sequences for pump control boxes, and grid integration adjustment strategies for photovoltaic power generation. The optimization results are then fed back in real time to the grid dispatch system, achieving bidirectional interaction of energy flow and information flow, thereby ensuring that the Agricultural Energy Internet system maintains globally optimal operation under dynamic conditions.

The power grid flow optimization sub-model in the Agricultural Energy Internet monitoring system not only undertakes the critical tasks of energy flow optimization and power supply security assurance but also through its collaboration with the deep reinforcement learning control module, achieves a deep integration of the aquaponics system in terms of power economy, environmental stability, and system intelligence, providing a solid algorithmic foundation and engineering basis for the adaptive optimization and intelligent decision-making of subsequent Agricultural Energy Internet systems.

6.5.3 Deep reinforcement learning model

In the design of control strategies for the Agricultural Energy Internet monitoring system, the construction of deep reinforcement learning models is a key step in achieving intelligent load regulation and optimal energy allocation. The system establishes separate deep reinforcement learning models for aquaculture and agriculture to address the differences in load dynamics in fish farming and plant cultivation scenarios, enabling the optimization of adaptive control strategies based on environmental conditions.

Specifically, the first step is to construct a deep reinforcement learning model for fisheries, as shown in Figure 6.19. The system combines water environment parameters, cultivation environment parameters, warning information, meteorological information, controllable electricity load module states, and energy storage device power states to form the state space S. That is, S = {S | S =[meteorological information, water quality parameters, warning information, fishery load states, current power of energy storage devices]}, which can comprehensively describe the current environmental state and operational characteristics of the fishery system. In the action space A, the system takes the start-stop power adjustment of electrical devices such as ultra-low temperature air energy water heaters, precision oxygen machines, blowers, ultraviolet lamps, water pumps, feed dispensers, variable-frequency circulation water pumps, and carbon dioxide generators, as well as the charge and discharge power regulation schemes of energy storage devices, as selectable actions, denoted as A = {a1, a2, a3, a4 . . . }. The system constructs the

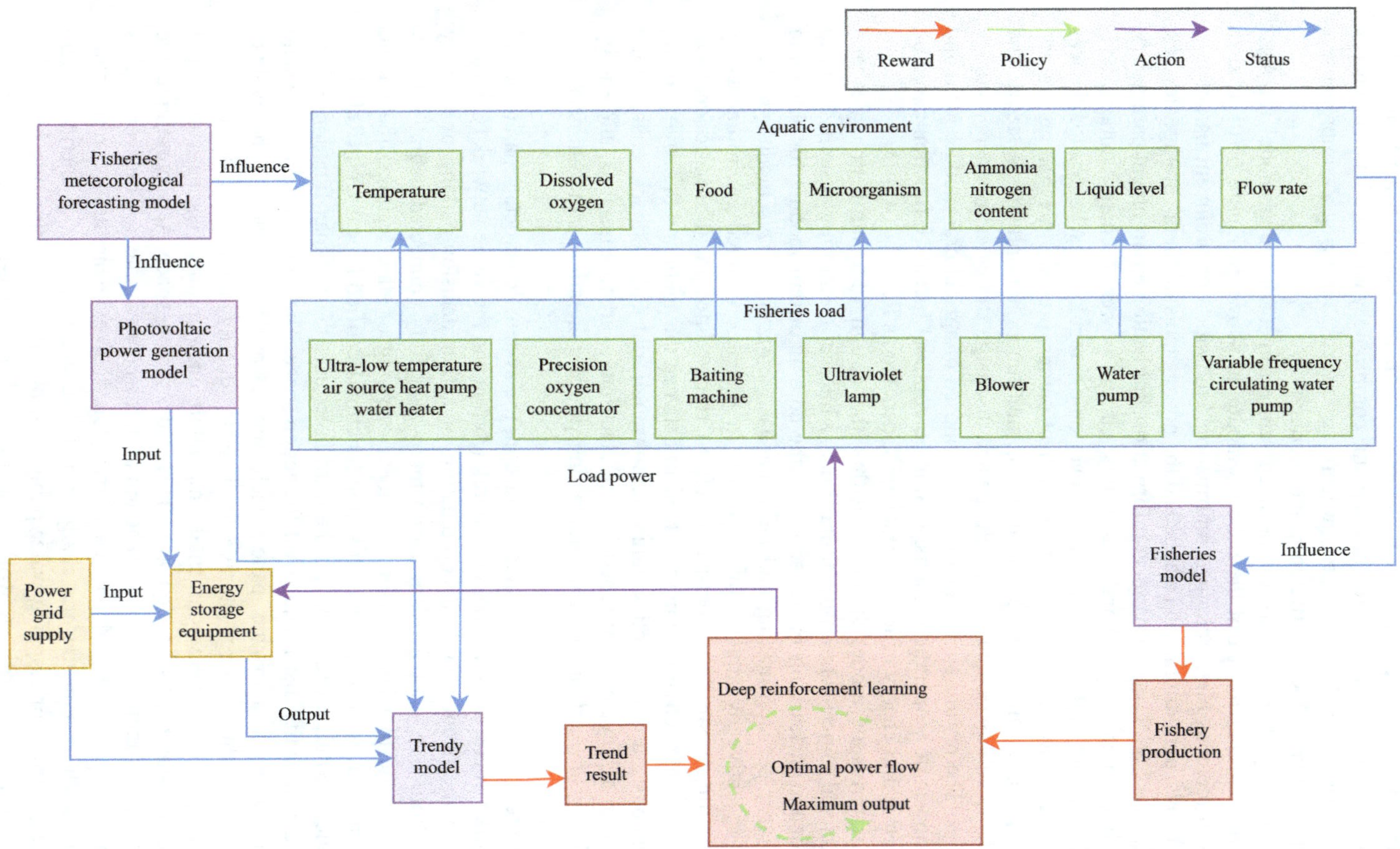

Figure 6.19 Fisheries deep reinforcement learning model

reward space R based on grid flow optimization results and fishery yield, and employs a cost-driven reward mechanism, combining yield benefits with energy consumption costs for reinforcement learning rewards, achieving a comprehensive balance between energy efficiency and productivity.

Similarly, the structure of the deep reinforcement learning model for agriculture is similar to that of the fisheries model, as shown in Figure 6.20. Its state space S includes meteorological information, cultivation environment parameters, alert levels, agricultural load status, and the power status of energy storage devices, that is, S = { S | S [meteorological information, cultivation environment parameters, alert information, agricultural load status, current power of energy storage devices]}. The action space A includes power-on/off regulation schemes for devices such as ultra-low temperature air source heat pumps, precision oxygen generators, water pumps, carbon dioxide generators, multifunctional spraying systems, and supplemental lighting, as well as charge and discharge strategies for energy storage devices. The reward space is similarly composed of agricultural module yield and grid power flow operation results, forming an optimal incentive function based on economic and energy efficiency trade-offs, denoted as R.

In terms of model structure, within the design of control strategies and operational logic for the Agricultural Energy Internet monitoring system, the deep reinforcement learning model serves as the core algorithm module for achieving intelligent decision-making and autonomous optimization. This module realizes multi-objective dynamic optimal control in complex coupled environments by constructing a policy-gradient-based Actor-Critic architecture. The basic idea is to utilize the interaction learning mechanism between the agent and the environment, continuously updating policy parameters through trial and feedback, enabling the system to achieve optimal scheduling of electricity, carbon emissions, and energy efficiency under dynamic load, fluctuating power generation, and nonlinear coupling conditions.

In terms of network architecture design, the system constructs separate neural network structures for the policy network (Actor) and the value network (Critic) to achieve the decoupling and coordination of action decision-making and policy evaluation. The actor network uses a three-layer fully connected neural network structure: the first layer performs a linear mapping of state features, the second layer introduces the nonlinear activation function Rectified Linear Unit (ReLU) for nonlinear feature extraction, and the third layer uses the Softmax function to generate a probability distribution over the action space. The network input is the state vector s_t in the system state space S, including key parameters such as photovoltaic power generation, energy storage level, water quality regulation load, real-time electricity price, and environmental temperature and humidity. The network output is the probability distribution $\pi(a|s_t)$ over the action space A, representing the probability of taking each control action a_t in the current state s_t. The agent makes sampling decisions based on this distribution, thereby generating control commands, such as adjusting the charging and discharging power of energy storage devices, starting or stopping water pump control boxes, or optimizing load scheduling plans. The critic network structure is also a three-layer fully connected

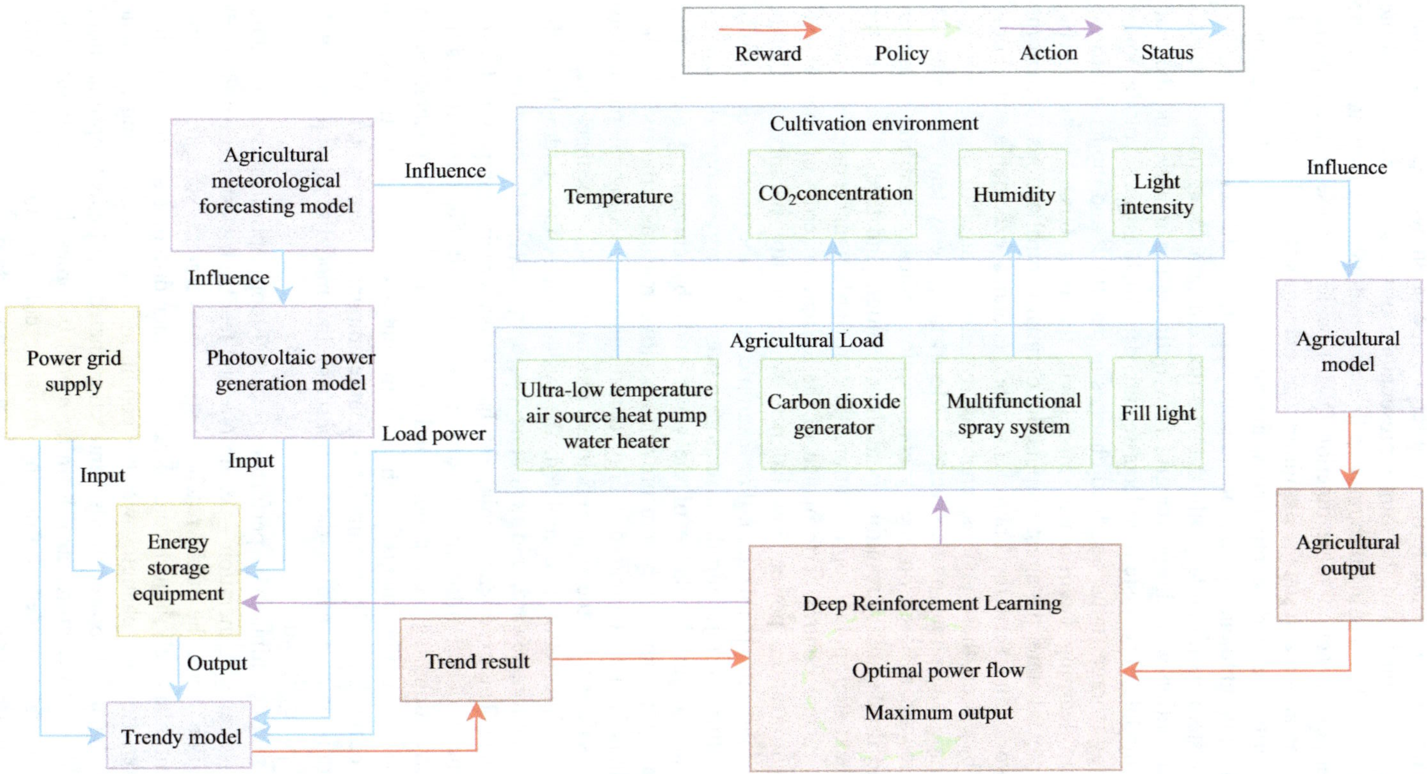

Figure 6.20 Agricultural deep reinforcement learning model

neural network, used to evaluate the long-term value of an agent taking a certain action. The input consists of the joint representation of the state vector s_t and the action vector a_t. After passing through two layers of nonlinear transformations, the output generates the action-value function $Q(s|a)$, which represents the expected return of taking action a in state s. By minimizing the temporal difference error (TD Error), the critic network continuously approximates the true value function, thereby providing gradient signals for the policy update of the actor network.

At the system operation level, the deep reinforcement learning model achieves real-time interaction with the power grid flow optimization module and the load forecasting module. The system collects feature data such as photovoltaic power output, energy storage status, and load changes in real time, and inputs it into the reinforcement learning environment model via the data bus. After receiving the environmental state, the agent outputs the optimal control actions through the policy network, and then the control module generates executable scheduling commands based on these actions, enabling dynamic control of energy storage devices, pumps, and photovoltaic inverters. Meanwhile, the system calculates the immediate reward r_t based on energy consumption indicators, carbon reduction benefits, and flow stability resulting from the execution, and feeds it back to the agent for next-step policy optimization. The reward signal comprehensively reflects the current control performance of the system in terms of energy consumption optimization, power flow balance, water quality stability, and plant growth environment optimization. For example, when the model successfully reduces peak load by adjusting the discharge power of energy storage devices, or maintains the ecological balance of the water body by controlling pumps and oxygenation equipment, the system provides positive rewards; conversely, when control actions lead to increased energy consumption or ecological parameters deviate from safe ranges, the system delivers punitive negative feedback.

After each interaction is completed, the system stores the quadruple of the current state, action, reward, and next state $\{(s_t, a_t, r_t, s_{t+1})\}$ into the experience replay buffer D. The experience replay mechanism is a crucial component for the stable training of deep reinforcement learning models, as it effectively breaks the temporal correlations between samples and enhances the model's generalization ability in complex environments. As the system continues to operate, the experience replay buffer gradually accumulates a large amount of representative historical interaction data, providing data support for the model's subsequent batch training and policy iteration.

The system continuously performs a closed-loop optimization of "state awareness — strategy generation — environmental feedback — strategy updating" during operation. Once the model reaches a stable convergence state, the resulting strategies can adaptively adjust control actions under different seasons, varying load characteristics, and changing weather conditions, thereby achieving efficient energy use, balanced power loads, and coordinated ecological optimization. Ultimately, the deep reinforcement learning model becomes the core algorithmic support for autonomous decision-making, dynamic control, and intelligent scheduling in the Agricultural Energy Internet monitoring system.

Through this deep reinforcement learning model, the Agricultural Energy Internet monitoring system can achieve intelligent control features such as self-learning, self-adaptation, and self-optimization, enabling the system to continuously optimize in complex dynamic environments and providing key technological support for achieving multi-energy complementarity and carbon neutrality goals.

In the model update phase, the critic network is first parameter-tuned. The main task of the critic network is to evaluate the performance of the current policy, and its update process is achieved by minimizing the mean squared error between the target *Q*-value and the predicted *Q*-value. Specifically, the definition of the target *Q*-value is shown in (6.10).

$$y_i = r_i + \gamma Q_{target}\left(s_{i+1}, \pi_{target}\left(s_{i+1}\right)\right) \tag{6.10}$$

In this formula, γ is the discount factor, used to balance immediate rewards with long-term rewards; Q_{target} and π_{target} represent the target critic network and target actor network, respectively, which are used to improve training stability.

The loss function of the critic network is shown in (6.11).

$$L_{critic} = \frac{1}{N}\sum_{i=1}^{N}\left(y_i - Q(s_i, a_i)\right)^2 \tag{6.11}$$

In this formula, N is the number of samples.

The system optimizes the critic network parameters through gradient descent, allowing the predicted *Q* values to gradually approach the target *Q* values.

Subsequently, the system performs a policy update on the actor network. The goal of the actor network is to maximize the long-term expected reward through gradient ascent, with its objective function shown in (6.12).

$$J(\theta) = \frac{1}{N}\sum_{i=1}^{N} Q(s_i, a_i)\log\pi(a_i|s_i) \tag{6.12}$$

where, θ represents the parameters of the actor network. By computing the gradient of this objective function and updating the parameters θ, continuous improvement of the control policy is achieved. In other words, under the guidance of the critic network, the actor network continuously adjusts its action decisions to achieve an optimal balance among multiple objectives (minimizing energy consumption, minimizing carbon emissions, and maximizing water quality stability).

The entire training process is carried out in a cyclical manner, with the system continuously sampling from the experience replay pool for batch learning and parameter updates until the set maximum training steps or policy convergence criteria are met. Ultimately, the trained deep reinforcement learning model acquires the ability to adaptively regulate the complex nonlinear dynamic environment of the Agricultural Energy Internet, automatically adjusting energy usage structures and environmental control strategies under different seasons and weather conditions, thereby achieving maximum energy efficiency and intelligent sustainable operation of the agricultural system.

During the system's operation, it continuously performs iterative interactions of state–action–reward. Through experience replay and target network mechanisms, it reduces sample correlation and training oscillations, enabling the model to converge stably. Ultimately, when the number of training steps reaches the set threshold, both the fisheries and agriculture deep reinforcement learning models can output the optimal load control strategies, achieving the comprehensive goals of minimizing electricity consumption, optimizing power grid flow, stabilizing aquatic and cultivation environments, and maximizing the yield of the aquaponics system.

6.6 Conclusion

This chapter provides a comprehensive explanation and technical implementation of the construction process of the Agricultural Energy Internet monitoring system from multiple perspectives, including overall architecture, functional modules, database design, system implementation, and key algorithms. The system centers on the integrated concept of "Smart Agriculture, Energy Internet, and Carbon Management" and addresses the needs for multi-source data monitoring, energy consumption analysis, and carbon emission accounting in aquaculture and plant cultivation systems within aquaponics scenarios. It proposes a comprehensive platform with real-time monitoring, visual display, and intelligent decision-making capabilities, achieving a fundamental shift in agricultural production from "manual experience management" to "intelligent data-driven management."

From an overall architecture perspective, the system implements dynamic visualization and interactive display, while the back end, built on Spring Boot and MyBatis Plus, provides an efficient data processing and service support framework. Through a combined MySQL and Redis database architecture, the system achieves hot and cold data separation and a real-time response mechanism. The system uses WebSocket technology to realize real-time push of monitoring data and automatic refresh of the front-end interface, ensuring high-frequency sampling and second-level update responses in complex aquaculture and cultivation environments. Under this architecture, the platform enables multi-module collaborative operation, including water quality and environmental monitoring, energy consumption management, carbon emission accounting, and intelligent control strategies, forming a stable and efficient energy Internet operation system.

In terms of functional design, the system revolves around the ecological logic of the aquaponics system, establishing nine major functional modules, including the fishery water quality monitoring subsystem, cultivation environment monitoring subsystem, historical data management subsystem, and energy consumption and carbon emission visualization subsystem, achieving a full-chain closed-loop from underlying perception to upper-level decision-making. The fishery monitoring module supports dynamic collection and threshold alerts for key parameters; the cultivation monitoring module can display indicators in real time; the energy module enables comparison of photovoltaic power generation and purchased electricity, energy consumption statistics, and trend forecasting; the carbon accounting module constructs models to perform multidimensional calculations and visualizations of carbon emissions, carbon absorption, and carbon reduction. Through data linkage between modules, the system achieves unified scheduling and visual presentation of the "energy flow–information flow–material flow."

In terms of database design and data management, the system adopts a dual-layer architecture with Redis and MySQL to achieve efficient data caching and persistent storage. Redis is primarily used for caching high-frequency real-time monitoring data and simulated sensor inputs, storing the real-time status of different breeding ponds, cultivation areas, and energy nodes using a hash structure, which significantly improves data retrieval efficiency and front-end response speed. MySQL is responsible for long-term storage of structured data, including historical water quality records, carbon emission logs, and energy consumption statistics, ensuring data integrity and traceability. The collaborative mechanism of the dual databases ensures system stability and efficiency under high-concurrency access, providing a solid data foundation for real-time monitoring and historical analysis in the Agricultural Energy Internet.

In terms of algorithm implementation, the system centers on deep reinforcement learning to construct an intelligent regulation model that integrates the load characteristics of both fisheries and agriculture. By dynamically perceiving multi-source information such as photovoltaic output, energy storage power, water quality parameters, ambient temperature, humidity, and carbon emission data, the model can identify the system's operational state in real time and autonomously generate optimal control strategies. On this basis, the system separately establishes a deep reinforcement learning model for fisheries and another for agriculture to address the differences in load characteristics, energy consumption patterns, and environmental requirements between fish farming and plant cultivation, achieving scenario-specific self-learning and strategy optimization. The model takes the power regulation of various controllable devices (such as water pumps, oxygenators, heating devices, supplemental lights, and energy storage systems) as the action space input. Through continuous sampling of environmental states and reward feedback, thereby achieving coordinated control of energy flows and environmental parameters. The system focuses on optimizing energy costs, carbon emission intensity, grid flow stability, and production efficiency, constructing a multi-objective reward mechanism that balances economic and ecological considerations, enabling the model to automatically mediate between energy saving and production output during operation. During model training, the system continuously optimizes strategies through experience replay and policy iteration mechanisms. The agent accumulates a large amount of state–action–reward data while interacting with the environment and continuously refines the control strategy through batch learning, maintaining stable control performance under varying weather conditions, load changes, and external disturbances. After sufficient training, the model can autonomously identify the optimal control scheme in complex nonlinear environments, achieving coordinated scheduling of photovoltaic output and energy storage systems, dynamic balancing of water quality and environmental parameters, and maximization of electricity utilization efficiency. Ultimately, the deep reinforcement learning model becomes the intelligent hub of the Agricultural Energy Internet monitoring system, endowing the system with self-learning, self-adaptive, and self-optimizing capabilities. It can achieve the collaborative goal of energy optimization and carbon emission control in a variable production environment, supporting the green and efficient operation of aquaponics systems with intelligent algorithms, and providing a sustainable technological pathway toward low-carbon and intelligent agricultural energy systems.

At the level of visualization and user interaction, the system utilizes ECharts graphical components and the Element Plus interface library to implement various interactive methods, including water quality parameter trend curves, carbon emission month-on-month charts, energy consumption comparison bar charts, and dynamic displays of carbon absorption. Users can perform data queries, threshold settings, and multidimensional comparative analysis within a unified interface. The system intuitively reflects water quality and environmental health levels through colors, icons, and a scoring mechanism. When a parameter exceeds the preset threshold, the system automatically triggers an alert, changing interface elements to red or yellow and pushing real-time warning information to the user side, enabling immediate response and intervention for abnormal situations.

In the carbon accounting module, the system innovatively introduces a bidirectional calculation logic for both carbon emissions and carbon absorption. By dynamically tracking purchased electricity and photovoltaic power generation, the system can calculate carbon emissions and reductions per time period; at the same time, it estimates total carbon absorption based on plant growth models and displays the annual carbon balance through a visual interface. The module supports statistics and comparisons across multiple time scales – hourly, daily, monthly, and yearly – helping managers comprehensively understand the low-carbon benefits of system operations. This design fully aligns with China's "dual carbon" strategy goals and provides data support for carbon accounting systems in the fisheries and agriculture sectors.

In terms of system implementation and deployment, the platform adheres to the principles of high availability, high scalability, and high security. The backend services are designed with modular configuration and layered interfaces, allowing for independent maintenance and flexible upgrades; the frontend uses a responsive design to accommodate multiple devices, supporting synchronized display on PCs, tablets, and mobile devices, enabling farmers regulate operations. On the security front, the system ensures information safety and operational compliance through user authentication, encrypted data transmission, and anti-injection mechanisms.

Through multidimensional integration and optimization, this system demonstrates significant innovation and engineering value in the following aspects: it introduces the concept of the "Energy Internet" into aquaponics ecosystems for the first time, establishing a multi-level monitoring and control system that integrates energy, material, and information flows; it proposes an agricultural energy monitoring framework based on a carbon accounting model, enabling visualized and quantitative management of fishery electricity consumption, photovoltaic power generation, and carbon balance; it constructs a reinforcement learning algorithm model for intelligent control, achieving dynamic optimization of load-side devices; and it implements a unified platform design from water quality monitoring to energy and carbon management, providing a scalable model for agricultural digitalization and green transformation.

The system, with its scalable technical architecture, intelligent algorithm models, and user-friendly human–computer interaction interface, demonstrates a feasible path for the deep integration of smart fisheries and green agriculture, providing a technical demonstration and practical basis for the future networked, intelligent, and low-carbon development of agriculture.

Looking ahead, the system will be further optimized and expanded in the following directions: at the algorithm level, introducing federated learning and transfer learning technologies to build cross-regional shared models, enhancing adaptability to various climates and water quality environments; at the model level, integrating digital twin technology to create virtual simulation scenarios for predictive maintenance and energy-carbon scheduling verification; at the application level, extending to a larger-scale Agricultural Energy Internet ecosystem, such as collaborative control across multiple farming areas and integration with cross-regional energy-carbon scheduling and carbon trading systems; at the international level, improving multi-language support and regional parameter adaptation mechanisms, promoting the implementation of agricultural digitalization and carbon neutrality applications in countries.

In summary, the Agricultural Energy Internet monitoring system designed and implemented in this chapter not only achieves multidimensional information integration and intelligent decision-making control in engineering practice but also provides a new paradigm for agricultural low-carbon development and energy internetization in theoretical research. The establishment of this system validates the feasibility and sustainability of a smart agriculture system driven by data and guided by carbon management, and it holds significant scientific significance and promotional value for the modernization of agriculture in China.

6.7 Study questions

In the process of research and implementation of the Agricultural Energy Internet monitoring system, although this project has initially built an integrated platform that combines environmental monitoring, energy consumption management, carbon accounting, and intelligent regulation, there are still numerous scientific issues and technical challenges worthy of in-depth exploration from the perspectives of scientific research and engineering applications. Studying these issues is not only related to improving system performance and enhancing intelligence levels but also holds significant importance for the improvement of the theoretical framework of the Agricultural Energy Internet. Based on the preceding analysis, this chapter summarizes the following core research directions.

On the issue of agricultural scenario modeling under the energy Internet system. Current aquaponics system models are still relatively idealized in their coupling representation of energy, material, and information flows, and insufficiently consider the dynamic impacts of natural disturbances, environmental fluctuations, and equipment aging. This leads to a decline in model accuracy and deviations from optimal control strategies during long-term operation. Future research should explore:

1. How to build system models with greater physical interpretability and environmental adaptability?
2. How can dynamic constraints (such as water cycle, energy flow balance, and greenhouse effect parameters) be incorporated to make the model more closely reflect the actual production environment?
3. How can model parameters be self-corrected and updated online to cope with dynamic changes in environmental conditions and device performance? In complex multivariable agricultural systems, reinforcement learning models are prone to policy oscillations, slow convergence, or overfitting. Therefore, the following scientific issues need to be addressed in the future:
4. How can we design deep reinforcement learning architectures that are interpretable and robust, enabling the model to maintain stable decision-making in dynamic and nonlinear environments?
5. How can multi-agent reinforcement learning be used to achieve coordinated regulation between fisheries and agriculture modules, reducing strategy conflicts between different subsystems?
6. How can environmental prior knowledge and physical constraints be incorporated into algorithms to improve model training efficiency and sample utilization?
7. Is it possible to build a multi-level control architecture based on Hierarchical Reinforcement Learning (Hierarchical RL) to balance both local responses and global optimization? Regarding the issue of multi-source heterogeneous data fusion and trustworthy sensing, the current system's data sources mainly include water quality sensors, weather stations, energy consumption measurement devices, and photovoltaic monitoring equipment. However, these data exhibit significant differences in sampling frequency, format, and accuracy, which affect the unified learning performance of the model. Therefore, future research needs to focus on addressing the following questions:
8. How can a unified multi-source heterogeneous data fusion framework be established to achieve effective collaboration between data of different modalities and different time scales?
9. How can methods such as Graph Neural Networks (GNN) and spatiotemporal attention mechanisms be used to mine potential correlation features from multidimensional data?
10. How can local data preprocessing and anomaly detection be implemented at edge computing nodes to enhance the overall system's real-time performance and reliability?

11. In the face of data privacy and security risks, how can distributed modeling and protection of agricultural energy data be achieved under the federated learning framework?

Regarding the issues of energy-carbon coupling and carbon benefit assessment. The current carbon accounting system is mainly based on calculations of electricity consumption and emission reduction factors from photovoltaic power generation, but a complete carbon lifecycle analysis system has not yet been established. In practice, the relationship between carbon emissions, carbon absorption, and energy dispatch is very complex, and there are time delays and spatial coupling characteristics. Future research should focus on the following issues:

1. How can a unified energy-carbon coupling model be constructed to achieve full-process mapping from electricity input to carbon emission output?
2. How can we quantify the differences in carbon emissions and carbon efficiency under different energy structures and modes of equipment operation?
3. How can digital twin technology be used to establish a hybrid virtual-physical carbon flow dynamic simulation system to achieve real-time monitoring and predictive regulation?
4. How can the carbon trading mechanism be integrated with system energy consumption optimization strategies to form a new model of "economic–energy–carbon emission"' coordinated optimization?

Regarding system security and sustainable operation issues. As the complexity of the Agricultural Energy Internet increases, issues related to system security and sustainable operation become more prominent. Data interactions between distributed sensor nodes, edge computing devices, and cloud collaborative platforms occur frequently, increasing the risks of security vulnerabilities and abnormal propagation. In the future, the following key issues need to be studied:

1. How can we build a multi-layered security defense system for the Agricultural Energy Internet, achieving full lifecycle protection of data links, model parameters, and control commands?
2. How can blockchain technology be used to ensure data traceability and tamper-resistance?
3. How can the system achieve rapid recovery and self-repair in the event of network abnormalities, equipment failures, or data attacks?
4. Is it possible to establish trusted computing and autonomous diagnostic mechanisms to enable the system with fault tolerance and self-healing capabilities, ensuring long-term stable operation?

Regarding the issues of system scalability and cross-scenario migration. Existing platforms are mainly developed for specific aquaponics system scenarios, and their algorithms, data models, and control logic still need to be adapted for different types of agriculture (such as recirculating aquaculture, vertical farming, greenhouse irrigation, etc.). Future research should address:

1. How can a modular and pluggable system architecture be built to support the rapid integration of different production units?
2. How can models be applied across different regions and crop types based on transfer learning and domain adaptation methods?
3. How can we, while ensuring accuracy, establish unified data interface standards and system communication protocols to achieve cross-regional collaborative operation of the Agricultural Energy Internet?

On decision intelligence and multi-objective optimization problems. In Agricultural Energy Internet systems, energy conservation, emission reduction, yield, and ecological benefits often conflict. How to achieve dynamic balance among multiple objectives is a key scientific issue for intelligent decision-making. Future research directions include:

1. How can an integrated optimization model based on multi-objective reinforcement learning (MORL) be established to achieve unified scheduling of energy efficiency, economic benefits, and ecological stability?
2. How can auxiliary methods such as fuzzy decision-making and the Analytic Hierarchy Process (AHP) be introduced to provide a scientific basis for the system's multi-objective weight allocation?
3. How can evolutionary algorithms (such as NSGA-II) be combined with reinforcement learning to achieve dynamic optimization of the non-dominated solution set?
4. Is it possible to establish a self-learning weight adjustment mechanism that allows the system to automatically adapt to different optimization priorities during operation?

Regarding the issues of theoretical systems and standard development. As an emerging interdisciplinary field, the Agricultural Energy Internet currently lacks a unified theoretical framework and engineering standards. Different research teams show significant differences in model construction, data interfaces, and carbon accounting methods, which affect the comparability and reproducibility of results.

1. How can we build a systematic theoretical framework for the Agricultural Energy Internet and clarify its core concepts, hierarchical relationships, and evaluation indicators?
2. How can technical standards and interface specifications be developed for agricultural carbon accounting, data connectivity, and intelligent control?
3. How can the standardization, industrialization, and internationalization of Agricultural Energy Internet technology be promoted under the national "dual carbon" policy?

In summary, research on Agricultural Energy Internet monitoring systems is still in a stage of continuous deepening and systematic improvement. From data-driven modeling to energy-carbon coordinated optimization, and from intelligent control algorithms to system security mechanisms, these research issues not only hold significant theoretical value but also directly relate to the practical outcomes of agricultural modernization and green low-carbon transformation. Future research will focus on these issues, continuously enhancing the system's intelligence level, decision-making accuracy, and operational safety, thereby providing scientific support and a technological foundation.

Funding

This study was supported by the National Key Research and Development Program of China (Project No. 2023YFD2400401).

References

[1] Zhu Z, Wu D, and Jiang Q. Chinese freshwater aquaculture: A comparative analysis of the competitiveness on regional aquaculture industries. *Aquaculture and Fisheries*, 2024, 9(5): 860–870.

[2] Aljehani F, N'doye I, and Laleg-Kirati T M. Feeding control and water quality monitoring on bioenergetic fish growth modeling: Opportunities and challenges. *Aquacultural Engineering*, 2024, 104: 102511.

[3] Verma D K, Satyaveer, Maurya N K, Kumar P, and Jayaswa R. Important water quality parameters in aquaculture: An overview. *Agriculture and Environment*, 2022, 3(3): 24–29.

[4] Wu Y, Duan Y, Wei Y, An D, and Liu J. Application of intelligent and unmanned equipment in aquaculture: A review. *Computers and Electronics in Agriculture*, 2022, 199: 107201.

[5] Munguia-Fragozo P, Alatorre-Jacome O, Rico-Garcia E, *et al*. Perspective for aquaponic systems: "Omic" technologies for microbial community analysis. *BioMed Research International*, 2015, 2015: 480386.

[6] Elia E, Hodoşan C, Nistor L, Dumitrache F, and Udroiu N A. System Cycling Stage On Aquaponic Systems As Required Prerequisite For Soilless Agriculture. *Scientific Papers. Series D. Animal Science*, 2015, 58: 381–384.

[7] Kyaw T Y, and Ng A K. Smart aquaponics system for urban farming. *Energy Procedia*, 2017, 143: 342–347.

[8] Rodgers D, Won E, Timmons M B, and Mattson N. Complementary nutrients in decoupled aquaponics enhance basil performance. *Horticulturae*, 2022, 8(2): 111.

[9] Chen W, and Gao S. Current status of industrialized aquaculture in China: a review. *Environmental Science and Pollution Research*, 2023, 30(12): 32278–32287.

[10] Sun J, Li M, Zhai N, Mu H, Li W, and Liu H. Enhancing carbon neutrality through blue carbon fisheries: Spatial analysis in Shandong. *Regional Studies in Marine Science*, 2025, 85: 104125.

[11] Li H, and Li J. Water quality parameter monitoring system in multi-region based on IOT. *Proceedings of the 2nd International Conference on*

Automation, Mechanical Control and Computational Engineering (AMCCE 2017). Atlantis Press, 2017: 271–274.

[12] Amraouiyine A E, Britel M R, Ouahabi M E, *et al.* Design of real-time monitoring system for aquaculture sea-cages with GPRS and embedded systems. *Sensors and Transducers*, 2020, 240(1): 26–35.

[13] Singh Y, and Walingo T. Smart water quality monitoring with IOT wireless sensor networks. *Sensors*, 2024, 24(9): 2871.

[14] Yu J, Wang G, Li X, and Li D. SDYOLO-Tracker: An efficient multi-fish hypoxic behavior recognition and tracking method. *Computers and Electronics in Agriculture*, 2025, 232: 110079.

[15] Liang Z, Shi Y, Yi Y, and Fan Y. Online water quality monitoring system based on NB-IoT. *International conference on internet of things, communication and intelligent technology*. Singapore: Springer Nature Singapore, 2024: 277–289.

[16] Rasheed A F, Hagem R M, and Khidhir A S M. Design and implementation of an interactive embedded system as a low-cost remotely operated vehicle for underwater applications. *Diyala Journal of Engineering Sciences*, 2024, 17(3): 173–198.

[17] Peng Z, Chen Y, Zhang Z, Qiu Q, and Han X. Implementation of water quality management platform for aquaculture based on big data. *2020 international conference on computer information and big data applications (CIBDA)*. IEEE, 2020: 70–74.

[18] Xie X, Hua J, Ding J, *et al.* An aquaculture big data platform for unmanned boats. *2025 32nd international conference on geoinformatics*. IEEE, 2025: 1–5.

[19] Liu T, Liu J, Wang J, and Xu J. Optimization of the intelligent sensing model for environmental information in aquaculture waters based on the 5G smart sensor network. *Journal of Sensors*, 2022, 2022(1): 6409046.

[20] Quiroz-Martinez M A, Perez-Vitonera A, Gómez-Rios M, Castro-Arias S, and Leyva-Vazquez M. Architecture design for the implementation of a water quality prediction system in aquaculture systems with big data. *International conference on applied technologies*. Cham: Springer Nature Switzerland, 2024: 178–190.

[21] Punt A E, Smith A D M, and Cui G. Evaluation of management tools for Australia's South East Fishery. 2. How well can management quantities be estimated? *Marine and Freshwater Research*, 2002, 53: 631–644.

[22] King J R, and McFarlane G A. Marine fish life history strategies: applications to fishery management. *Fisheries Management and Ecology*, 2003, 10: 249–264.

Index

www.ingramcontent.com/pod-product-compliance
Lightning Source LLC
LaVergne TN
LVHW012322171026
850499LV00003B/68

* 9 7 8 1 8 0 7 0 5 0 8 9 4 *